Graduate Texts in Contemporary Physics

Series Editors:

Joseph L. Birman
Helmut Faissner
Jeffrey W. Lynn

Graduate Texts in Contemporary Physics

Michio Kaku

Strings, Conformal Fields, and Topology

An Introduction

With 40 Illustrations in 50 Parts

Springer-Verlag
New York   Berlin   Heidelberg   London
Paris   Tokyo   Hong Kong   Barcelona

Michio Kaku
Department of Physics
City College of the City University
of New York
New York, NY 10031, USA

Series Editors

Joseph L. Birman
Department of Physics
The City College of the
City University of New York
New York, NY 10031, USA

H. Faissner
Physikalisches Institut
RWTH Aachen
5100 Aachen
Federal Republic of Germany

Jeffrey W. Lynn
Department of Physics and Astronomy
University of Maryland
College Park, MD 20742, USA

Library of Congress Cataloging-in-Publication Data
Kaku, Michio,
 Strings, conformal fields, and topology: an introduction / Michio Kaku.
 p. cm.
 Includes bibliographical references.
 ISBN 0-387-97496-2 (New York). -- ISBN 3-540-97496-2 (Berlin)
 1. String models. 2. Superstring theories. 3. Quantum field
theory. 4. Algebraic topology. I. Title.
QC794.6.S85K36 1991   90-23741
539.7'2--dc20

Printed on acid-free paper.

Photocomposed pages prepared from the author's TeX file.
Printed and bound by R.R. Donnelley & Sons, Harrisonburg, Virginia.
Printed in the United States of America.

9 8 7 6 5 4 3 2 1

ISBN 0-387-97496-2 Springer-Verlag New York Berlin Heidelberg
ISBN 3-540-97496-2 Springer-Verlag Berlin Heidelberg New York

This book is dedicated to my parents.

Preface

String theory, because it is the leading candidate for a theory of all fundamental physical forces, has advanced at an astonishing rate in the last few years. Accordingly, the purpose of this book is to acquaint the reader with the most active topics of research in string theory. After reading this book, a student will hopefully understand the main areas of current progress in string theory, and may be able to engage directly in research. The primary focus, therefore, is to place the reader at the forefront of current string research.

This book is complementary to my previous book, *Introduction to Superstrings,* which gives the reader a firm foundation in the fundamentals of string theory. It contains new material not covered in that book, such as the classification of conformal field theories, the nonpolynomial closed string field theory, the matrix models, and topological field theory.

Although it would be helpful to read *Introduction to Superstrings,* it is not absolutely necessary. The overlap between this book with the previous one is minimal, but in some chapters, such as One and Ten, I have reviewed the necessary background material where needed so this book would be self-contained. However, readers are encouraged to consult the previous book for its glossary and appendix which contains a brief introduction to group theory, supergravity, supersymmetry, the theory of forms, and general relativity.

Strings, Conformal Fields, and Topology will be a success if it conveys some of the vitality and vigor of current activity in string theory to the reader and prepares him or her for research.

Michio Kaku

Acknowledgements

I would like to thank the hospitality of the Institute for Advanced Study at Princeton, where this book was written. I would especially like to thank Drs. E. Witten and S. Adler for inviting me to come to the Institute.

I would like to thank Dr. B. Sakita and Dr. J. Birman and the faculty of the City College of New York for their constant encouragement and support. I would like to acknowledge support from the National Science Foundation, Department of Energy, and CUNY-FRAP.

I would like to thank L. Alvarez Gaumé, who once again has given detailed and productive comments throughout the entire draft. Both the presentation and content of this book has greatly benefited from his insightful comments. I would also like to thank D. Karabali, O. Lechtenfeld, L. Hua, P. Huet, B. Grossman, and A. Jevicki for reading various chapters of this book and making many valuable comments.

Contents

Part II   Nonperturbative Methods

Chapter 9   Beyond the Planck Length   285

Chapter 10   String Field Theory   315

Chapter 11   Nonpolynomial String Field Theory   354

Chapter 12   Geometric String Field Theory   389

Part I

Conformal Field Theory and Perturbation Theory

Chapter 1

Introduction to Superstrings

1.1. Introduction

During this century, two great physical theories have emerged. Each theory remains unchallenged within its respective domain.

The first, quantum theory, has given us a theory of the microcosm. At the subatomic level, quantum mechanics has unraveled the secrets of matter and energy. Quantum mechanics has given us the language by which we can unite three of the four fundamental forces, the strong and weak forces and electromagnetic interactions.

The second, the general theory of relativity, has given us a theory of the macrocosm. At the cosmic level, relativity theory has given us an unrivaled description of creation and cosmology. Black holes, warped space–time, and the expanding universe are all consequences of this theory of gravitation.

No experimental deviation from either theory has been seen in the laboratory. The entire body of physical knowledge amassed by physicists over the last two millennia is based on these two physical theories, without exception.

The unification of these two theories, however, has eluded some of the greatest thinkers of our time. Wolfang Pauli, Werner Heisenberg, and Albert Einstein wrestled with the problem and ultimately failed. At the center of this puzzle is the realization that the theories are based on different assumptions, obey different physical principles, and use different mathematics. Whenever attempts are made at merging these two theories, disastrous infinities emerge that render the hybrid theory meaningless.

In particular, these two theories have fundamentally different viewpoints regarding the meaning of a force. Quantum theory reduces all forces to the exchange of tiny discrete packets of energy, called "quanta." General relativity, however, considers forces to be an apparent effect, the consequence of the smooth distortion of space and time.

At present, superstring theory [1–7] has emerged as the leading candidate for the unification of these two theories and hence all known forces. No other theory can make the claim of unifying both general relativity and quantum mechanics into a simple, self-consistent formalism. Calculations carried out to the 8th-loop order, and some calculations to all loop orders, show that the potential divergences inherent in other quantum field theories miraculously cancel in string theory due to the enormously powerful symmetries built into the theory. From a technical point of view, superstrings theory seems to be totally free of quantum anomalies and divergences, which riddle all known point particle theories of gravity and matter. Its successes can be summarized as follows:

1.   The theory explains the origin of particles and resonances. The myriad particles found in nature can be viewed as the vibrations of a string, in much the same way that the notes found in music can be explained as the modes of a vibrating string. Pursuing this analogy, the basic particles of our world correspond to the musical notes of the superstring, the laws of physics correspond to the harmonies that these notes obey, and the universe itself corresponds to a symphony of superstrings.

2.   The theory explains the unification of gravity with matter. The simplest vibration of a closed string in curved space, in fact, corresponds to a massless spin-two object. The gauge symmetries of string theory show that this spin-two particle has all the properties of the graviton. Moreover, when the string executes its motions, it actually forces space–time to curl up around it, yielding the complete set of Einstein's equations of motion. Thus, the string naturally merges the two divergent pictures of a force: the modes of vibration are quantized, but the string can only self-consistently vibrate in a curved space–time consistent with Einstein's equations of motion.

3.   The theory explains how to remove the divergences found in quantum gravity. In quantum field theory, point particles interact via Feynman graphs, which badly diverge when the graph is "pinched," that is, when one of the legs of the graph shrinks to zero. When the string moves, it repeatedly splits and reforms, thereby tracing the topology of two-dimensional sheets or Riemann surfaces, such as a doughnut. However, since it is difficult to pinch or stretch a doughnut, one can show that the string graphs are actually ultraviolet finite. Thus, the topology of the string removes the divergences of quantum gravity.

4.   The theory shows how to simply incorporate the symmetries of the standard model. In the heterotic string, for example, one set of vibrations exists in 26 dimensions. However, since the superstring exists in 10 dimensions, we let 16 unwanted dimensions curl up like a ball. If we compactify on the root lattice of a rank 16 Lie algebra, such as $E_8 \otimes E_8$, then we have a theory that is large enough to accomodate $SU(3) \otimes SU(2) \otimes U(1)$. (By contrast, the string's nearest rival, supergravity, has a maximum Lie group of $SO(8)$, which is too small to accomodate the standard model.)

In spite of the simplicity of string theory and the success of string phenomenology, there are, of course, several formidable obstacles, both experimental and theoretical, that must be overcome before the theory can be accepted.

Experimentally, unification in superstring theory takes place at the Planck length (10^{19} GeV), and hence experimental verification of this theory remains problematic. Even the largest particle accelerators or observations from cosmic ray detectors and satellites will, at best, be able to probe only indirect signals emerging from the Planck length. One hope is that the superconducting super collider (SSC) will observe supersymmetric partners of the lower lying particles and hence give us indirect clues to the superstring itself. However, to probe energies at the Planck length is impossible with conceivable technology.

The origin of this problem is that superstring theory (or any theory that claims to unite all four forces) is inherently a theory of creation, where all four forces were united into a single superforce. Thus, to experimentally verify superstring theory means recreating creation in the laboratory.

Although this seemingly insurmountable experimental barrier is usually cited as the chief criticism of superstring theory, our philosophy is very different. We feel that the main problem facing superstring theory is theoretical, not experimental. If we could fathom the principles underlying the theory, then we would be able to calculate the physics of our low-energy world, and settle, once and for all, whether superstring theory is correct or not. Since the basic equations of superstring theory (at least perturbatively) are well known, it means that we are not clever enough, with our limited mathematical skills, to solve the theory.

If superstring theory, as some have claimed, is 21st-century physics that fell accidentally into the 20th century, then the fundamental problem facing us is that 21st-century mathematics powerful enough to solve 21st-century physics has not been discovered.

As Richard Feynman once said, one of the goals of a theoretical physicist is to "prove yourself wrong as soon as possible." Thus, the real problem facing us, in our opinion, is to theoretically settle the following question as quickly as possible: what is the true vacuum (ground state) of superstring theory? Since the ground state should correspond to our physical universe, if the true vacuum could be discovered, we might be able to decisively settle whether superstring theory is a theory of the universe or just the latest in a series of failed efforts to discover the Holy Grail of physics, the unified field theory.

Once the true ground state of the superstring theory is found, we should be able to calculate the quantum numbers and physical properties of the various particles found in nature and compare them with actual experimental data. In theory, starting from the superstring action and extracting its ground state, we should be able to calculate all the basic parameters of the universe from first principles.

The search for the true vacuum of string theory is therefore the central theme of this book. It is the theme that binds the various chapters together and determines the structure of this book. It is also the theme that drives almost all current research in superstring theory.

As a consequence, there are two parts to this book:

1.   perturbation theory and conformal field theory and
2.   nonperturbative methods.

In the first part, we focus on the perturbative vacuum, which can be described using conformal field theory. The basic strategy for the first part of this book is therefore to classify all possible conformal field theories. Although a complete classification of conformal field theories that reveals the deeper relationship between perturbative vacuums still lies beyond our grasp, we will review the enormous progress made in this direction.

One of the successes of the conformal field theory approach is that, with only a few mild constraints, one finds reasonably acceptable candidates for practical phenomenology. For example, the heterotic string is the leading superstring theory, containing the gauge group $E_8 \otimes E_8$. By making a few reasonable assumptions about its broken phase, it is possible to break this group down to $E_8 \otimes E_6$ and finally to E_6, which contains the standard model's gauge group, $SU(3) \otimes SU(2) \otimes U(1)$. The basic fermion multiplet naturally occurs in the **27** multiplet of E_6, which is consistent with known grand unified theory (GUT) phenomenology.

Thus, it is surprising that, with very few minimal assumptions, we are naturally led to the following symmetry-breaking scheme:

$$E_8 \otimes E_8 \to E_8 \otimes E_6 \to E_6 \to SU(3) \otimes SU(2) \otimes U(1) \qquad (1.1.1)$$

Not only is the theory internally self-consistent after unifying quantum mechanics with general relativity, the theory also goes far beyond the usual GUT description of quarks and leptons. The gauge hierarchy problem, for example, which is impossible to solve within the framework of the standard GUT, is naturally explained by the supersymmetry (SUSY) of the superstring. Furthermore, the generation problem, which has plagued all GUTs, is naturally explained in string theory via topological arguments. For example, it is now relatively easy to construct pheonomenological models that contain only the three experimentally observed fermion generations.

In spite of the early successes of the conformal field theory approach to the perturbative vacuum, there are several fundamental flaws with this approach that force us to study nonperturbative effects.

1.   Millions of conformal field theories have been discovered, each one representing a possible (perturbative) ground state of the theory. Which one is correct? No single scheme has been found that can classify them in a coherent fashion. In Chapter 7, we will study some of the more powerful classification schemes so far proposed, but all fall far short of an all-embracing classification scheme. This, in turn, has made it difficult, if not impossible, to decide which of the millions of conformal field theories describes our known universe.

2. String phenomenology, although surprisingly successful, suffers from a fatal defect. We can show to all orders in perturbation theory that supersymmetry is unbroken. However, our low-energy world does not manifest supersymmetry, which must therefore be dynamically broken. The low-energy spectrum of string theory thus cannot give us a correct description of the physical spectrum of particles. For example, it includes massless particles, such as the dilaton, which have not been seen in nature.

3. More important, 10-dimensional space–time also seems perfectly stable in perturbation theory. At present, the breaking of 10-dimensional space down to 4 dimensions is purely an ad hoc assumption that is built into conformal field theory from the very beginning, not a consequence of string dynamics. Until this fundamental problem is solved nonperturbatively, all the predictions of perturbative string theory are suspect.

4. Conformal field theories may also be unstable at the Planck energy level. Studies of the higher loops of bosonic theory show that the theory is not Borel summable, and hence, the theory is not stable when power is expanded perturbatively around a conformal field theory. [We should note that quantum electrodynamics (QED), quantum chromodynamics (QCD), and gauge theories in general are also not Borel summable, but we implicitly assume that they can be embedded into a higher theory that is finite. However, superstring theory, making the claim of being the final theory, cannot therefore be embedded into a higher theory.] This instability seems to be independent of the existence of the tachyon and hence may persist even in the entire superstring theory.

If perturbation theory is flawed because it fails to precisely describe our low-energy universe and is unstable, then we must resort to nonperturbative calculations, which takes us to the second part of this book.

Although nonperturbative calculations are notoriously difficult to perform, even in point particle theories, in the second part of this book we review what is known about nonperturbative approaches to string theory. Although most superstring research is presently confined to perturbation theory, the future of superstring research in the coming years may lie in the nonperturbative realm.

We review all the major approaches to nonperturbative string theory. At present, string field theory is the leading formalism for this type of calculation. String field theory successfully unifies all known information about string theory in a simple, second quantized action, rather than appealing to folklore and rules of thumb.

String field theory, like other nonperturbative approaches, is still too difficult to solve. Systems with an infinite number of degrees of freedom have traditionally been difficult to solve, and string field theory is no exception. Recent research on much simpler "toy" models, with finite degrees of freedom, has enjoyed some success.

Some of the most important systems with finite degrees of freedom are topological models. In fact, topology will prove to be one of the most powerful tools are our disposal to calculate the properties of the vacuum of string theory. Topology enters into string theory in at least three ways.

First, topological arguments are crucial in determining the phenomenological predictions of the theory once we have compactified the theory down to four dimensions. In particular, topological invariants defined on six dimensional manifolds are the key ingredients necessary to solve for the properties of nature. For example, the Euler number is related to the generation number found in GUT theories.

Second, topology is useful in classifying the possible vacuums of the theory, that is, the various conformal field theories. In particular, knot theory, which until recently had no practical application in theoretical physics, is now known to be crucial in classifying the so-called rational conformal field theories. We will find that the "Clebsch-Gordan coefficients" for the conformal group can be determined by the braiding operations performed on knots, and that we can calculate the known knot invariants and even infinite classes of new ones directly from physics.

Third, topology has played a key role in solving string theory in $D \leq 1$. Certain topological field theories, in fact, can be shown to be equivalent to the matrix model formulation of string theory in low dimensions, which give us a complete non-perturbative solution to the Green's functions to all orders in the genus. In these $D \leq 1$ models, the theory has only a finite number of degrees of freedom, and hence topological arguments can be used to solve them exactly.

Our interest, however, in topology arises from the fact that studies of the high-energy, high-temperature phase of perturbative string theory show that new, unexpected symmetries emerge associated with systems of finite degrees of freedom in that exotic region. Since the "natural home" of string theory lies beyond the Planck scale, perhaps topology will give us a key tool by which to probe string theory in its natural state. Perhaps topology will give us a clue to solve the central problem of superstring theory, the search for its true vacuum state.

We begin the first part of this book with a discussion of conformal field theory, which has emerged as a powerful tool by which to probe the perturbative phenomenology of the theory. It is conjectured that each conformal field theory corresponds to a possible string vacuum. Thus, by classifying all possible conformal field theories, we hope to exhaust all possible vacuums allowed by superstring theory. Perhaps one of them describes the physical universe.

This chapter is self-contained and will hopefully serve as a brief introduction to superstring theory. Students with a familiarity of elementary string theory may skip this chapter. However, the reader is encouraged to consult *Introduction to Superstrings* for certain concepts that may be explained at greater length and depth.

1.2. Quantizing the Relativistic String

To begin our discussion of string theory and conformal field theory, let us study the first quantized action of the relativistic string. If a point particle moves in space–time, it sweeps out a one-dimensional world line. Likewise, as a string moves in space–time, it sweeps out a two-dimensional world sheet. Let $X_\mu(\sigma, \tau)$ represent a vector defined in D-dimensional space–time that begins at the origin of our coordinate system and ends at some point along the two-dimensional string world sheet, labeled by $\xi^a = \{\sigma, \tau\}$. Let $\eta_{\mu\nu} = (-, +, +, + \cdots)$ be the flat metric in D-dimensional space–time, where $\mu = 0, 1, 2, ..., D - 1$. [We will use Roman letters $(a, b, c, ...)$ to represent two-dimensional world sheet indices, and Greek letters $(\mu, \nu, ...)$ to represent D-dimensional space–time indices.]

Let us take our action to be the area of the world sheet swept out by the string [8–9]. Because the area of the world sheet is independent of the two-dimensional coordinates used to measure the area, it must be a generally covariant scalar in two dimensions.

There are several ways in which we can write the area of the string world sheet. The simplest is to introduce the tensor g^{ab}, which represents a two-dimensional metric defined on the surface.

Our action can be written as [10]:

$$S = -\frac{1}{4\pi\alpha'} \int d^2\xi \, \sqrt{g} g^{ab} \, \partial_a X_\mu \, \partial_b X_\nu \, \eta^{\mu\nu} \qquad (1.2.1)$$

where $\alpha' = 1/2$ for open strings and $\alpha' = 1/4$ for closed strings. The action is manifestly reparametrization invariant. If we reparametrize the two-dimensional world sheet according to:

$$\sigma \to \tilde\sigma(\sigma, \tau), \qquad \tau \to \tilde\tau(\sigma, \tau) \qquad (1.2.2)$$

then the action is invariant under this two-dimensional general coordinate transformation if:

$$\bar{g}^{ab}(\bar{x}) = \left(\frac{\partial \bar{x}^a}{\partial x^c}\right)\left(\frac{\partial \bar{x}^b}{\partial x^d}\right) g^{cd}(x) \qquad (1.2.3)$$

where $x^a = \{\sigma, \tau\}$. Under this transformation, the action is manifestly coordinate invariant (because the transformation of $\sqrt{g}$ cancels against the transformation of the two-dimensional measure).

If we take an infinitesimal transformation, then the transformation of the fields becomes:

$$\begin{aligned} \delta g^{ab} &= \epsilon^c \, \partial_c g^{ab} - g^{ac} \, \partial_c \epsilon^b - g^{bc} \, \partial_c \epsilon^a \\ \delta X_\mu &= \epsilon^a \, \partial_a X_\mu \end{aligned} \qquad (1.2.4)$$

The action is also trivially invariant under local scale transformations:

$$g^{ab} \rightarrow e^{\phi} g^{ab} \tag{1.2.5}$$

(When we quantize the system, we will find that the classical scale invariance of the system actually breaks down, and the metric tensor obeys the equations of Liouville theory. This conformal anomaly disappears in 26 dimensions for bosonic strings and 10 dimensions for the superstring. In later chapters, we will discuss how to quantize the two-dimensional gravitational system in dimensions other than 10 and 26, where the Liouville mode plays a crucial role.)

The two-dimensional metric in the action does not have any derivatives, and hence we may classically eliminate it via its equations of motion. Then, we find:

$$g_{ab} \sim \partial_a X_\mu \, \partial_b X^\mu \tag{1.2.6}$$

Reinserting this value of the metric tensor back into the action, we find the Nambu–Goto action, a nonlinear action written totally in terms of the string variable [8, 9]:

$$S = \frac{1}{2\pi\alpha'} \int d^2\xi \, \sqrt{(\dot{X}^\mu)^2 (X^{\mu\prime})^2 - (\dot{X}_\mu X^{\mu\prime})^2} \tag{1.2.7}$$

where $\dot{X}^\mu$ equals $\partial_\tau X^\mu$ and X'^μ equals $\partial_\sigma X^\mu$. Notice that the above expression is proportional to the area of the world sheet, given by the integral of $\sqrt{\det g_{ab}}$.

The action is now written as the determinant of the metric tensor of the world sheet, which shows that the action is proportional to the two-dimensional area swept out by the string. It is remarkable that string theory, which provides a comprehensive scheme in which to unite general relativity with quantum mechanics and all known physical forces, begins with this simple statement: the action is proportional to the area of the string world sheet.

In order to extract the physical predictions of the theory, we must first quantize the theory and extract its physical states. The quantization of the string action is nontrivial, however, because the system has a powerful symmetry and is hence highly redundant in its degrees of freedom.

Over the years, three equivalent quantization schemes have been developed: (1) Gupta–Bleuler quantization, (2) light cone quantization, and (3) Becchi–Rouet–Stora–Tyupin (BRST) quantization. Each method has its own advantages and disadvantages.

Gupta–Bleuler Quantization

Because the action possesses local reparameterization invariance (with two parameters) and scale invariance (with one parameter), we are allowed to impose a total of three constraints on the metric tensor. This will break the reparameterization invariance and will allow ghosts to propagate in the theory because of the wrong sign of the time component of the string field in the propagator. However, these ghost states will be eliminated by directly applying constraints on the states of the theory.

Let us choose the conformal gauge, in which all components of the metric tensor are set to constants:

$$\text{conformal gauge} : \quad g^{ab} = \delta^{ab} \tag{1.2.8}$$

Then, our Lagrangian linearizes to the following [11]:

$$L = \frac{1}{4\pi\alpha'} \left[(\dot{X}_\mu)^2 + (X'_\mu)^2 \right] = \frac{1}{2\pi} \partial_z X_\mu \partial_{\bar{z}} X^\mu \tag{1.2.9}$$

where:

$$z = \sigma + i\tau \tag{1.2.10}$$

where we have performed a Wick rotation so the system is conformally invariant and where the equations of motion become:

$$\left(\frac{\partial^2}{\partial^2 \sigma} + \frac{\partial^2}{\partial^2 \tau} \right) X_\mu = 0 \tag{1.2.11}$$

The equations of motion can now be trivially solved in terms of functions of

$$X_\mu(\sigma \pm i\tau) \tag{1.2.12}$$

which in turn can be decomposed into cosine modes (for open strings) or into cosine and sine modes (for closed strings).

Notice that the action, while no longer locally reparameterization invariant, is still globally invariant under *conformal transformations* [11]:

$$z \to f(z) \tag{1.2.13}$$

Under conformal transformations, the string transforms as:

$$\delta X_\mu(z, \bar{z}) = \epsilon(z) \partial_z X_\mu + \epsilon(\bar{z}) \partial_{\bar{z}} X_\mu \tag{1.2.14}$$

Written in the conformal gauge, the action can be trivially quantized because it has been reduced to a set of free harmonic oscillators. If we define the canonical momentum to the string variable as:

$$\begin{aligned} [P_\mu(\sigma), X_\nu(\sigma')] &= -i\eta_{\mu\nu}\delta(\sigma - \sigma') \\ P_\mu &= \delta L / \delta \dot{X}^\mu \end{aligned} \tag{1.2.15}$$

then we can decompose the string variable into harmonic oscillators [12]:

$$\begin{aligned} X^\mu(\sigma) &= x^\mu + i \sum_{n=1}^{\infty} \frac{1}{\sqrt{n}} (a_n^\mu - a_{-n}^\mu) \cos n\sigma \\ P^\mu(\sigma) &= \frac{1}{\pi} \left[p^\mu + \sum_{n=1}^{\infty} \sqrt{n} (a_n^\mu + a_{-n}^\mu) \cos n\sigma \right] \end{aligned} \tag{1.2.16}$$

where the commutation relations are satisfied if:

$$[a_{n\mu}, a_{m\nu}] = \delta_{n,-m}\eta_{\mu\nu} \tag{1.2.17}$$

Then, the spectrum of string theory is given by the eigenvalues of the Hamiltonian:

$$H = \int_0^\pi d\sigma (P_\mu \dot{X}^\mu - L) = \sum_{n=1}^\infty na_{-n\mu}a_n^\mu + \alpha' p_\mu^2 \tag{1.2.18}$$

Since the Hamiltonian is just the sum over an infinite set of uncoupled, free harmonic oscillators, the spectrum consists of the sum total of all possible products of harmonic oscillator states [12]:

$$\prod_{n,\mu}\{a_{-n,\mu}\}|0\rangle \tag{1.2.19}$$

To analyze the symmetries of the system, let us calculate the energy-momentum tensor corresponding to the action. The energy-momentum tensor is defined as:

$$T_{ab} = -4\pi\alpha' \frac{1}{\sqrt{g}} \frac{\delta L}{\delta g^{ab}} \tag{1.2.20}$$

This, in turn, can be shown to equal:

$$T_{ab} = \partial_a X_\mu \, \partial_b X^\mu - \frac{1}{2} g_{ab} g^{cd} \, \partial_c X^\mu \, \partial_d X_\mu \tag{1.2.21}$$

We notice several important features of the energy-momentum tensor, that is, it satisfies:

$$\partial_b T^{ab} = 0; \qquad \mathrm{Tr}\, T_{ab} = 0 \tag{1.2.22}$$

The first statement simply states that the energy-momentum tensor is conserved. The second statement states that it is traceless, that is, it is scale invariant.

For us, perhaps the most important feature of the energy-momentum tensor is that it forms an algebra that closes. Let us first define the *Virasoro generators* L_n as the Fourier moments of the energy-momentum tensor [13]:

$$\begin{aligned}
L_m &= \frac{1}{4\pi\alpha'} \int_0^\pi d\sigma \left[e^{im\sigma}(T_{00} + T_{01}) + e^{-im\sigma}(T_{00} - T_{01}) \right] \\
&= \frac{1}{8\pi\alpha'} \int_{-\pi}^\pi d\sigma \left[e^{im\sigma}(\dot{X} + X')^2 \right] \\
&= \frac{1}{2} \sum_{n=-\infty}^\infty \alpha_{m-n}\alpha_n
\end{aligned} \tag{1.2.23}$$

These generators, L_n, which will appear throughout this book, in turn form a closed algebra, the Virasoro algebra (which generates the conformal group):

$$[L_n, L_m] = (n - m)L_{n+m} + \frac{c}{12}n(n + 1)(n - 1)\delta_{n,-m} \qquad (1.2.24)$$

where c is a constant.

In the Gupta–Bleuler quantization scheme, we allow ghosts to propagate in the theory. However, unitarity is reestablished by applying the gauge constraints directly on the Fock space. Thus, we apply:

$$L_n|R\rangle = 0, \quad n > 0$$
$$(L_0 - 1)|R\rangle = 0 \qquad (1.2.25)$$

where the second condition is the mass-shell condition. The solution of these constrains is the set of real states $|R\rangle$ of the theory. (The actual proof, however, that these conditions are sufficient to eliminate all ghost states is quite involved [14, 15]. The proof shows that string theory is ghost free if the dimension of space–time is 26 and the intercept, or the highest spin of the massless sector the theory, is equal to 1 for open strings and 2 for closed strings.)

Spurious states are those that do not couple to the real states, that is,

$$\langle S|R\rangle = 0 \qquad (1.2.26)$$

This equation, in turn, implies that any spurious state contains Virasoro generators:

$$|S\rangle = L_{-k_1}L_{-k_2}, \cdots, L_{-k_n}|R\rangle \qquad (1.2.27)$$

The advantage of the Gupta–Bleuler method is that Lorentz covariance is maintained throughout. However, the proof that unphysical states are eliminated by the Virasoro generators is highly nontrivial. Hence, Lorentz covariance is manifest in the Gupta–Bleuler formalism but unitarity is not.

We now turn to another quantization scheme where only the physical states are present and unitarity is manifest.

Light Cone Quantization

The advantage of the light cone quantization method, as in the case of ordinary point particle gauge theories, is that the theory is manifestly ghost free, and hence, only physical states propagate. All gauge constraints have been eliminated explicitly, leaving only the physical Hilbert space of transverse modes. The essence of the light cone quantization method lies in eliminating the redundant longitudinal modes by gauge fixing two degrees of freedom and solving the constraints.

We will define the light cone coordinates as:

$$X^+ = \frac{1}{\sqrt{2}}\left(X^0 + X^{D-1}\right)$$
$$X^- = \frac{1}{\sqrt{2}}\left(X^0 - X^{D-1}\right) \qquad (1.2.28)$$

and fix the gauge as [16]:

$$X^+(\sigma,\tau) = p^+\tau \tag{1.2.29}$$

The momentum canonical to X_μ is P^μ, whose components are not all independent. If we take Eq. (1.2.7) as our action and then take derivatives with respect to $\dot{X}_\mu$ to form P_μ, we find that the resulting momenta, which are complicated nonlinear objects, are not independent but obey the following simple identities:

$$P_\mu^2 + \frac{1}{(2\pi\alpha')^2}X_\mu'^2 \equiv 0 \tag{1.2.30}$$
$$P_\mu X'^\mu \equiv 0$$

The light cone quantization program begins by solving these constraints to eliminate the redundant longitudinal modes:

$$P^-(\sigma) = \frac{\pi}{2p^+}\left(P_i^2 + \frac{X_i'^2}{\pi^2}\right)$$
$$X^-(\sigma) = \int_0^\sigma d\sigma' \frac{\pi}{p^+}(P_iX_i') \tag{1.2.31}$$

The Hamiltonian in the light cone gauge reduces to:

$$H = \frac{\pi}{2}\int_0^\pi \left(P_i^2 + \frac{X_i^2}{\pi^2}\right)d\sigma \tag{1.2.32}$$

so that the physical Hilbert space corresponds to the set of all transverse harmonic oscillator states.

The disadvantage of the light cone formalism, of course, is that Lorentz covariance is broken and, in fact, must be reestablished at each step of the way. We check Lorentz invariance by constructing the Lorentz generators:

$$M^{\mu\nu} = \int_0^\pi d\sigma\left(X^\mu P^\nu - X^\nu P^\mu\right)$$
$$= x^\mu p^\nu - x^\nu p^\mu - i\sum_{n=1}^\infty \frac{1}{n}\left(\alpha_{-n}^\mu \alpha_n^\nu - \alpha_{-n}^\nu \alpha_n^\mu\right) \tag{1.2.33}$$

Checking Lorentz invariance now amounts to reinserting the constrained values of $X^\pm$ and $P^\pm$ into the Lorentz generators. The only difficult commutator is given by the following [16]:

$$[M^{-i}, M^{-j}] = -\frac{1}{p^{+2}}\sum_{n=1}^\infty \left(\alpha_{-n}^i \alpha_n^j - \alpha_{-n}^j \alpha_n^i\right)\Delta_n \tag{1.2.34}$$

where:

$$\Delta_n = \frac{n}{12}(26 - D) + \frac{1}{n}\left(\frac{D-26}{12} + 2 - 2a\right) \tag{1.2.35}$$

where a is the intercept. In order to have Lorentz invariance, we must set Δ_n equal to zero, that is,

$$D = 26; \quad a = 1 \tag{1.2.36}$$

In the Gupta–Bleuler quantization, conformal invariance alone fixes the dimension of space–time to 26. In the light cone case, because Lorentz invariance and conformal invariance are now mixed nontrivially together, Lorentz invariance fixes $D = 26$.

BRST Quantization

The BRST quantization method [17, 18] combines the best aspects of each of the previous quantization methods. Like the Gupta–Bleuler quantization scheme, it is Lorentz invariant; and like the light cone quantization, we can easily extract the physical states (and the $D = 26$ constraint).

The BRST method begins with the conformal gauge, but then uses the Faddeev–Popov method to introduce ghost states. These ghost states allow us to maintain covariance throughout the calculation; the final theory remains unitary because these ghosts will cancel against unphysical states.

Let us rewrite the gauge transformation of the metric as:

$$\delta g_{ab} = g_{ac} \partial_b \delta v^c + \partial_a \delta v^c g_{cb} - \delta v^c \partial_c g_{ab} = \nabla_a \, \delta v_b + \nabla_b \, \delta v_a \tag{1.2.37}$$

The Faddeev–Popov determinant [19] arises because we cannot simply insert the conformal gauge [Eq. (1.2.8)] into the functional integral because it will contribute an incorrect measure. In fact, the correct insertion of the conformal gauge is given by the number one:

$$1 = \delta(g_{ab} - \delta_{ab})\Delta_{FP} \tag{1.2.38}$$

This determinant Δ_{FP} can be shown to equal the determinant of the variation of Eq. (1.2.37) with respect to δv_a. This, however, is just the determinant of the operator ∇_a.

The Faddeev–Popov determinant can thus be rewritten as:

$$\Delta_{FP} = \det(\nabla_a) = \det \nabla_z \det \nabla_{\bar{z}} \tag{1.2.39}$$

We now use the fact that any determinant can be rewritten as an exponential integral. Normally, the functional integral over a Gaussian yields a determinant in the denominator. Because the Faddeev–Popov determinant appears in the numerator, rather than the denominator, we must introduce functional integration over Grassmann variables, denoted by b, c, which are the Faddeev–Popov ghosts.

We can rewrite the Faddeev–Popov determinant as [19]:

$$\Delta_{FP} = \int Db D\bar{b} Dc D\bar{c} \, e^{i \int L^{bc} d^2 z} \tag{1.2.40}$$

where:

$$L^{bc} = \frac{1}{\pi}\left(b\,\partial_{\bar{z}}c + \bar{b}\,\partial_z\bar{c}\right) \tag{1.2.41}$$

This action, in turn, has an additional symmetry associated with it, called the BRST symmetry. This symmetry is a global one, so no fields can be eliminated through this symmetry. The generator of this global symmetry is given by [20]:

$$Q = \sum_{n=-\infty}^{\infty} : c_{-n}\left(L_n^X + \frac{1}{2}L_n^{bc} - a\delta_{n,0}\right)$$

$$= c_0(L_0 - a) + \sum_{n=1}^{\infty}\left(c_{-n}L_n + L_{-n}c_n\right) \tag{1.2.42}$$

$$-\frac{1}{2}\sum_{n,m=-\infty}^{\infty} : c_{-m}c_{-n}b_{n+m} : (m-n)$$

where:

$$\{c_n, b_m\} = \delta_{n,-m} \tag{1.2.43}$$

and:

$$Q^2 = \frac{1}{2}\sum_{m=-\infty}^{\infty}\left[\frac{D}{12}(m^3 - m) + \frac{1}{6}(m - 13m^3) + 2am\right]c_m c_{-m} \tag{1.2.44}$$

which vanishes only if $D = 26$ and $a = 1$, as before. (The double dots indicate normal ordering, that is, the creation oscillators with negative indices appear to the left, and the annihilation oscillators with positive indices appear to the right.)

Although the derivation of the BRST operator Q came from the conformal gauge, its actual origin is quite general, independent of any gauge. For example, given any Lie algebra $[\tau_a, \tau_b] = f_{ab}^c \tau_c$, with generators τ_a, it is possible to construct a nilpotent operator Q such that $Q^2 = 0$ by introducing anticommuting variables b, c. Notice that [21]:

$$Q = \sum_{n=-\infty}^{\infty} c_{-n}\left(\tau_n - \frac{1}{2}f_{nm}^p c_{-m}b_p\right) \tag{1.2.45}$$

satisfies the identity $Q^2 = 0$ if the Jacobi identity is satisfied for the algebra.

The addition of the ghost states has vastly increased the Fock space of the theory, which now consists of all possible products over the string creation oscillators and ghost oscillators:

$$\prod_{n,m,p,\mu} \{\alpha_{-n}^\mu\}\{b_{-m}\}\{c_{-p}\}|0\rangle \tag{1.2.46}$$

The last step in the BRST quantization program is to eliminate the unphysical states by applying the operator Q onto states:

$$Q|\psi\rangle = 0 \qquad (1.2.47)$$

We don't count the states that vanish trivially, that is, those that can be written as $|\psi\rangle = Q|\lambda\rangle$ for some $|\lambda\rangle$.

Thus, the criterion for physical states is:

$$Q|\text{physical}\rangle = 0$$
$$|\text{physical}\rangle \neq Q|\lambda\rangle \qquad (1.2.48)$$

Mathematically speaking, we say that the physical states lie within the *cohomology* of the BRST operator Q. They are given by the kernel of the operator Q, divided by the image of Q:

$$|R\rangle = \frac{\{\ker Q\}}{\{\operatorname{im} Q\}} \qquad (1.2.49)$$

1.3. Scattering Amplitudes

This completes the discussion of the free string. In order to discuss how to construct scattering amplitudes for strings, let us quickly review the development of the interacting string.

String theory developed quite by accident when Veneziano [22] and Suzuki [23] stumbled across the Euler beta function, which seemed to satisfy all the properties of an S matrix (scattering matrix) for hadronic scattering. The one property that the beta function failed to satisfy was unitarity.

Then, Kikkawa, Sakita, and Virasoro (KSV) [24] postulated that the beta function should be treated as the lowest order Born term in a Feynman-like perturbation series involving multiloops. This conjecture was verified when Kaku, Yu, Lovelace, and Alessandrini [25–27] actually constructed the multiloop amplitudes by sewing together tree diagrams. The integrand of these higher order amplitudes was given by the solution to Laplace's equation defined on a Riemann surface of genus g.

This perturbation series in terms of Riemann surfaces was given an elegant interpretation in terms of path integrals. Hsue, Sakita, and Virasoro [28] showed that the entire perturbation series could be written as a path integral summed over all conformally inequivalent Riemann surfaces of genus g (see Fig. 1.1.):

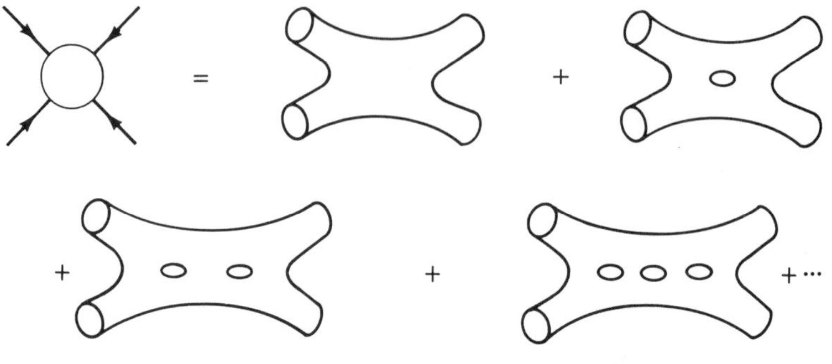

Fig. 1.1.

$$A_N = \sum_{\text{Topologies}} \int DX_\mu \int d\mu \exp\left[i\int d^2z L(z) + \sum_{i=1}^{N} ik_{i,\nu}X^\nu(z_i)\right]$$

$$= \sum_{\text{Topologies}} \int d\mu \left\langle \prod_{i=1}^{N} e^{ik_i X_i}\right\rangle$$

$$(1.3.1)$$

where $d\mu$ is a conformal measure. Because of the simplicity of the action (which is that of a free theory), the N-point functions are all exactly calculable.

There are two direct ways in which to actually calculate the N-point function from the functional integral. The first is to calculate the functional integral over the complex plane, and the second is to use harmonic oscillators.

The first method performs the functional integral by shifting the X_μ variable by a classical solution. For open strings, the region of the complex plane over which we wish to integrate is an infinite horizontal strip, sitting on the x axis, with a width of π. The points z_i where momenta p_i enter into the strip, are located on the real axis. By an exponential conformal transformation, we can map this horizontal strip to the upper half plane. (For closed strings, the strip will have width 2π and will be mapped to the entire complex plane. The points z_i will be located throughout the complex plane.)

Let us shift the integration variable by a solution to the classical equation:

$$X_\mu \to X_{\mu,\text{classical}} + X_\mu \qquad (1.3.2)$$

where the classical solution is determined via the Green's function for Laplace's equation on the upper half plane:

$$X_{\mu,\text{classical}} = -i\alpha' \int G(z,z')J(z')\,dz'$$

$$G(z,z') = \ln|z - z'| + \ln|z - z^{*'}|$$

(1.3.3)

The integral is easy to perform, since it is just a Gaussian in the string variable, so we find:

$$A_N = \int \prod_{i=3}^{N-1} dz_i \prod_{2 \le i < j \le N} |z_i - z_j|^{k_i k_j}$$

(1.3.4)

where we have used projective invariance $SL(2,R)$ to fix $\infty = z_1 \ge z_2 = 1 \ge z_3 \cdots z_{N-1} \ge z_N = 0$.

The other equivalent way is to convert path integrals to harmonic oscillators by taking vertical "time slices" along the horizontal strip. The Hamiltonian on the conformal surface is related to L_0, so the propagator becomes:

$$D = \int_0^\infty e^{-\tau(L_0-1)}d\tau = \frac{1}{L_0 - 1}$$

(1.3.5)

while the vertex function becomes:

$$V(k) = \; : e^{ik_\mu X_\mu} : \; = \exp\left(k \cdot \sum_{n=1}^\infty \frac{\alpha_{-n}}{n}\right) \exp\left(-k \cdot \sum_{n=1}^\infty \frac{\alpha_n}{n}\right)$$

(1.3.6)

The N-point function becomes:

$$A_N = \langle 0, k_1 | V(k_2)DV(k_3)\cdots V(k_{N-1})|0, k_N\rangle$$

(1.3.7)

When this is explicitly evaluated (using, for example, coherent states), we find the previous expression in Eq. (1.3.4). For $N = 4$, this becomes the celebrated Veneziano formula:

$$B_4(s,t) = \int_0^1 dx\, x^{-s/2-2}(1-x)^{-t/2-2} = \frac{\Gamma[-\alpha(s)]\Gamma[-\alpha(t)]}{\Gamma[-\alpha(s)-\alpha(t)]}$$

(1.3.8)

where $\alpha(s) = 1 + \frac{1}{2}s$, $\alpha(t) = 1 + \frac{1}{2}t$, $s = -(k_1 + k_2)^2$, and $t = -(k_2 + k_3)^2$.

The accidental discovery of this formula in 1968 by Veneziano and Suzuki [22, 23], who were trying to describe the scattering matrix for hadronic interactions, marked the birth of what eventually became superstring theory.

We should also emphasize that strings come in two types, open and closed. Closed strings are described in almost exactly the same terms as open strings, except the number of oscillators is doubled, and their amplitudes are defined in the entire complex plane, not just the upper half plane.

We can decompose the string variable in terms of two sets of commuting harmonic oscillators:

$$X_\mu(\sigma) = x_\mu + \left(\frac{\alpha'}{2}\right)^{1/2} \sum_{n=1}^{\infty} \frac{1}{\sqrt{n}} \Big(a_n e^{-in\sigma} + \tilde{a}_n e^{in\sigma}$$

$$+ a_n^\dagger e^{in\sigma} + \tilde{a}_n^\dagger e^{-in\sigma}\Big)_\mu$$

$$P_\mu(\sigma) = \frac{p_\mu}{2\pi} + \frac{1}{2\pi\sqrt{2\alpha'}} \sum_{n=1}^{\infty} \sqrt{n} \Big(-i\tilde{a}_n e^{-in\sigma} - ia_n e^{in\sigma}$$

$$+ ia_n^\dagger e^{in\sigma} + i\tilde{a}_n^\dagger e^{-in\sigma}\Big)_\mu$$

(1.3.9)

where $X_\mu(0) = X_\mu(2\pi)$.

The Hamiltonian now also has doubled the number of oscillators:

$$H = \pi \int_0^{2\pi} d\sigma \left(\alpha' P_\mu^2 + \frac{X_\mu'^2}{4\pi^2\alpha'}\right) = \sum_{n=1}^{\infty} (n a_n^\dagger a_n + n \tilde{a}_n^\dagger \tilde{a}_n) + \alpha' p_\mu^2 \quad (1.3.10)$$

The important feature of the closed string spectrum is that it possesses a massless spin-two particle, that is, the graviton. In fact, this is a general feature of string theory; the graviton and hence relativity are unavoidable parts of the spectrum. This, in fact, is perhaps the most attractive, and most mysterious, feature of string theory, that relativity is an essential part of the theory. While other point particle theories try to avoid including the graviton, string theory views gravity as an inseparable part of its formulation.

The spectrum, at its lowest level, now consists of the tachyon, represented by the vacuum $|0\rangle$ and the graviton $a_1^{\mu\dagger} \tilde{a}_1^{\nu\dagger}|0\rangle$. We will, by convention, set the slope α' to be $\frac{1}{2}$ and the slope of the closed strings to be $\frac{1}{4}$. This means that the tachyon appears at $s = -8$.

The propagator for closed strings is similar to the open string propagator, except for one difference: there is an extra rotation factor that guarantees that the final result is not dependent on the origin of the parameterization. Thus, the propagator is:

$$\frac{1}{L_0 + \tilde{L}_0 - 2} P \tag{1.3.11}$$

where:

$$P = \int_0^{2\pi} d\theta \, e^{i\theta(L_0 - \tilde{L}_0)} \tag{1.3.12}$$

The propagator can be written in an equivalent way:

$$D = \frac{1}{2\pi} \int_{|z|\leq 1} z^{L_0-2} \bar{z}^{\tilde{L}_0-2} d^2 z = \frac{\sin\pi(L_0 - \tilde{L}_0)}{\pi(L_0 - \tilde{L}_0)} \frac{1}{L_0 + \tilde{L}_0 - 2} \tag{1.3.13}$$

Notice that the operator P can also be interpreted as a projection operator, which forces $L_0 - \tilde{L}_0$ acting on the states to be zero. Thus, the spectrum of the closed string model is now determined by the following constraints (for $n > 0$):

$$L_n|\phi\rangle = \tilde{L}_n|\phi\rangle = 0$$

$$(L_0 + \tilde{L}_0 - 2)|\phi\rangle = 0 \tag{1.3.14}$$

$$(L_0 - \tilde{L}_0)|\phi\rangle = 0$$

where the last constraint is due to the fact that the states should be independent of where we chose the origin of our parameterization.

The N-point function can now be calculated in two equivalent ways, by the oscillator method and the functional method. The oscillator expression for the N-point function is still of the form:

$$A_N = \sum_{\text{perm}} \langle k_1, 0|V(k_2)DV(k_3)\cdots DV(k_{N-1})|k_N, 0\rangle \tag{1.3.15}$$

where the new feature is that we have to sum over all permutations of the order of the external lines (since all of them are defined on the sphere).

For the functional method, the integral for N-point functions is now defined over the entire complex plane, rather than just the upper half plane. Performing the functional integral, we obtain:

$$A_N = \int d\mu \prod_{2\leq i<j\leq N} |z_i - z_j|^{(1/2)k_i \cdot k_j} \tag{1.3.16}$$

where the measure is given by:

$$d\mu = |z_a - z_b|^2|z_b - z_c|^2|z_c - z_a|^2 \frac{\prod_{i=1}^N d^2 z_i}{d^2 z_a \, d^2 z_b \, d^2 z_c} \tag{1.3.17}$$

Setting $N = 4$, we find the amplitude $A(s,t,u)$ for the Shapiro–Virasoro model [29, 30]:

$$A_4(s,t,u)$$
$$= \frac{\Gamma[-\alpha(s)/2]\Gamma[-\alpha(t)/2]\Gamma[-\alpha(u)/2]}{\Gamma\{-[\alpha(s)+\alpha(u)]/2\}\Gamma\{-[\alpha(t)+\alpha(u)]/2\}\Gamma\{-[\alpha(s)+\alpha(t)]/2\}} \tag{1.3.18}$$

In contrast to the open string case, where the amplitude $A(s,t)$ only had poles in two channels at a time, the amplitude $A(s,t,u)$ has poles in all three channels simultaneously, that is, whenever z_i comes close to z_j, the amplitude has a pole.

Last, we mention that although KSV introduced the multiloop interpretation of string theory in order to obtain unitarity, the perturbation theory was not manifestly unitary. Furthermore, it was not clear how to determine the weights of each of these multiloop diagrams. The origin of this problem was that the multiloop series was not derived from a Hermitian Hamiltonian, but was simply postulated via Eq. (1.3.1).

Mandelstam [31] gave the solution to this problem by making a conformal transformation on the Riemann surfaces of genus g, demonstrating that in the light cone gauge the world sheet was equivalent to a string picture in which strings split into smaller strings or joined to form larger ones. Because this formalism eliminated all redundant modes from the very start and could be shown to be Lorentz invariant, the theory was unitary from the beginning.

Then, Kaku and Kikkawa [32] showed that the theory could be expressed as a genuine *field theory of strings*, where unitarity was trivially implemented by an explicit interacting Hamiltonian. Not only did the field theory of strings solve the problem of unitarity and fix the weights of the diagrams appearing in the perturbation series, it gave the possibility of writing a nonperturbative formalism in which certain symmetries (supersymmetry, 10-dimensional Lorentz invariance, etc.) could be broken down to obtain realistic phenomenology. We will elaborate more on the field theory of strings in the second part of this book.

1.4. Supersymmetry

The bosonic string, as we have seen, can be derived by postulating a simple set of assumptions: that the first quantized theory is given by the two-dimensional area swept out by the string and that we sum over all conformally inequivalent topologies [which generates a set of Feynman diagrams defined on Riemann manifolds].

However, the physical spectrum of the bosonic theory cannot accomodate fermions. One crucial feature of the string model is that it can be generalized to include fermion fields defined along the string, which in turn allows us to define a new symmetry, *supersymmetry*, between the bosons and fermions. Indeed, the discovery of supersymmetry took place first in the string model.

We must be careful, however, to stress that there are two types of supersymmetry on the string. The first is *world sheet supersymmetry*, which is an unphysical supersymmetry between fermions and bosons defined in two-dimensional space. World sheet supersymmetry is manifest in the *Neveu–Schwarz–Ramond model* [33, 34].

The second is *space–time sypersymmetry*, which is defined in 10-dimensional space–time, which corresponds to the physical supersymmetry defined in space-time. This physical symmetry is manifest in the *Green–Schwarz model* [35].

We will begin with world sheet supersymmetry in the Neveu–Schwarz–Ramond model and discuss space–time supersymmetry later.

Let us introduce a new fermion field ψ_μ, which is a vector in space–time but transforms as a two-dimensional spinor in the two-dimensional world sheet. Then, Gervais and Sakita showed that the Neveu–Schwarz–Ramond (NS–R) model could be derived from a new symmetry, called supersymmetry. They introduced the Lagrangian [36]:

$$L = -\frac{1}{2\pi}\left(\partial_a X_\mu \, \partial^a X^\mu - i\bar{\psi}^\mu \rho^a \, \partial_a \psi_\mu\right) \qquad (1.4.1)$$

where:

$$\rho^0 = \begin{pmatrix} 0 & -i \\ i & 0 \end{pmatrix}, \qquad \rho^1 = \begin{pmatrix} 0 & i \\ i & 0 \end{pmatrix} \qquad (1.4.2)$$

and:

$$\psi^\mu = \begin{pmatrix} \psi_0^\mu \\ \psi_1^\mu \end{pmatrix}, \qquad \bar{\psi}^\mu = \psi^\mu \rho^0 \qquad (1.4.3)$$

with the metric $\{\rho^a, \rho^b\} = -2\eta^{ab}$, where η is given by (-1, +1).
Written explicitly, this equals:

$$L = \frac{1}{2\pi}\left[\dot{X}\dot{X} - X'X' + i\psi_0(\partial_\tau + \partial_\rho)\psi_0 + i\psi_1(\partial_\tau - \partial_\rho)\psi_1\right] \qquad (1.4.4)$$

This Lagrangian is explicitly invariant under the following:

$$\delta X^\mu = \bar{\epsilon}\psi^\mu, \qquad \delta\psi^\mu = -i\rho^a \, \partial_a X^\mu \, \epsilon \qquad (1.4.5)$$

The energy-momentum tensor can be written as:

$$T_{ab} = \partial_a X_\mu \, \partial_b X^\mu + \frac{i}{4}\bar{\psi}^\mu \rho_a \, \partial_b \psi_\mu + \frac{i}{4}\bar{\psi}^\mu \rho_b \, \partial_a \psi_\mu - (\text{Trace}) \qquad (1.4.6)$$

Generally, in field theory, there is a conserved current associated for every symmetry given by:

$$J^{\mu a} = \frac{\delta L}{\delta \, \partial_\mu \phi}\frac{\delta\phi}{\delta\epsilon^a}, \qquad \partial_\mu J^{\mu a} = 0. \qquad (1.4.7)$$

For our case, the current associated with the two-dimensional superconformal symmetry of Eq. (1.4.5) is given by:

$$J_a = \frac{1}{2}\rho^b \rho_a \psi^\mu \, \partial_b X_\mu \qquad (1.4.8)$$

We can rewrite the superconformal current J_a as:

$$T_F = -\frac{1}{2}\psi_\mu \, \partial X^\mu \qquad (1.4.9)$$

and its Fourier moments as:

$$G_n = 2 \oint \frac{dz}{2\pi i} z^{n+(1/2)} T_F(z) \qquad (1.4.10)$$

To quantize the system, we find that the fermion is self-conjugate and that:

$$\{\psi_a^\mu(\sigma,\tau), \psi_b^\nu(\sigma',\tau)\} = \pi\delta_{ab}\delta(\sigma - \sigma')\eta^{\mu\nu} \qquad (1.4.11)$$

With these commutation relations, the superconformal algebra becomes:

$$[L_m, L_n] = (m - n)L_{m+n} + \frac{\hat{c}}{8}(m^3 - m)\delta_{m+n,0}$$

$$[L_m, G_r] = \left(\frac{m}{2} - r\right)G_{m+r} \tag{1.4.12}$$

$$\{G_r, G_s\} = 2L_{r+s} + \frac{\hat{c}}{2}\left(r^2 - \frac{1}{4}\right)\delta_{r+s,0}$$

where $\hat{c} = 2c/3$ and where, if G_r is integral moded, we have the Ramond (R) algebra, and if G_r is half-integral moded, then we have the Neveu–Schwarz (NS) algebra.

This is most easily implemented by forcing the fermion field to have periodic (R) or antiperiodic (NS) boundary conditions:

$$R: \quad \psi_0(\pi, \tau) = \psi_1(\pi, \tau)$$
$$NS: \quad \psi_0(\pi, \tau) = -\psi_1(\pi, \tau) \tag{1.4.13}$$

With these boundary conditions, the harmonic oscillator decomposition is given by:

$$R: \quad \psi_{0,1}^\mu = \frac{1}{\sqrt{2}} \sum_{n=-\infty}^{\infty} d_n^\mu e^{-in(\tau \pm \sigma)}$$

$$NS: \quad \psi_{0,1}^\mu = \frac{1}{\sqrt{2}} \sum_{r \in \mathbf{Z}+1/2}^{\infty} b_r^\mu e^{-ir(\tau \pm \sigma)} \tag{1.4.14}$$

where we associate 0 (1) with the $+$ (-1) sign, and where we have the anticommutation relation among oscillators:

$$R: \quad \{d_n^\mu, d_m^\nu\} = \eta^{\mu\nu}\delta_{n,-m}$$
$$NS: \quad \{b_r^\mu, b_s^\nu\} = \eta^{\mu\nu}\delta_{r,-s} \tag{1.4.15}$$

The states (including ghosts) of the theory, as usual, are given by the complete set of harmonic oscillator states:

$$R: \quad \{a_{-n}^\mu\}\{d_{-r}^\nu\}|0\rangle u_\alpha$$
$$NS: \quad \{a_{-n}^\mu\}\{b_{-r}^\nu\}|0\rangle \tag{1.4.16}$$

where u_α is a 10-dimensional (32-component) Dirac spinor.

With this decomposition in terms of oscillators, we can now give an explicit representation of the generators of the superconformal group. For the NS sector, the generators are given by:

$$L_m = \frac{1}{2} \sum_{n=-\infty}^{\infty} : \alpha_{-n}\alpha_{m+n} : + \frac{1}{2} \sum_{r=-\infty}^{\infty} \left(r + \frac{1}{2}m\right) : b_{-r}b_{m+r} :$$

$$G_r = \sum_{n=-\infty}^{\infty} \alpha_{-n}b_{r+n} \tag{1.4.17}$$

For the R sector, the generators are given by:

$$L_m = \frac{1}{2} \sum_{n=-\infty}^{\infty} : \alpha_{-n}\alpha_{m+n} : + \frac{1}{2} \sum_{n=-\infty}^{\infty} \left(n + \frac{1}{2}m\right) : d_{-n}d_{m+n} :$$

$$G_m = \sum_{n=-\infty}^{\infty} \alpha_{-n}d_{m+n}$$

(1.4.18)

Finally, let us define the operator Q, which can be easily derived by using the previous expression for Q in Eq. (1.2.45) in terms of any Lie algebra. We find that the Faddeev–Popov ghost factor can be written in terms of two commuting ghosts β, γ as:

$$L = \frac{1}{\pi}\left(\beta\, \partial_{\bar{z}}\gamma + c.c.\right)$$

(1.4.19)

where $c.c. =$ complex conjugate. The superconformal generators must therefore be rewritten to include this new factor coming from the b, c and β, γ ghosts:

$$L_m^{\text{ghost}} = \sum_{n=-\infty}^{\infty} (m+n) : b_{m-n}c_n + \sum_{n=-\infty}^{\infty} \left(\frac{1}{2}m + n\right) : \beta_{m-n}\gamma_n :$$

$$G_m^{\text{ghost}} = -2 \sum_{n=-\infty}^{\infty} b_{-n}\gamma_{m+n} + \sum_{n=-\infty}^{\infty} \left(\frac{1}{2}n - m\right) c_{-n}\beta_{m+n}$$

(1.4.20)

Finally, Q can be written as:

$$Q = \sum_{n=-\infty}^{\infty} (L_{-n}c_n + G_{-n}\gamma_n) - \frac{1}{2}\sum_{m,n=-\infty}^{\infty} (m-n) : c_{-m}c_{-n}b_{m+n} :$$

$$+ \sum_{m,n=-\infty}^{\infty} \left(\frac{3n}{2} + m\right) c_{-n}\beta_{-m}\gamma_{m+n} + \sum_{m,n=-\infty}^{\infty} \gamma_{-m}\gamma_{-n}b_{m+n} - ac_0$$

(1.4.21)

As usual, we can check for the vanishing of Q^2, and we find the constraints:

$$D = 10, \qquad a = \begin{cases} \frac{1}{2} & \text{(NS)} \\ 0 & \text{(R)} \end{cases}$$

(1.4.22)

1.5. 2D SUSY versus 10D SUSY

In previous sections, we presented the gauge-fixed NS–R action, where the action was only invariant under global, not local, superconformal invariance. This was to show how the theory could be quantized and to show the nature of its spectrum.

We will now present the full NS–R action, with all its invariances intact. The key will be to introduce a zweibein e_a^α and its supersymmetric partner χ_α, which is a two-dimensional world sheet spinor (whose indices we shall suppress). We shall use a, b, c to denote flat two-dimensional indices and α, β, γ to denote curved two-dimensional indices. Then, the action is [37, 38]:

$$
L = -\frac{1}{4\pi\alpha'}\sqrt{g}\left(g^{\alpha\beta}\partial_\alpha X^\mu \partial_\beta X_\mu - i\bar{\psi}_\mu \rho^\alpha \nabla_\alpha \psi^\mu \right.
$$
$$
\left. + 2\bar{\chi}_\alpha \rho^\beta \rho^\alpha \psi^\mu \partial_\beta X_\mu + \frac{1}{2}\bar{\psi}_\mu \psi^\mu \bar{\chi}_\alpha \rho^\beta \rho^\alpha \chi_\beta \right)
\tag{1.5.1}
$$

where ρ^α is not a constant matrix, but is multiplied by the two-dimensional zweibein:

$$
\rho^\alpha = e_a^\alpha \rho^a
\tag{1.5.2}
$$

The action is invariant under:

$$
\begin{aligned}
\delta X_\mu &= \bar{\epsilon}\psi^\mu \\
\delta\psi^\mu &= -i\rho^\alpha \epsilon(\partial_\alpha X^\mu - \bar{\psi}^\mu \chi_\alpha) \\
\delta e_\beta^a &= -2i\bar{\epsilon}\rho^a \chi_\beta \\
\delta\chi_\alpha &= \nabla_\alpha \epsilon
\end{aligned}
\tag{1.5.3}
$$

as well as under local Weyl rescaling:

$$
\begin{aligned}
\delta X_\mu &= 0 \\
\delta\psi^\mu &= -\frac{1}{2}\sigma\psi^\mu \\
\delta e_\beta^a &= \sigma e_\beta^a \\
\delta\chi_\alpha &= \frac{1}{2}\sigma\chi_\alpha
\end{aligned}
\tag{1.5.4}
$$

as well as:

$$
\delta\chi_\alpha = i\rho_\alpha \eta, \qquad \delta e_\beta^a = \delta\psi_\mu = \delta X_\mu = 0
\tag{1.5.5}
$$

We have enough gauge constraints to place the following conditions on the zweibein and the gravitino:

$$
e_a^\alpha = \delta_a^\alpha, \qquad \chi_\alpha = 0
\tag{1.5.6}
$$

With these constraints, the action reduces to the linear NS–R action studied earlier.

Although the NS–R formalism is both simple and elegant and can be effortlessly quantized both covariantly and canonically, there is one extreme drawback: it lacks genuine 10-dimensional space–time supersymmetry. Notice that the NS–R action only has supersymmetry of the world sheet, that is, it interchanges X_μ with ψ_μ, but it does not interchange space–time bosons with space–time fermions.

By hand, one can impose the Gliozzi–Scherk–Olive (GSO) projection [39] on the states of the NS–R model, and the resulting fermionic and bosonic states become space–time supersymmetric. The superstring, therefore, possesses two-dimensional superconformal symmetry at the Lagrangian level and space–time supersymmetry at the Fock space level. However, the origin of this space–time supersymmetry is very obscure and still not well understood.

Space–time supersymmetry is a physical symmetry that lies at the heart of many of the near-miraculous properties of the string model, including its finiteness and lack of anomalies, so we need another formalism in which space–time supersymmetry is manifest. This second formalism, equivalent to the NS–R formalism after the GSO projection, is called the Green–Schwarz (GS) formalism [35] and explicitly contains space–time supersymmetry by introducing a 32-component, 10-dimensional space–time spinor. The advantage of the GS formalism is that space–time supersymmetry is built-in from the start and can be used to analyze the cancellation of divergences and anomalies in the theory.

The GS action introduces two space–time spinors θ^A, $A = 1, 2$ (we will suppress the spinorial index of the spinor). The action is:

$$S = -\frac{1}{4\alpha'\pi} \int d\sigma \, d\tau \{ \sqrt{g} g^{\alpha\beta} \Pi_\alpha \cdot \Pi_\beta + 2i\epsilon^{\alpha\beta} \partial_\alpha X^\mu (\bar{\theta}^1 \Gamma_\mu \partial_\beta \theta^1$$
$$- \bar{\theta}^2 \Gamma_\mu \partial_\beta \theta^2) - 2\epsilon^{\alpha\beta} \bar{\theta}^1 \Gamma_\mu \partial_\alpha \theta^1 \bar{\theta}^2 \Gamma^\mu \partial_\beta \theta^2 \} \tag{1.5.7}$$

where:

$$\Pi_\alpha^\mu = \partial_\alpha X^\mu - i\bar{\theta}^A \Gamma^\mu \partial_\alpha \theta^A \tag{1.5.8}$$

where Γ_μ are 10-dimensional Dirac spinors, and α, β are local, two-dimensional world sheet indices. The A index, however, labels two distinct world sheet scalars and not a two-component world sheet spinor.

The action is explicitly invariant under:

$$\delta\theta^A = \epsilon^A, \qquad \delta X_\mu = i\bar{\epsilon}^A \Gamma_\mu \theta^A \tag{1.5.9}$$

In the proof of the invariance of the action, we will use the following identity for a 10-dimensional spinor ψ:

$$\Gamma_\mu \psi_{[1} \bar{\psi}_2 \Gamma^\mu \psi_{3]} = 0 \tag{1.5.10}$$

which is only true for the following:

1. $D = 3$ and ψ is a Majorana fermion,
2. $D = 4$ and ψ is a Majorana or Weyl fermion,
3. $D = 6$ and ψ is a Weyl fermion, and
4. $D = 10$ and ψ is both a Majorana and a Weyl fermion.

We will, of course, take the case when $D = 10$, so that the spinors are both Majorana and Weyl.

In general, a 10-dimensional complex Dirac spinor has 32 complex components. Imposing that it be real (Majorana) and that it have a definite eigenvalue under the chiral operator:

$$\frac{1}{2}(1 \pm \Gamma_{D+1}), \quad \Gamma_{D+1} \equiv \Gamma_0 \Gamma_1 \Gamma_2 \cdots \Gamma_{D-1} \qquad (1.5.11)$$

reduces the number of components by half each time, so that a Majorana–Weyl spinor in 10 dimensions has 16 real components.

To show invariance under space–time supersymmetry, we need the following definitions. Let us define a spinorial parameter $\kappa^{A\alpha a}$, where $A = 1, 2$, α is a local, two-dimensional world sheet index, and a is a spinorial index, which we will often suppress. Also, we now introduce the operator $P_\pm^{\alpha\beta}$, which projects onto self-dual and anti-self-dual pieces of a two-dimensional vector:

$$P_\pm^{\alpha\beta} = (1/2)(g^{\alpha\beta} \pm \epsilon^{\alpha\beta}/\sqrt{g}) \qquad (1.5.12)$$

which has the following important properties:

$$\begin{aligned}
P_\pm^{\alpha\beta} g_{\beta\gamma} P_\pm^{\gamma\delta} &= P_\pm^{\alpha\delta} \\
P_\pm^{\alpha\beta} g_{\beta\gamma} P_\mp^{\gamma\delta} &= 0 \\
\kappa^{1\alpha} &= P_-^{\alpha\beta} \kappa_\beta^1 \\
\kappa^{2\alpha} &= P_+^{\alpha\beta} \kappa_\beta^2
\end{aligned} \qquad (1.5.13)$$

Then, the GS action is invariant under:

$$\begin{aligned}
\delta\theta^A &= 2i\Gamma \cdot \Pi_\alpha \kappa^{A\alpha} \\
\delta X^\mu &= i\bar{\theta}^A \Gamma^\mu \delta\theta^A \\
\delta(\sqrt{g} g^{\alpha\beta}) &= -16\sqrt{g}\,(P_-^{\alpha\gamma} \bar{\kappa}^{1\beta} \partial_\gamma \theta^1 + P_+^{\alpha\gamma} \kappa^{2\beta} \partial_\gamma \theta^2)
\end{aligned} \qquad (1.5.14)$$

Although the GS action is symmetric under a wide variety of invariances, including space–time supersymmetry, the problem is that covariant quantization of this action is exceedingly difficult. This is because the theory is highly nonlinear, and the momenta associated with coordinates do not have simple properties. For example, one might naively construct the momenta corresponding to the spinor:

$$\pi_A = \frac{\delta L}{\delta \dot{\theta}^A} \sim i\Gamma^\mu \partial_\sigma X_\mu \theta^A \qquad (1.5.15)$$

We find that the commutation relations are proportional to the inverse of the constraints. Thus, the commutators diverge, and the naive approach makes no sense.

The problem is that the first and second constraints of the theory are mixed together in a way that cannot be separated. The reason for this is actually quite easy to see.

We saw that a Majorana–Weyl spinor has 16 real components. This is the smallest spinorial representation of the 10-dimensional Lorentz group $SO(9,1)$. However, if we were to apply all the covariant constraints on the spinor, we would have to place one additional constraint, reducing the spinor to 8 components. *But, there are no 8-dimensional spinorial representations of $SO(9,1)$.* Thus, naive quantization of the theory is impossible. Only recently, with advances in the covariant quantization method of Batalin–Vilkovisky (BV), has the covariant gauge-fixed GS action been quantized (see Appendixes 1 and 2 for a more complete discussion of this important point).

However, although the covariant quantization of the theory is quite difficult, the light cone quantization of the theory is quite easy (although covariance is completely lost). Let us apply the light cone gauge conditions:

$$\Gamma^+ \theta^{1,2} = 0$$
$$\Gamma^\pm = 2^{-1/2}(\Gamma^0 \pm \Gamma^9) \tag{1.5.16}$$

so that the Dirac matrices satisfy:

$$(\Gamma^+)^2 = (\Gamma^-)^2 = 0 \tag{1.5.17}$$

Then the nonlinear terms in the action in Eq. (1.5.7) disappear, leaving us with the simple action:

$$S = -\frac{1}{4\pi\alpha'} \int d\sigma \, d\tau \left(\partial_a X^i \, \partial^a X^i - i\bar{S}\rho^b \, \partial_b S \right) \tag{1.5.18}$$

where a, b are flat world sheet indices and i, j are 8-dimensional transverse indices and where we have substituted $\sqrt{p^+}\theta$ with S. Notice that the two space–time spinors θ^A, which were scalars on the world sheet (and not world sheet spinors), have strangely transformed into genuine world sheet spinors S.

The light cone action has a very simple global space–time supersymmetry associated with it. The supersymmetry transformation is given by:

$$\delta S^a = (2p^+)^{1/2}\eta^a$$
$$\delta X^i = 0$$
$$\delta S^a = -i\sqrt{p^+}\rho^b \, \partial_b X^i (\gamma^i)_{a\dot{a}} \epsilon^{\dot{a}} \tag{1.5.19}$$
$$\delta X^i = (p^+)^{-1/2}(\gamma^i)_{a\dot{a}} \bar{\epsilon}^{\dot{a}} S^a$$

where $(\gamma^i)_{a\dot{a}}$ are the Dirac matrices that generate a representation of $SO(8)$, and the parameters of the symmetry are given by η^a and $\epsilon^{\dot{a}}$.

To calculate the generators of this symmetry, let us first quantize the model. Since the action is now linear, this is trivial. We find:

$$\left[S^{Aa}(\sigma,\tau), S^{Bb}(\sigma',\tau) \right] = \pi\delta^{ab}\delta^{AB}\delta(\sigma - \sigma') \tag{1.5.20}$$

and there is only one choice of boundary conditions for open strings:

$$S^{1a}(0,\tau) = S^{2a}(0,\tau), \qquad S^{1a}(\pi,\tau) = S^{2a}(\pi,\tau) \tag{1.5.21}$$

so that the strings have the same $SO(8)$ chirality.

The generator of these symmetries is given by:

$$Q^a = (2p^+)^{1/2} S_0^a$$

$$Q^{\dot{a}} = (p^+)^{-1/2}(\gamma^i)_{\dot{a}a} \sum_{n=-\infty}^{\infty} S_{-n}^a \alpha_n^i \tag{1.5.22}$$

The generators, in turn, form the supersymmetric algebra:

$$\{Q^a, Q^b\} = 2p^+ \delta^{ab}$$
$$\{Q^a, Q^{\dot{a}}\} = \sqrt{2}(\gamma^i)_{a\dot{a}} p^i \tag{1.5.23}$$
$$\{Q^{\dot{a}}, Q^{\dot{b}}\} = 2H \delta^{\dot{a}\dot{b}}$$

where:

$$H = \frac{1}{p^+}\left[\sum_{n=1}^{\infty} (\alpha_{-n}^i \alpha_n^i + n S_{-n}^a S_n^a) + \frac{1}{2}p_i^2\right] \tag{1.5.24}$$

In contrast to the covariant quantization of the GS string, which is quite difficult and involved, the light cone quantization of the GS string yields a simple, free supersymmetric theory.

1.6. Types of Strings

At this point, we may ask what are the various types of string theories one can write that are supersymmetric, ghost free, and anomaly free. The easiest way to catalog the various possibilities is through the light cone quantization of the GS string, since all ghosts have been removed and the theory is globally supersymmetric in space–time.

The list of totally self-consistent string theories consists of:

1. type I,
2. type IIA,
3. type IIB, and
4. heterotic.

It may seem surprising that there are so few self-consistent string theories, while there are an infinite number of possible theories of point particles. The reason for this is that the Feynman diagrams of a point particle are based on one-dimensional graphs, upon which we can impose any number of Lorentz covariant vectors and spinors (e.g., Feynman's rules). However, the Feynman diagrams of string theory are two-dimensional manifolds, obeying strict self-consistency constraints, so it is not surprising that we only find four self-consistent string theories.

Type I

The first string theory is called type I, in which, as we have seen, the spinors S^1 and S^2 have the same $SO(8)$ chirality. The theory of open strings, by itself, however, is incomplete (because the nonplanar one-loop diagram has a pole, which corresponds to a closed string). Because the closed string emerges as a bound state of open string graphs, we must add the closed string sector to the open string in order to maintain unitarity.

Gauge invariance can be added into the theory by multiplying the N-point function with appropriate traces over the generators of some Lie algebra (called Chan–Paton factors). The gauge group must be $SO(32)$ in order to cancel all anomalies.

Type IIA

For closed strings, there is a choice as to how to choose the chiralities of S^1 and S^2. If we choose them to be of opposite chirality, then we have the type IIA string. Type IIA closed string theory is appealing because it has no chiral anomalies from the very beginning (since the two chiral sectors cancel against each other). In the zero slope limit, when only the massless sector of the theory survives, the theory reduces to the point particle $N = 1$, $D = 10$ supergravity theory.

Type IIB

For closed strings, if we take the choice where S^1 and S^2 have the same chirality, then we have the type IIB superstring. However, in the zero slope limit, when we analyze the massless sector, we find that there does not exist any known covariant version of this theory. Its light cone reduction is well defined, but its covariant precursor apparently cannot be written. (This may be because of our limited understanding of how to construct point particle supersymmetric theories in 10 dimensions.)

At present, it seems, however, that the type II string cannot describe the physical $SU(3) \otimes SU(2) \otimes U(1)$ symmetry of our low-energy universe. By compactifying from 10 dimensions to 4 dimensions, the type II string can introduce a wide array of symmetries, but none of them seems to fit the description of our world.

Heterotic String

The string theory that holds the most promise of describing the physical world is that of the heterotic string [40]. While the type I string uses Chan–Paton factors to introduce isospin symmetry, the heterotic string uses the 16 dimensions left from the compactification of 26 dimensions down to 10 dimensions to introduce a rank 16 Lie group. Since E_8 is a rank eight

Lie group, the heterotic string can be compactified so that its spectrum is $E_8 \otimes E_8$ [or $Spin(32)/Z_2$], which is certainly large enough to permit a serious phenomenological investigation.

The heterotic string, however, achieves this compactification in an unorthodox fashion. We recall that the closed string has right-moving and left-moving oscillator modes, which, for the most part, do not interact. The heterotic string splits these modes apart. The left-moving modes are purely bosonic and live in a 26-dimensional space which, has been compactified to 10-dimensions, leaving us with an $E_8 \otimes E_8$ isospin symmetry. However, the right-moving modes only live in a 10-dimensional space and contain the supersymmetric GS or NS-R theory. When the left-moving half (containing the isospin) and the right-moving half (containing the supersymmetry) are put together, they produce a self-consistent, ghost-free, anomaly-free, one-loop finite theory, the heterotic string (meaning "hybrid vigor").

The action for the heterotic string is therefore:

$$S = -\frac{1}{4\pi\alpha'} \int d\tau \int d\sigma \left[\partial_a X^i \, \partial^a X^i + \sum_{I=1}^{16} \partial_a X^I \, \partial^a X^I + i\bar{S}\gamma^-(\partial_\tau + \partial_\sigma)S \right]$$

(1.6.1)

where $I = 1, 2, \ldots, 16$ and is an isospin index and where we enforce the constraints:

$$(\partial_\tau - \partial_\sigma)X^I = 0, \qquad \gamma^+ S = \frac{1}{2}(1 + \gamma_{11})S = 0 \qquad (1.6.2)$$

where $\gamma^+ = 2^{-1/2}(\gamma^0 + \gamma^9)$. (Some have criticized the heterotic string for being artificial and contrived because of the way it splits the left- and right-moving oscillators, indicating that perhaps the heterotic string, in turn, is a broken version of an even higher string. However, attempts to embed the heterotic string into a larger string theory have not been particularly successful.) As we shall see in later chapters, with a mild set of assumptions, we can obtain surprisingly realistic string theories that contain the $SU(3) \otimes SU(2) \otimes U(1)$ low-energy theory of our world.

To understand these compactifications, in the first part of this book, we will turn to a discussion of conformal field theory, which will hopefully give us a classification of all possible vacuums of the theory. Perhaps one of the millions of conformal field theories that have been discovered describes our universe.

However, this perturbative approach alone can never yield totally realistic results. Supersymmetry and 10-dimensional space–time seem to be preserved to all orders in perturbation theory, so perturbation theory by itself can never break the symmetries of the superstring in order to yield realistic phenomenology. In the second part of this book, we will concentrate on nonperturbative approaches to superstring theory, especially string field theory and matrix models.

1.7. Summary

Superstring theory has emerged as the leading candidate for all known forces. Not only has the theory enough symmetries to include the four fundamental forces as subsets of its symmetries, it is also the only theory that can claim to yield a finite quantum theory of gravity.

The bosonic Lagrangian for the string is given by:

$$L = -\frac{1}{4\pi\alpha'}\sqrt{g}\,g^{ab}\,\partial_a X_\mu\,\partial_b X_\nu\,\eta^{\mu\nu} \tag{1.7.1}$$

Notice that the action is manifestly reparameterization invariant. If we reparameterize the two-dimensional world sheet according to:

$$\sigma \to \tilde{\sigma}(\sigma,\tau), \qquad \tau \to \tilde{\tau}(\sigma,\tau) \tag{1.7.2}$$

then the action is invariant under this two-dimensional general coordinate transformation if:

$$\bar{g}^{ab}(\bar{x}) = \left(\frac{\partial \bar{x}^a}{\partial x^c}\right)\left(\frac{\partial \bar{x}^b}{\partial x^d}\right)g^{cd}(x) \tag{1.7.3}$$

where $x^a = \{\sigma,\tau\}$. The theory is also scale invariant under:

$$g_{ab} \to e^\phi g_{ab} \tag{1.7.4}$$

There are enough symmetries in the theory to select out the conformal gauge:

$$g_{ab} = \delta_{ab} \tag{1.7.5}$$

so the action becomes manifestly conformally invariant:

$$L = \frac{1}{2\pi}\partial_z X_\mu\,\partial_{\bar{z}} X^\mu \tag{1.7.6}$$

In the first quantized formalism, interactions are introduced by the functional integral:

$$A_N = \sum_{\text{Topologies}} \int DX_\mu \int d\mu \, \exp\left[i\int d^2 z\, L(z) + \sum_{i=1}^{N} ik_{i,\nu} X^\nu(z_i)\right]$$

$$= \sum_{\text{Topologies}} \int d\mu \left\langle \prod_{i=1}^{N} e^{ik_i X_i} \right\rangle \tag{1.7.7}$$

We must sum over all conformally distinct surfaces, including Riemann surfaces of arbitrary genus, which means that the first quantized formalism is necessarily perturbative. This is one of the fundamental deficiencies of the first quantized system.

In the genus zero limit, we can calculate the full spectrum of the theory. To do this, we first define the energy-momentum tensor:

$$T_{ab} = -4\pi\alpha' \frac{1}{\sqrt{g}} \frac{\delta L}{\delta g^{ab}} \tag{1.7.8}$$

This, in turn, can be shown to equal:

$$T_{ab} = \partial_a X_\mu \, \partial_b X^\mu - \frac{1}{2} g_{ab} g^{cd} \, \partial_c X^\mu \, \partial_d X_\mu \tag{1.7.9}$$

We define the *Virasoro generators* L_n as the moments of the energy-momentum tensor:

$$\begin{aligned} L_m &= \frac{1}{4\pi\alpha'} \int_0^\pi e^{im\sigma}(T_{00} + T_{01}) + e^{-im\sigma}(T_{00} - T_{01}) \, d\sigma \\ &= \frac{1}{8\pi\alpha'} \int_{-\pi}^\pi e^{im\sigma}(\dot{X} + X')^2 \, d\sigma \\ &= \frac{1}{2} \sum_{n=-\infty}^\infty \alpha_{m-n}\alpha_n \end{aligned} \tag{1.7.10}$$

These generators, L_n, in turn form a closed algebra, the Virasoro algebra (which generates the conformal group):

$$[L_n, L_m] = (n - m)L_{n+m} + \frac{c}{12} n(n + 1)(n - 1)\delta_{n,-m} \tag{1.7.11}$$

where c is a constant.

In the Gupta–Bleuler quantization scheme, we allow ghosts (due to the negative sign in the Lorentz metric) to propagate in the theory. We eliminate them by applying the Virasoro generators on the states:

$$\begin{aligned} L_n|R\rangle &= 0, \quad n > 0 \\ (L_0 - 1)|R\rangle &= 0 \end{aligned} \tag{1.7.12}$$

where the second condition is the mass-shell condition.

In the alternative BRST formalism, we allow Faddeev–Popov b, c ghosts to circulate in the theory and then cancel them by requiring the physical states to satisfy:

$$Q|\psi\rangle = 0 \tag{1.7.13}$$

where Q is nilpotent.

Fermions are introduced into the theory by adding supersymmetry (which was first discovered in string theory).

Let us introduce a new fermion field ψ_μ, which is a vector in space–time but transforms as a two-dimensional spinor on the two-dimensional world sheet. The Neveu–Schwarz–Ramond action is:

$$L = -\frac{1}{2\pi}\left(\partial_a X_\mu \, \partial^a X^\mu - i\bar{\psi}^\mu \rho^a \, \partial_a \psi_\mu\right) \tag{1.7.14}$$

where:

$$\rho^0 = \begin{pmatrix} 0 & -i \\ i & 0 \end{pmatrix}; \qquad \rho^1 = \begin{pmatrix} 0 & i \\ i & 0 \end{pmatrix} \tag{1.7.15}$$

and:

$$\psi^\mu = \begin{pmatrix} \psi_0^\mu \\ \psi_1^\mu \end{pmatrix}, \qquad \bar{\psi}^\mu = \psi^\mu \rho^0 \tag{1.7.16}$$

with the metric $\{\rho^a, \rho^b\} = -2\eta^{ab}$, where η is given by $(-1, +1)$.

The $N = 1$ superconformal algebra is constructed by taking the moments of the energy-momentum tensor and the supercurrent:

$$[L_m, L_n] = (m - n)L_{m+n} + \frac{\hat{c}}{8}(m^3 - m)\delta_{m+n,0}$$

$$[L_m, G_r] = \left(\frac{m}{2} - r\right)G_{m+r} \tag{1.7.17}$$

$$\{G_r, G_s\} = 2L_{r+s} + \frac{\hat{c}}{2}\left(r^2 - \frac{1}{4}\right)\delta_{r+s,0}$$

where, if G_r is integral moded, we have the Ramond algebra, and if G_r is half-integral moded, then we have the Neveu–Schwarz algebra. By a similar analysis, the NS–R model is ghost free in 10 dimensions.

The NS–R model's main problem, however, is lack of space–time supersymmetry, which can only be restored by truncating the Fock space. An equivalent formalism, which is manifestly space–time supersymmetric, is the Green–Schwarz string:

$$S = -\frac{1}{4\alpha'\pi} \int d\sigma d\tau \left[\sqrt{g} g^{\alpha\beta} \Pi_\alpha \cdot \Pi_\beta + 2i\epsilon^{\alpha\beta} \partial_\alpha X^\mu \left(\bar{\theta}^1 \Gamma_\mu \partial_\beta \theta^1 \right.\right.$$
$$\left.\left. - \bar{\theta}^2 \Gamma_\mu \partial_\beta \theta^2\right) - 2\epsilon^{\alpha\beta} \bar{\theta}^1 \Gamma_\mu \partial_\alpha \theta^1 \bar{\theta}^2 \Gamma^\mu \partial_\beta \theta^2\right] \tag{1.7.18}$$

where:

$$\Pi_\alpha^\mu = \partial_\alpha X^\mu - i\bar{\theta}^A \Gamma^\mu \partial_\alpha \theta^A \tag{1.7.19}$$

where Γ_μ are 10-dimensional Dirac spinors, α, β are local two-dimensional world sheet indices, and $A = 1, 2$. This A index, however, labels two distinct world sheet scalars, not a two-component world sheet spinor.

Superstrings are enormously constrained because of the large symmetry group and because interactions are defined on manifolds. (By contrast, point particle theories are defined on graphs, and hence an infinite number of them can be written.)

So far, only four superstrings have been discovered.

1. For type I superstrings, we combine open and closed superstrings. The theory is anomaly free for the gauge group $SO(32)$.
2. For type IIA superstrings, we have a closed string theory in which the spinors have opposite chirality.
3. For type IIB superstrings, the spinors are of the same chirality.

4.   For heterotic superstrings, we have a closed superstring theory in which the left-moving sector is bosonic and lives in 26-dimensional space and the right-moving sector is supersymmetric in 10 dimensions. By compactifying the 26-dimensional left-moving sector to 10 dimensions, we obtain the gauge group $E_8 \otimes E_8$ or $\mathrm{Spin}(32)/Z_2$.

The action for the heterotic string is therefore:

$$S = -\frac{1}{4\pi\alpha'} \int d\tau \int d\sigma \left[\partial_a X^i \, \partial^a X^i + \sum_{I=1}^{16} \partial_a X^I \, \partial^a X^I + i\bar{S}\gamma^-(\partial_\tau + \partial_\sigma)S \right]$$

$$(1.7.20)$$

where I labels the $E_8 \otimes E_8$ symmetry and where we enforce the constraints:

$$(\partial_\tau - \partial_\sigma)X^I = 0, \qquad \gamma^+ S = \frac{1}{2}(1 + \gamma_{11})S = 0 \qquad (1.7.21)$$

where $\gamma^+ = 2^{-1/2}(\gamma^0 + \gamma^9)$.

The central theme of this book, and also the most pressing problem in superstring research, is the search for the true vacuum of the theory. Notice that our discussion has been mainly perturbative. However, because perturbation theory can never compactify space–time to four dimensions or break supersymmetry, we must turn to nonperturbative formalisms, which are discussed in the second half of this book.

We have written the book in two parts. In the first part, we discuss conformal field theories, which give us the complete set of possible perturbative vacuums of string theory. However, the true vacuums of the theory must necessarily break supersymmetry and compactify space–time to four dimensions, so in the second part of this book, we discuss nonperturbative approaches to superstring theory, in particular string field theory and matrix models.

References

For introductions to string theory, see Refs. 1 and 2.

1. M. Kaku, *Introduction to Superstrings*, Springer–Verlag, Berlin and New York (1989).
2. M. B. Green, J. H. Schwarz, and E. Witten, *Superstring Theory*, Cambridge Univ. Press, Vols. 1 and 2, London and New York (1987).

For reviews of the older, dual resonance model, see Refs. 3 to 7.

3. J. H. Schwarz, *Phys. Rep.* **89**, 223 (1982).
4. M. Jacob, ed., *Dual Theory*, North-Holland Publ., Amsterdam (1974).
5. J. H. Schwarz, ed., *Superstrings: the First 15 Years of Superstring Theory*, World Scientific, Singapore (1985).
6. J. Scherk, *Rev. Mod. Phys.* **47**, 1213 (1975).
7. P. Frampton, *Dual Resonance Models*, Benjamin, New York (1974).
8. Y. Nambu, *Lectures at the Copenhagen Summer Symposium* (1970).
9. T. Goto, *Prog. Theor. Phys.* **46**, 1560 (1971).
10. A. M. Polyakov, *Phys. Lett.* **103B**, 207, 211 (1981).

11. J.L. Gervais and B. Sakita, *Nucl. Phys.* **B34**, 632 (1971); *Phys. Rev.* **D4**, 2291 (1971); *Phys. Rev. Lett.* **30**, 716 (1973).
12. S. Fubini, D. Gordon, and G. Veneziano, *Phys. Lett.* **29B**, 679 (1969).
13. M. A. Virasoro, *Phys. Rev.* **D1**, 2933 (1970).
14. P. Goddard and C. B. Thorn, *Phys. Lett.* **40B**, 235 (1972).
15. R. C. Brower and K. A. Friedman, *Phys. Lett.* **D7**, 535 (1973).
16. P. Goddard, J. Goldstone, C. Rebbi, and C. B. Thorn, *Nucl. Phys.* **B56**, 109 (1973).
17. C. Becchi, A. Rouet, and R. Stora, *Ann. Phys.* **98**, 287 (1976).
18. I. V. Tyupin, Lebedev preprint, FIAN No. 39 (1975), unpublished.
19. L. D. Faddeev and V. N. Popov, *Phys. Lett.* **25B**, 29 (1967).
20. M. Kato and K. Ogawa, *Nucl. Phys.* **B212**, 443 (1983).
21. E. S. Fradkin and G. A.Vilkoviski, *Phys. Lett.* **55B**, 224 (1975).
22. G. Veneziano, *Nuovo Cim.* **57A**, 190 (1968).
23. M. Suzuki, unpublished.
24. K. Kikkawa, B. Sakita, and M. A. Virasoro, *Phys. Rev.* **184**, 1701 (1969).
25. M. Kaku and L. P. Yu, *Phys. Lett.* **33B**, 166 (1970); *Phys. Rev.* **D3**, 2992, 3007, 3020 (1971); M. Kaku and J. Scherk, *Phys. Rev.* **D3**, 430 (1971); *Phys. Rev.* **D3**, 2000 (1971).
26. C. Lovelace, *Phys. Lett.* **32B**, 703 (1970); *Phys. Lett.* **34B**, 500 (1971).
27. V. Alessandrini, *Nuovo Cim.* **2A**, 321 (1971).
28. C. S. Hsue, B. Sakita, and M. A. Virasoro, *Phys. Rev.* **D2**, 2857 1970.
29. M. A. Virasoro, *Phys. Rev.* **177**, 2309 (1970).
30. J. Shapiro, *Phys. Lett.* **33B**, 361 (1970).
31. S. Mandelstam, *Nucl. Phys.* **B64**, 205 (1973); *Nucl. Phys.* **B69**, 77 (1974).
32. M. Kaku and K. Kikkawa, *Phys. Rev.* **D10**, 1110, 1823 (1974).
33. A. Neveu and J. H. Schwarz, *Nucl. Phys.* **B31**, 86 (1971).
34. P. Ramond, *Phys. Rev.* **D3**, 2415 (1971).
35. M. Green and J. H. Schwarz, *Phys. Lett.* **136B**, 367 (1984); *Nucl. Phys.* **B198** 252, 441 (1982).
36. J. L. Gervais and B. Sakita, *Nucl. Phys.* **B34**, 632 (1971).
37. L. Brink, P. Di Vecchia, and P. Howe, *Phys. Rev.* **D5**, 988 (1972).
38. S. Deser and B. Zumino, *Phys. Lett.* **65B**, 369 (1976).
39. F. Gliozzi, J. Scherk, and D. Olive, *Nucl. Phys.* **B122**, 253 (1977).
40. D. Gross, H. A. Harvey, E. Martinec, and R. Rohm, *Phys. Rev. Lett.* **54**, 502 (1985); *Nucl. Phys.* **B256**, 253 (1986); **B267**, 75 (1986).

Chapter 2

BPZ Bootstrap and Minimal Models

2.1. Conformal Symmetry in D Dimensions

In the original pioneering paper of Belavin, Polyakov, and Zamolodchikov (BPZ) [1], two questions were asked. Is conformal invariance by itself sufficiently restrictive to uniquely determine all Green's functions of a conformal field theory? If not, then what additional conditions are necessary before we can solve for Green's functions?

These questions are not as outlandish as they may seem. For a typical quantum field theory in higher dimensions, the space–time symmetries of the system are not strong enough to uniquely determine all Green's functions. However, we know that the case of two dimensions is special: the number of generators of the conformal group is infinite. As a result, the restriction of conformal invariance creates an infinite number of conserved currents, which are often sufficient to solve a two-dimensional quantum field theory [1, 2].

In general, because of the explosion of conformal field theory solutions found within the last few years [3–10], we know that the conformal bootstrap method requires new constraints that must be imposed, such as modular invariance (as we will see in Chapter 4), to determine the correlation functions. There is a class of conformal field theories, called the "minimal series," for which Green's functions can be computed using only the constraint of unitarity and a finite number of primary fields.

Let us first begin with a general discussion of the conformal group in D dimensions, and then single out why the case $D = 2$ is so special. We define conformal transformations as those that leave the metric invariant up to a scale change:

$$g_{\mu\nu}(x) \to g'_{\mu\nu}(x') = \Omega(x)g_{\mu\nu}(x) \tag{2.1.1}$$

Notice that this transformation preserves angles, that is, the angle

$$\frac{x_\mu g^{\mu\nu} y_\nu}{\sqrt{x^2 y^2}} \tag{2.1.2}$$

between x_μ and y_μ is preserved under a conformal transformation. Let us represent this with a small infinitesimal transformation, $x^\mu \rightarrow x^\mu + \epsilon^\mu$. Then, the infinitesimal distance ds^2 transforms as:

$$ds^2 \rightarrow ds^2 + (\partial_\mu \epsilon_\nu + \partial_\nu \epsilon_\mu)\, dx^\mu\, dx^\nu \qquad (2.1.3)$$

Next, we place constraints on ϵ_μ so that we can make it compatible with a scale transformation on the metric. This means that the right-hand side of the previous equation must be proportional to $\eta_{\mu\nu}$, so that

$$\partial_\mu \epsilon_\nu + \partial_\nu \epsilon_\mu = \frac{2}{D}(\partial \cdot \epsilon)\eta_{\mu\nu} \qquad (2.1.4)$$

Let us take the trace of both sides of the equation and then compare it with the Ω found in Eq. (2.1.1). We find:

$$\Omega = 1 + (2/D)(\partial \cdot \epsilon) \qquad (2.1.5)$$

so that:

$$\left[\eta_{\mu\nu}\, \partial_\lambda\, \partial^\lambda + (D-2)\, \partial_\mu\, \partial_\nu\right] \partial \cdot \epsilon = 0 \qquad (2.1.6)$$

Notice that the constraint on ϵ_μ for $D = 2$ is different than for $D > 2$.

For D greater than 2, let us now tabulate the components of ϵ_μ for the conformal group:

$\epsilon^\mu = a^\mu$ generates constant *translations*,

$\epsilon^\mu = \omega^\mu_\nu x^\nu$ generates *Lorentz transformations* for antisymmetric ω^μ_ν,

$\epsilon^\mu = \lambda x^\mu$ generates *scale transformations*, and

$\epsilon^\mu = b^\mu x^2 - 2x^\mu b_\nu x^\nu$ generates *proper conformal transformations*.

The first two are the familiar transformations of the Poincaré group. The third is a scale transformation, and the fourth is a combination of an inversion and a translation. To see this, we can write the last transformation in a more transparent way:

$$\frac{x^{\mu\prime}}{x^{\prime 2}} = \frac{x^\mu}{x^2} + b^\mu \qquad (2.1.7)$$

that is, proper conformal transformations correspond to an inversion followed by a translation.

If D is the dimension of space–time and $D > 2$, then the total number of parameters in the conformal group is equal to $\frac{1}{2}(D+1)(D+2)$. The conformal group is thus isomorphic to the orthogonal group on $(D+2) \times (D+2)$ matrices.

For finite, rather than infinitesimal transformations, we have the following transformations:

$$x^\mu \rightarrow x^{\mu\prime} = x^\mu + a^\mu$$

$$x^\mu \rightarrow x^{\mu\prime} = \Lambda^\mu_\nu x^\nu$$

$$x^\mu \rightarrow x^{\mu\prime} = \lambda x^\mu \qquad (2.1.8)$$

$$x^\mu \rightarrow x^{\mu\prime} = \frac{x^\mu + b^\mu x^2}{1 + 2b \cdot x + b^2 x^2}$$

(Λ^ν_μ parameterizes Lorentz transformations.)

For the special case of 4 dimensions, the conformal group is easily constructed from the 6 generators of the Lorentz group, 4 generators for translations, 4 generators for proper conformal transformations, and one generator for scale transformations, for a total of 15 generators. The conformal group in 4 dimensions is therefore:

$$SO(2,4) \sim SU(2,2) \qquad (2.1.9)$$

For the special case of $D = 2$, we find that the finite number of parameters in the conformal group becomes infinite. In this case, the infinitesimal conformal transformation becomes:

$$\partial_1 \epsilon_1 = \partial_2 \epsilon_2, \quad \partial_1 \epsilon_2 = -\partial_2 \epsilon_1 \qquad (2.1.10)$$

that is, we have precisely the Cauchy–Riemann equations for a two-dimensional transformation. If we define $\epsilon(z) = \epsilon^1 + i\epsilon^2$, $\bar\epsilon(\bar z) = \epsilon^1 - i\epsilon^2$, $z(\bar z) = x^1 + (-1)ix^2$, then the conformal transformation becomes:

$$z \rightarrow z + \epsilon(z), \quad \bar z \rightarrow \bar z + \bar\epsilon(\bar z) \qquad (2.1.11)$$

If we make the following infinitesimal change given by: $\epsilon(z) = -z^{n+1}$ and $\bar\epsilon(\bar z) = -\bar z^{m+1}$, then it is easy to compute the generators of this transformation:

$$L_n = -z^{n+1}\partial_z, \quad \bar L_n = -\bar z^{n+1}\partial_{\bar z} \qquad (2.1.12)$$

which obey the algebraic relations:

$$[L_n, L_m] = (n-m)L_{m+n}, \quad [\bar L_n, \bar L_m] = (n-m)\bar L_{m+n} \qquad (2.1.13)$$

This is called *Witt algebra*. When the central term is added, it becomes the familiar Virasoro algebra [11].

In two dimensions, there is a qualitative change in the conformal group because the number of generators has suddenly become infinite. This means that many two-dimensional models are actually exactly soluble, contrary to the situation in higher dimensions, because of the presence of an infinite number of conserved currents. Enormous simplifications of the correlation functions occur only in two dimensions, sufficient to make a wide variety of models exactly soluble.

2.2 Conformal Group in Two Dimensions

Let us now discuss how two-dimensional conformal fields transform. We say that a conformal field has weight $h_1 + h_2$ and conformal spin $h_1 - h_2$ if it transforms as:

$$\phi(z, \bar{z}) = \left(\frac{dz}{dz'} \right)^{-h_1} \left(\frac{d\bar{z}}{d\bar{z}'} \right)^{-h_2} \phi(z', \bar{z}') \qquad (2.2.1)$$

under a conformal transformation.

The full conformal transformation on $\phi(z, \bar{z})$ is actually the product of two copies of the conformal group, acting on each z and $\bar{z}$ individually. [Since a function $\phi(z, \bar{z})$ transforms under the product of two commuting conformal algebras, we will often delete the dependence on $\bar{z}$. It is an easy matter to reinsert the dependence on $\bar{z}$ into all the equations.]

If we power expand this infinitesimally, we find:

$$\delta\phi(z) = \epsilon(z)\,\partial_z\phi(z) + h\,\partial_z\epsilon(z)\phi(z) \qquad (2.2.2)$$

A field that transforms in such a manner is called a *primary field* with conformal weight or dimension h [1].

Notice that the conformal weight of the product of two fields ϕ_n and ϕ_m (with conformal weights n and m) is equal to the sum of the conformal weights, given by $n + m$, that is,

$$\phi_n\phi_m = \phi_{m+n} \qquad (2.2.3)$$

Also, notice that if a field has conformal weight 1, then its integral is actually an invariant:

$$\delta \int dz\,\phi(z) = \int dz\,\partial_z[\epsilon(z)\phi(z)] = 0 \qquad (2.2.4)$$

Example: Free Boson Field

To illustrate these methods, let us study the simplest of all conformal systems, a single free boson field ϕ. We begin with the Lagrangian:

$$L = \frac{1}{2\pi}\,\bar{\partial}\phi\,\partial\phi \qquad (2.2.5)$$

from which we can naively define the energy-momentum tensor for the free boson field as:

$$T(z) = -\frac{1}{2}\,\partial_z\phi(z)\,\partial_z\phi(z) \qquad (2.2.6)$$

When we proceed to the quantum theory, however, this previous expression has no meaning. There are two fields defined at the same point in space–time, which is formally divergent. To make some sense out of this, we therefore must define a method by which to subtract the divergent piece, while maintaining conformal invariance.

We can define the *normal ordered product* in several ways by subtracting its divergent piece:

$$T(z) = -\frac{1}{2} : \partial\phi(z)\,\partial\phi(z) :$$

$$= -\frac{1}{2} \lim_{z \to w} \left[\partial\phi(z)\,\partial\phi(w) + \frac{1}{(z-w)^2} \right] \qquad (2.2.7)$$

$$= -\frac{1}{2} \lim_{z \to w} \left[\partial\phi(z)\,\partial\phi(w) + \left\langle \partial\phi(z)\,\partial\phi(w) \right\rangle \right]$$

where:

$$\left\langle \phi(w)\phi(z) \right\rangle = -\log(w - z) \qquad (2.2.8)$$

Yet another way to define the normal ordered product is to simply reshuffle the operators contained within the product of two fields so that the creation (annihilation) operators appear on the left (right), so that the resulting expression has a zero vacuum expectation value. Because of the problem of potential divergences of two fields multiplied at the same point, we will adopt the following conventions in this book. If two fields are multiplied at the same point, we will tacitly assume that the product is normal ordered. Also, when taking the correlation function of two fields, we will tacitly assume that they are "radially ordered," which is the counterpart of time ordering found in ordinary point particle quantum theory. Radial ordering means that the products of all fields are ordered according to their distance from the origin of the complex plane. Let us now calculate the conformal weight of the derivative of a free boson field:

$$T(w)\,\partial\phi(z) \sim -\frac{1}{2} : \partial\phi(w)\partial\phi(w) : \partial\phi(z)$$

$$\sim (-2)\frac{1}{2}\,\partial\phi(w)\left\langle \partial\phi(w)\,\partial\phi(z) \right\rangle$$

$$\sim \partial\phi(w)\frac{1}{(w-z)^2} \qquad (2.2.9)$$

$$\sim \left[\partial\phi(z) + (w-z)\,\partial^2\phi(z) \right] \frac{1}{(w-z)^2}$$

Comparing with our previous expression for the transformation of a field of weight h, we find that $\partial\phi$ has weight 1 when ϕ has weight 0. Next, let us compute the value of the central term for the free boson field. The operator product expansion of two energy-momentum tensors, for the scalar field, is given by:

$$T(w)T(z) \sim 2\left(-\frac{1}{2}\right)^2 \left(\langle \partial_w \phi \, \partial_z \phi \rangle\right)^2$$

$$+ 4\left(-\frac{1}{2}\right)^2 \partial_w \phi \langle \partial_w \phi \, \partial_z \phi \rangle \, \partial_z \phi$$

$$\sim \frac{1}{2}\frac{1}{(w-z)^4} + \partial_w \phi \frac{-1}{(w-z)^2} \partial_z \phi \qquad (2.2.10)$$

$$\sim \frac{1}{2}\frac{1}{(w-z)^4} + \frac{2}{(w-z)^2}\left[-\frac{1}{2}(\partial_z \phi)^2\right]$$

$$+ \frac{1}{(w-z)} \partial_z \left[-\frac{1}{2}(\partial_z \phi)^2\right]$$

To understand this expression, let us write the general expression for the operator product expansion of the energy-momentum tensor:

$$T(z)T(w) \sim \frac{1}{2}\frac{c}{(z-w)^4} + \frac{2}{(z-w)^2}T(w) + \frac{1}{(z-w)} \partial_w T(w) + \cdots \quad (2.2.11)$$

where c is the central charge of the Virasoro algebra.

Comparing the two expressions, we see that the second term in the expansion shows that $T(z)$ has conformal weight 2 (although it is not a primary field because of the presence of the central charge), and that the free boson field has $c = 1$.

This trivial example is important because the string can be viewed as 26 free bosons added together:

$$T(z) = -\frac{1}{2} \partial X_\mu \, \partial X^\mu \qquad (2.2.12)$$

so the central term is just equal to 26.

Example: Free Fermion Field

The next simplest example is a free fermion field, which has $c = \frac{1}{2}$. Let us begin with the free fermion action:

$$L = \frac{1}{2\pi} \psi(z) \, \partial \psi(z) \qquad (2.2.13)$$

which yields the following energy-momentum tensor:

$$T(z) = \frac{1}{2} : \psi(z) \, \partial \psi(z) : \qquad (2.2.14)$$

The vacuum expectation value of two fermion fields is therefore given by:

$$\left\langle \psi(w)\psi(z) \right\rangle = \frac{1}{(w-z)} \qquad (2.2.15)$$

We can now repeat all the previous steps, replacing the free boson field by the fermion field. We find that:

$$T(w)\psi(z) \sim \frac{1/2}{(w-z)^2}\psi(z) + \cdots \tag{2.2.16}$$

so the fermion field has conformal weight equal to $1/2$. Furthermore, the operator product expansion of two energy-momentum fields now yields:

$$T(w)T(z) \sim \frac{1/4}{(w-z)^4} + \cdots \tag{2.2.17}$$

so the central term for the free fermion field is given by $c = 1/2$.

In summary, we found two representations of the conformal group given by a free boson and a free fermion with:

$$\begin{aligned} \text{free boson}: \quad & c = 1, \quad && h = 0 \\ \text{free fermion}: \quad & c = 1/2, \quad && h = 1/2 \end{aligned} \tag{2.2.18}$$

Last, let us analyze the transformation properties of the energy-momentum tensor. The generator of conformal transformations will be the Virasoro generators, which in turn are the moments of the energy-momentum tensor. It is straightforward to calculate the product of T with the conformal field $\phi_h(z)$ with weight h:

$$T(z)\phi_h(w) \sim \frac{h}{(z-w)^2} + \frac{1}{z-w}\partial_w\phi_h(w) \tag{2.2.19}$$

The Virasoro generators [Eq. (1.2.23)] emerge when we take the moments of the energy-momentum operator:

$$\begin{aligned} L_n &= \oint \frac{dz}{2\pi i} z^{n+1}T(z) \\ T(z) &= \sum_{n=-\infty}^{\infty} z^{-n-2}L_n \end{aligned} \tag{2.2.20}$$

Let us rewrite these equations in perhaps a more familiar form, in terms of commutators. Let us define the generator of conformal transformations as T_ϵ:

$$T_\epsilon = \oint \epsilon(z)T(z)\,dz \tag{2.2.21}$$

Then, we can write the variation of the field $\phi_h(z)$ as a commutator:

$$\begin{aligned} \delta\phi_h(z) &= [T_\epsilon, \phi_h(z)] \\ [L_m, \phi_h(z)] &= z^{m+1}\partial\phi_h(z) + h(m+1)z^m\phi_h(z) \end{aligned} \tag{2.2.22}$$

while the variation of the energy-momentum tensor becomes:

$$[T_\epsilon, T(z)] = \epsilon(z)T(z)' + 2\epsilon(z)'T(z) + \frac{1}{12}c\epsilon(z)''' \tag{2.2.23}$$

The infinitesimal transformation of the energy-momentum tensor can be integrated, giving us the finite transformation under $z \to f(z)$ as follows:

$$T(z) \to (\partial f)^2 T[f(z)] + \frac{c}{12} S(f, z) \qquad (2.2.24)$$

where the last term is called the Schwartzian derivative:

$$S(f, z) \equiv \frac{\partial f \, \partial^3 f - (3/2)(\partial^2 f)^2}{(\partial f)^2} \qquad (2.2.25)$$

This expression will be useful when we discuss modular invariance in Chapter 4.

Last, using these techniques, we mention that we can construct the full operator product expansion of the superconformal algebra [Eq. (1.4.12)] for the NS–R model [see Eq. 1.4.10)]:

$$T_B(w)T_B(z) \sim \frac{3\hat{c}/4}{(w-z)^4} + \frac{2}{(w-z)^2}T_B(z) + \frac{1}{w-z}\partial_z T_B(z)$$

$$T_B(w)T_F(z) \sim \frac{3/2}{(w-z)^2}T_F(w) + \frac{1}{w-z}\partial_z T_F(z) \qquad (2.2.26)$$

$$T_F(w)T_F(z) \sim \frac{\hat{c}/4}{(w-z)^4} + \frac{1/2}{w-z}T_B(z)$$

2.3. Representations of the Conformal Group

Now, let us try to classify the representations of the conformal group in much the same way that we classify the representations of an ordinary Lie group. For the familiar example of $SU(2)$, we know that representations are constructed by taking ladder operators L_+ and acting on the eigenstate $|l, -l\rangle$, in which m (the eigenvalue of L_z) has its lowest value. (This state is often called the "highest weight state.") In general, the series generated by all products of these ladder operators creates the universal enveloping algebra of the system, which in turn contains the various representations of the group:

$$(L_+)^n |l, -l\rangle \qquad (2.3.1)$$

Representations of higher groups, such as $SU(3)$, are created in the same way. Here, there are three sets of ladder operators U_+, V_+, and T_+, which then act on an eigenstate with the lowest values of the quantum numbers. Within this series we find the octets, decuplets, etc. Thus, although the number of ladder operators that hit the highest weight state is unlimited, the dimension of each representation of $SU(3)$ is finite.

In much the same way, we construct representations of the conformal group, except for several crucial differences. We choose as our ladder operators the set L_{-n}, where n is positive. The highest weight state is specified by two quantum numbers, h and c, such that:

$$L_0|h, c\rangle = h|h, c\rangle$$
$$L_n|h, c\rangle = 0, \quad n = 1, 2, \dots$$

(2.3.2)

The eigenvalue of L_0 is called the level number, and c is the central charge of the Virasoro algebra. Then, the universal enveloping algebra is created by all products of the ladder operators acting on the highest weight state [see Eqs. (1.2.25)–(1.2.27)]:

$$|\{n\}\rangle = L_{-n_1}, L_{-n_2}, \cdots, L_{-n_k}|h, c\rangle$$

(2.3.3)

It is easy to see that the enveloping algebra does, in fact, form a representation of the algebra. For example, if we hit an element of the enveloping algebra with L_{-n}, then it obviously transforms this element into another element of the enveloping algebra. Also, if we hit this state with L_n, for n positive, then we can use the commutation relations of the original Virasoro algebra to shove this operator to the right, until it annihilates on the highest weight state, thereby giving us a new element of the representation.

The dimension of this collection of products is now infinite, in contrast to the Lie algebra case. This representation, constructed out of the enveloping algebra, is called a *Verma module* [13].

Let us now introduce an operator language for this representation. Let $\phi_h(z)$ represent a conformal field of weight h. Then, let us define the vacuum state:

$$\{L_0, L_1, L_{-1}\}|0\rangle = 0$$

(2.3.4)

The three generators of $SL(2, R)$ given by $\{L_1, L_0, L_{-1}\}$ vanish on this vacuum. Then, we can show:

$$|h, c\rangle = \phi_h(0)|0\rangle$$

(2.3.5)

where c is the central term of the Virasoro generator. Let us now apply the Virasoro generators on this state. Let $\phi_n(z)$ be a primary field that satisfies Eq. (2.2.2). Then, its commutators with L_n are given by Eq. (2.2.22). By fixing the value of z, it is easy to show that:

$$L_n\phi_h(0)|0\rangle = L_n|h, c\rangle = 0; \quad n > 0$$
$$L_0\phi_h(0)|0\rangle = L_0|h, c\rangle = h|h, c\rangle$$

(2.3.6)

In other words, given the fact that ϕ_h transforms as a primary field of weight h, the state $|h, c\rangle$ is a highest weight state of weight h.

Conformal fields that do not transform as primary fields are called *secondary fields*. For example, the derivatives of primary fields, which may have complicated transformation rules under the conformal group, are usually secondary fields. To see how secondary fields are constructed from a primary field, let us define:

$$L_{-k}(z) = \oint \frac{dw\, T(w)}{(w-z)^{k+1}} \qquad (2.3.7)$$

Let us also define:

$$\phi_h^{-k_1,-k_2,\ldots,-k_n}(z) = L_{-k_1}(z), L_{-k_2}(z) \cdots L_{-k_n}(z)\phi_h(z) \qquad (2.3.8)$$

which can be rewritten as:

$$|\{k\}\rangle = \phi_h^{-k_1,-k_2,\ldots,-k_n}(0)|0\rangle = L_{-k_1}(0)L_{-k_2}(0) \cdots L_{-k_n}(0)|h,c\rangle \qquad (2.3.9)$$

We see, therefore, that the fields $\phi_h^{\{k\}}(z)$ are secondary fields that are descendants of the original, primary field $\phi_h(z)$. These secondary fields are constructed from derivatives of the original primary fields, which in turn can be composed from Virasoro generators acting on the primary field.

2.4. Fusion Rules and Correlation Functions

In general, there may be an infinite number of primary fields, with each primary field in turn having an infinite series of descendant secondary fields associated with it. Let the symbol $[\phi_n]$ represent the conformal family containing the primary field ϕ_n and all its secondary fields created by acting on it with L_{-k}, as in Eq. (2.3.3). Our task is to determine, for a fixed value of c, all possible conformal families $[\phi_n]$ and their correlation functions.

At first, it seems that the categorization of all representations of a conformal field theory seems hopeless. However, these conformal operators fortunately must satisfy a large set of identities that often make it possible to completely solve the theory.

To see this, let us recall that in ordinary quantum field theory two fields ϕ_1 and ϕ_2 have the following Wilson operator product expansion when the fields are close to each other:

$$\phi_1(x)\phi_2(y) \sim \sum_i C_i(x-y)O_i(y) \qquad (2.4.1)$$

where the O_i's are a complete set of operators and the C_i's are singular numerical coefficients.

Similarly, the same can be said for conformal fields, except that we can place more constraints on the right-hand side. For example, by equating the scaling dimension on both sides of the operator product expansion, we can calculate the singularity structure of the C_i's as follows:

$$C_i \sim \frac{1}{(x-y)^{h_1+h_2-h_i}} \qquad (2.4.2)$$

where the h's are the scaling dimensions of the various fields.

Now, let us take the operator product expansion of the two conformal fields ϕ_n and ϕ_m, with conformal weights given by δ_n and δ_m. We claim that the operator product of the two primary fields is given by:

$$\phi_n(z)\phi_m(w) \sim \sum_k C_{nmk}(z-w)^{\delta_k-\delta_n-\delta_m} O_k(w) \qquad (2.4.3)$$

for some constants C_{nmk}. (The power of $z - w$ is easily determined by examining the dimensions of the left- and right-hand sides of the equation.)

Actually, conformal invariance places even more constraints on the operator product expansion. We note that the right-hand side can, in turn, be represented by a complete set of primary and secondary fields, denoted $\phi_p^{\{k\}}$, where $\{k\}$ indexes the various elements of the Verma module.

Now, let us write the full operator product relation, including all dependence on the complex variables $z, \bar{z}$:

$$\phi_n(z,\bar{z})\phi_m(0) \sim \sum_p \sum_{\{k\},\{\bar{k}\}} C_{nm}^p \beta_{nm}^{p;\{k\}} \beta_{nm}^{p;\{\bar{k}\}}$$
$$\times z^{[\delta_p-\delta_n-\delta_m+\sum_i \{k\}]} \bar{z}^{[\bar{\delta}_p-\bar{\delta}_n-\bar{\delta}_m+\sum_i \{\bar{k}\}]} \phi_p^{\{k\},\{\bar{k}\}}(0) \qquad (2.4.4)$$

The matrix C_{nm}^p expresses the "Clebsch–Gordan" coefficient found in the tensor product decomposition of two primary fields, with weights n and m, into another set of fields labeled by p. (Strictly speaking, this is not a Clebsch–Gordan coefficient in the usual sense because the value of c for all primary fields is the same. For a normal Clebsch–Gordan coefficient, the values of the c's are additive as we multiply different representations.)

Conformal invariance is so powerful that we can often determine the precise numerical values of all the coefficients appearing on the right-hand side of the equation. The numerical calculation is straightforward, but rather lengthy. We simply multiply both sides of the equation by $|0\rangle$ and act on both sides of the equation with L_k. When L_k acts on a primary field, it transforms, as in Eq. (2.2.19). However, when L_k acts on a secondary field, it creates many other secondary fields as L_k commutes past L_{-q}, until it annihilates on the vacuum. By equating the terms on the left with the terms on the right, we can find an iterative procedure to calculate all values of C_{nm}^p and $\beta_{nm}^{p,\{k\}}$. Thus, conformal invariance alone is sometimes sufficient to determine the value of all C_{nm}^p [1].

We wish to express this rather lengthy equation in shorthand. We simply write [1]:

$$[\phi_n] \times [\phi_m] = \sum_k C_{nm}^k [\phi_k] \qquad (2.4.5)$$

where we suppress the presence of all secondary fields by putting everything in brackets. We implicitly assume that the infinite series of coefficients that we have suppressed in the above equation can be numerically calculated by hitting both sides with a conformal transformation and then equating terms, order by order.

In principle, if we knew the values of the "structure constants" C_{nm}^k, then we would actually know everything about the representation of the conformal field theory. Most of the work in solving conformal field theory reduces to determining, for a fixed value of c, the number of conformal families $[\phi_n]$ and the structure constants C_{nm}^p created by taking products between them. (Once the C_{nm}^p are known, then conformal invariance alone will determine the $\beta_{nm}^{p,\{k\}}$.)

In general, conformal invariance alone is not powerful enough to determine the fusion rules among the primary fields. Outside input is required. To see this, let us first construct the correlation functions between primary fields. For the two-point function, conformal invariance alone is sufficient to determine the correlation function up to a constant:

$$\langle \phi_{n_1}(z_1)\phi_{n_2}(z_2)\rangle = \frac{k}{(z_1 - z_2)^{\delta_1 + \delta_2}}\delta_{n_1,n_2} \qquad (2.4.6)$$

This is proven by taking the conformal transformation of both sides of the equation. The transformation under $\epsilon(z) = $ const forces the right-hand side to be a function of $z_{12} = z_1 - z_2$, and the transformation under $\epsilon(z) = z^2$ fixes the conformal weights δ_i.

Similarly, using only the transformation rules of the conformal group, we can determine the correlation function of three primary fields, up to a constant, in terms of $z_{ij} = z_i - z_j$:

$$\left\langle \prod_{i=1}^3 \phi_{n_i}(z_i) \right\rangle = kC_{123}z_{12}^{\delta_3-\delta_1-\delta_2}z_{23}^{\delta_1-\delta_2-\delta_3}z_{13}^{\delta_2-\delta_1-\delta_3} \qquad (2.4.7)$$

Conformal invariance does not fix the value of the structure constant itself, however.

For products of four or higher fields, the situation gets worse. For example, for the four-point function, we have:

$$\left\langle \prod_{i=1}^4 \phi_{n_i}(z_i) \right\rangle = f(x)\prod_{i<j} z_{ij}^{-\delta_i-\delta_j+\delta} \qquad (2.4.8)$$

where $\delta = \sum_{i=1}^4 \delta_i/3$ and:

$$x = z_{12}z_{34}/z_{13}z_{24} \qquad (2.4.9)$$

Conformal invariance can reduce the correlation function to functions of z_{ij}, but it is not sufficient to determine the function $f(x)$.

However, we can exploit one more condition, that the product of four primary fields is associative, that is, we can take the pairwise contraction of primary fields in two different ways and get the same answer. Thus, by pairing the four primary fields in two different fashions, we have yet another constraint on the correlation function. In Fig. 2.1, we see how the fusion rules may be used in two different ways, by pairing different sets of primary fields within the same correlation function. Since the final answer is independent of the way in which the pairing takes place, we have a new restriction on the functions appearing in the correlation function.

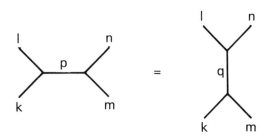

Fig. 2.1.

For simplicity, let us first fix the values of the z's to be $z_1 = \infty, z_2 = 1, z_3 = x, z_4 = 0$. Then, the full four-point function, as a function of both x and $\bar{x}$, becomes:

$$G_{nm}^{lk}(x, \bar{x}) = \left\langle \phi_k(z_1, \bar{z}_1)\phi_l(z_2, \bar{z}_2)\phi_n(z_3, \bar{z}_3)\phi_m(z_4, \bar{z}_4) \right\rangle$$
$$= \left\langle k|\phi_l(1, 1)\phi_n(x, \bar{x})|m \right\rangle \tag{2.4.10}$$

Let us now perform the contractions by pairing the primary fields and then using the fusion rules. For example, by pairing the nth and mth fields, we find:

$$G_{nm}^{lk}(x, \bar{x}) = \sum_p C_{nm}^p C_{klp} F_{nm}^{lk}(p|x)\bar{F}_{nm}^{lk}(p|\bar{x}) \tag{2.4.11}$$

where:

$$F_{nm}^{lk}(p|x) = x^{\delta_p - \delta_n - \delta_m} \sum_{\{k\}} \beta_{nm}^{p,\{k\}} x^{\sum_i k_i}$$
$$\times \frac{\left\langle k|\phi_l(1, 1)L_{-k_1}L_{-k_2}\dots L_{-k_N}|p \right\rangle}{\left\langle k|\phi_l(1, 1)|p \right\rangle} \tag{2.4.12}$$

The function $F_{nm}^{lk}(p|x)$ is called a *conformal block* [1], because higher order correlation functions can always be written in terms of such blocks.

They are the building blocks by which we can write arbitrary correlation functions.

Now, the key step is to pair the primary fields in different fashions using the fusion rules and then compare the results of different sets of pairings. Since the final result must be the same, we have a new set of identities. Taking the pairwise contraction in two different ways, we find:

$$\sum_p C^p_{nm} C_{lkp} F^{lk}_{nm}(p|x) \bar{F}^{lk}_{nm}(p|\bar{x}) = \sum_q C^q_{nl} C_{mkq} F^{mk}_{nl}(q|1-x) \bar{F}^{mk}_{nl}(q|1-\bar{x})$$

(2.4.13)

which expresses the fact that the operator product expansion is associative.

This still leaves the problem of how to actually solve for the conformal blocks $F^{lk}_{nm}(p|x)$, using our knowledge of conformal invariance. One way of tackling this problem is to insert a $T(z)$ operator into the correlation function. Because the commutation relation between a $T(z)$ operator and a primary field gives us back a primary field (without the T insertion), we can write Ward-like identities for the correlation functions. These identities relate the correlation function where T is inserted and a differential equation on a correlation function where T has disappeared.

Let us now insert T_ϵ into the following correlation function:

$$\left\langle \oint \frac{dz}{2\pi i} \epsilon(z) T(z) \phi_1(w_1) \cdots \phi_n(w_n) \right\rangle$$

$$= \sum_{j=1}^{n} \left\langle \phi_1(w_1) \cdots \left[\oint \frac{dz}{2\pi i} \epsilon(z) T(z) \phi_j(w_j) \right] \cdots \phi_n(w_n) \right\rangle \quad (2.4.14)$$

$$= \sum_{j=1}^{n} \left\langle \phi_1(w_1) \cdots \delta_\epsilon \phi_j(w_j) \cdots \phi_n(w_n) \right\rangle$$

By inserting the value of the variation of a primary field, we find our final result:

$$\langle T(z) \phi_1(w_1) \phi_2(w_2) \cdots \phi_n(w_n) \rangle$$

$$= \sum_{i=1}^{n} \left[\frac{\Delta_i}{(z-w_i)^2} + \frac{1}{z-w_i} \frac{\partial}{\partial w_i} \right] \langle \phi_1(w_1) \cdots \phi_n(w_n) \rangle \quad (2.4.15)$$

where we have repeated Eq. (2.2.19):

$$T(z) \phi_i(w_i) \sim \frac{\Delta_i}{(z-w)^2} \phi_i(w_i) + \frac{1}{z-w} \partial_{w_i} \phi(w_i) \quad (2.4.16)$$

In general, the differential equations for the conformal blocks are too difficult to solve, especially if there are an infinite number of primary fields.

Unfortunately, this is the case for string theory, which has an infinite number of primary fields given by the real states $|R\rangle$, which satisfy Eq. (1.2.25) or Eq. (2.3.6). Reinterpreted from the point of view of conformal field theory, we see that the spurious states $|S\rangle$ in (1.2.27) form the secondary states of the Verma module labeled by $|R\rangle$. Thus, the Fock space of string theory can be decomposed in terms of an infinite family of Verma modules. Each module consists of a real state $|R\rangle$ (which is a primary field) and the infinite number of spurious states $|S\rangle$ associated with each primary state, which are the singular secondary states.

Although the correlation functions for the full string model are too difficult to solve exactly for all Verma modules, we will be interested in a subclass of conformal field theories that have only a finite number of primary fields, which we will show is exactly solvable. Although the conformal field theories with a finite number of primary fields are not very physical, they will give us a theoretical laboratory in which to test many of our ideas concerning conformal field theory.

It can be shown that if there are a finite number of primary fields, then the values of h and c take on rational values. These are called *rational conformal field theories* and will be studied in this and the next chapter in connection with Kac–Moody algebras. The simplest of these rational conformal field theories are called *minimal models*, which we will now study.

2.5. Minimal Models

One important question to ask of any representation is whether it is reducible or not. For ordinary Lie groups, we can construct the scalar product between the various elements $|\alpha_i\rangle$ of a representation and treat it as a matrix $\langle\alpha_i|\alpha_j\rangle$. Then, the representation is reducible if the determinant of this matrix is zero.

Similarly, we can determine whether a Verma module is reducible or not by taking its elements $|\{n\}\rangle$ in Eq. (2.3.3) and forming the scalar product between them. Then, the representation is reducible if the determinant of this matrix is zero, that is, if

$$\det\langle\{m\}|\{n\}\rangle = 0 \tag{2.5.1}$$

This is not a trivial task, because the elements within a Verma module grows rapidly, as the partition of the level. For example, at the first level, we only have one element, given by $L_{-1}|h\rangle$, so the matrix has only one element:

$$\langle h,c|L_1 L_{-1}|h,c\rangle = 2\langle h,c|L_0|h,c\rangle = 2h \tag{2.5.2}$$

However, at the second level, we have two members of the Verma module: $L_{-1}^2|h,c\rangle$ and $L_{-2}|h,c\rangle$. To determine whether they are truly linearly independent, we must form the 4×4 matrix given by:

$$\det M_2 = \begin{pmatrix} \langle h|L_2 L_{-2}|h\rangle & \langle h|L_1^2 L_{-2}|h\rangle \\ \langle h|L_2 L_{-1}^2|h\rangle & \langle h|L_1^2 L_{-1}^2|h\rangle \end{pmatrix}$$

$$= \begin{pmatrix} 4h + c/2 & 6h \\ 6h & 4h(1+2h) \end{pmatrix} \tag{2.5.3}$$

The determinant of this matrix, in turn, can be written in terms of h, which factorizes nicely:

$$\det M_2 = 2(16h^3 - 10h^2 + 2h^2 c + hc)$$
$$= 32 \left[h - h_{1,1}(c)\right]\left[h - h_{1,2}(c)\right]\left[h - h_{2,1}(c)\right] \tag{2.5.4}$$

where:

$$h_{1,1}(c) = 0$$
$$h_{1,2}(c) = \frac{1}{16}(5 - c) \mp \frac{1}{16}\sqrt{(1-c)(25-c)} = h_{2,1} \tag{2.5.5}$$

We see that the determinant conveniently factorizes into a product of factors, which vanishes if h equals one of the h_{rs}. Although it seems like a hopeless task to generalize this equation to all levels, we now use a remarkable formula due to Kac [12–15], which states that the determinant at the nth level is given by:

$$\det M_n = \prod_{k=1}^{n} \psi_k(h, c)^{p(n-k)} \tag{2.5.6}$$

where:

$$\psi_k(h, c) = \prod_{pq=k} \left[h - h_{p,q}(c)\right]$$
$$h_{p,q}(c) = \frac{[(m+1)p - mq]^2 - 1}{4m(m+1)} \tag{2.5.7}$$

where $p(n)$ is the partition of the integer n (that is, the number of ways in which n can be written as the sum of smaller integers) and where the parameter c is related to m via:

$$c = 1 - \frac{6}{m(m+1)}$$
$$m = -\frac{1}{2} \pm \frac{1}{2}\sqrt{\frac{25 - c}{1 - c}} \tag{2.5.8}$$

where p and q are positive integers. This formula is one of the most important tools that we have at our disposal for understanding the representations of the conformal group. We will refer to this formula throughout the first part of this book.

Ordinarily, if $h \neq h_{p,q}$, then the Kac determinant does not vanish, and the representation is irreducible. In general, however, this case is exceedingly difficult to analyze. However, some of the more interesting cases occur when the Fock space is reducible, that is, when it contains linear relations between the various secondary states. We will study these models because many of them can be solved exactly.

If $h = h_{p,q}$, then the Verma module is reducible, but one can (by suitable truncation of the secondary states) extract a smaller subspace of the reducible Verma module that is irreducible. The advantage of this truncation is that, although the number of primary states ϕ_n may in general be infinite, for these reducible models, the number of primary fields actually becomes finite.

For this reason, these models with a finite number of primary fields are solvable and hence provide us insight into the more difficult (and physically relevant) case of an infinite number of primary fields.

If $h = h_{p,q}$ for some p and q, then the Kac determinant is zero, the representation is reducible, and there exists a state $|\chi\rangle$ at the Nth level whose matrix elements with other states in the module at level N vanish. However, the matrix elements between states with levels N and M ($N \neq M$) vanish (since we must have an equal number of creation and annihilation operators sandwiched between $\langle 0|$ and $|0\rangle$). Thus, $|\chi\rangle$ is a null state that has vanishing matrix elements with all members $\langle\phi|$ of the Verma module, that is,

$$\langle\phi|\chi\rangle = 0, \qquad \langle\chi|\chi\rangle = 0 \qquad (2.5.9)$$

The level of $|\chi\rangle$ is equal to $h_{p,q} + pq$. The important point, however, is that $|\chi\rangle$ can be a primary field. To see this, notice that each secondary field $\langle\phi|$ appearing in the above equation consists of products of L_k. We see that L_k therefore annihilates $|\chi\rangle$, that is,

$$L_n|\chi\rangle = 0, \qquad L_0|\chi\rangle = (h_{p,q} + pq)|\chi\rangle \qquad (2.5.10)$$

The fact that reducible Verma modules contain secondary null states $|\chi\rangle$ that are by themselves primary states means that there is a "smaller" Verma module contained within the larger one with $|\chi\rangle$ as its highest weight state. This new Verma module, contained within the larger reducible Verma module, is called $[\phi_{p,q}]$.

(This smaller Verma module, however, may also be reducible. There may be, in turn, null states within this module. However, as we shall see in Chapter 4, we can systematically extract these null states within $[\phi_{p,q}]$ until the final result is irreducible.)

Example: Null States

It is instructive to construct some of these null states explicitly. For example, the secondary state at level 1,

$$|\chi\rangle = L_{-1}|h, c\rangle \qquad (2.5.11)$$

is a null state if $h = 0$, for any value of c, because its norm is equal to $2h$, as in Eq. (2.5.2). By applying L_n on this null state, we see that it is also a primary state.

[There is another, more transparent way of seeing this. We know that L_{-1} can be represented in z space as the operator ∂, so the state in question is $\partial\phi_h$. This state, for arbitrary h, is not a primary state [because the derivative acts on the Jacobian appearing in Eq. (2.2.1). However, if $h = 0$, then there is no Jacobian factor in Eq. (2.2.1), and $\partial\phi_h$ is a primary field. Thus, for a reducible module, a secondary field has become a null primary field. This is analogous to the general theory of relativity, where the derivative of a vector $\partial_\mu \phi_\nu$ is not a covariant tensor, but the derivative of a scalar $\partial_\mu \phi$ is a genuine vector.]

At the second level, let us try:

$$|\chi\rangle = (L_{-2} + aL_{-1}^2)|h, c\rangle \qquad (2.5.12)$$

Demanding that this state be annihilated by L_1 and L_2 fixes the following: $a = (3/2)(2h+1)$ and $c = 2h(5 - 8h)/(2h + 1)$.

Now, let us analyze the unitarity properties of minimal models. If we are looking for conformal models with a finite number of fields, then BPZ [1] found that the minimal models are labeled by two numbers, m and m', which are relatively prime positive integers, such that

$$\text{minimal series}: \quad \begin{cases} c & = 1 - 6(m - m')^2/mm' \\ h_{p,q} & = (4mm')^{-1}\left[(pm' - qm)^2 - (m - m')^2\right] \end{cases}$$
$$(2.5.13)$$

The areas of most interest for us, however, are those states that are unitary, that is, those representations where the Kac determinant has only positive eigenvalues. Let us now analyze the unitary representations of the Virasoro algebra (that is, representations that have a positive norm). Analyzing the Kac formula, we find that there are three regions of interest. In the region $c > 1$, $h \geq 0$, the Kac determinant has no zeros at all and hence the representations are irreducible. For the region $1 < c < 25$, however, m is not real, and the $h_{p,q}$'s have an imaginary part or, for $p = q$, are negative. For $c \geq 25$, we can choose the branch $-1 < m < 0$, and all $h_{p,q}$'s are negative.

We can also show that the representations are unitary in this region. For very large h, the diagonal elements along the Kac matrix dominate the matrix, and they are all positive. Thus, the matrix has positive eigenvalues for large h. But, since the determinant never vanishes for $c > 1$, $h \geq 0$, all of the eigenvalues must stay positive in this region, and hence the representation is unitary. For $c = 1$, the determinant vanishes at $h = n^2/4$ and never becomes negative. So, there is no obstacle to being unitary.

For $0 < c < 1$, $h > 0$, the situation is rather delicate. Naively, we can show that this region is nonunitary. Let us draw the curves formed by $h = h_{p,q}(c)$ in the h, c plane. For each set of integers p, q, we have a curve in the h, c plane. Then, one can show, by graphical methods, that any point in the region $0 < c < 1$, $h > 0$, can be connected to the $c > 1$ region by a path that crosses a single curve of the Kac determinant. This shows that the determinant reverses sign passing through the curve, which proves the existence of negative norm states and hence unitary representations are excluded from this region.

There is, however, a loophole in this demonstration. It may turn out that the determinant vanishes along the curves. We know, however, where the determinant vanishes, and that is given by Eqs. (2.5.6)–(2.5.8). We find that the representations are unitary for the following discrete set of values [16]:

$$\text{unitary series}: \quad c = 1 - \frac{6}{m(m+1)}, \quad m = 3, 4, \ldots \quad (2.5.14)$$

(If we set $m' = m + 1$ in the minimal series in Eq. (2.5.13), we obtain the unitary series.) Notice that $h_{p,q}$ has certain symmetries that enable us to establish some of the structure of the representation space. In particular, notice that it is symmetric under:

$$p \leftrightarrow m - p, \quad q \leftrightarrow m + 1 - p \quad (2.5.15)$$

Thus, if we allow q to range from $1 \le q \le m$, then there are a total of $m(m-1)$ values of $h_{p,q}$, each appearing twice. It is sometimes convenient to display the allowed values of $h_{p,q}$ in a grid. We choose p to label the horizontal axis (increasing from left to right) and q to label the vertical axis (increasing from bottom to top). Then, the allowed values of $h_{p,q}$ for $m = 3, 4, 5$ are:

$$m = 3 \rightarrow \begin{pmatrix} 1/2 & 0 \\ 1/16 & 1/16 \\ 0 & 1/2 \end{pmatrix}, \quad m = 4 \rightarrow \begin{pmatrix} 3/2 & 7/16 & 0 \\ 3/5 & 3/80 & 1/10 \\ 1/10 & 3/80 & 3/5 \\ 1/10 & 3/80 & 3/5 \\ 0 & 7/16 & 3/2 \end{pmatrix}$$

$$m = 5 \rightarrow \begin{pmatrix} 3 & 7/5 & 2/5 & 0 \\ 13/18 & 21/40 & 1/40 & 1/8 \\ 2/3 & 1/15 & 1/15 & 2/3 \\ 1/8 & 1/40 & 21/40 & 13/8 \\ 0 & 2/5 & 7/5 & 3 \end{pmatrix} \quad (2.5.16)$$

Example: The Ising Model

The case $m = 3$ is most interesting because it will correspond to the critical point of the two-dimensional Ising model, which has been extensively studied in the field of statistical mechanics. At the critical point of the Ising model, the correlation lengths become infinite, and the theory loses all reference to any scale, that is, it becomes scale invariant. Thus, we expect to find the Ising model at the critical point among the list of various conformal models.

We recall that we can write down null states at the second level for various values of h and c. In particular, for the Ising model, we have three primary fields, with conformal weights 0, $\frac{1}{2}$, $\frac{1}{16}$.

If we have left–right symmetric fields, then the field $(\frac{1}{16}, \frac{1}{16})$ is called the "order parameter" σ, and the field $(\frac{1}{2}, \frac{1}{2})$ is called the "energy operator" ϵ. (This will be discussed further in Chapter 6.)

In conformal field theory language, the energy operator can be written as two free fermion fields $\psi\bar{\psi}$. The order parameter, however, cannot be written in terms of the fermion field ψ, since we cannot construct a field with weight $\frac{1}{16}$ starting with a field with weight $\frac{1}{2}$. (In the next chapter, we will show that the σ field can be written as a spin field S using the mechanism of bosonization.) In addition to σ, there is also the "disorder" parameter μ in the Ising model at criticality. The disorder parameter has the same conformal weight as σ (but has different product ordering with ψ).

These fields, in turn, allow us to construct null states for the $m = 3$ minimal series. From Eqs. (2.5.12) and (2.5.13), we have:

$$
\begin{aligned}
\left(L_{-2} - \frac{3}{4}L_{-1}^2\right)\Big|h = \frac{1}{2}\Big\rangle \\
\left(L_{-2} - \frac{4}{3}L_{-1}^2\right)\Big|h = \frac{1}{16}\Big\rangle
\end{aligned}
\tag{2.5.17}
$$

If we insert these null states into any matrix element, then it is sure to vanish because these states have zero matrix elements with all elements of the Verma module. Thus, given the fact that Virasoro operators can be converted into partial derivatives, we can derive differential equations that are satisfied by the correlation functions. These differential equations, in turn, allow us to completely calculate many of the lower order correlation functions.

Let us define $G^{(2M,2N)}$ to be:

$$
\langle\sigma(z_1, \bar{z}_1)\cdots\sigma(z_{2M}, \bar{z}_{2M})\mu(z_{2M+1}, \bar{z}_{M+1})\cdots\mu(z_{2M+2N}, \bar{z}_{2M+2N})\rangle
\tag{2.5.18}
$$

Because of the differential equation satisfied by correlation functions with null states, we find the following differential equation for $G^{(2M,2N)}$:

$$
\left\{\frac{4}{3}\frac{\partial^2}{\partial z_i^2} - \sum_{j\neq i}^{2M+2N}\left[\frac{(1/16)}{(z_i - z_j)^2} + \frac{1}{z_i - z_j}\frac{\partial}{\partial z_j}\right]\right\}G^{(2M,2N)} = 0
\tag{2.5.19}
$$

Let us now consider the case when $M = 2$ and $N = 0$. Then, the differential equation simplifies to the following equation:

$$
\left[x(1 - x)\frac{\partial^2}{\partial x^2} + \left(\frac{1}{2} - x\right)\frac{\partial}{\partial x} + \frac{1}{16}\right]f(x, \bar{x}) = 0
\tag{2.5.20}
$$

where:

$$G^{(4,0)}(z_i) = \left[(z_1 - z_3)(z_2 - z_4)(\bar{z}_1 - \bar{z}_3)(\bar{z}_2 - \bar{z}_4)\right]^{-1/8} Y(x, \bar{x}) \qquad (2.5.21)$$

and:

$$x = \frac{(z_1 - z_2)(z_3 - z_4)}{(z_1 - z_3)(z_2 - z_4)} \qquad (2.5.22)$$

and $Y = [x\bar{x}(1 - x)(1 - \bar{x})]^{-1/8} f(x, \bar{x})$.

There are two independent solutions to this equation given by:

$$f_{\pm}(x) = (1 \pm \sqrt{1 - x})^{1/2} \qquad (2.5.23)$$

The complete solution for Y thus contains four possibilities $f_{\pm}(x)f_{\pm}(\bar{x})$. Therefore, the method of null states is quite powerful, often giving us an explicit solution to the correlation functions.

Notice that most of our discussion in this chapter has been quite general and often did not depend in any way on the particular model being studied. Thus, any two-dimensional theory, which becomes conformally invariant, must have representations given by the above analysis for various values of h and c. In particular, the minimal conformal field theories can be shown to describe certain integrable (solvable) models found in statistical mechanics, such as the Ising model. Because a second-order phase transition in an integral model corresponds to an infinite correlation length, the theory loses all references to any scale, that is, we have a scale invariant theory at the critical point. By comparing the critical exponents of various statistical models, we find the following one-to-one correspondence between minimal models and various integrable statistical mechanical models at criticality [17, 18]:

$$
\begin{aligned}
m &= 3 \rightarrow \text{Ising model} \\
m &= 4 \rightarrow \text{Tricritical Ising model} \\
m &= 5 \rightarrow 3 - \text{state Potts model} \\
m &= 6 \rightarrow \text{Tricritical 3} - \text{state Potts model} \\
m &\text{ arbitrary} \rightarrow \text{RSOS models}
\end{aligned}
\qquad (2.5.24)
$$

This relationship between statistical mechanical models at criticality and conformal field theory will be explored more fully in Chapter 6.

2.6. Fusion Rules for Minimal Models

We now wish to use the various identities that we have established to calculate the fusion rules for the minimal models, which determine the operator product expansion of all the primary fields.

Our first goal is to find the fusion rules for the product of a minimal field $\phi_{1,2}$ and some arbitrary primary field ϕ_Δ with weight Δ:

$$\phi_{1,2}(z)\phi_\Delta(w) = \mathrm{const}(z - w)^k[\phi_{\Delta'} + (z - w)\phi_{\Delta'}^{(-1)} + \cdots] \qquad (2.6.1)$$

Our task is to first determine the possible values for Δ' on the right-hand side of the equation and then to generalize both $\phi_{1,2}$ and ϕ_Δ to become arbitrary minimal primary fields.

In general, it is impossible to determine the values for Δ' on the right-hand side of the equation for an arbitrary conformal field theory without more information. We need extra input, which will be that the fields are primary fields for the minimal model.

Our strategy is to take the matrix element between a null field χ and a product of ordinary fields ϕ_i. Because the resulting correlation function is equal to zero and because the null field χ can be decomposed in terms of Virasoro operators, we then arrive at a differential equation involving the correlation function.

The key assumption that we will use is that the primary field $\phi_{1,2}$ is a null field and can be written explicitly, as in Eq. (2.5.12), where the values of the coefficient a and c are given in Eq. (2.5.13). Because the L_n operators can all be written in terms of differential operators, we find that this null field, via (2.5.12), can be written as:

$$\chi_{\delta+2} = \phi_\delta^{(-2)} + \frac{3}{2(2\delta + 1)} \frac{\partial^2}{\partial z^2} \phi_\delta \qquad (2.6.2)$$

$\chi_{\delta+2}$ has conformal weight $\delta + 2$, where δ can be solved by inserting Eq. (2.5.8) into Eq. (2.5.13):

$$\delta = \frac{1}{16}\left[5 - c \pm \sqrt{(c - 1)(c - 25)}\right] \qquad (2.6.3)$$

The term $\phi_\delta^{(-2)}$ contains the operator L_{-2}, which in turn can be written as a differential operator. Anytime a secondary or descendant field enters a correlation function, we can extract the energy-momentum tensor and hence write the correlation function as a differential equation:

$$\langle \phi_n^{-k_1, -k_2, \ldots, -k_M}(z)\phi_1(z_1) \cdots \phi_N(z_N)\rangle$$
$$= \hat{L}_{-k_M}(z, z_i)\hat{L}_{-k_{M-1}}(z, z_i)\langle \phi_n(z)\phi_1(z_1) \cdots \phi_N(z_N)\rangle \qquad (2.6.4)$$

where:

$$\hat{L}_{-k}(z, z_i) = \sum_{i=1}^{N}\left[\frac{(1 - k)\Delta_i}{(z - z_k)^k} - \frac{1}{(z - z_i)^{k-1}}\frac{\partial}{\partial z_i}\right] \qquad (2.6.5)$$

If we take the matrix element of this null state $\phi_{2,1}$ and a product of several fields $\phi_i(z_i)$, then the result must be zero. However, by writing the null vector in terms of Virasoro operators and then converting Virasoro operators into differential operators, we find the following differential equations:

$$\left[\frac{3}{2(2\delta+1)} \frac{\partial^2}{\partial z^2} - \sum_{i=1}^{N} \frac{\Delta_i}{(z-z_i)^2} - \sum_{i=1}^{N} \frac{1}{z-z_i} \frac{\partial}{\partial z_i} \right]$$
$$\times \langle \phi_{1,2}(z)\phi_1(z_1)\cdots\phi_N(z_N) \rangle = 0 \tag{2.6.6}$$

Take the most singular term as $z \to z_1$. Using Eq. (2.6.1) and (2.6.6), we have

$$\frac{3\kappa(\kappa-1)}{2(2\delta+1)} - \Delta + \kappa = 0; \quad \kappa = \Delta' - \Delta - \delta \tag{2.6.7}$$

Solving, we find two solutions:

$$\Delta'_{(1)} = \Delta_0 + \frac{1}{4}(\alpha+\alpha_\pm)^2 \equiv \hat{\delta}(\alpha+\alpha_\pm)$$
$$\Delta'_{(2)} = \Delta_0 + \frac{1}{4}(\alpha-\alpha_\pm)^2 \equiv \hat{\delta}(\alpha-\alpha_\pm) \tag{2.6.8}$$

where:

$$\Delta_0 = \frac{(c-1)}{24}$$
$$\alpha_\pm = \frac{\sqrt{1-c} \pm \sqrt{25-c}}{\sqrt{24}} \tag{2.6.9}$$

and $\hat{\delta}(\alpha) \equiv \Delta_0 + \frac{1}{4}\alpha^2$. This is the equation that we want. We now have the possible values of Δ' that appear on the right-hand side of Eq. (2.6.1).

Thus, the fusion rules give us:

$$\phi_{1,2}\phi_\alpha = [\phi_{\alpha-\alpha_+}] + [\phi_{\alpha+\alpha_+}] \tag{2.6.10}$$

where the conformal fields in the Verma modules on the right-hand side have conformal weight $\hat{\delta}(\alpha \pm \alpha_+)$.

Now, let us gradually generalize the left-hand side of Eq. (2.6.1). If we replace $\phi_{1,2}$ with $\phi_{n,m}$, then the fusion rules can also be calculated using the same techniques, that is, replace $\phi_{n,m}$ with a null field, take its matrix element with a product of fields, and then rewrite the Virasoro generators as differential operators. Then, the fusion rules give:

$$[\phi_{n,m}] \times [\phi_\alpha] = \sum_{l=1-m}^{1+m} \sum_{k=1-n}^{1+n} [\phi_{\alpha+l\alpha_+ + k\alpha_+}] \tag{2.6.11}$$

where the height weight fields on the right-hand side have conformal weight $\hat{\delta}(\alpha + l\alpha_- + k\alpha_+)$.

It is now a straightforward process to generalize ϕ_α as a minimal field, which would then give us the fusion rules for all minimal fields. After a bit

of work, we find that all the fusion rules for the minimal fields are given by
[1]:

$$[\phi_{p_1,q_1}] \times [\phi_{p_2,q_2}] = \sum_{p_3=|p_1-p_2|+1}^{\substack{\min[p_1+p_2-1, \\ 2m-1-(p_1+p_2)]}} \sum_{q_3=|q_1-q_2|+1}^{\substack{\min[q_1+q_2-1, \\ 2m+1-(q_1+q_2)]}} [\phi_{p_3,q_3}] \qquad (2.6.12)$$

2.7. Superconformal Minimal Series

Now that we have examined the conformal properties of the minimal models, let us make a few remarks about the superconformal generalization of the minimal models.

The calculation of the superconformal minimal series proceeds in much the same way as in the conformal minimal series, so let us quickly review how we obtained them. First, we construct the Verma modules, created by the action of the generators of the algebra on some vacuum state. Then, we contract these Verma modules, creating the Kac determinant in Eqs. (2.5.6)–(2.5.8). For certain values of h and c, the Kac determinant does not vanish, and then we have an irreducible representation of the algebra. However, there are usually an infinite number of primary fields associated with this representation, so we are more interested in the values of the Kac determinant that do vanish.

When the determinant vanishes, there is a zero norm state, which in turn can be used as the primary field of its own module. However, this module, in turn, contains zero norm states as well. After all extraneous null states and their secondaries are extracted, we find an irreducible module $[\phi_{p,q}]$. The fusion rules close on a finite number of such primary fields. This allows us to extract a finite number of primary fields, giving us the minimal series.

To analyze the superconformal series, we will find it convenient to introduce the Grassmann variable θ so we can combine the bosonic energy-momentum tensor with the fermionic superconformal current into one superfield:

$$T(z,\theta) = T_F(z) + \theta T_B(z) = \sum_n z^{-n-3/2} T_{F,n} + \theta z^{-n-2} T_{B,n} \qquad (2.7.1)$$

where T_B is the usual bosonic current, with conformal weight 2, and T_F is the superconformal current, with conformal weight $\frac{3}{2}$.

Then, the superconformal algebra [Eq. (2.2.26)] can be deduced from just one operator expansion:

$$T(z_1, \theta_1)T(z_2, \theta_2) \sim \frac{1}{4}\hat{c}z_{12}^{-3} + \left(\frac{3}{2}\theta_{12}z_{12}^{-2} + \frac{1}{2}z_{12}^{-1}D_2 + \theta_{12}z_{12}^{-1}\,\partial_2\right)T(z_2, \theta_2)$$

$$(2.7.2)$$

where:

$$z_{12} = z_1 - z_2 - \theta_1\theta_2, \qquad \theta_{12} = \theta_1 - \theta_2, \qquad D = \partial_\theta + \theta\,\partial_z \qquad (2.7.3)$$

The central charge of the Virasoro algebra is now normalized to $c = 3\hat{c}/2$. A free scalar superfield now consists of a scalar field with $c = 1$ and a Majorana fermion with $c = 1/2$, combined together in a superfield with $\hat{c} = 1$.

A conformal superfield $\phi(z, \theta)$ transforms as:

$$T(z_1, \theta_1)\phi(z_2, \theta_2) \sim h\theta_{12}z_{12}^{-2}\phi + \frac{1}{2}z_{12}^{-1}D_2\phi + \theta_{12}z_{12}^{-1}\partial_2\phi \qquad (2.7.4)$$

which generalizes Eq. (2.2.19).

This can be used to generate the commutation relation:

$$[T_v, \phi(z, \theta)] = v\,\partial\phi + \frac{1}{2}(Dv)D\phi + h(\partial v)\phi \qquad (2.7.5)$$

The key to the construction of the minimal series is the Kac determinant formula. Let us construct Verma modules for the NS sector, whose elements are given by ladder operators G_{-n} acting on the highest weight vacuum state:

$$G_{-n_1}G_{-n_2}\cdots G_{-n_N}|h, c\rangle \qquad (2.7.6)$$

Notice that we do not have to have to add the usual Virasoro generators, since the superconformal generator G_{-n} is the "square root" of the Virasoro generator, as can be seen from the anticommutation relations.

To construct the vacuum for the NS–R theory, let us define the state $|0\rangle$, which is annihilated by all five generators: L_{-1}, L_0, L_1, $G_{1/2}$, and $G_{-1/2}$, that is, it is invariant under the action of the group $Osp(2|1)$. Then, a highest weight state with conformal weight h can be constructed from a field $\phi_h(z, \theta)$ as follows:

$$|h\rangle = \phi_h(0, 0)|0\rangle \qquad (2.7.7)$$

Notice that the highest weight vacuum $|h\rangle$ is annihilated by all generators with positive indices.

Let us first analyze the Neveu–Schwarz sector. The determinant of this matrix, at level n, is given by [13,19–21]:

$$\det(M_n) = \prod_{p,q} \left[h - h_{p,q}(\hat{c})\right]^{p_{NS}(n-pq/2)} \qquad (2.7.8)$$

where the product is over positive p, q subject to the constraint that $pq/2 \leq n$ and $p - q$ is even; p_{NS} is given indirectly by taking the coefficients of the following power expansion:

$$\sum_{k=0}^{\infty} t^k p_{NS}(k) = \prod_{k=1}^{\infty} \frac{(1 + t^{k-1/2})}{(1 - t^k)} \tag{2.7.9}$$

To define $h_{p,q}$, let us first introduce $\hat{m}$, defined by:

$$\hat{c}(\hat{m}) = 1 - \frac{8}{\hat{m}(\hat{m} + 2)} \tag{2.7.10}$$

Then, we define $h_{p,q}$ as follows:

$$h_{p,q} = \frac{\left[(\hat{m} + 2)p - \hat{m}q\right]^2 - 4}{8\hat{m}(\hat{m} + 2)} + \frac{1}{32}\left[1 - (-1)^{p-q}\right] \tag{2.7.11}$$

Let us analyze this formula in the same way we analyzed the conformal determinant. For $\hat{c} \geq 1$ and $h \geq 0$, all representations are unitary. For $\hat{c} < 1$, however, we have the possibility that there is a discrete series of unitary representations.

The minimal unitary representations of the conformal algebras with $\hat{c} < 1$ are given by:

$$\begin{aligned} \hat{c} &= \hat{c}(\hat{m}), & \hat{m} &= 2, 3, 4, \ldots \\ h &= h_{p,q}(\hat{m}), & 1 \leq p &< \hat{m}, & 1 \leq q < \hat{m} + 2 \end{aligned} \tag{2.7.12}$$

The determinant formula for the Ramond sector is a bit more delicate. The vacuum $|0\rangle$, we saw, belonged to the NS sector of the theory. To create a fermionic vacuum, we need to multiply the bosonic vacuum $|0\rangle$ by a *spin field* $S^{\pm}(z)$, that is,

$$|h^{\pm}\rangle = S^{\pm}(0)|0\rangle, \qquad |h^-\rangle = G_0|h^+\rangle \tag{2.7.13}$$

We will give an explicit representation of the spin field $S(z)$ in the next chapter. However, for our purposes, we only need to know its transformation properties, not how to construct it.

The vacuum state $|h^{\pm}\rangle$ is actually a fermion. Thus, there is a chirality operator $\Gamma = (-1)^F$, where F is the fermion number, which splits the vacuum into two pieces:

$$\Gamma|h^{\pm}\rangle = \pm|h^{\pm}\rangle \tag{2.7.14}$$

where:

$$\{\Gamma, G_n\} = [\Gamma, L_n] = 0 \tag{2.7.15}$$

For the lowest state, we find that the determinants are different for opposite chirality (due to the central term in the 0–0 anticommutator in the algebra, that is, $G_0^2 = L_0 - \hat{c}/16$):

$$\det(M_0^+) = 1, \qquad \det(M_0^-) = (h - \hat{c}/16) \tag{2.7.16}$$

For higher levels, they are the same:

$$\det(M_n^+) = \det(M_n^-) = \left(h - \frac{\hat{c}}{16}\right)^{p_R(n)/2} \prod_{p,q} [h - h_{p,q}(\hat{c})]^{p_R(n-pq/2)}$$

$$(2.7.17)$$

where the product over p, q is over all positive integers p, q subject to the constraint that $pq/2 \leq n$ and $p - q$ is odd. In turn, $p_R(k)$ is given by taking the coefficients of the power expansion:

$$\sum_{k=0}^{\infty} t^k p_R(k) = \sum_{k=1}^{\infty} \frac{(1+t^k)}{(1-t^k)}$$

$$(2.7.18)$$

The superconformal minimal Ramond series is then the same as the one found for the Neveu–Schwarz space.

One interesting fact is that the first member of the superconformal minimal series is given by:

$$\hat{m} = 3, \quad \hat{c} = 7/15, \quad c = 7/10$$

$$(2.7.19)$$

which is precisely the same value found for the second member of the conformal minimal series in Eq. (2.5.14). In fact, this is the only value for which the conformal and superconformal minimal series coincide.

The tricritical Ising model, therefore, actually has a superconformal representation as well as a conformal one. (This means that, experimentally speaking, it is possible to find a superconformal representation in nature. The adsorbing of helium-4 on krypton-plated graphite provides the first known example of a realization of a superconformal theory.)

The tricritical model has a $\mathbf{Z}_2$ symmetry, which flips the order operators, the Ising spins, and the disorder operators. The even sector of the tricritical model then corresponds to the NS sector of the $N = 1$ superconformal theory. The odd sector, then, corresponds to the Ramond sector. The allowed values of h for the two sectors are given by:

$$\begin{aligned} \text{NS}: \quad & h_{1,1} = 0, \quad && h_{2,2} = \frac{1}{10} \\ \text{R}: \quad & h_{1,2} = \frac{3}{80}, \quad && h_{2,1} = \frac{7}{16} \end{aligned}$$

$$(2.7.20)$$

Since the states of the tricritical Ising model can be represented either in terms of the standard Virasoro theory or with the superconformal theory (with different values of c), we can then see the correspondence between states defined in one representation expressed as sums of states in the other. For example, the following NS states can be expressed as sums over bosonic Virasoro states:

$$\begin{aligned} |h = 0\rangle_{\text{NS}} &= |h = 0\rangle_{\text{VIR}} \oplus \left|h = \frac{3}{2}\right\rangle_{\text{VIR}} \\ \left|h = \frac{1}{10}\right\rangle_{\text{NS}} &= \left|h = \frac{1}{10}\right\rangle_{\text{VIR}} \oplus \left|h = \frac{6}{10}\right\rangle_{\text{VIR}} \end{aligned}$$

$$(2.7.21)$$

We can use the same techniques used to calculate the correlation functions of the minimal series to calculate the correlation functions for the superconformal one. Specifically, we note that a null state is given by:

$$\left[G_{-3/2} - \left(\frac{5}{3}\right) L_{-1} G_{-1/2}\right] |h = \frac{1}{10}\rangle \qquad (2.7.22)$$

As expected, coupling this null state to a product of superfields gives zero, which in turn gives us differential equations for the correlation functions. For example, the correlation function of four superfields can be calculated. Let Φ be a superfield:

$$\Phi(z, \bar{z}, \theta, \bar{\theta}) = \epsilon + \theta\psi + \bar{\theta}\bar{\psi} + \theta\bar{\theta}t \qquad (2.7.23)$$

with conformal weight given by $(h, \bar{h}) = \left(\frac{1}{10}, \frac{1}{10}\right)$, while the conformal weights of the various fields are given by $\left(\frac{1}{10}, \frac{1}{10}\right)$ for the ϵ field, $\left(\frac{6}{10}, \frac{6}{10}\right)$ for the t field, and $\left(\frac{6}{10}, \frac{1}{10}\right)$ for the ψ field. Then, its correlation function is given by [21]:

$$\left\langle \Phi(z_1, \theta_1)\Phi(z_2, \theta_2)\Phi(z_3, \theta_3)\Phi(z_4, \theta_4) \right\rangle = |z_{12}z_{23}z_{34}z_{41}|^{-15}(|f|^2 + A|g|^2) \qquad (2.7.24)$$

where:

$$
\begin{aligned}
f &= \left[1 + \left(\xi\eta\frac{d}{d\eta}\right)\right]\left\{[\eta(1-\eta)]^{-1/10} F_1\left(\frac{1}{5}, -\frac{2}{5}, \frac{2}{5}, \eta\right)\right\} \\
&\quad + \frac{3}{35}\xi[\eta(1-\eta)]^{9/10} F_1\left(\frac{8}{5}, \frac{11}{5}, \frac{12}{5}, \eta\right) \\
g &= \left[1 + \left(\xi\eta\frac{d}{d\eta}\right)\right]\left\{[\eta(1-\eta)]^{12} F_1\left(\frac{7}{5}, \frac{4}{5}, \frac{8}{5}, \eta\right)\right\} \qquad (2.7.25) \\
&\quad + \frac{1}{5}\xi[\eta(1-\eta)]^{-1/2} F_1\left(-\frac{6}{5}, -\frac{3}{5}, -\frac{2}{5}, \eta\right) \\
A &= \frac{(4/9)\Gamma(4/5)\Gamma(2/5)^3}{\Gamma(1/5)\Gamma(3/5)^3}
\end{aligned}
$$

where:

$$\eta = z_{12}z_{34}/z_{13}z_{24}, \qquad \xi = 1 - \eta - z_{14}z_{23}/z_{13}z_{24} \qquad (2.7.26)$$

Lastly, we note some more equivalences between the superconformal minimal series and known statistical models. We noted before that the $\hat{m} = 3$ superconformal theory is identical to the minimal bosonic $m = 4$ theory. We also note that the $\hat{m} = 4$ theory is equivalent to a special case of the Ashkin-Teller model, which we shall study in Chapter 6. Also, the $\hat{m} = 6$ superconformal theory is equivalent to a critical point of the $\mathbf{Z}_6$ Ising model.

In summary, the power of our formalism is that we can solve for all Green's functions of the minimal models using the differential equations that they satisfy. Although the minimal models are unrealistic, with only a finite number of primary fields, they give us an invaluable laboratory in which to test many of our ideas concerning the full string theory, such as supersymmetry and modular invariance.

2.8. Summary

All perturbative vacuums of string theory possess conformal symmetry. Thus, it is important to search for a classification scheme for conformal field theories to determine the physics behind string theory. A two-dimensional conformal field theory, however, is special in that it has an infinite number of conserved currents. Thus, it is often possible to solve them exactly.

We say that a conformal field has weight $h_1 + h_2$ and conformal spin $h_1 - h_2$ if it transforms as:

$$\phi(z, \bar{z}) = \left(\frac{dz}{dz'}\right)^{-h_1} \left(\frac{d\bar{z}}{d\bar{z}'}\right)^{-h_2} \phi(z', \bar{z}') \tag{2.8.1}$$

under a conformal transformation. (We will often drop the $\bar{z}$ dependence in this book because we have two exact copies of the same algebra. It can always be restored later.)

If we power expand this infinitesimally, we find:

$$\delta\phi(z) = \epsilon(z) \, \partial_z \phi(z) + h \, \partial_z \epsilon(z) \phi(z) \tag{2.8.2}$$

A field that transforms in such a manner is called a *primary field*.

One example of a weight zero field is a free boson, with:

$$L = \frac{1}{2\pi} \bar{\partial}\phi \, \partial\phi$$
$$T(z) = -\frac{1}{2} \partial_z\phi(z) \, \partial_z\phi(z) \tag{2.8.3}$$

It obeys the operator product expansion:

$$T(z)T(w) \sim \frac{1}{2} \frac{c}{(z-w)^4} + \frac{2}{(z-w)^2} T(w) + \frac{1}{(z-w)} \partial_w T(w) + \cdots \tag{2.8.4}$$

where c is the central charge of the Virasoro algebra; $c = 1$ for the free boson and $\frac{1}{2}$ for a free fermion.

Representations of the conformal group are constructed out of a highest weight state, labeled by two numbers h, c:

$$L_0|h, c\rangle = h|h, c\rangle$$
$$L_n|h, c\rangle = 0, \quad n = 1, 2, \ldots \tag{2.8.5}$$

Then, the universal enveloping algebra is created by all products of the ladder operators acting on the highest weight state:

$$|\{n\}\rangle = L_{-n_1} L_{-n_2} \dots L_{-n_k} |h, c\rangle \qquad (2.8.6)$$

The elements of this set are called *Verma modules*.

One of the most important tools in conformal field theory is the Kac determinant equation. When it is not zero, the Verma module is irreducible. The determinant is:

$$\det M_n = \prod_{k=1}^{n} \psi_k(h, c)^{P(n-k)} \qquad (2.8.7)$$

where:

$$\psi_k(h, c) = \prod_{pq=k} [h - h_{p,q}(c)] \qquad (2.8.8)$$

and

$$h_{p,q}(c) = \frac{[(m+1)p - mq]^2 - 1}{4m(m+1)} \qquad (2.8.9)$$

where the parameter c is related to m via:

$$\begin{aligned} c &= 1 - \frac{6}{m(m+1)} \\ m &= -\frac{1}{2} \pm \frac{1}{2}\sqrt{\frac{25-c}{1-c}} \end{aligned} \qquad (2.8.10)$$

where p and q are positive integers.

Usually, the representations in which we are interested have an infinite number of primary fields. However, when the Kac determinant is zero, the representation is reducible, and one can truncate the Verma modules until one obtains a finite number of primary fields.

Unitary representations with a finite number of primary fields occur for $c < 1$. For minimal models, the operator product expansions, correlation functions, partition functions, etc., can be solved exactly.

The operator product expansion of two arbitrary primary fields with weights n and m yields the following series:

$$\begin{aligned} \phi_n(z, \bar{z})\phi_m(0) &\sim \sum_p \sum_{\{k\},\{\bar{k}\}} C_{nm}^p \beta_{nm}^{p;\{k\}} \beta_{nm}^{p;\{\bar{k}\}} \\ &\times z^{\delta_p - \delta_n - \delta_m + \sum_i k} \bar{z}^{\bar{\delta}_p - \bar{\delta}_n - \bar{\delta}_m + \sum_i \bar{k}} \phi_p^{\{k\},\{\bar{k}\}}(0) \end{aligned} \qquad (2.8.11)$$

$$[\phi_n] \times [\phi_m] = C_{nm}^p [\phi_p]$$

This is called the fusion rule. By acting on both sides of the equation with a conformal transformation, one obtains an infinite series of equations. Solving them gives an explicit solution for the coefficients appearing in the equation.

Correlation functions can be solved exactly. Correlation functions obey the following Ward-like identity:

$$\left\langle T(z)\phi_1(w_1)\phi_2(w_2)\cdots\phi_n(w_n)\right\rangle$$

$$= \sum_{i=1}^{n}\left[\frac{\delta_i}{(z-w_i)^2}+\frac{1}{z-w_i}\frac{\partial}{\partial w_i}\right]\left\langle\phi_1(w_1)\cdots\phi_n(w_n)\right\rangle \qquad (2.8.12)$$

which is derived by inserting T into the correlation function and then commuting T past the ϕ's. This leads to a differential equation that can be solved for minimal models.

For example, the correlation function:

$$G_{nm}^{lk}(x) = \sum_{p} C_{nm}^{p} C_{klp} F_{nm}^{lk}(p|x) \qquad (2.8.13)$$

where:

$$F_{nm}^{lk}(p|x) = x^{\delta_p-\delta_n-\delta_m}\sum_{\{k\}}\beta_{nm}^{p,\{k\}}x^{\sum_i k_i}$$

$$\times\frac{\left\langle k|\phi_l(1,1)L_{-k_1}L_{-k_2}\ldots L_{-k_N}|p\right\rangle}{\left\langle k|\phi_l(1,1)|p\right\rangle} \qquad (2.8.14)$$

contains the function $F_{nm}^{lk}(p|x)$, which is called a *conformal block*, because higher order correlation functions can always be written in terms of such blocks. They are the building blocks by which we can write arbitrary correlation functions. They can be solved explicitly in the minimal models.

Last, we note that superconformal theories also have unitary minimal models. Their central charge is given by:

$$\hat{c}(\hat{m}) = 1 - \frac{8}{\hat{m}(\hat{m}+2)} \qquad (2.8.15)$$

Thus, minimal models, because they only have a finite number of primary fields, have correlation functions that can be solved exactly, and therefore, they give us valuable insights into the structure of string theory, which necessarily has an infinite number of primary fields.

References

1. A. A. Belavin, A. M. Polyakov, and A. B. Zamolodchikov, *Nucl. Phys.* **B241**, 333 (1984).
2. D. Friedan, E. Martinec, and S. Shenker, *Nucl. Phys.* **B271** , 93 (1986).
   For reviews, see Refs. 3 to 10.
3. L. Alvarez-Gaumé, C. Gomez, and G. Sierra, "Topics in Conformal Field Theory," in *Physics and Mathematics of Strings*, L. Brink, D. Friedan and A. M. Polyakov, eds., World Scientific, Singapore (1990); G. Moore and N. Seiberg, *Lectures on RCFT*, 1989 Trieste Summer School.
4. M. Peskin, 1986 Santa Cruz TASI lectures, SLAC-PUB-4251.
5. T. Banks, 1987 Santa Cruz TASI lectures, SCIPP 87/111.

6. D. Friedan in *Unified String Theories*, M. Green and D. Gross, eds., World Scientific, Singapore (1986).
7. T. Eguchi, Inst. Phys. Lectures, Taipei (1986).
8. J. Cardy in *Phase Transitions* 11, Academic Press, San Diego (1987).
9. P. Ginsparg and J. L. Cardy in *Fields, Strings, and Critical Phenomena*, 1988 Les Houches School, E. Brezin and J. Zinn-Justin, eds., Elsevier Science Publ., Amsterdam (1989).
10. Vl. S. Dotsenko, *Lectures on Conformal Field Theory*, Adv. Studies in Pure Math. 16 (1988).
11. M. A. Virasoro, *Phys. Rev.* D1, 2933 (1970).
12. V. Kac, *Infinite Dimensional Lie Algebras*, Birkhaeuser, Basel (1983).
13. V. Kac in *Lec. Notes in Phys.* 94, Springer-Verlag, Berlin and New York (1979).
14. B. L. Feigin and D. B Fuchs, *Func. Anal. Appl.* 16 (1982).
15. C.B. Thorn, *Nucl. Phys.* B248, 551 (1984).
16. D. Friedan, Z. Qiu, and S. H. Shenker in *Vertex Operators in Mathematics and Physics*, Springer-Verlag, Berlin and New York (1985).
17. D. A. Huse, *Phys. Rev.* B30, 3908 (1984).
18. G. E. Andrews, R. J. Baxter, and P. J. Forrester, *J. Stat. Phys.* 35, 193 (1984).
19. A. Meurman and A. Rocha-Caridi, MSRI preprint.
20. C. Thorn, *Nucl. Phys.* B248, 551 (1984).
21. D. Friedan, Z. Qiu, and S. Shenker *Phys. Lett.* 151B, 37 (1984).

Chapter 3

WZW Model, Cosets, and Rational Conformal Field Theory

3.1. Compactification and the WZW Model

In the previous chapter, we emphasized the importance of conformal invariance as a stringent requirement that allowed us to calculate many of the simpler Green's functions from first principles. Because the conformal group has an infinite number of generators, a surprisingly large number of mathematical results flow from the requirement of conformal invariance alone.

Unfortunately, conformal invariance alone cannot determine the correlation functions that we desire, nor can't lead to a realistic string theory. In reality, conformal invariance alone cannot explain the rich diversity of particles found in nature, which includes particles transforming under a gauge group and perhaps supersymmetry.

In the next two chapters, therefore, we explore additional constraints that will define the model and give us a more realistic phenomenology. In this chapter, we introduce the concept of compactification, that is, the curling up of some of the unwanted dimensions into a compact manifold, leaving us with a physical, four-dimensional theory.

Because there may be symmetries associated with the compactified space, we will introduce new symmetries in the theory, which are described by Kac–Moody algebras [1]. In this regard, string theory has revived an old trick due to Kaluza, introduced in 1919 [2].

Kaluza's idea was to embed both Maxwell's equations and Einstein's theory of gravity into a single field, the metric tensor g_{AB} in five dimensions. Let us decompose the 5D metric tensor as follows:

$$\begin{pmatrix} g_{\mu\nu} & g_{\mu 5} \\ g_{5\nu} & g_{55} \end{pmatrix} = \begin{pmatrix} g_{\mu\nu} + \kappa^2 A_\mu A_\nu & \kappa A_\mu \\ \kappa A_\nu & \phi \end{pmatrix} \tag{3.1.1}$$

Let us assume that the unseen fifth dimension has compactified into a circle, that is,

$$x_5 = x_5 + R \tag{3.1.2}$$

so the fifth dimension is periodic. If we take R small enough, derivatives with respect to the fifth dimension will be small and can be neglected.

With this reduction, the variation of $g_{\mu 5}$ field yields:

$$\delta A_\mu \sim \partial_\mu \xi_5 \qquad (3.1.3)$$

and the final action is the sum of the Einstein action and the Maxwell action:

$$L = -\frac{1}{2\kappa^2} R - \frac{1}{4} \sqrt{g}\, g^{\mu\nu} g^{\alpha\beta} F_{\mu\alpha} F_{\nu\beta} + \cdots \qquad (3.1.4)$$

To explain why the fifth dimension was never seen, Kaluza speculated that the fifth dimension had curled up into a small circle, too small to be experimentally observed. Thus, although Kaluza's idea gave great elegance and beauty to a unification of gravity with light, he had no idea why the fifth dimension had curled up or what size it was.

This process can be duplicated for higher dimensions. If we take Einstein's theory in higher dimensions beyond the fifth, we can compactify the unwanted dimensions on a manifold that, like Kaluza's circle, has certain symmetries or isometries associated with it. This symmetry can be represented by a Lie group. Not surprising, Einstein's theory in N dimensions then reduces to Yang–Mills theory coupled to four-dimensional gravity.

String theory must necessarily incorporate Kaluza's compactification scheme if it is to become a realistic theory. This is both its strength and weakness. With relatively few assumptions, one can show that a compactification of string theory yields remarkably realistic phenomenological models. This is the advantage of compactification.

The weakness of this compactification, of course, is that we still cannot answer the questions raised by Kaluza 70 years ago, for example, why did the universe compactify in this manner?

Now we wish to generalize this discussion to a bosonic string propagating in curved or compactified space–time. Specifically, we wish the first quantized action to contain the term:

$$L = \frac{1}{\pi} G_{\mu\nu}(X)\, \partial_a X^\mu\, \partial_b X^\nu g^{ab} + \cdots \qquad (3.1.5)$$

where we have now explicitly added the curvature of space–time through the metric tensor $G_{\mu\nu}$. In general, this action is much too difficult to solve exactly, so we will make simplifications. Depending on which assumption we make about the background metric, we will arrive at different conformal field theories.

Assume, for the moment, that the string is propagating on a manifold specified by a Lie group, that is, a group manifold. Let G be a semisimple Lie group and let g be an element of this group. We will exploit the similarity between string theory and the sigma model, so our first guess for a string propagating on this group manifold may be:

$$L = \frac{1}{\pi} \operatorname{tr}\left(\partial_a g^{-1}\, \partial^a g\right) \qquad (3.1.6)$$

where g is a function of the string field X_μ. In this form, we can calculate $G_{\mu\nu}$ in terms of the g field. By differentiating, we find that $\partial_a g = \partial_a X_\mu f^{a\mu}(X)$ for some function $f^{a\mu}$. Then, the metric $G_{\mu\nu}$ can be expressed in terms of $f^{a\mu}$.

It turns out, however, that the naive choice of our action is incorrect. Treating the model as a σ model, it can be shown that it is not conformally invariant. In order to have a fully conformally invariant model, let us modify our naive choice and add a new term to the previous action:

$$S = \frac{1}{4\lambda^2} \int \mathrm{tr}\left(\partial_a g^{-1} \partial^a g\right) d^2\xi + k\Gamma(g) \tag{3.1.7}$$

where the new term is called a Wess–Zumino term:

$$\Gamma(g) = \frac{1}{24\pi} \int d^3 X \epsilon^{\alpha\beta\gamma} \mathrm{tr}\left[(g^{-1}\partial_\alpha g)(g^{-1}\partial_\beta g)(g^{-1}\partial_\gamma g)\right] \tag{3.1.8}$$

The Wess–Zumino term is integrated over a three-dimensional disk whose boundary is two-dimensional space-time.

For $k = 0$, this theory reduces to the familiar σ model, which is known to be asymptotically free and massive. Thus, conformal symmetry is violated, and the model is not suitable for our purposes. However, for $k = 1, 2, ...$, the theory becomes effectively massless and possesses an infrared-stable fixed point at:

$$\lambda^2 = 4\pi/k \tag{3.1.9}$$

Therefore, at these special values of k, we have a conformally invariant σ model where the theory is defined on a group manifold. We will call the action at this value the Wess–Zumino–Witten WZW model [3, 4]. In addition to conformal invariance, the remarkable feature of this model is that it is also invariant under the following transformation:

$$\begin{aligned} g(\xi) &\to \Omega(z)g(\xi)\bar\Omega^{-1}(\bar z) \\ z &= \xi^1 + i\xi^2 \\ \bar z &= \xi^1 - i\xi^2 \end{aligned} \tag{3.1.10}$$

One can show that the action is invariant under this symmetry using the identity:

$$S(gh^{-1}) = S(g) + S(h) + \frac{k\,\mathrm{tr}}{16\pi} \int \left(g^{-1}\partial_{\bar z} g h^{-1} \partial_z h\right) d^2\xi \tag{3.1.11}$$

To analyze this new symmetry further, let us now extract the generators of this symmetry, which will turn out to represent an infinite set of currents [3, 5]:

$$\begin{aligned} J &= -\frac{1}{2}k\,\partial_z g\, g^{-1} = J^a t^a \\ \bar J &= -\frac{1}{2}k g^{-1} \partial_{\bar z} g = \bar J^a t^a \end{aligned} \tag{3.1.12}$$

where:

$$\partial_{\bar{z}} J = 0; \quad \partial_z \bar{J} = 0 \tag{3.1.13}$$

where t^a are the generators of a Lie algebra.

Let us decompose the generator J in terms of its moments:

$$J(z) = \sum_{n=-\infty}^{\infty} J_n z^{-n-1} \tag{3.1.14}$$

The J's generate an algebra given by:

$$[J_n^a, J_m^b] = f^{abc} J_{n+m}^c + \frac{1}{2} k n \delta^{ab} \delta_{n+m,0} \tag{3.1.15}$$

This is a special case of what is called a *Kac–Moody algebra* [1, 6, 7], and it effectively smears the generators of an ordinary Lie algebra around a circle or string. Notice that for $n = m = 0$, we retrieve a classical Lie algebra.

Since the conformal anomaly vanishes at the fixed point of the WZW theory, the theory is conformally invariant, and it should be possible to give the explicit form of the energy-momentum tensor $T(z)$ in terms of these currents. In fact, we find the Sugawara form of the energy-momentum tensor:

$$T(z) = \frac{1}{2\kappa} : J^a(z) J^a(z) := \frac{1}{2\kappa} \sum_{n,m} : J_{n-m}^a J_m^a : z^{-n-2} \tag{3.1.16}$$

where:

$$\kappa = -\frac{1}{2}(c_v + k), \qquad f^{abc} f^{bcd} = c_v \delta^{ad} \tag{3.1.17}$$

where c_v is called the second Casimir of the adjoint representation of the Lie algebra.

Written in component form, we have the Virasoro generators written as:

$$L_n = -\frac{1}{c_v + k} \sum_{m=-\infty}^{\infty} : J_m^a J_{n-m}^a : \tag{3.1.18}$$

If we commute two generators of the Virasoro algebra written in Sugawara form, we find [see (1.2.24)]:

$$c = kD/(c_v + k) \tag{3.1.19}$$

where D is the dimension of the group.

Last, we find that the two algebras can be spliced together by taking the semidirect sum of their generators:

$$[L_n, J_m^a] = -m J_{n+m}^a \tag{3.1.20}$$

Written in terms of conformal operators, we have:

$$T(z)J^a(z') \sim \frac{1}{(z-z')^2}J^a(z') + \frac{1}{z-z'}\partial J^a(z')$$

$$J^a(z)J^b(z') \sim \frac{(1/2)k\delta^{ab}}{(z-z')^2} + \frac{f^{abc}}{z-z'}\partial J^c(z') \qquad (3.1.21)$$

Example: O(D) and SU(N)

Now consider, for the moment, the following tensor field composed out of fermion fields:

$$J_{\mu\nu}(z) = i\psi_\mu\psi_\nu \qquad (3.1.22)$$

where we use:

$$\psi_\mu(z)\psi_\nu(w) \sim \delta_{\mu\nu}/(z-w) \qquad (3.1.23)$$

Notice that $J_{\mu\nu}$ satisfies the commutation relations of the Kac–Moody algebra $O(D)$. Thus, fermion fields give us a simple realization of both a conformal field theory as well as a Kac–Moody algebra.

We could also have taken:

$$J^a(z) = \psi(z)t^a\psi(z) \qquad (3.1.24)$$

where $\psi(z)$ transforms in the vector representation of $SO(D)$. Then, we find that the Kac–Moody algebra of $SO(D)$ with $k = 1$ is satisfied, with:

$$c_{SO(N),k=1} = \frac{(1/2)D(D-1)}{1+(D-2)} = \frac{D}{2} \qquad (3.1.25)$$

The central term is thus consistent with D free fermion fields.

Let us say that we now have complex fermions, transforming under $SU(N)$, such that:

$$J^a(z) = \psi^*(z)t^a\psi(z) \qquad (3.1.26)$$

Then, it is easy to check that the affine $SU(N) \otimes U(1)$ is realized, so that the central term is

$$c_{U(1)} + c_{SU(N)} = 1 + \frac{(N^2-1)}{1+N} = N \qquad (3.1.27)$$

which is consistent with N free complex fermion fields.

Example: Bosonization

Let us now introduce two new techniques, bosonization and external charges, which will prove invaluable in constructing explicit representations of Kac–Moody algebras and conformal field theories. We will use these two techniques repeatedly throughout this book.

In two dimensions, because Lorentz transformations have only one generator, the distinction between a boson and a fermion is not that great. In fact, the main distinction lies in their statistics. Thus, it is possible to exponentiate a boson and (after normal ordering) obtain a fermion field, similar to the way that we exponentiate the string variable in order to obtain a vertex function for the Veneziano model.

Let us now calculate the conformal weight of this vertex operator. A straightforward calculation yields:

$$
\begin{aligned}
T(w) : e^{q\phi(z)} : &\sim -\frac{1}{2}\Big[\langle\partial_w\phi(w)q\phi(z)\rangle\Big]^2 : e^{q\phi(z)} : \\
&\quad -\frac{1}{2}2\,\partial_w\phi(w)\langle\partial_w\phi(w)q\phi(z)\rangle : e^{q\phi(z)} : \\
&\sim \frac{-q^2/2}{(w-z)^2} : e^{q\phi(z)} : +\frac{q\,\partial_w\phi}{w-z} : e^{q\phi(z)} :
\end{aligned}
\tag{3.1.28}
$$

so that the vertex operator has conformal weight equal to $-q^2/2$. (We will often use the fact that a vertex $e^{iq\phi}$ has conformal weight $q^2/2$.)

We also find that:

$$
\left\langle e^{i\alpha_1\phi(z_1)}e^{i\alpha_2\phi(z_2)}\dots e^{i\alpha_N\phi(z_N)}\right\rangle = \prod_{i<j}(z_i-z_j)^{\alpha_i\alpha_j}
\tag{3.1.29}
$$

where the α_i sum to zero. (It is easy to check that this formula has the correct conformal weight. The left-hand side has conformal weight $\sum_i \alpha_i^2/2$, while the right-hand side has weight $-\sum_{i<j}\alpha_i\alpha_j$. These two expressions, however, are equal, which can easily be seen by squaring the sum of the α_i, which is zero.)

One of the most important uses of this formalism is to create fermion operators out of boson operators (which is only possible in two dimensions). Notice that the vertex operator has anticommutation relations with itself, so that we can consider it to be a fermion. For example, for $\alpha = 1$, we have a fermion with a conformal weight equal to $\frac{1}{2}$, as expected.

Example: External Charges

Let us examine the case of the free boson with the energy-momentum tensor given by:

$$
T(z) = \frac{\epsilon}{2}\big[\partial_z\phi(z)\big]^2 - \frac{Q}{2}\,\partial_z^2\phi(z)
\tag{3.1.30}
$$

Notice that we have made several changes in the usual form for the energy-momentum tensor. First, we have put in a factor of $\epsilon = \pm 1$ in T. In order to satisfy the usual operator product expansion of T in the presence of ϵ, we must alter Eq. (2.2.8) to read:

$$
\phi(w)\phi(z) \sim \epsilon \ln(w-z)
\tag{3.1.31}
$$

Note the presence of ϵ in front of the logarithm.

Second, we have added a term proportional to $\partial^2\phi$. Because of the presence of this last term, we have altered the conformal properties of the free boson. Let us now calculate the contribution of this last term to the central term. We find:

$$T(w)T(z) \sim 2 \left(\frac{\epsilon}{2}\right)^2 \langle \partial_w \phi \, \partial_z \phi \rangle^2 + \left(\frac{Q^2}{4}\right) \langle \partial_w^2 \phi \, \partial_z^2 \phi \rangle + \ldots$$

$$\sim \frac{1}{2} \frac{1}{(w-z)^4} - \left(\frac{3\epsilon}{2}\right) Q^2 \frac{1}{(w-z)^4} + \cdots \tag{3.1.32}$$

so that the central term is equal to:

$$c = 1 - 3\epsilon Q^2 \tag{3.1.33}$$

For $Q = 0$, we arrive back at the free boson. We will, however, use the case $Q \neq 0$ extensively. For example, the conformal weight of $e^{q\phi}$ can also be calculated, and it is now given by:

$$\frac{1}{2}\epsilon q(q + Q) \tag{3.1.34}$$

We will use this expression for the conformal weight of the vertex operator throughout this book.

3.2. Frenkel–Kac Construction

Let us now use the methods developed in the last section, such as bosonization, to write explicit representations of Kac–Moody algebras. This is the Frenkel–Kac construction [8–10], which in turn is based on the Cartan–Weyl representation of an ordinary Lie algebra. The Frenkel–Kac construction of the Kac–Moody algebra is perhaps the most commonly used representation.

A Lie algebra is usually written as:

$$[\tau_a, \tau_b] = f_{ab}^c \tau_c \tag{3.2.1}$$

Although concise and elegant, this representation tells us very little about the structure of the Lie algebra, which is hidden within the structure constants. Thus, we will sometimes find more convenient the Cartan–Weyl representation, which displays the structure of the algebra in a more transparent way. The Cartan–Weyl construction is based on the fact that, within an ordinary Lie algebra, we have two types of generators: the generators H_i, which mutually commute among themselves (forming the Cartan subalgebra), and the set E_α of all other generators.

In general, the number of generators within the Cartan subalgebra is called the *rank* r of the algebra. Thus, $SU(2)$ has rank 1 because L_z is usually singled out as the generator of the Cartan subalgebra. For $SU(3)$, the rank is 2, because T_3 and Y are usually singled out as the mutually commuting operators.

The other elements of the algebra E_α are labeled by the vectors α, called the *root vectors*, which live in an r-dimensional space. The number of elements within the Cartan subalgebra and the number of vectors α obviously equals the number of parameters of the group, called the *dimension*.

The complete commutation relations of the Lie algebra can now be rewritten in terms of the Cartan subalgebra and the root vectors as:

$$
\begin{aligned}
[H_i, H_j] &= 0 \\
[H_i, E_\alpha] &= \alpha_i H_i \\
[E_\alpha, E_\beta] &= N_{\alpha,\beta} E_{\alpha+\beta} \\
[E_\alpha, E_{-\alpha}] &= \alpha_i H_i
\end{aligned}
\tag{3.2.2}
$$

where the $N_{\alpha,\beta}$ are the structure constants of the algebra.

We would now like to find a representation of the Cartan–Weyl basis, generalized to the case of the Kac–Moody algebra, based on free boson fields. Let us begin by writing the generators of $O(2N)$. We begin with N free boson fields ϕ_i with weight zero. The simplest representation of the Cartan subalgebra, with conformal weight one, is, therefore:

$$
H_i(z) = \partial \phi_i(z) \tag{3.2.3}
$$

Because the ϕ_i are free, the generators obviously commute with each other.

We now need a representation of the other elements E_α. Our first guess might be something like $: \exp(\phi_i + \phi_j) :$, which has conformal spin one. This naive choice almost works, but it has the wrong commutation relations.

To remedy this, let us define e_i to be a unit vector in the ith direction. (We suppress the index labeling the isospin space.) Then, ϕ is also a vector in this space. Now, define:

$$
\begin{aligned}
f_{\pm e_i}(z) &= \exp(\pm e_i \cdot \phi) c_{\pm e_i} \\
E_{\pm i \pm j} &= f_{\pm e_i} f_{\pm e_j} = -E_{\pm j \pm i}, \quad j \neq i
\end{aligned}
\tag{3.2.4}
$$

The numbers $c_{\pm e_i}$ are constants, called the *cocycles*, which must be added to the definition in order to get the correct statistics for the generators of the algebra. A convenient choice of these cocycles is:

$$
c_i = (-1)^{N_1 + N_2 + \cdots + N_{i-1}} \tag{3.2.5}
$$

where N_i is the fermion number for the ith fermion. (Actually, there exists a wide variety of choices for the cocycle.)

With this definition, it is now straightforward to check [8, 11]:

$$
\begin{aligned}
H_j(z) H_k(w) &\sim \frac{\delta_{jk}}{(z-w)^2} \\
H_j(z) E_\alpha &\sim \frac{\alpha_j E_\alpha}{z-w} \\
E_{j \pm k}(z) E_{-j \mp k}(w) &\sim -\frac{1}{(z-w)^2} + \frac{[H_j(w) \pm H_k(w)]}{z-w} \\
E_{\pm i \pm j}(z) E_{\mp j \pm k}(w) &\sim \frac{E_{\pm i \pm k}(w)}{z-w}, \quad i \neq k \\
E_{\pm i \pm j}(z) E_{\pm k \pm l}(w) &\sim \text{finite}
\end{aligned}
\tag{3.2.6}
$$

Thus, we have now successfully represented the generators of the Lie algebra for affine $O(2N)$ in terms of N free boson fields, ϕ_i.

As we mentioned earlier, we can also represent the $O(2N)$ Kac–Moody algebra in terms of fermion fields ψ_A.

Let:

$$f_{\pm e_j} = \frac{1}{\sqrt{2}}\left(\psi_{2j-1} \mp i\psi_{2j}\right) \tag{3.2.7}$$

Then, the generators of $O(2N)$ can be written in terms of these fermion fields as:

$$J_{MN} = i\psi_M(z)\psi_N(z) \tag{3.2.8}$$

Then, the following algebra is satisfied:

$$
\begin{aligned}
J_{MN}(z)J_{PQ}(w) \sim &\ \frac{k(\delta_{MP}\delta_{NQ} - \delta_{MQ}\delta_{NP})}{(z-w)^2} \\
&- i\frac{(\delta_{MP}J_{NQ} - \delta_{MQ}J_{NP} - \delta_{NP}J_{MQ} + \delta_{NQ}J_{MP})(w)}{z-w}
\end{aligned}
\tag{3.2.9}
$$

We also have the relation:

$$J_{MN}(z)\psi_P(w) \sim -i\frac{\delta_{MP}\psi_N - \delta_{NP}\psi_M}{z-w} + \text{finite} \tag{3.2.10}$$

In calculating this operator product expansion, we have used the convenient formula (which is easily derived from the Baker–Hausdorff formula $e^A e^B = e^{A+B+(1/2)[A,B]+\cdots}$) [11]:

$$
\begin{aligned}
O^\lambda(z)O^{\lambda'}(w) \sim &\ (z-w)^{\lambda\cdot\lambda'}\exp[i\pi(\lambda\cdot M\lambda')] \\
&\times \exp[\lambda\cdot\phi(z) + \lambda'\cdot\phi(w)]c_{\lambda+\lambda'} \\
\sim &\ (z-w)^{\lambda\cdot\lambda'}\exp[i\pi(\lambda\cdot M\lambda')]O^{\lambda+\lambda'} \\
&\times \left\{1 + (z-w)\lambda\cdot\partial\phi + \frac{1}{2}(z-w)^2 \right. \\
&\left. \times [\lambda\cdot\partial^2\phi + (\lambda\cdot\partial\phi)^2] + \cdots\right\}(w)
\end{aligned}
\tag{3.2.11}
$$

where:

$$O^\lambda(z) = e^{\lambda\cdot\phi}c_\lambda \tag{3.2.12}$$

The advantage of this bosonized formalism is that we can now write an explicit representation of the space–time spinors S_α in Eq. (2.7.13) occurring in the Ramond sector of the superstring in terms of free bosons. In this way, we can now represent both the NS and R fields in terms of a more basic set of conformal fields.

In $2N$ dimensions, we want to construct a 2^N component spinor. To do this, let us first write the following row matrix (with N entries):

$$A = (\pm, \pm, \ldots \pm, \pm)/2 \qquad (3.2.13)$$

which can take on 2^N possible values. Our first guess for a spinor might be:

$$e^{A \cdot \phi} \qquad (3.2.14)$$

but this has the wrong conformal weight.

We recall from Eqs. (3.1.30)–(3.1.33) that $: e^{a\phi} :$ has conformal weight $\frac{1}{2}a^2$ for $\epsilon = +1$ and $Q = 0$, so that a spinor composed in this fashion would have weight $N/8$. In 10 dimensions, this spinor would have weight $\frac{5}{8}$, so we need a new field with weight $\frac{3}{8}$ to give us the desired weight of 1.

We thus introduce a new field, ϕ_6 (with $\epsilon = -1$ and $Q = 2$) so that:

$$T(z) : e^{\lambda \cdot \phi + q\phi_6} :\sim \left[\frac{1}{2}\lambda \cdot \lambda - 1 - \frac{1}{2}q(q+2) \right] (z-w)^2 : e^{\lambda \cdot \phi + q\phi_6} : \quad (3.2.15)$$

that is, the ϕ_6 field contributes $- \left(\frac{1}{2} \right) q(q+2)$ to the conformal weight (that is, it is defined with a background charge). If we choose $q = -\frac{1}{2}$, then the spinor has unit conformal weight:

$$\frac{5}{8} + \frac{3}{8} = 1 \qquad (3.2.16)$$

as desired.

The operator product expansions of the spin field S^A with the Lorentz generators and the anticommuting vector ψ_M are:

$$\begin{aligned} J_{MN} S^A(w) &\sim -\frac{1}{2} i \frac{(\Gamma_{MN})^A{}_B S^B(w)}{z-w} \\ \psi_M S^A(w) &\sim \frac{1}{\sqrt{2}} \frac{(\Gamma_M)^A{}_B S^B(w)}{\sqrt{z-w}} \end{aligned} \qquad (3.2.17)$$

The constant factors appearing in the operator product expansion, called Γ matrices, can be shown to satisfy the properties of the usual Dirac matrices:

$$\{\Gamma_M, \Gamma_N\} = 2\delta_{MN} \qquad (3.2.18)$$

so we will simply define them to be the Dirac matrices for $O(2N)$.

The operator product expansion therefore gives us an explicit representation of these Dirac matrices. They can be given as follows. Let us define the usual Pauli spin matrices:

$$\sigma^1 = \begin{pmatrix} 0 & 1 \\ 1 & 0 \end{pmatrix}, \quad \sigma^2 = \begin{pmatrix} 0 & -i \\ i & 0 \end{pmatrix}, \quad \sigma^3 = \begin{pmatrix} 1 & 0 \\ 0 & -1 \end{pmatrix} \qquad (3.2.19)$$

and σ^0 is the unit matrix. Then, the Dirac matrices can be given as:

$$\Gamma^{2j-1} = (-1)^{(j-1)/2}(\sigma^3\otimes)^{j-1}\sigma^1(\otimes\sigma^0)^{N-j}, \qquad j \text{ odd}$$
$$\Gamma^{2j-1} = -(-1)^{j/2}(\sigma^3\otimes)^{j-1}\sigma^2(\otimes\sigma^0)^{N-j}, \qquad j \text{ even}$$
$$\Gamma^{2j} = (-1)^{(j-1)/2}(\sigma^3\otimes)^{j-1}\sigma^2(\otimes\sigma^0)^{N-j}, \qquad j \text{ odd} \tag{3.2.20}$$
$$\Gamma^{2j} = (-1)^{j/2}(\sigma^3\otimes)^{j-1}\sigma^1(\otimes\sigma^0)^{N-j}, \qquad j \text{ even}$$

Notice that we can define the vector $\{\lambda, q\}$ to span a six-dimensional space. Thus, in this completely bosonized formalism, each element of the algebra can be uniquely specified by fixing the value of $\{\lambda, q\}$. This gives us a *lattice* representation of all fields occurring in the affine $O(2N)$ construction.

In the lattice construction, all conformal operators appearing in the theory can be expressed as a point in this lattice space via bosonization.

3.3. GKO Coset Construction

In our search for representations of the conformal group, we were aided by the fact that the Kac–Moody algebra gave us, via the Sugawara construction [Eq. (3.1.16)], an explicit representation of the Virasoro algebra for certain values of c. The value of c obtained by the Sugawara construction is always greater or equal to 1.

For $SU(2)_k$, for example, from Eq. (3.1.19), we find:

$$c_{SU(2)} = \frac{3k}{k+2} \tag{3.3.1}$$

For an arbitrary Lie group G, we find:

$$\text{rank}\, G \le c_G \le \dim G \tag{3.3.2}$$

However, we can use a trick, called the Goddard, Kent and Olive (GKO) "coset construction," which allows us an explicit representation of all minimal models (as well as possibly all rational conformal field theories, which have rational values of h, c) [12].

Let us say that the group G contains a subgroup H. Now, let us construct the generators of conformal transformations in terms of the current J_G^a transforming under G as well as the current J_H^a transforming under H. Our goal is to construct the conformal generator associated with G/H.

Using the Sugawara construction for T_G in terms of J_G^a in Eq. (3.1.16), we can calculate [see Eq. (3.1.21)]:

$$T_G(z)J_H^a(w) \sim \frac{J_H^a(w)}{(z-w)^2} + \frac{\partial J_H^a(w)}{z-w} \tag{3.3.3}$$

But, we also know:

$$T_H(z)J_H^a(w) \sim \frac{J_H^a(w)}{(z-w)^2} + \frac{\partial J_H^a(w)}{z-w} \tag{3.3.4}$$

Notice that the right-hand side of both equations is the same. Thus, if we subtract the two equations, the right-hand side will equal zero.

If we write:

$$T_G = (T_G - T_H) + T_H \equiv T_{G/H} + T_H \qquad (3.3.5)$$

then we also have:

$$[T_{G/H}, T_H] = 0 \qquad (3.3.6)$$

This last equation means that T_G can be split into two mutually commuting pieces, $T_{G/H}$ and T_H, both of which generate representations of the conformal algebra (but with different values of c). If we now calculate the operator product expansion for T_G, we find:

$$T_G(z)T_G(w) \sim \frac{1}{2}\frac{c_{G/H} + c_H}{(z-w)^4} + \cdots \qquad (3.3.7)$$

In other words, we now have the final expression [see Eq. (3.1.19)]:

$$c_{G/H} = c_G - c_H = \frac{k_G|G|}{k_G + \tilde{h}_G} - \frac{k_H|H|}{k_H + \tilde{h}_H} \qquad (3.3.8)$$

where $\tilde{h}$ is the second Casimir of the adjoint representation of the group, and the vertical bars represent the dimension of the group.

This is the desired result. Because T_G has been decomposed into two mutually commuting pieces, the central term of the coset algebra generated by $T_{G/H}$ is given by the difference of the central terms for T_G and T_H. Obviously, $c_{G/H}$ can be less than one.

A simple example is given by the following:

$$G/H = SU(2)_k \otimes SU(2)_1/SU(2)_{k+1} \qquad (3.3.9)$$

The value of the central term is therefore given by:

$$c_{G/H} = \frac{3k}{k+2} + 1 - \frac{3(k+1)}{(k+1)+2} = 1 - \frac{6}{(k+2)(k+3)} \qquad (3.3.10)$$

which is precisely the discrete sequence of the minimal unitary models for $m = k+2 = 3, 4, 5, \ldots$ as in Eq. (2.5.8). Thus, we have the correspondence:

$$\text{unitary series} \leftrightarrow SU(2)_k \otimes SU(2)_1/SU(2)_{k+1} \qquad (3.3.11)$$

Yet another sequence is given by

$$G/H = SU(2)_k \otimes SU(2)_2/SU(2)_{k+2} \qquad (3.3.12)$$

for which we have:

$$c_{G/H} = \frac{3k}{k+2} + \frac{3}{2} - \frac{3(k+2)}{(k+2)+2} = \frac{3}{2}\left[1 - \frac{8}{(k+2)(k+4)}\right] \qquad (3.3.13)$$

We immediately recognize this as generating the superconformal $N = 1$ discrete series for $m = k + 2$ in Eq. (2.7.10). Thus, we have the correspondence:

$$N = 1 \text{ unitary series } \leftrightarrow SU(2)_k \otimes SU(2)_2/SU(2)_{k+2} \qquad (3.3.14)$$

The GKO coset construction, of course, gives us the power to generate much larger representations of the conformal group. It is believed that all rational conformal field theories, not just the minimal ones, can be constructed in this fashion. In fact, the GKO construction gives us one of the most powerful methods of unifying conformal field theories.

Although the coset construction has great power in unifying conformal field theories, it has a fundamental weakness. To understand how to construct the specific tensor representations involved in G/H, we must know the corresponding representations in G and H, which are in general not known. Thus, although the coset construction is one of the most general procedures yet found to unify conformal field theories, in actual practice, it is sometimes not very useful for specific calculations.

3.4. Conformal and Current Blocks

As in the case of the minimal models, where we could solve for the correlation functions by solving certain differential equations involving the conformal blocks, we find a similar situation with regard to the WZW model. Again, we will find that the lower order correlation functions can be determined in terms of hypergeometric functions by exploiting their transformation properties alone.

Consider the following operator product expansions between a field ϕ_l which is primary under *both* the generators of the Kac–Moody and Virasoro algebras [see Eq. (2.2.19)]:

$$T(w)\phi_l(z, \bar{z}) \sim \frac{\Delta_l}{(w - z)^2} \phi_l(z, \bar{z}) + \frac{1}{w - z} \partial_z \phi_l(z, \bar{z}) + \cdots$$

$$J^a(w)\phi_l(z, \bar{z}) \sim \frac{t^a}{w - z} \phi_l(z, \bar{z}) + \cdots \qquad (3.4.1)$$

where the first relation states that the field has conformal weight Δ_l, and the second states that the current J^a acts as a generator of isospin transformations on the field, which transforms under some representation of the affine group. (We omit the parallel transformation properties involving $\bar{z}$).

As before, we can use these relations to construct differential equations for correlation functions. Let us insert T and J^a in an N-point correlation function. Then, we arrive at:

$$\Big\langle T(z)\phi_1(z_1,\bar{z}_1)\cdots\Big\rangle = \sum_{j=1}^{N}\left[\frac{\Delta_j}{(z-z_j)^2}+\frac{1}{z-z_j}\frac{\partial}{\partial z_j}\right]\langle\phi_1(z_1,\bar{z}_1)\cdots\rangle$$

$$\Big\langle J^a(z)\phi_1(z_1,\bar{z}_1)\cdots\Big\rangle = \sum_{j=1}^{N}\frac{t_j^a}{z-z_j}\langle\phi_1(z_1,\bar{z}_1)\cdots\rangle$$

$$\text{(3.4.2)}$$

Let us now observe that the generators T and J^a are regular at infinity, which means that as $z \to \infty$ these operators behave as:

$$T(z) \sim z^{-4}, \qquad J^a(z) \sim z^{-2} \qquad \text{(3.4.3)}$$

If we now impose these asymptotic limits on the previous correlation function relations, then:

$$\sum_{j=1}^{N}\left[z_j^{n+1}\frac{\partial}{\partial z_j}+(n+1)\Delta_j z_j^n\right]\langle\phi_1(z,\bar{z}_1)\cdots\phi_N(z_N,\bar{z}_N)\rangle = 0 \qquad \text{(3.4.4)}$$

for $n = -1, 0, +1$ and:

$$\sum_{j=1}^{N}t_j^a\langle\phi_1(z_1,\bar{z}_1)\cdots\phi_N(z_N,\bar{z}_N)\rangle = 0 \qquad \text{(3.4.5)}$$

Now, we would like to generalize this discussion and calculate the Green's functions for the WZW model. We now replace the ϕ appearing in the Green's function with $g(z,\bar{z})$, and use the operator product expansion:

$$J^a(w)t^a g(z,\bar{z}) = \frac{c_g}{w-z}g(z,\bar{z})+\sum_{n=1}^{\infty}(w-z)^{n-1}t^a J_{-n}^a g(z,\bar{z}) \qquad \text{(3.4.6)}$$

where $t^a t^a = c_g I$. The term proportional to $w-z$ raised to the zeroth power coincides with the operator ∂_z multiplied by some constant, which we call κ.

Repeating many of the steps we used for the Green's function of a conformal field in Eq. (3.4.2), we find the differential equation satisfied for the Green's function of the WZW model:

$$t_i^a\Big\langle J^a(z)g(z_1,\bar{z}_1)...g(z_1,\bar{z}_N)\Big\rangle$$
$$= \left(\frac{c_g}{z-z_i}+\sum_{j\neq i}^{N}\frac{t_i^a t_j^a}{z-z_j}\right)\langle g(z_1,\bar{z}_1)\cdots g(z_1,\bar{z}_N)\rangle \qquad \text{(3.4.7)}$$

Now, insert Eq. (3.4.6) into Eq. (3.4.7) and take the limit as $z \to z_i$. We arrive at the desired relationship:

$$\left(\kappa\frac{\partial}{\partial z_i} - \sum_{j\neq i}^{N}\frac{t_i^a t_j^a}{z_i - z_j}\right)\langle g(z_1,\bar{z}_1)...g(z_N,\bar{z}_N)\rangle = 0 \qquad (3.4.8)$$

This is the Knizhnik–Zamolodchikov (KZ) relation [13], which is useful in actually calculating explicit expressions for the correlation functions. (Another way of deriving this important relation is to insert a certain null state into the matrix element. We notice from Eq. (3.1.18) that $L_{-1} - (\frac{1}{2}\kappa) : J_m^a J_{-1-m}^a :$ must be zero. When this expression acts on a primary state, the only terms that survive are $L_{-1} - (\frac{1}{2}\kappa) : J_1^a J_0^a :$. Inserting null states constructed out of this operator into a correlation function, we find immediately the KZ relationship.)

Example: Four-Point Correlation Function

As an example to illustrate these techniques, consider, for example, calculating the four-point correlation function:

$$G(z_i,\bar{z}_i) = \left\langle g(z_1,\bar{z}_1)g^{-1}(z_2,\bar{z}_2)g^{-1}(z_3,\bar{z}_3)g(z_4,\bar{z}_4)\right\rangle \qquad (3.4.9)$$

By the usual conformal arguments, we know that this expression can be reduced to a function of the anharmonic ratios:

$$x = \frac{(z_1 - z_2)(z_3 - z_4)}{(z_1 - z_4)(z_3 - z_2)} \qquad (3.4.10)$$

Thus, we find that we can reexpress this correlation function of the g's, with weight Δ, in terms of a function of x and $\bar{x}$:

$$G(z_i,\bar{z}_i) = \left[(z_1 - z_4)(z_2 - z_3)(\bar{z}_1 - \bar{z}_4)(\bar{z}_2 - \bar{z}_3)\right]^{-2\Delta}G(x,\bar{x}) \qquad (3.4.11)$$

This is all that conformal invariance can tell us, because we have not yet specified the representation to which the fields belong. To be specific, let the group $G = SU(N)$ and let g transform in the fundamental representation of $SU(N)\otimes SU(N)$. Then, we can insert the explicit form of the indices for the group elements into g:

$$g(z_i,\bar{z}_i) \rightarrow g_{\alpha_i}^{\beta_i}(z_i,\bar{z}_i) \qquad (3.4.12)$$

Then, the $G(x,\bar{x})$ matrix can be written as:

$$G(x,\bar{x}) = \sum_{A,B=1,2}(I_A)(\bar{I}_B)G_{AB}(x,\bar{x}) \qquad (3.4.13)$$

where:

$$I_1 = \delta_{\alpha_1}^{\alpha_2}\delta_{\alpha_3}^{\alpha_4}, \quad \bar{I}_1 = \delta_{\beta_2}^{\beta_1}\delta_{\beta_3}^{\beta_4}, \quad I_2 = \delta_{\alpha_1}^{\alpha_4}\delta_{\alpha_3}^{\alpha_2}, \quad \bar{I}_2 = \delta_{\beta_2}^{\beta_4}\delta_{\beta_3}^{\beta_1} \qquad (3.4.14)$$

Using the KZ relation, let us now contract over the indices and reduce the equations to a set of two 2×2 matrix equations:

$$\frac{\partial G}{\partial x} = \left[\frac{1}{x}P + \frac{1}{x-1}Q\right]G, \qquad \frac{\partial G}{\partial \bar{x}} = G\left[\frac{1}{\bar{x}}P^T + \frac{1}{\bar{x}-1}Q^T\right] \qquad (3.4.15)$$

where T represents taking the transpose of the matrices, which are defined by [13]:

$$P = \frac{1}{2N\kappa}\begin{bmatrix} N^2-1 & N \\ 0 & -1 \end{bmatrix}, \qquad Q = \frac{1}{2N\kappa}\begin{bmatrix} -1 & 0 \\ N & N^2-1 \end{bmatrix} \qquad (3.4.16)$$

where $\kappa = -\frac{1}{2}(N+k)$.

Fortunately, this system of equations is completely solvable. We can find a solution in terms of hypergeometric functions. Let us decompose $G(x,\bar{x})$ as follows:

$$G_{AB}(x,\bar{x}) = \sum_{p,q=0,1} U_{pq} F_A^{(p)}(x) F_B^{(q)}(\bar{x}) \qquad (3.4.17)$$

for some constant matrix U. Then, the complete solution is given by [13]:

$$F_1^{(0)}(x) = x^{-2\Delta}(1-x)^{\Delta_1-2\Delta}F\left(-\frac{1}{2\kappa},\frac{1}{2\kappa},1+\frac{N}{2\kappa};x\right)$$

$$F_2^{(0)}(x) = -(2\kappa+N)^{-1}x^{1-2\Delta}(1-x)^{\Delta_1-2\Delta}$$
$$\times F\left(1-\frac{1}{2\kappa},1+\frac{1}{2\kappa},2+\frac{N}{2\kappa};x\right)$$

$$F_1^{(1)}(x) = x^{\Delta_1-2\Delta}(1-x)^{\Delta_1-2\Delta}F\left(-\frac{N-1}{2\kappa},-\frac{N+1}{2\kappa},1-\frac{N}{2\kappa};x\right)$$

$$F_2^{(1)}(x) = -Nx^{\Delta_1-2\Delta}(1-x)^{\Delta_1-2\Delta}F\left(-\frac{N-1}{2\kappa},-\frac{N+1}{2\kappa},-\frac{N}{2\kappa};x\right)$$
$$(3.4.18)$$

where:

$$\Delta = \frac{N^2-1}{2N(N+k)}, \qquad \Delta_1 = \frac{N}{N+k} \qquad (3.4.19)$$

In summary, we find that large classes of correlation functions for the WZW model could be completely solved using the KZ relation.

3.5. Racah Coefficients for Rational Conformal Field Theory

In Chapter 2, we saw that, in general, there were an infinite number of primary fields for certain values of h and c. However, we also saw that the representations of the conformal group simplified enormously if we had a finite number of primary fields. In this case, the minimal models were unique, and we could also calculate exactly their correlation functions.

When we enlarge our discussion to include fields primary under the Kac–Moody algebra, the minimal models are no longer the only models possessing a finite number of primaries. In general, for an affine Lie algebra to have a finite number of primaries, the values of c and h must be rational numbers. Hence, the goal of this chapter is to study what are called *rational conformal field theories* [14].

Just a few of the examples of rational conformal field theories include the following

1. The WZW models have central charges given by Eq. (3.1.19):

$$c = \frac{k \dim G}{c_v + k} \qquad (3.5.1)$$

where c_v is the dual Coxeter number, or the Casimir of the adjoint representation.

2. The unitary discrete series has a central charge given by Eq. (2.5.8):

$$c = 1 - \frac{6}{m(m+1)}, \qquad m = 3, 4, \ldots \qquad (3.5.2)$$

3. The parafermion models (to be discussed in Chapter 5), have central charge:

$$c = \frac{2(N-1)}{N+2} \qquad (3.5.3)$$

4. The rational toroidal compactification, has central charge:

$$c = d \qquad (3.5.4)$$

where d is the dimension of the torus.

5. The various coset constructions (some of the most phenomenologically interesting ones are constructions based on "free fermions" and "free bosons") [15–18] can be represented by lattices when we bosonize all the fields.

6. The allowed orbifolds of the above exist when there is a discrete symmetry [19]. (Orbifolds are not manifolds; they have singularities, which arise when we divide a manifold by the action of some discrete symmetry.)

7. The tensor products of the above are another example when the central charge is the sum of the individual central charges:

$$c_{\text{total}} = \sum_i c_i \qquad (3.5.5)$$

The advantage of studying the rational conformal field theories, like the simpler minimal models, is that their modular invariant partition functions can usually be computed exactly. Thus, they serve as a laboratory to study the more complicated compactifications found in string theory. (The proof that rational conformal field theories have a finite number of primary fields is given in Chapter 4.)

Because of the close resemblance between a conformal field theory and an ordinary Lie algebra, it is possible to take tensor products of representations, which in turn yield other representations. In this fashion, we can construct the analog of Clebsch–Gordan coefficients and Racah $6j$ symbols by taking repeated tensor products.

However, there is a slight, but important, complication. In general, we wish to take the tensor product of two Verma modules $[\phi^i]$ and $[\phi^j]$ that have the same central charge c. If we carelessly define the product of these two representations, we will find that the resulting representation has a central charge $2c$, which is the sum of the two individual central charges. However, we wish to find product representations in which the central charge is still c.

As in ordinary Lie algebra theory, the goal is to construct a series of self-consistency conditions by taking tensor products of representations in different orders, decomposing them, and setting them equal at the end. In this way, we will find a series of polynomial consistency equations in terms of the conformal blocks.

There are two operations that we wish to perform on these conformal blocks, called B (corresponding to braiding or interchanging two points) and F (corresponding to pinching the graph, that is, making an s-channel graph into a t-channel graph).

On the space of conformal blocks, we may now write a representation of the B-twist operation on a four-point function, representing the interaction of conformal fields labeled by i, j, k, l.

Let us begin with the standard four-point function:

$$\langle 0|\phi^i(\infty)\phi^j(z_1)\phi^k(z_2)\phi^l(0)|0\rangle = \langle i|\phi^j(z_1)\phi^k(z_2)|l\rangle \qquad (3.5.6)$$

where we are taking the correlation function of four primary fields.

In Eq. (2.4.13), we obtained identities by contracting the fields in different orders and then setting them equal. For example, we may first contract the i, j fields and the k, l fields. The second way is to contract the i, k and the j, l fields. Since the final answer must be the same, we have established a relationship between the conformal blocks.

If we select the pth element in the sum, we find that the two ways of contracting the matrix element via the braiding operator B yields the following relationship:

$$\Phi^j_{ip}(z_1)\Phi^k_{pl}(z_2) = \sum_q B_{pq}\begin{bmatrix} j & k \\ i & l \end{bmatrix}\Phi^k_{iq}(z_2)\Phi^j_{ql}(z_1) \qquad (3.5.7)$$

Viewing Fig. 3.1(a), it is easy to see that the effect of the B operation is to interchange z_1 and z_2. Notice that the sum over q corresponds to summing over all intermediate states.

If we write the B matrix as a mapping from the space of one set of three-point couplings to another, it can be symbolically represented as:

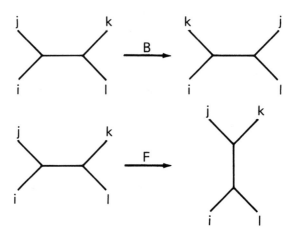

Fig. 3.1.

$$B_{pq}\begin{bmatrix} j & k \\ i & l \end{bmatrix}: \qquad V_{jp}^i \otimes V_{kl}^p \to V_{kq}^i \otimes V_{jl}^q \qquad (3.5.8)$$

Similarly, we can introduce the operation F in Fig. 3.1(b), which pinches a four-point graph [see Eq. (2.4.13)]. On the space of conformal blocks, it has the specific representation:

$$\Phi_{lp}^i(z_1)\Phi_{pr}^j(z_2) = \sum_q F_{pq}\begin{bmatrix} i & j \\ l & r \end{bmatrix}\Phi_{lr}^q(z_2)\Phi_{qj}^i(z_1 - z_2) \qquad (3.5.9)$$

(where, for convenience, we have suppressed the summation over descendant fields). Symbolically, this can be represented as:

$$F_{pq}\begin{bmatrix} j & k \\ i & l \end{bmatrix}: \qquad V_{jp}^i \otimes V_{kl}^p \to V_{ql}^i \otimes V_{jk}^q \qquad (3.5.10)$$

These identities are quite interesting because, by iterating a series of manipulations with the B and F operators, we can deduce a large number of nontrivial relations. These identities, in turn, are analogous to the identities found among $3j$ Clebsch–Gordan and $6j$ Racah coefficients in the usual $SU(2)$ theory of addition of angular momentum.

We can also introduce a new operator Ω, which simply twists two external legs around an internal leg. Actually, Ω is not a new operator, but can be represented by setting one of the legs of the B matrix to the identity, that is,

$$B\begin{bmatrix} i & j \\ 0 & k \end{bmatrix} \qquad (3.5.11)$$

defines the map:

$$\Omega^i_{jk} : \quad V^i_{jk} \rightarrow V^i_{kj} \qquad (3.5.12)$$

that is, Ω twists the j, k lines.

An explicit representation for Ω can be found because it satisfies a number of constraints. We find:

$$\Omega^i_{jk} = \xi^i_{jk} e^{\pi i(\Delta_j + \Delta_k - \Delta_i)}, \qquad (\Omega^i_{jk})^2 = e^{2\pi i(\Delta_j + \Delta_k - \Delta_i)} \qquad (3.5.13)$$

where $\xi = \pm 1$ depending on which conformal field theory we are analyzing.

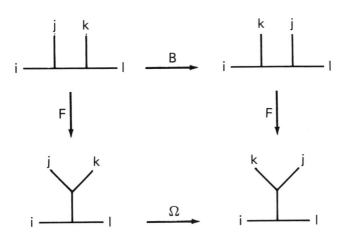

Fig. 3.2.

We can now deduce nontrivial identities among the F, B, Ω matrices. In Fig. 3.2, we see how a four-point block can be deformed in two ways, by the operation of BF or $F\Omega$, leading to the same result. Equating the two operations, we obtain a representation of this figure by:

$$\sum_{p'} B_{pp'} \begin{bmatrix} j & k \\ i & l \end{bmatrix} (\epsilon) F_{p'q} \begin{bmatrix} k & j \\ i & l \end{bmatrix} = F_{pq} \begin{bmatrix} j & k \\ i & l \end{bmatrix} e^{-i\pi\epsilon(\Delta_k + \Delta_j - \Delta_q)} \qquad (3.5.14)$$

where ϵ represents the sense of the braiding.

In order to obtain more identities, we also note that the B and F operations can be pictorially represented as in Fig. 3.3. Since both B and F transform four-point functions into other four-point functions, in Fig. 3.3 we have simply sandwiched these two four-point functions back to back. For each B or F, there is a unique representation in terms of the diagrams in Fig. 3.3.

With this new identification for B and F, we can write an identity corresponding to Fig. 3.4:

$$F_{p_1 q_1} \begin{bmatrix} j_2 & j_3 \\ j_1 & p_2 \end{bmatrix} B_{p_2 q_2} \begin{bmatrix} q_1 & j_4 \\ j_1 & j_5 \end{bmatrix} (\epsilon)$$

$$= \sum_s B_{p_2 s} \begin{bmatrix} j_3 & j_4 \\ p_1 & j_5 \end{bmatrix} (\epsilon) B_{p_1 q_2} \begin{bmatrix} j_2 & j_4 \\ j_1 & s \end{bmatrix} (\epsilon) F_{s q_1} \begin{bmatrix} j_2 & j_3 \\ q_2 & j_5 \end{bmatrix}$$

(3.5.15)

(We can read this identity by tracing the indices in Fig. 3.4. In the first diagram in Fig. 3.4, the F operation appears in the lower left, and the B operation appears in the upper right. In the second diagram in Fig. 3.4, the F operation appears on the upper right, and the two B operations appear on the left and lower right.)

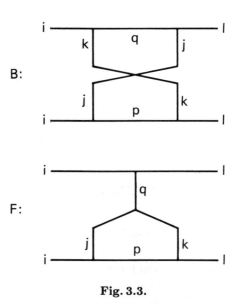

Fig. 3.3.

Now set $j_5 = 0$. Because the B matrix reduces to a twist Ω, we can compare the previous two equations and find an identity between the B and F matrices:

$$F_{pq} \begin{bmatrix} j & k \\ i & l \end{bmatrix} = e^{-i\pi\epsilon(\Delta_i + \Delta_k - \Delta_p - \Delta_q)} B_{pq} \begin{bmatrix} j & l \\ i & k \end{bmatrix} (\epsilon)$$

(3.5.16)

Now let us analyze Fig. 3.5, where we have two sequences of operations. Notice that the beginning and final result is the same. By explicitly representing each of the five manipulations show in the picture, we will be able to represent the "pentagon relations" [14].

By carefully labeling all the legs and decomposing each of the manipulations in terms of the F and B matrices, it is each to show an identity between BBF and FB, given explicitly by:

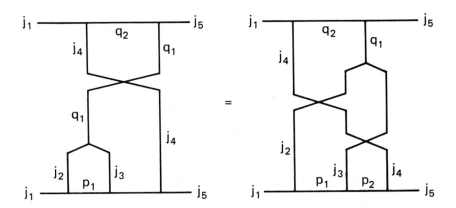

Fig. 3.4.

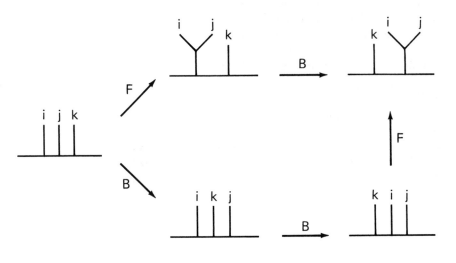

Fig. 3.5.

$$\sum_s F_{p_2 s} \begin{bmatrix} j & k \\ p_1 & b \end{bmatrix} F_{p_1 l} \begin{bmatrix} i & s \\ a & b \end{bmatrix} F_{sr} \begin{bmatrix} i & j \\ l & k \end{bmatrix}$$
$$= F_{p_1 r} \begin{bmatrix} i & j \\ a & p_2 \end{bmatrix} F_{p_2 l} \begin{bmatrix} r & k \\ a & b \end{bmatrix}$$

(3.5.17)

Similarly, we can also construct the "hexagon relations" by tracing the manipulations shown in Fig. 3.6. We equate the action of $F\Omega F$ and $\Omega F\Omega$. Explicitly, the hexagon identity becomes:

$$\Omega_{lk}^m(\epsilon)F_{mn}\begin{bmatrix} j & k \\ i & l \end{bmatrix}\Omega_{jk}^i(\epsilon) = \sum_r F_{mr}\begin{bmatrix} j & l \\ i & k \end{bmatrix}\Omega_{kr}^i(\epsilon)F_{rn}\begin{bmatrix} k & j \\ i & l \end{bmatrix} \quad (3.5.18)$$

At the genus 0 level, one can show that one pentagon identity and two hexagon identities are enough to completely specify all possible higher order identities. This means that octagon (and higher) identities for tree graphs can be reduced to combinations of pentagon and hexagon identities. (For the higher loop graphs, one must introduce two more operators, which we define in the next chapter, to completely specify all possible polynomial equations.)

One interesting observation is that we have now reduced rational conformal field theory to a set of *finite* polynomial equations. This, in itself, is remarkable because conformal field theory is usually defined over an infinite dimensional space. By reducing everything to the finite dimensional space of primary fields, we have a remarkable way in which to specify any rational conformal field theory strictly through their pentagon and hexagon graphs.

In some sense, this may even serve as an alternative, finite dimensional definition of rational conformal field theory.

Example: Ising Model

One of the simplest examples of this construction is the Ising model, which is equal to the first minimal model at criticality. We recall that for $c = \frac{1}{2}$, the allowed values of h for the primary fields were $\frac{1}{2}$ and $\frac{1}{16}$.

Let us represent the Verma modules associated with each primary field as $[\psi]$ and $[\sigma]$. Then, the fusion rules can be represented as:

$$[\psi] \times [\psi] = [1]$$
$$[\psi] \times [\sigma] = [\sigma] \quad (3.5.19)$$
$$[\sigma] \times [\sigma] = [1] + [\psi]$$

Let us now represent the B and F matrices in terms of these primary fields. There is a certain freedom in choosing the values of these matrices, so we can always choose a gauge by setting:

$$F\begin{bmatrix} \sigma & \psi \\ \sigma & \psi \end{bmatrix} = F\begin{bmatrix} \psi & \sigma \\ \psi & \sigma \end{bmatrix} = F\begin{bmatrix} \psi & \psi \\ \sigma & \sigma \end{bmatrix} = F\begin{bmatrix} \sigma & \sigma \\ \psi & \psi \end{bmatrix} = 1 \quad (3.5.20)$$

By solving the polynomial equations, we arrive at [14]:

$$F\begin{bmatrix} \psi & \psi \\ \psi & \psi \end{bmatrix} = 1, \qquad F\begin{bmatrix} \sigma & \sigma \\ \sigma & \sigma \end{bmatrix} = \frac{1}{\sqrt{2}}\begin{pmatrix} 1 & 1 \\ 1 & -1 \end{pmatrix}$$

$$F\begin{bmatrix} \sigma & \psi \\ \psi & \sigma \end{bmatrix} = -1, \qquad B\begin{bmatrix} \sigma & \sigma \\ \sigma & \sigma \end{bmatrix} = \frac{e^{-i\pi/8}}{\sqrt{2}}\begin{pmatrix} 1 & i \\ i & 1 \end{pmatrix} \qquad (3.5.21)$$

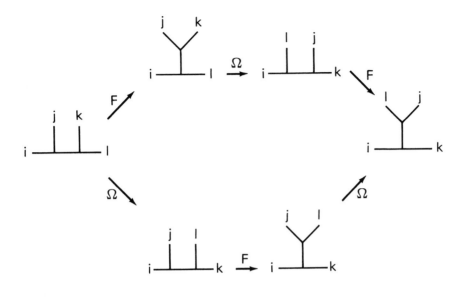

Fig. 3.6.

3.6. Summary

Systems with a finite number of primary fields give us a theoretical laboratory for testing our ideas about string vacuums. In particular, the minimal
models exhaust all possible unitary representations of the conformal group
with a finite number of primary fields, and they have correlation functions
and fusion rules that can be calculated exactly.

To go beyond minimal models, we study Kac–Moody algebras. It is
possible to find new representations with a finite number of fields that are
primary with respect to both the Virasoro and Kac–Moody algebras.

One way to find representations of these algebras is to take the string
and let it move on some curved manifold:

$$L = \frac{1}{\pi} G_{\mu\nu}(X)\, \partial_a X^\mu\, \partial_b X^\nu\, g^{ab} + \cdots \qquad (3.6.1)$$

Specifically, we will take the following σ model action:

$$S = \frac{1}{4\lambda^2} \int \operatorname{tr}\left(\partial_a g^{-1}\, \partial^a g\right) d^2\xi + k\Gamma(g) \qquad (3.6.2)$$

where the last term is a Wess–Zumino term:

$$\Gamma(g) = \frac{1}{24\pi} \int d^3 X \, \epsilon^{\alpha\beta\gamma} \text{tr}\left[(g^{-1}\partial_\alpha g)(g^{-1}\partial_\beta g)(g^{-1}\partial_\gamma g)\right] \qquad (3.6.3)$$

where the last term is integrated over a three-dimensional disk whose boundary is two dimensional space-time.

For $k = 1, 2, \dots$, the theory becomes effectively massless and possesses an infrared-stable fixed point at:

$$\lambda^2 = 4\pi/k \qquad (3.6.4)$$

The currents for the WZW model are:

$$J = -\frac{1}{2}k\,\partial_z g\, g^{-1} = J^a t^a, \qquad \bar{J} = -\frac{1}{2}kg^{-1}\partial_{\bar{z}}g = \bar{J}^a t^a \qquad (3.6.5)$$

The J's generate an algebra given by:

$$[J_n^a, J_m^b] = f^{abc} J_{n+m}^c + \frac{1}{2}kn\delta^{ab}\delta_{n+m,0} \qquad (3.6.6)$$

One of the most important representations of the Virasoro generators is obtained by

$$T(z) = \frac{1}{2\kappa} : J^a(z)J^a(z) := \frac{1}{2\kappa} \sum_m : J_{n-m}^a J_m^a : z^{-n-2} \qquad (3.6.7)$$

where:

$$\kappa = -\frac{1}{2}(c_v + k), \qquad f^{abc}f^{bcd} = c_v\delta^{ab} \qquad (3.6.8)$$

where c_v is called the second Casimir of the adjoint representation of the Lie algebra.

Written in component form, we have the Virasoro generators written as:

$$L_n = -\frac{1}{c_v + k} \sum_{m=-\infty}^{\infty} : J_m^a J_{n-m}^a : \qquad (3.6.9)$$

If we commute two generators of the Virasoro algebra written in Sugawara form, we find:

$$c = \frac{kD}{c_v + k} \qquad (3.6.10)$$

where D is the dimension of the group.

One of the most powerful methods of generating new conformal field theories is the GKO coset method, which uses the representations of a group G and a subgroup H.

Using the Sugawara construction for T_G in terms of J_G^a, we can calculate:

$$T_G(z)J_H^a(w) \sim \frac{J_H^a(w)}{(z-w)^2} + \frac{\partial J_H^a(w)}{z-w} \qquad (3.6.11)$$

But, we also know:

$$T_H(z)J_H^a(w) \sim \frac{J_H^a(w)}{(z-w)^2} + \frac{\partial J_H^a(w)}{z-w} \qquad (3.6.12)$$

Notice that the right-hand side of both equations is the same. Thus, if we subtract the two equations, the right-hand side will equal zero.

If we write:

$$T_G = (T_G - T_H) + T_H \equiv T_{G/H} + T_H \qquad (3.6.13)$$

then we also have:

$$[T_{G/H}, T_H] = 0 \qquad (3.6.14)$$

This last equation means that T_G can be split into two mutually commuting pieces, $T_{G/H}$ and T_H, both of which generate representations of the conformal algebra (but with different values of c). If we now calculate the operator product expansion for T_G, we find:

$$T_G(z)T_G(w) \sim \frac{1}{2}\frac{c_{G/H} + c_H}{(z-w)^4} + \cdots \qquad (3.6.15)$$

In other words, we now have the final expression:

$$c_{G/H} = c_G - c_H = \frac{k_G|G|}{k_G + \tilde{h}_G} - \frac{k_H|H|}{k_H + \tilde{h}_H} \qquad (3.6.16)$$

A simple example of how the coset method can generate conformal field theories is given by the following:

$$G/H = SU(2)_k \otimes SU(2)_1/SU(2)_{k+1} \qquad (3.6.17)$$

The value of the central term is therefore given by:

$$c_{G/H} = \frac{3k}{k+2} + 1 - \frac{3(k+1)}{(k+1)+2} = 1 - \frac{6}{(k+2)(k+3)} \qquad (3.6.18)$$

which is precisely the discrete sequence of the minimal unitary models for $m = k + 2 = 3, 4, 5, ...$

Thus, we have the correspondence:

$$\text{unitary series} \leftrightarrow SU(2)_k \otimes SU(2)_1/SU(2)_{k+1} \qquad (3.6.19)$$

Yet another sequence is given by

$$G/H = SU(2)_k \otimes SU(2)_2/SU(2)_{k+2} \qquad (3.6.20)$$

for which we have:

$$c_{G/H} = \frac{3k}{k+2} + \frac{3}{2} - \frac{3(k+2)}{(k+2)+2} = \frac{3}{2}\left(1 - \frac{8}{(k+2)(k+4)}\right) \qquad (3.6.21)$$

We immediately recognize this as generating the superconformal $N = 1$ discrete series for $m = k + 2$. Thus, we have the correspondence:

$$N = 1 \text{ unitary series} \leftrightarrow SU(2)_k \otimes SU(2)_2/SU(2)_{k+2} \tag{3.6.22}$$

For rational conformal field theories, like the minimal models, one can construct Wardlike identities on the correlation functions which allow us to solve them exactly. If we insert J into a correlation function and then commute it past the fields, we find:

$$\left(\kappa \frac{\partial}{\partial z_i} - \sum_{j \neq i}^{N} \frac{t_i^a t_j^a}{z_i - z_j} \right) \langle g(z_1, \bar{z}_1) \cdots g(z_N, \bar{z}_N) \rangle = 0 \tag{3.6.23}$$

This is the Knizhnik–Zamolodchikov relation, which is useful in actually calculating explicit expressions for the correlation functions.

Finally, it can be shown that if a conformal field theory has a finite number of primary fields then h, c are rational. These are called rational field theories and include the WZW and minimal models as subsets. It is believed all known rational conformal field theories can be generated by the coset method.

Rational conformal field theories can be treated much like ordinary Lie algebras, where we calculate Clebsch–Gordan coefficients and Racah coefficients. On the space of conformal blocks arising from four-point correlators, we can define two operators, B and F, which have well-defined geometric interpretations. Starting with an initial conformal block, there is more than one way in which to use the B and F operators to deform the original block into a new one. By equating the various ways in which this is done, we therefore obtain a series of identities, such as the hexagon identity, which have a direct counterpart in classical Lie algebras.

The advantage of this technique for rational conformal field theories is that we can find a unique prescription for calculating all such blocks and hence calculate all correlation functions. Like the minimal models for conformal theory, the rational conformal field theories for the Kac–Moody theory can be completely solved by these techniques.

References

1. For a review of Kac–Moody algebras, see P. Goddard and D. Olive, *Int. J. of Mod. Phys.* **A1**, 303 (1986).
2. Th. Kaluza, *Sitz. Preuss. Akad. Wiss.* **K1**, 966 (1921).
3. E. Witten, *Comm. Math. Phys.* **92**, 455 (1986).
4. S. P. Novikov, *Usp. Mat. Nauk.* **37**, 3 (1982).
5. D. Gepner and E. Witten, *Nucl. Phys.* **B278**, 493 (1986).
6. V. Kac, *Infinite Dimensional Lie Algebras*, Birkhauser, Basel (1983).
7. R. V. Moody, *Bull. A. M.S.* **73**, 217 (1974).
8. I. Frenkel and V. G. Kac, *Inven. Math.* **62**, 23 (1980).
9. M. Halpern, *Phys. Rev.* **D12**, 1684 (1975).
10. G. Segal, *Comm. Math. Phys.* **80**, 301 (1981).
11. V. A. Kostelecky, O. Lechtenfeld, W. Lerche, S. Samuel, and S. Watamura, *Nucl. Phys.* **B288**, 173 (1987).

12. P. Goddard, A. Kent, and D. Olive, *Comm. Math. Phys.* **103**, 105 (1986); *Comm. Math. Phys.* **103**, 105 (1986).
13. V. Knizhnik and A. B. Zamolodchikov, *Nucl. Phys.* **B247**, 83 (1984).
14. G. Moore and N. Seiberg, *Lectures on RCFT*, 1989 Trieste Summer School; *Phys. Lett.* **212B**, 451 (1988); *Nucl. Phys.* **B313**, 16 (1989); *Comm. Math. Phys.* **123**, 77 (1989); *Phys. Lett.* **220B** 422 (1989).
15. K. S. Narain, *Phys. Lett.* **169B**, 41 (1986).
16. W. Lerche, D. Lust, and A. N. Schellekens, *Nucl. Phys.* **B287**, 477 (1987).
17. H. Kawai, D. C. Lewellen, and S.-H. Tye, *Phys. Rev. Lett.* **57**, 1832 (1986); *Phys. Rev.* **D34**, 3794 (1986); *Nucl. Phys.* **B288**, 1 (1987).
18. I. Antoniadis, C. P. Bachas, and C. Kounnas, *Nucl. Phys.* **B289**, 87 (1987).
19. L. Dixon, J. Harvey, C. Vafa, and E. Witten, *Nucl. Phys.* **B261**, 651, (1985); **B274**, 285 (1986).

Chapter 4

Modular Invariance and the A–D–E Classification

4.1. Dehn Twists

Up to now, we have seen that conformal invariance by itself is not sufficiently restrictive to give us the fusion rules and conformal blocks of any conformal field theory. Thus, we need to add an extra constraint to fix the theory. The next constraint that we will impose is *modular invariance* [1], a powerful tool which will allow us to fix many of the features of a conformal field theory and which yields many new surprises.

To understand why we want strings to be modular invariant, recall that the string amplitude is given as a path integral sum [Eq. (1.3.1)] over all possible conformally distinct Riemann surfaces [2, 3], so that we do not introduce overcounting into the path integral. However, we should distinguish between two types of transformations.

The familiar class of conformal transformations are those that can smoothly be deformed back to the identity. By smoothly changing the parameters that typify a conformal transformation, we can slowly change it back to the identity.

However, there are also other kinds of conformal transformations that we must subtract, which are those that cannot be smoothly deformed back to the identity, the global transformations. To visualize these global transformations, notice that there are only two ways to draw circles on a torus, called the a cycle and b cycle, such that the circles are not contractible to a point. If we slice the torus around the a cycle and b cycle and unravel the torus, we find that we have a parallelogram, such that points on opposite sides are identified with each other. Thus, a torus is topologically equivalent to a parallelogram with properly identified sides.

The simplest global transformation for a torus is given by a *Dehn twist*, which involves slicing a torus along an a cycle, rotating one of the open ends by 360°, and then resealing the cut. Equivalently, we could have sliced the torus along a b cycle, rotated one end of the slice by 360°, and then resealed the cut.

Notice that if we trace the motion of the various points on the torus we can see that these two transformations cannot be smoothly mapped back to the identity. Moreover, these two transformations can be viewed from the perspective of the parallelogram.

Let us place the parallelogram onto the complex z plane as in Fig. 4.1. Notice that we have arbitrarily fixed the bottom leg to have a length equal to one. The parallelogram can then be uniquely fixed by placing the upper left corner at point τ, a complex number. Given a fixed value of τ, we have uniquely specified a torus.

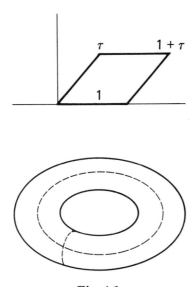

Fig. 4.1.

However, under the action of Dehn twists, the torus is mapped back into itself, so we must demand that all our amplitudes be modular invariant, that is, invariant under the action of Dehn twists. Specifically, if we make the transformation called T,

$$T: \quad \tau \to \tau + 1 \tag{4.1.1}$$

we see that upper left-hand corner moves to the right by one unit, but this just corresponds to a Dehn twist.

A bit more delicate is the other Dehn twist, because we have arbitrarily fixed the length of the bottom leg to be one. The other Dehn twist corresponds to interchanging the a cycle and b cycle, or flipping the parallelogram onto its side. However, if we place the parallelogram on its side, then we have to rescale the entire parallelogram so that its bottom side has length one. Then, if we now calculate the position of its upper left corner, we find that it has made the following transformation:

$$S: \quad \tau \to -1/\tau \qquad (4.1.2)$$

Thus, we demand that all loop amplitudes of the string be invariant under the T and S transformations (or else we will be counting the same torus an infinite number of times).

The group generated by repeatedly taking S and T transformations is called the modular group $SL(2, \mathbf{Z})$ [4–6], which is specified by the following:

$$\tau \to \frac{a\tau + b}{c\tau + d}, \qquad (a, b, c, d \in \mathbf{Z}, \qquad ad - bc = 1) \qquad (4.1.3)$$

(The transformation remains the same if we simultaneously reverse the sign of a, b, c, d. We can thus take the smaller group $PSL(2, \mathbf{Z}) = SL(2, \mathbf{Z})/\mathbf{Z}_2$ as the modular group.)

The parameters of the group are integers, not real numbers, which accounts for the difficulty in writing the representations of this group.

In ordinary Lie group theory, we wish to write the *character*, a function that immediately tells us how large a representation of an algebra is. For a Verma module [7, 8], we wish to write a function of x such that its nth Taylor coefficient tells us how many elements the representation has at the nth level.

Recall that the Verma module, for example, at the third level, is given by:

$$L_{-1}^3|0\rangle, \qquad L_{-1}L_{-2}|0\rangle, \qquad L_{-3}|0\rangle \qquad (4.1.4)$$

that is, the number of elements of an irreducible Verma module at the nth level (with no null states) is given by the number of ways in which n can be broken down into integers. This number is called $p(n)$, or the partition of the integer n.

Thus, we can define a function of x such that the coefficient of x^n tells us the number of elements in the Verma module V_n at the nth level:

$$\mathrm{Tr}\, x^{L_0} = \sum_{n=0}^{\infty} x^n \dim V_n = \sum_{n=0}^{\infty} x^n p(n) \qquad (4.1.5)$$

However, let us use the formula:

$$\prod_{n=1}^{\infty} (1 - x^n)^{-1} = \sum_{n=0}^{\infty} x^n p(n) \qquad (4.1.6)$$

Let us also define the Dedekind η function (as a function of $q = e^{2\pi i \tau}$ or of τ) as:

$$\eta(q) = q^{1/24} \prod_{n=1}^{\infty} (1 - q^n) = q^{1/24} \left(\mathrm{Tr}\, q^{L_0}\right)^{-1} \qquad (4.1.7)$$

which transforms as follows under the modular group:

$$\eta(-1/\tau) = (-i\tau)^{1/2}\eta(\tau)$$
$$\eta(\tau + 1) = e^{\pi i/12}\eta(\tau) \qquad (4.1.8)$$

where we have rewritten the Dedekind function as a function of τ.

We now write the character of an irreducible Verma module (over complex q) as

$$Z(q, \bar{q}) = q^{-c/24}\bar{q}^{-c/24} \operatorname{Tr} q^{L_0}\bar{q}^{\bar{L}_0} \qquad (4.1.9)$$

(The origin of the $c/24$ term is a bit tricky. In the trace, notice that we have implicitly made a conformal transformation from the complex plane to a cylinder, given by $w \rightarrow z = e^w$. However, as we saw in Chapter 2, the energy-momentum tensor does not transform homogeneously under a conformal transformation, but picks up a quantity proportional to the Schwartzian [Eq. (2.2.24)], that is,

$$T_{\text{cylin}}(w) = \left(\frac{\partial z}{\partial w}\right)^2 T(z) + \frac{c}{12}S(z, w) \qquad (4.1.10)$$

where the Schwartzian can be computed to equal $S(e^w, w) = -\frac{1}{2}$. This means that the L_0 defined on the cylinder is not the L_0 that we defined on the complex plane:

$$(L_0)_{\text{cylin}} = L_0 - (c/24) \qquad (4.1.11)$$

Normally, this term can be thrown away, since we are only interested in counting the states at each level. However, when we compute the effect of modular transformations, this factor becomes crucial in proving modular invariance. Traditionally, this factor comes from zeta-function regularization, but this obscures the conformal nature of its origin.)

The next step is to calculate the transformation properties of these characters under modular transformations in order to calculate functions that are invariant under the modular group [9- 11]. This will, in turn, place nontrivial restrictions on the conformal field theories that we have been studying, including the $N = 1$ superconformal theories [12, 13]. The input of modular invariance, therefore, will prove to be a powerful tool by which to put restrictions on the vast number of conformal field theories and their representations that we have found.

4.2. Free Fermion and Boson Characters

Let us now try to calculate the modular functions corresponding to a more complicated models, a free fermion with $c = \frac{1}{2}$ and a free boson on a torus with $c = 1$.

Example: Free Fermion

There is a trick we will use when calculating fermionic partition functions. If ψ_{-n} is a Fourier mode of $\psi(z)$, corresponding to a creation operator, then we know that the Fock space spanned by this oscillator is trivial, that is, only one ψ_{-n} can act on the vacuum at any given time, since it is a Grassmann odd variable, that is, $\psi_{-n}^2|0\rangle = 0$.

The trace over the nth fermionic oscillator mode is therefore easy to perform, since the trace consists of only two elements, the vacuum and $\psi_{-n}|0\rangle$.

If F equals the fermion number, then we have:

$$\text{Tr}\, q^{n\psi_{-n}\psi_n} = 1 + q^n$$
$$\text{Tr}\,(-1)^F q^{n\psi_{-n}\psi_n} = 1 - q^n \qquad (4.2.1)$$

[The insertion of the factor $(-1)^F$ into the trace converts the $1 + q^n$ into a $1 - q^n$ term.]

The entire Fock space, however, consists of monomials one can create out of various products of ψ_{-n} for different values of n, so we can convert the trace, which is a sum over states, into a product over different Fock spaces:

$$\text{Tr}(q)^{L_0} = \text{Tr}(q)^{\sum_n n\psi_{-n}\psi_n} = \text{Tr} \prod_n (q)^{n\psi_{-n}\psi_n} = \prod_n (1 + q^n) \qquad (4.2.2)$$

where the sum n is over positive integers for the NS fermions and positive half-integers for the R fermions.

However, there is a complication in the sum due to the boundary conditions on the fermions, that is, we can take different periodic and antiperiodic boundary conditions on the torus. If we take the trace over q^{L_0}, we must first specify whether we are tracing (in the τ direction) over Ramond fermions (which are periodic) or Neveu–Schwarz fermions (which are antiperiodic). We will denote this by Tr_P and Tr_A. But, we must also specify the boundary conditions in the σ direction as well as in the τ direction. Since the boundary conditions can be either periodic or antiperiodic in the τ or σ directions, we have a total of four possible boundary conditions and hence four possible traces.

The four different boundary conditions on the torus or parallelogram define what is called the *spin structure* on that surface.

We will use the symbol $\chi(A, P)$, for example, to denote the trace over antiperiodic (periodic) boundary conditions in the σ (τ) direction. In the case of $\chi(A, P)$, we will simply trace over Ramond fermions: $q^{-1/48}\,\text{Tr}_P\, q^{L_0}$. Notice that the trace over Ramond fermions automatically specifies antiperiodic boundary conditions in the σ direction.

If we wish, however, to calculate $\chi(P, P)$, which has periodic conditions in the σ direction, then we must insert the operator $(-1)^F$, where F is the fermion number, in order to reverse the periodicity.

Given these rules, it is now a simple matter to write out in detail the four possible traces corresponding to the four possible spin structures for the torus. Because all trace operations can now be converted into products over different Fock spaces labeled by n, we now have:

$$\chi(A, A) = q^{-1/48} \operatorname{Tr}_A q^{L_0} = q^{-1/48} \prod_{n=0}^{\infty} (1 + q^{n+1/2}) = \sqrt{\frac{\vartheta_3}{\eta}}$$

$$\chi(P, A) = q^{-1/48} \operatorname{Tr}_A (-1)^F q^{L_0} = q^{-1/48} \prod_{n=0}^{\infty} (1 - q^{n+1/2}) = \sqrt{\frac{\vartheta_4}{\eta}}$$

$$\chi(A, P) = \frac{1}{\sqrt{2}} q^{-1/48} \operatorname{Tr}_P q^{L_0} = \frac{1}{\sqrt{2}} q^{1/24} \prod_{n=0}^{\infty} (1 + q^n) = \sqrt{\frac{\vartheta_2}{\eta}} \qquad (4.2.3)$$

$$\chi(P, P) = \frac{1}{\sqrt{2}} q^{-1/48} \operatorname{Tr}_P (-1)^F q^{L_0} = \frac{1}{\sqrt{2}} q^{1/24} \prod_{n=0}^{\infty} (1 - q^n) = 0$$

where the $\vartheta_i = \vartheta_i(0, q)$ are the usual Jacobi theta functions (taken with one variable set to zero). These functions can be defined through the above equations as infinite products, or as infinite sums via:

$$\vartheta_1(\nu, q) = i \sum_{n=-\infty}^{\infty} (-1)^n q^{1/2[n-(1/2)]^2} e^{i\pi(2n-1)\nu}$$

$$\vartheta_2(\nu, q) = \sum_{n=-\infty}^{\infty} q^{1/2[n-(1/2)]^2} e^{i\pi(2n-1)\nu}$$

$$\vartheta_3(\nu, q) = \sum_{n=-\infty}^{\infty} q^{n^2/2} e^{2\pi in\nu} \qquad (4.2.4)$$

$$\vartheta_4(\nu, q) = \sum_{n=-\infty}^{\infty} (-1)^n q^{n^2/2} e^{2\pi in\nu}$$

We should also mention that the ϑ functions can also be written as infinite products:

$$\vartheta_1(\nu, q) = 2q_0 q^{1/8} \sin \pi\nu \prod_{i=1}^{\infty} (1 - 2q^n \cos \pi\nu + q^{2n})$$

$$\vartheta_2(\nu, q) = 2q_0 q^{1/8} \cos \pi\nu \prod_{n=1}^{\infty} (1 + 2q^n \cos 2\pi\nu + q^{2n})$$

$$\vartheta_3(\nu, q) = q_0 \prod_{n=1}^{\infty} \left[1 + 2q^{n-(1/2)} \cos 2\pi\nu + q^{2n-1} \right] \qquad (4.2.5)$$

$$\vartheta_4(\nu, q) = q_0 \prod_{n=1}^{\infty} \left[1 - 2q^{n-(1/2)} \cos \pi\nu + q^{2n-1} \right]$$

where $q_0 = \prod_{n=1}^{\infty}(1 - q^n)$. (Our value of q is the square root of the value quoted in Ref. 14.)

We now would like to reanalyze the partition functions for the $c = \frac{1}{2}$ fermion, rearranging the characters according to the irreducible representations of the conformal group, that is, according to h, c. Comparing the characters found above with the characters found in the usual Ising model, where we also have $c = \frac{1}{2}$, we find from Eq. (2.5.16) that there are three primary fields with weights given by

$$\{h_{1,1}, h_{2,1}, h_{2,2}\} = \left\{0, \frac{1}{2}, \frac{1}{16}\right\} \tag{4.2.6}$$

Our task is now to rearrange the partition functions found earlier so that we have only the characters of irreducible representations. To do this, we note that the Verma modules are built up by multiplying the vacuum by L_{-n}, which does not change the overall fermion number of the state. Thus, the Verma module with $h = 0$ is built of states with even fermion numbers and integer eigenvalues of L_0, while the module with $h = \frac{1}{2}$ is built of states with odd fermion numbers and half-integral eigenvalues of L_0.

Thus, instead of arranging the character depending on whether the boundary conditions are periodic or antiperiodic, we will now rearrange the module according to whether the module has an even or odd fermion number. The way to do this is to insert $\frac{1}{2}[1 \pm (-1)^F]$ into the trace, which selects the states with definite fermion numbers.

Given this decomposition according to the fermion number, the characters in this representation can be defined via:

$$\chi_0 = q^{-1/48} \operatorname{Tr}_{h=0} q^{L_0} = q^{-1/48} \operatorname{Tr}_A \frac{1}{2}\left[1 + (-1)^F\right] q^{L_0}$$

$$\chi_{\frac{1}{2}} = q^{-1/48} \operatorname{Tr}_{h=1/2} q^{L_0} = q^{-1/48} \operatorname{Tr}_A \frac{1}{2}\left[1 - (-1)^F\right] q^{L_0} \tag{4.2.7}$$

$$\chi_{\frac{1}{16}} = q^{-1/48} \operatorname{Tr}_{h=1/16} q^{L_0} = q^{-1/48} \operatorname{Tr}_P \frac{1}{2}\left[1 \pm (-1)^F\right] q^{L_0}$$

(In the last expression for $h = \frac{1}{16}$, we used the fact that the trace over $(-1)^F q^{L_0}$ equals zero due to a cancellation between equal numbers of states with different fermion numbers.) However, since we know how to evaluate all the traces in the above expression in terms of the four periodic and antiperiodic characters we wrote before, we now can write the complete expression for each of the three Verma modules [4]:

$$\chi_0 = \frac{1}{2}\left[\chi(A,A) + \chi(P,A)\right] = \frac{1}{2}\left(\sqrt{\frac{\vartheta_3}{\eta}} + \sqrt{\frac{\vartheta_4}{\eta}}\right)$$

$$\chi_{\frac{1}{2}} = \frac{1}{2}\left[\chi(A,A) - \chi(P,A)\right] = \frac{1}{2}\left(\sqrt{\frac{\vartheta_3}{\eta}} - \sqrt{\frac{\vartheta_4}{\eta}}\right) \qquad (4.2.8)$$

$$\chi_{\frac{1}{16}} = \frac{1}{\sqrt{2}}\left[\chi(A,P) \pm \chi(P,P)\right] = \frac{1}{\sqrt{2}}\sqrt{\frac{\vartheta_2}{\eta}}$$

The point of this discussion, of course, is to calculate modular invariant combinations of the χ. Let us now calculate, therefore, how each of these χ functions change under a modular transformation.

Under the operation T, we see that the top of the parallelogram is shifted one unit to the right, as in Fig. 4.1. We can now simply determine how the spin structures change under this operation. For example, we can start with a torus with the (A,A) spin structure and apply the operation T. When we move from the origin to the point τ, we pick up a factor of (-1). When we now move from the origin to the point $\tau + 1$, we pick up a factor of $(-1)(-1) = +1$, which is periodic. By moving in the σ direction on the new torus, we now pick up a factor of $+1$, so it is now periodic in this direction. So the spin structure, under the T operation, has now changed to (P,A).

Likewise, we can, by simply moving from the origin to the point $\tau + 1$, determine how all four spin structures change. By explicit calculation with the known values of χ, we find:

$$T : \begin{cases} \chi(A,A) \to e^{-i\pi/24}\chi(P,A) \\ \chi(P,A) \to e^{-i\pi/24}\chi(A,A) \\ \chi(A,P) \to e^{i\pi/12}\chi(A,P) \end{cases} \qquad (4.2.9)$$

We can also calculate how the spin structures change under the operation S (which reverses the τ and σ directions). Using the simple-minded rule given above, we find the following transformations of the traces under S:

$$S : \begin{cases} \chi(A,A) \to \chi(A,A) \\ \chi(A,P) \to \chi(P,A) \\ \chi(P,A) \to \chi(A,P) \end{cases} \qquad (4.2.10)$$

Notice that none of the χ are modular invariant by themselves. To obtain modular invariant functions, we must also include complex conjugates of the χ. We can satisfy T and S invariance by multiplying the various characters with their complex conjugates (which eliminates the phase factor introduced via T transformations) and then choosing the correct combination of absolute values of characters to get complete modular invariance.

The goal of this discussion is to formulate a modular invariant partition function for the $c = \frac{1}{2}$ fermions, which we see are now given by:

$$
\begin{aligned}
Z_{\text{Ising}} &= \frac{1}{2}\Big(|\vartheta_3/\eta| + |\vartheta_4/\eta| + |\vartheta_2/\eta| \pm |\vartheta_1/\eta| \Big) \\
&= \chi_0 \bar{\chi}_0 + \chi_{\frac{1}{2}} \bar{\chi}_{\frac{1}{2}} + \chi_{\frac{1}{16}} \bar{\chi}_{\frac{1}{16}}
\end{aligned}
\tag{4.2.11}
$$

In general, when we discuss increasingly more complicated characters for the Virasoro and Kac–Moody representations, we will take the *ansatz* that the final modular invariant partition function will be a bilinear sum over both holomorphic and antiholomorphic representations, each corresponding to the various primary fields, that is,

$$
Z(q, \bar{q}) = \sum_{h, \bar{h}} N_{h\bar{h}} \chi_h(q) \chi_{\bar{h}}(\bar{q})
\tag{4.2.12}
$$

We will make modular transformations on this bilinear combination in order to obtain constraint equations on the coefficients $N_{h\bar{h}}$, thus yielding the invariant solution.

These methods developed for the characters of free fermions can be carried directly over to the partition function over the free boson defined on the torus, such that $X = X + 2\pi r$.

Example: Free Boson on a Torus

The trace we are interested in is:

$$
\chi = (q\bar{q})^{-c/24} \operatorname{Tr} q^{L_0} \bar{q}^{\tilde{L}_0}
\tag{4.2.13}
$$

Special care, however, has to be given to the zero mode sector in the trace. Because of periodic boundary conditions, momentum is quantized on the torus, and because the string can wind around the torus, we have to introduce the winding number. Thus, two integers are required to describe compactification of a closed string. We thus have:

$$
\alpha_0 = p_{\mathrm{L}} = \frac{m}{2r} + nr, \qquad \bar{\alpha}_0 = p_{\mathrm{R}} = \frac{m}{2r} - nr
\tag{4.2.14}
$$

for integer n and m. This means that we have to sum over states $|m, n\rangle$, indexed by two integers, and then multiply by arbitrary numbers of creation operators.

The eigenvalues of L_0 and $\tilde{L}_0$ on these states are easily calculated. Let:

$$
|\{n_i\}, \{m_j\}, m, n\rangle \equiv \Big(\prod_i a_{n_i}^{\dagger} \Big) \Big(\prod_j \bar{a}_{m_j}^{\dagger} \Big) |m, n\rangle
\tag{4.2.15}
$$

for some collection of integers n_i and m_j.

Then, we have:

$$L_0 |\{n_i\}, \{m_j\}, n, m\rangle = \left[\sum_i n_i + \frac{1}{2}\left(\frac{m}{2r} + nr\right)^2\right] |\{n_i\}, \{m_j\}, n, m\rangle$$

(4.2.16)

(The equation for $\tilde{L}_0$ is the same, except that the term $+nr$ is changed to $-nr$ on the right-hand side, and we replace n_i with m_j.)

The partition function therefore splits into two pieces. The first piece simply records the number of ways in which we can create states multiplied by monomials in the creation oscillators $a_n^\dagger$. This part is easy to compute and gives us $\prod_n (1 - q^n)^{-1}$. This contributes a factor of $1/\eta\bar{\eta}$.

The second part is the summation over the zero modes, which in turn are indexed by two numbers. Putting everything together, we now have:

$$\chi = \frac{1}{\eta\bar{\eta}} \sum_{n,m=-\infty}^{\infty} q^{1/2(m/2r+nr)^2} \bar{q}^{1/2(m/2r-nr)^2}$$

(4.2.17)

This expression is modular invariant by itself. (If we make the transformation $\tau \to \tau + 1$, the $\eta\bar{\eta}$ is invariant because they only change by a phase. The zero mode part picks up a phase $\exp \pi i(p_L^2 - p_R^2)$, which equals unity when we plug in the values for the momenta. If we make the transformation $\tau \to -1/\tau$, the calculation is a bit more difficult, but can also be performed by reversing the boundary conditions on the torus.)

4.3. GSO and Supersymmetry

Let us now apply some of these techniques to the $D = 10$ superstring, investigating the surprising link between supersymmetry and modular invariance. We recall that the NS–R model by itself is not space–time supersymmetric. However, we can recover space–time supersymmetry by imposing the GSO projection [15].

Let us define the G-parity operator by:

$$G = (-1)^{\sum_{n=1/2} b_{-r}b_r}$$

(4.3.1)

We will now take the even G-parity sector of the theory. This truncation of the NS–R sector has several important implications.

First, it eliminates the troublesome tachyon that appears in the bosonic theory. Second, it restores space–time supersymmetry. To see this, it is most convenient to work with the light cone quantized NS–R string, where we can prove that the number of states of the fermionic sector equals the number of states in the bosonic sector.

In the NS sector, the even G-parity sector is equal to the trace over the following:

$$P_{NS} = q^{-1/2} \, \mathrm{Tr} \left[\frac{1}{2}(1+G)q^R \right]$$

$$R = \sum_{n=1}^{\infty} n a_n^\dagger a_n + \sum_{r=1/2}^{\infty} r b_r^\dagger b_r \qquad (4.3.2)$$

Using the previous techniques, we can show that this partition function equals:

$$P_{NS} = \frac{1}{2} q^{-1/2} \prod_{n=1}^{\infty} (1-q^n)^{-8} \left[\prod_{n=1}^{\infty} (1+q^{n-1/2})^8 - \prod_{n=1}^{\infty} (1-q^{n-1/2})^8 \right]$$

$$(4.3.3)$$

Next, we can set up the Ramond sector partition function as:

$$P_R = 8 \, \mathrm{Tr}(q^R)$$

$$R = \sum_{n=1}^{\infty} n(a_n^\dagger a_n + d_n^\dagger d_n) \qquad (4.3.4)$$

(where the 8 comes from the fact that, in the light cone gauge, only these components of the spinor survives). This can be shown to equal:

$$P_R = 8 \prod_{n=1}^{\infty} (1-q^n)^{-8}(1+q^n)^8 \qquad (4.3.5)$$

It was recognized in 1829 by Jacobi that these two expressions are equal:

$$P_{NS} = P_R \qquad (4.3.6)$$

This formula has great implications for the superstring. Not only does it show that the NS–R model is space–time sypersymmetric after the GSO projection, but it also shows the power of modular invariance. Notice that we inserted the projection operator $\frac{1}{2}[1 + (-1)^F]$ into the trace, which is precisely the summing over the (A, P) and (P, A) sectors. Modular invariance means invariance under the interchange and transformation of the homology cycles, which is precisely what the insertion of the GSO operator performs.

For the superstring theory, it means that, in some sense, modular invariance, the GSO projection, and space–time supersymmetry are all interdependent. Since modular invariance is necessary for a unitary theory, this implies that space–time supersymmetry is necessary for the internal consistency of the theory.

Space–time supersymmetry, far from being a luxury for the theory, now appears absolutely essential for the self-consistency of the entire theory. The importance of this fact will become even more important in the next chapter, where we will discuss different supersymmetric compactification schemes.

4.4. Minimal Model Characters

Let us proceed to the more complicated case of the minimal models. Let us calculate the character in two different ways. First, we will take the above expressions for the character of an irreducible representation and generalize it to the case where null states are present. We will derive the character formula by carefully subtracting out all the null states, leaving us with the character of the minimal model. Second, we will use the coset construction of the previous chapter and calculate the character of the minimal models by reexpressing it in terms of the affine $SU(2)_k$ via Eq. (3.3.10). Finally, let us recall that the Ising model at criticality is equivalent to the minimal model, with $m = 3$ in Eq. (2.5.24). So, our new results on the character of the minimal models should give us an independent check on the previous formula.

We begin by noting that, if there are no null states at all, then the character is given by:

$$\chi_h(q) = q^{-c/24+h} \sum_n p(n)q^n = q^{-(c-1)/24} \eta(q)^{-1} q^h \qquad (4.4.1)$$

since the number of states at level n for this irreducible case equals the partition of that integer $p(n)$. (We have inserted a normalization factor q^h, which will simplify our discussion.) Let us now generalize this formula for the minimal model. Recall that the weights of the fields in the minimal models are labeled by two integers and are given by Eq. (2.5.7):

$$h_{rs} = \frac{[r(m+1) - sm]^2 - 1}{4m(m+1)}, \qquad (1 \le s \le r \le m-1) \qquad (4.4.2)$$

If we analyze the Verma module ϕ_{rs}, we notice that it contains a null state χ at level rs, which therefore has weight given by $h_{rs} + rs$. We must, therefore, explicitly remove from the character the states given by the module generated by χ.

Let us now analyze the character χ_{rs} of the mth minimal model. If we subtract the contribution from the null states, we find:

$$\chi_{rs}(q) = q^{-(c-1)/24} \eta(q)^{-1} q^{h_{rs}} (1 - q^{rs} + \cdots) \qquad (4.4.3)$$

However, we cannot stop here. It turns out that the module $[\chi]$ itself contains a null state, at level $(m+r)(m+1-s)$ because:

$$h_{rs} + rs = h_{m+r,m+1-s} \qquad (4.4.4)$$

so it would be overcounting to simply remove the module χ. Instead, we must carefully subtract this new null module from the first one.

Subtracting this new null module, we find:

$$\chi_{rs} = q^{-(c-1)/24}\eta(q)^{-1}\{q^{h_{rs}} - q^{h_{r,-s}}[1 - q^{(m+r)(m+1-s)} + \cdots]\}$$
$$= q^{-(c-1)/24}\eta(q)^{-1}(q^{h_{rs}} - q^{h_{r,-s}} + q^{h_{2m+r,s}} - \cdots) \tag{4.4.5}$$

Not surprisingly, this process continues forever, with null states within the module generated by the previous null state. Therefore, an infinite succession of subtraction of these factors is necessary. The final result can be found by summing the various subtractions:

$$\chi_{rs} = q^{-(c-1)/24}\eta(q)^{-1}\sum_{k=-\infty}^{\infty}(q^{h_{2mk+r,s}} - q^{h_{2mk+r,-s}}) \tag{4.4.6}$$

This is the desired character for the mth minimal model.

Now let us calculate the transformation properties of this character under S and T. It is straightforward to show that:

$$T: \quad \chi_h(q) \rightarrow e^{2\pi i(h-c/24)}\chi_h(q) \tag{4.4.7}$$

A bit more difficult is the calculation of the character under an S transformation:

$$S: \quad \chi_{rs}(q) \rightarrow \sum_{pq} S^{pq}_{rs}\chi_{pq}(q) \tag{4.4.8}$$

where:

$$S^{pq}_{rs} = \left[\frac{8}{m(m+1)}\right]^{1/2}(-1)^{(r+s)(p+q)}\sin\frac{\pi rp}{m}\sin\frac{\pi sq}{m+1} \tag{4.4.9}$$

This is a rather remarkable formula, because encoded within the S matrix is a finite dimensional representation of the modular group. However, notice also that χ_{rs} by itself is not modular invariant. It transforms "covariantly" under the modular group, and hence, it is not an invariant. To form a genuine invariant, we must take various combinations of bilinear sums of representations in order to obtain an invariant character as in Eq. (4.2.12).

One set of invariants can be calculated by observing that the matrix obeys $S^2 = 1$, with real elements. Then, it is easy to show that the following diagonal form for the N matrix is modular invariant:

$$N_{h\bar{h}} = \delta_{h\bar{h}} \tag{4.4.10}$$

As a check on our results, we take the case $m = 3$, which corresponds to the Ising model, with three primary fields. Notice that the diagonal N matrix yields the following form for the invariant:

$$Z = |\chi_0|^2 + |\chi_{1/2}|^2 + |\chi_{1/16}|^2 \tag{4.4.11}$$

which is precisely the form of the modular invariant found earlier in Eq. (4.2.11) when analyzing the $c = \frac{1}{2}$ fermion system.

We can find other invariants by using some tricks. Notice that for m odd, we can show that the S matrix obeys:

$$S^{r's'}_{rs} = (-1)^{s'-1} S^{r',m+1-s'}_{rs} = (-1)^{s-1} S^{r's'}_{r,m+1-s} \qquad (4.4.12)$$

This, in turn, implies that we can form modular invariants out of the combination:

$$\chi_{rs} + \chi_{r,m+1-s} \qquad (4.4.13)$$

Specifically, we can show that the following is also an invariant [5]:

$$Z = \frac{1}{2} \sum_r \sum_{s \text{ odd}} |\chi_{rs} + \chi_{r,m+1-s}|^2 \qquad (4.4.14)$$

Thus, we have constructed two infinite series of modular invariant partition functions. Similarly, it can be shown that there are only a finite number of other possibilities, corresponding to $m = 11, 12, 17, 18, 29, 30$. Later, we will see how they give rise to two infinite series and three finite ones, which exhaust all possible modular invariant partition functions for the minimal series.

However, we would like to find a systematic way in which to construct these invariants rather than appealing to mathematical tricks. To gain some insight into this difficult question, we now turn to the characters over the Kac–Moody algebras and will compute the modular invariant functions for both the minimal model and $SU(2)_k$.

4.5. Affine Characters

It can be shown that the minimal series found earlier exhausts all possible unitary representations with a finite number of primary fields. However, we can enlarge the system to include Kac–Moody algebras. Then, there exist representations of affine Lie groups with finite numbers of primary fields where the fields are now primary with respect to both the conformal group and the affine Lie group. (These affine systems, however, can have an infinite number of primaries with respect to just the conformal group.)

To understand the characters of these Kac–Moody algebras, let us first calculate the character of the minimal models in another way, using the coset construction of the previous chapter. We will write the character of the minimal model in terms of the character of the simplest affine Lie group, $SU(2)_k$.

In general, modules over the affine Lie algebras will be more complicated than conformal modules because they are generated by isotopic ladder operators J^a_{-m} as well as L_{-n} operating on the highest weight state. To typify a state, we must calculate its eigenvalue under both the level operator L_0 and the Cartan subalgebra H^i_0.

Let us consider the level k representation built on the spin-j vacuum state $|j\rangle$ for $SU(2)_k$. The character will now depend on two variables, τ

(associated with the Virasoro operators appearing in the module) and θ (associated with the Kac-Moody operators). We define:

$$\chi_{(j)}^{k}(\theta, \tau) = q^{-c/24} \, \text{Tr}_{(j),k} \left(q^{L_0} e^{i\theta J_0^3} \right) \tag{4.5.1}$$

For a more general affine Lie algebra, the character now depends on the parameter τ as well as θ^i, where i ranges over the elements of the generators H_0^i of the mutually commuting Cartan subalgebra. We define:

$$\chi_{(\lambda)}^{k}(\theta^i, \tau) = q^{-c_G/24} \, \text{Tr}_{(\lambda),k} \left(q^{L_0} e^{i\theta^i H_0^i} \right) \tag{4.5.2}$$

Fortunately, almost all formulas in the theory of Lie algebras generalizes to the affine case, so we will need the Weyl character formula for an arbitrary Lie algebra and its generalization to the affine case: the Weyl–Kac character formula.

To understand the Weyl character formula for Lie algebras and the Weyl–Kac character formula for affine Lie algebras, we must first make a few definitions. In the Cartan–Weyl basis, we define α_i to be the root vectors. A vector ρ can be written in terms of these roots as $\rho = \sum_i c_i \alpha_i$. If the first nonzero coefficient c_i is positive, then we say that ρ is a *positive root*. (This definition is somewhat arbitrary, of course, since we can mix up the root vectors. But, once a fixed basis for the roots has been chosen, this convention is a useful one.)

For the group $SU(2)$, for example, we notice that the representations are indexed by the integral or half-integral l and contain elements that run from $-l$ to $+l$, that is, they are symmetric if we rotate them 180°, exchanging L_3 for $-L_3$. This symmetry under reflections is called a *Weyl reflection*. For higher Lie algebras, the representations (when plotted on a graph whose coordinates are the independent eigenvalues of the Cartan subalgebra) have a larger discrete symmetry, which can be generated by a group of reflections called W, the *Weyl group*. For example, if we take a root λ, we can rotate it as follows:

$$w_\alpha(\lambda) = \lambda - \alpha \langle \lambda, \alpha \rangle \tag{4.5.3}$$

where the Weyl rotation operator w_α within W is associated with the root α.

Let us also introduce the convenient notation that e^α, where α is a root, represents an operator that acts on an arbitrary root β as follows:

$$e^\alpha(\beta) \equiv e^{(\alpha,\beta)} \tag{4.5.4}$$

Then, the classical Weyl formula states that the character of a representation $L(\Lambda)$ (associated with a highest weight vector Λ) is given by [16]:

$$\text{ch} \, L(\Lambda) = \frac{\sum_{w \in W} \epsilon(w) e^{w(\Lambda+\rho)-\rho}}{\sum_{w \in W} \epsilon(w) e^{w(\rho)-\rho}} \tag{4.5.5}$$

where we sum over the elements of the Weyl group, and $\epsilon(w)$ is $+1$ (-1), depending on whether the member of the Weyl group w can be expressed in terms of an even (odd) number of reflections. Here, ρ is half the sum over all positive roots:

$$\rho = \frac{1}{2} \sum_{\alpha > 0} \alpha \tag{4.5.6}$$

Example: SU(3)

This formula can be used to calculate the dimension appearing in any representation R of a Lie group. Let Eq. (4.5.5) operate on a vector, called $\bar\rho$, and then let $\bar\rho$ go to zero. The limit as this arbitrary vector goes to zero can be easily computed, and we find the celebrated result of Weyl:

$$\dim R = \prod_{\alpha > 0} \frac{(\alpha, \Lambda + \rho)}{(\alpha, \rho)} \tag{4.5.7}$$

where the sum over the Weyl group has now been replaced by the product over the positive roots. This is a very powerful result and can be used to calculate the dimension of virtually all the representations found in classical Lie group theory.

For example, for the group $SU(3)$, we can calculate the dimension of a representation with Dynkin coefficients (m_1, m_2). Inserting this into the above expression, we find:

$$\dim R = \left(\frac{m_1 + 1}{1}\right) \left(\frac{m_2 + 1}{1}\right) \left(\frac{m_1 + m_2 + 2}{2}\right) \tag{4.5.8}$$

Inserting various values of m_1 and m_2 into the equation, we easily compute the dimension of the well-known representations of $SU(3)$.

Example: $SU(2)_k$

Now that we have treated the classical case, we wish to generalize this discussion to the affine case. For a Kac–Moody algebra, the Weyl–Kac formula is, remarkably enough, essentially the same as the Weyl formula, except that the definitions of a root vector and the Weyl reflection have to be generalized.

We must generalize our previous discussion because the Kac–Moody algebra differs from the usual Lie algebra in two essential ways. Besides the usual root vectors, we also describe states by the number operator (eigenvalue of L_0) and the c-number term appearing in the algebra. Thus, a root vector in the Kac–Moody case actually has three entries:

$$\lambda \equiv (\bar\lambda, k, n) \tag{4.5.9}$$

where $\bar\lambda$ is the classical root vector, k is the eigenvalue of the number operator, and n is a c number.

We take the scalar product and the bracket product of two vectors $\alpha = (\bar{\alpha}, k, n)$ and $\beta = (\bar{\beta}, k', n')$ in the following way:

$$(\alpha, \beta) \equiv (\bar{\alpha}, \bar{\beta}) + kn' + nk', \qquad \langle \alpha, \beta \rangle = \frac{2(\alpha, \beta)}{(\alpha, \alpha)} \qquad (4.5.10)$$

We define the Weyl reflection in the same way as before:

$$w_\alpha(\lambda) = \lambda - \alpha\langle\lambda, \alpha\rangle \qquad (4.5.11)$$

except for the important fact that, because the root vector now has three entries, the effect of a Weyl reflection consists of a classical Weyl reflection and a translation.

This translation is easy to see. If we let $\alpha = (\bar{\alpha}, 0, 1)$ and $\lambda = (\bar{\lambda}, k, n)$, then, inserting both expressions into the definition of a Weyl reflection, we have:

$$w_\alpha(\lambda) = \left\{ w_{\bar{\alpha}}(\bar{\lambda} + 2k\bar{\alpha}/\bar{\alpha}^2), k, n + \frac{1}{2k}[\bar{\lambda}^2 - (\bar{\lambda} + 2k\bar{\alpha}/\bar{\alpha}^2)^2] \right\} \qquad (4.5.12)$$

This Weyl reflection is easily split into two parts, the classical Weyl reflection (which we denote by $\dot{W}$) and a translation by the following vector:

$$t_\beta(\lambda) = \left\{ \bar{\lambda} + k\beta, k, n + \frac{1}{2k}[\bar{\lambda}^2 - (\bar{\lambda} + k\beta)^2] \right\} \qquad (4.5.13)$$

where $\beta = 2\bar{\alpha}/\bar{\alpha}^2$. (For example, for the simple case of $SU(2)_k$, this separation is trivial: the classical Weyl reflection just flips the root α into $-\alpha$, and the translation produces a shift by $j\alpha$, where j is an integer.)

Let us perform the sum over translation T first, thereby obtaining a Θ function, and sum over $\dot{W}$ later. When we perform this separation, we find:

$$\sum_{w\in W} \epsilon(w) e^{w(\rho)-\rho} = e^{-\rho} \sum_{w\in\dot{W}} \epsilon(w) \sum_{\alpha\in M} e^{t_\alpha[w(\rho)]}$$
$$= e^{-\rho+\bar{\rho}^2\delta/2g} \sum_{w\in\dot{W}} \epsilon(w)\Theta_{w(\rho)} \qquad (4.5.14)$$

where M is the lattice generated by translations, $\delta = (0,0,1)$, $\bar{\rho} = \rho - g(0,1,0)$, and the Θ function comes directly from summation over the translations:

$$\Theta_\lambda \equiv e^{-|\lambda|^2\delta/2k} \sum_{t\in M} e^{t(\lambda)} = e^{k(0,1,0)} \sum_{\gamma\in M+k^{-1}\bar{\lambda}} e^{-k|\gamma|^2(0,0,1)+k\gamma/2} \qquad (4.5.15)$$

In this form, the character can be written totally in terms of Θ functions as:

$$\operatorname{ch} L(\lambda) = e^{s_\Lambda \delta} \frac{\sum_{w \in \hat{W}} \epsilon(w) \Theta_{w(\Lambda+\rho)}}{\sum_{w \in \hat{W}} \epsilon(w) \Theta_{w(\rho)}} \equiv e^{s_\Lambda \delta} \frac{C_{\Lambda+\rho}}{C_\rho}$$

$$s_\Lambda = \frac{|\Lambda + \rho|^2}{2(k+g)} - \frac{|\rho|^2}{2g}$$

(4.5.16)

For the case $SU(2)_k$, we find a vast simplification in all our formulas. In this case, there is only one root α, where $\alpha^2 = 2$. The Weyl reflections simply flips the root, and the sum over translations equals the sum over $j\alpha$ for integer j. In particular, we find that the Θ function can be written as:

$$\Theta_{n,k}(u, \tau, z) = e^{-2\pi i u} \sum_{j \in \mathbf{Z} + n/2m} \exp(2\pi i \tau m j^2 + 2\pi i j z) \qquad (4.5.17)$$

(where we set $u = z = 0$). Then everything can be expressed in terms of

$$C_{\Lambda+\rho} \to C_{n,k} \equiv \Theta_{n,k} - \Theta_{-n,k} \qquad (4.5.18)$$

Simplifying the above results for our case, we find the final expression for the character [17]:

$$\chi_\lambda(\tau) = C_{2j+1,k+2}/C_{1,2}$$

$$= \eta^{-3}(\tau) \sum_{n=-\infty}^{\infty} \left[2n(k+2) + (2j+1) \right] \times e^{i\pi\tau[2n(k+2)+2j+1]^2/2(k+2)}$$

(4.5.19)

where $\lambda = 2j + 1$. This is our final result for the character of $SU(2)_k$. (To prove the last step, we used the identity $\eta^3(\tau) = \sum_m (4m+1)q^{(4m+1)^2/8}$.)

Under S and T, this formula transforms as:

$$T: \quad \chi_\lambda(\tau+1) = \exp\left[2i\pi \left(\frac{\lambda^2}{2N} - \frac{1}{8} \right) \right] \chi_\lambda(\tau)$$

$$S: \quad \chi_\lambda\left(-\frac{1}{\tau} \right) = \sqrt{\frac{2}{k+2}} \sum_{1 \le \tilde{\lambda} \le k+1} \sin\left(\frac{\pi \lambda \tilde{\lambda}}{k+2} \right) \chi_{\tilde{\lambda}}(\tau)$$

(4.5.20)

where $N = 2(k+2)$.

Using the modular transformations T and S for $SU(2)_k$, we can now read the explicit values for the matrices that generate modular transformations:

$$S_{jj'}^{(k)} = \left(\frac{2}{k+2} \right)^{1/2} \sin \frac{\pi(2j+1)(2j'+1)}{k+2}$$

$$T_{jj'}^{(k)} = \exp\left\{ 2i\pi \left[\frac{(2j+1)^2}{4(k+2)} - \frac{1}{8} \right] \right\} \delta_{j,j'}$$

(4.5.21)

with $j, j' = 0, ..., k/2$. It is straightforward to check that $S^2 = (ST)^3 = 1$, as they should.

Now that we have successfully calculated the characters for $SU(2)_k$, we are in a position to exploit this result and calculate the characters of the minimal models using the coset construction, giving us an independent check on the correctness of our formalism. In the last chapter, we showed that the minimal models are equivalent to the GKO construction for the coset, that is, $G/H = SU(2)_k \otimes SU(2)_1/SU(2)_{k+1}$. We now proceed by noticing that the energy-momentum tensor $T(z)_G$ splits into two commuting sectors $T(z)_{G/H}$ and $T(z)_H$. This allows us to write the Fock space of the theory in terms of the direct product of the two sectors. Concretely, it means that we can decompose the character of affine G in terms of the characters associated with $T_{G/H}$.

Symbolically, we can write:

$$\chi^{k_G}(\tau) = \chi_{G/H}\chi^{k_H} \tag{4.5.22}$$

Under a modular transformation, we have:

$$\chi^{k_G}(\tau') = S^{k_G}\chi^{k_G}(\tau) \tag{4.5.23}$$

This means that:

$$\chi^{k_G}(\tau') = \chi_{G/H}(\tau')S^{k_H}\chi^{k_H}(\tau) \tag{4.5.24}$$

We can now solve for the transformation of the character of the coset, so that we have the desired result:

$$\chi_{G/H}(\tau') = S^{k_G}\chi_{G/H}(\tau)(S^{k_H})^{-1} \tag{4.5.25}$$

This, in turn, gives us an independent way in which to confirm our previous formulas concerning the minimal model. Using the above formula for the characters of the cosets, we can calculate the character of the minimal models in terms of the characters of the affine $SU(2)_k$. The calculation is not difficult. We use Eqs. (4.5.25) and (3.3.10) to calculate the character of the minimal models and find exact agreement with Eq. (4.4.6), which was derived in an entirely different fashion, by subtracting the characters of null state Verma modules.

4.6. A–D–E Classification

Now that we have successfully computed the character $\chi_\lambda(\tau)$ for $SU(2)_k$, our next step is to construct a modular invariant combination of such representations. Because the $\chi_\lambda(\tau)$ are not modular invariant, we will assume that we can create a genuine modular invariant by analyzing the bilinear expression [9–11]:

$$Z = \sum_{\lambda,\lambda'} \chi_\lambda(\tau)^* N_{\lambda,\lambda'} \chi_{\lambda'}(\tau) \tag{4.6.1}$$

Fortunately, with some work, it is possible to write the complete list of solutions to the $N_{\lambda,\lambda'}$ matrix, giving us all modular invariant partition functions for affine $SU(2)_k$. For example, one trivial solution is given by the diagonal matrix:

$$N_{\lambda,\lambda'} \sim \delta_{\lambda,\lambda'} \tag{4.6.2}$$

Then, the sum over λ runs from 1 to $k+1$, and the invariant becomes:

$$\sum_{\lambda=1}^{k+1} |\chi_\lambda|^2 \tag{4.6.3}$$

The complete solution to the problem of constructing modular invariants for $SU(2)_k$, however, is more involved.

Under a general modular transformation, the characters transform as:

$$\chi_\lambda(\tau') = \sum_{\lambda'=0}^{N-1} U_{\lambda,\lambda'}(A)\chi_{\lambda'}(\tau) \tag{4.6.4}$$

where the $U(A)$ matrix generalizes the S and T matrices found earlier, and A is an element of the modular group. It satisfies:

$$U(A)U(A') = e^{i\phi(A,A')}U(AA'), \qquad U(A)U^\dagger(A) = \mathbf{1} \tag{4.6.5}$$

To prove the modular invariance of Z, it can be shown that this means the N matrix of Eq. (4.6.1) must satisfy:

$$NU(A) = U(A)N \tag{4.6.6}$$

Finding the general solution of this equation is rather difficult and not very transparent, but the result is quite elegant. What is remarkable is that the final classification is so simple, corresponding in a one-to-one fashion with the A–D–E classification of Lie algebras. Each solution to the above equation corresponds to one of the simply laced Lie groups [18–22].

Let us write these modular invariant characters. We use the symbol $Z(A, D, E)$ to represent the modular invariant that can be placed in correspondence with one of the simply laced Lie groups. Then, we find:

$$Z(A_{k+1}) = \sum_{\lambda=1}^{k+1} |\chi_\lambda|^2; \qquad k \geq 1$$

$$Z(D_{2\rho+2}) = \sum_{\lambda\,\mathrm{odd}=1}^{2\rho-1} |\chi_\lambda + \chi_{4\rho+2-\lambda}|^2 + 2|\chi_{2\rho+1}|^2; \qquad k = 4\rho, \quad \rho \geq 1$$

$$Z(D_{2\rho+1}) = \sum_{\lambda\,\mathrm{odd}=1}^{4\rho-1} |\chi_\lambda|^2 + |\chi_{2\rho}|^2$$

$$+ \sum_{\lambda\,\mathrm{even}=2}^{2\rho-2} (\chi_\lambda \chi_{4\rho-\lambda}^* + \mathrm{c.c.}); \qquad k = 4\rho - 2, \quad \rho \geq 2$$

$$Z(E_6) = |\chi_1 + \chi_7|^2 + |\chi_4 + \chi_8|^2 + |\chi_5 + \chi_{11}|^2; \qquad k + 2 = 12$$

$$Z(E_7) = |\chi_1 + \chi_{17}|^2 + |\chi_5 + \chi_{13}|^2$$
$$+ |\chi_7 + \chi_{11}|^2| + |\chi_9|^2; \qquad k + 2 = 18$$

$$Z(E_8) = |\chi_1 + \chi_{11} + \chi_{19} + \chi_{29}|^2 + |\chi_7 + \chi_{13}$$
$$+ \chi_{17} + \chi_{23}|^2; \qquad k + 2 = 30$$

$$(4.6.7)$$

The deeper reason why this elegant one-to-one correspondence exists between the modular invariants of $SU(2)_k$ and the simply laced Lie groups is still rather obscure and not well understood. (General arguments can be made to show that, given a simply laced Lie group, one can construct modular invariant combinations for $SU(2)_k$. However, this does not explain why all the modular invariant combinations should be generated in this way.)

For completeness, we now present the A–D–E classification of the modular invariants for the minimal series mentioned earlier, where we displayed two infinite series and mentioned the existence of three exceptional cases.

This series for the minimal theory can be placed in correspondence with pairs of simply laced algebras. For example, the three exceptional cases can be placed in correspondence with (A, E).

The complete set of modular invariants for the minimal series are then given by [18–22]:

$$Z(A_{p'-1}, A_{p-1}) = \frac{1}{2} \sum_{r=1}^{p'-1} \sum_{s=1}^{p-1} |\chi_{rs}|^2$$

$$Z(D_{2\rho+2}, A_{p-1}) = \frac{1}{2} \sum_{s=1}^{p-1} \left[\sum_{\substack{r \text{ odd}=1 \\ r \neq 2\rho+1}}^{4\rho+1} |\chi_{rs}|^2 + 2|\chi_{2\rho+1,s}|^2 \right.$$

$$\left. + \sum_{\substack{r \text{ odd}=1}}^{2\rho-1} (\chi_{rs}\chi_{r,p-s}^* + c.c.) \right]; \qquad p' = 4\rho + 2, \quad p \geq 1$$

$$Z(D_{2\rho+1}, A_{p-1}) = \frac{1}{2} \sum_{s=1}^{p-1} \left[\sum_{r \text{ odd}=1}^{4\rho-1} |\chi_{rs}|^2 + |\chi_{2\rho,s}|^2 \right.$$

$$\left. + \sum_{\substack{r \text{ even}=1}}^{2\rho-2} (\chi_{rs}\chi_{p'-r,s}^* + c.c.) \right]; \qquad p' = 4\rho, \quad p \geq 2$$

$$Z(E_6, A_{p-1}) = \frac{1}{2} \sum_{s=1}^{p-1} \left(|\chi_{1s} + \chi_{7s}|^2 + |\chi_{4s} + \chi_{8s}|^2 \right.$$

$$\left. + |\chi_{5s} + \chi_{11s}|^2 \right); \qquad p' = 12$$

$$Z(E_7, A_{p-1}) = \frac{1}{2} \sum_{s=1}^{p-1} \left\{ |\chi_{1s} + \chi_{17s}|^2 + |\chi_{5s} + \chi_{13s}|^2 + |\chi_{7s} + \chi_{11s}|^2 \right.$$

$$\left. + |\chi_{9s}|^2 + [(\chi_{3s} + \chi_{15s})\chi_{9s}^* + c.c.] \right\}; \qquad p' = 18$$

$$Z(E_8, A_{p-1}) = \frac{1}{2} \sum_{s=1}^{p-1} \left(|\chi_{1s} + \chi_{11s} + \chi_{19s} + \chi_{29s}|^2 \right.$$

$$\left. + |\chi_{7s} + \chi_{13s} + \chi_{17s} + \chi_{23s}|^2 \right); \qquad p' = 30$$

$$(4.6.8)$$

(For these modular invariants, we find that the central charge equals $c = 1 - 6(p - p')/pp'$. To construct modular invariant combinations, we can show that the matrix $N_{rs,r's'}$ factorizes in terms of sums over $N_{rr'}$ and $N_{ss'}$ with levels $k = p - 2$ and $k' = p' - 2$. Since p and p' are coprime, this means that they cannot both be even, so that one of modular invariant combinations must have $N = 1$, that is, it must be of the A-type. That is why the A series always appears in each pair of modular invariant combinations.)

4.7. Higher Invariants and Simple Currents

In the last section, although we achieved a complete classification of the modular invariant combinations for $SU(2)_k$, at this point it may seem prohibitive to generalize this calculation for the higher Kac-Moody algebras $SU(N)_k$.

Actually, there is a trick one can use which considerably cuts down the work necessary to generate the higher modular invariant combinations. In fact, we will only use part of the information contained within the fusion rules to calculate these modular invariant combinations.

Let us begin by defining a *simple current* J [23] as a primary field which has the following fusion rule with all other primary fields Φ:

$$J \times \Phi = \Phi' \qquad (4.7.1)$$

This differs from the usual fusion rules in an important way. The crucial observation is that just one primary field, rather than a sum, appears on the right hand side. We define the conjugate field J^c such that it has the fusion rule $J \times J^c = \mathbf{1}$.

Now let us multiply J repeatedly with itself n times:

$$J_n \equiv J \times J \times \ldots \times J \qquad (4.7.2)$$

Since the fusion rules are associative, J_n is also a simple current, that is, $J_n \times \Phi = \Phi_n$, where Φ_n is a single primary field. Soon or later, we will find that this process terminates because $J_N = \mathbf{1}$ for some N. Then the order N of J is the smallest integer N for which $J_N = \mathbf{1}$. The orbit created by repeated multiplication by J thus has N elements, such that $J_0 = J_N = \mathbf{1}$, $J_1 = J$, and $J_n^c = J_{N-n}$. Thus, by repeatedly using the fusion rules as a multiplication operator, we have generated an orbit of simple currents whose elements $\{J_n\}$ form the group $\mathbf{Z}_N$.

(The orbit of J may not exhaust all possible simple currents. In general, other primary fields may generate other orbits of order N_i, which in turn generate the group $Z_{N_1} \otimes Z_{N_2} \otimes \ldots \otimes Z_{N_k}$.)

Now let us analyze the monodromy properties of simple currents a bit more carefully. The fusion rules give us:

$$J(z)J(w) \sim (z - w)^{-\alpha} J_2(w) \qquad (4.7.3)$$

Let us now repeatedly multiply this fusion rule with J. Because $J_N = \mathbf{1}$, we can show that the exponent α must equal r/N for some integer r. We call this integer r the monodromy parameter, which will label the modular invariants we construct using this method.

Let the conformal weight of J under L_0 equal h_J. Since we know the conformal weight of J, then we can calculate the conformal weight h_n of the element J_n:

$$h_n = \frac{rn(N - n)}{2N} \bmod 1 \qquad (4.7.4)$$

for $r = 0, 1, \ldots, N - 1$ for odd N and $r = 0, 1, \ldots, 2N - 1$ for N even. (We have imposed charge conjugation, so that $h_n = h_{N-n}$).

Now that we have calculated the orbit of operators created by repeated multiplication by J, let us now calculate the properties of the orbit created by repeatedly fusing another primary field Φ with J. Its fusion rule gives us $J(z)\Phi(w) \sim (z-w)^{-t/N}\Phi_1(w)$ for some integer monodromy parameter t.

Using the same arguments as above, we can determine the conformal weight of Φ_n, denoted by $h(\Phi_n)$ in terms of the conformal weight $h(\Phi)$ of Φ:

$$h(\Phi_n) = h(\Phi) + \frac{rn(N-N)}{2N} - \frac{tn}{N} \mod 1 \qquad (4.7.5)$$

It is useful to introduce the concept of a conserved charge. The charge of Φ with respect to J is defined to be $Q(\Phi) = t/N \mod 1$. Then we have:

$$Q(\Phi_n) = \frac{t}{N} + \frac{rn}{N} \qquad (4.7.6)$$

We also define $Q_n(\Phi) = nQ \mod 1$ to be the charge of Φ with respect to the primary field J_n.

This charge is important for several reasons. First, if we take the fusion rule $\Phi_i \times \Phi_j = \sum_k C_{ijk}\Phi_k + ...$, we find that

$$Q(\Phi_i) + Q(\Phi_j) = Q(\Phi_k) \qquad (4.7.7)$$

for all fields which appear in the fusion rule. In other words, the charge in conserved under the fusion rule.

Secondly, the charge $Q(\Phi)$ appears when we carry a field around a twist field. If we define a twist field as $T(z,\bar{z}) = J(z)J^c(\bar{z})$, then we can calculate the effect that this twist operator has when we carry a primary field $\Phi_{ij}(z,\bar{z})$ (with left (right) moving conformal weight labeled by $i(j)$) around the twist field. In general, we pick up the phase $\exp(2\pi Q i)$ where the total charge Q is defined as $Q = Q(\Phi_i) + Q(\Phi_j)$.

So far, our discussion of simple currents has been rather formal. We now come to the heart of this construction. If we are given any modular invariant partition function (e.g. the trivial, diagonal one), we can form yet another modular invariant function in a simple way. First, remove all states in the diagonal sum which are not invariant under Q (that is, those which do not have integer charge) and then add all twisted sectors (all those obtained from the untwisted sector by acting with twist T).

Notice that this is a generalization of the usual trick used in constructing modular invariant partition functions on orbifold spaces, that is, we begin with a known modular invariant function on a given space, and then modify the sum by including contributions with different boundary conditions. In this way, new modular invariant combinations defined on orbifolds can be defined in terms of known modular invariants.

There are several advantages to this approach. First, we do not have to know the entire set of fusion rules, just the set of fusion rules for the orbits.

Second, the group structure of the orbits is trivial, given by $\mathbf{Z}_N$. Third, we can always generate new modular invariant combinations starting from a known invariant, such as the diagonal one. The resulting modular invariants will, in general, be non-diagonal.

When we employ this trick, we will find that if a particular conformal field theory has center $\mathbf{Z}_N$, then different modular invariant combinations are generated for every divisor of N if N is odd or if N and r are both even. In the case where N is even and r is odd, then this method generates different modular invariant combinations for every divisor of $N/2$. (The standard diagonal invariant $M_{ij} = \delta_{ij}$ corresponds to the divisor being 1.)

Let us examine how we turn a known modular invariant into another one by this procedure. For example, if $N = 9$ and $r = 3$, we can convert the diagonal modular invariant into a non-diagonal one.

For $\mathbf{Z}_1$, we have:

$$\sum_{\text{all orbits}} \sum_{i=0}^{8} |\chi_i|^2 \tag{4.7.8}$$

For $\mathbf{Z}_3$, we have:

$$\sum_{Q_3=0 \, \text{orbits}} |\chi_0 + \chi_3 + \chi_6|^2 + |\chi_1 + \chi_4 + \chi_7|^2 + |\chi_2 + \chi_5 + \chi_8|^2 \tag{4.7.9}$$

For $\mathbf{Z}_9$, we have:

$$\sum_{Q_3=0 \, \text{orbits}} |\chi_0 + \chi_3 + \chi_6|^2 + [(\chi_1 + \chi_4 + \chi_7)(\chi_2 + \chi_5 + \chi_8)^* + c.c.] \tag{4.7.10}$$

Let us now be more precise. Let J_n be of order N in $\mathbf{Z}_N$. We begin by postulating that the following combination is modular invariant: $\sum_{ij} M_{ij} \chi(\Phi_i)\chi^*(\Phi_j)$. Then we make the appropriate twists, so the modular invariant turns into the combination:

$$\frac{1}{N} \sum_{l=0}^{N-1} e^{2i\pi l[Q_n(\Phi_i)+Q_n(\Phi_j)]} \sum_{k=0}^{N-1} M_{ij} \, \chi(J_{nk}\Phi_i)\chi^*(J_{nk}^c \Phi_j) \tag{4.7.11}$$

Notice that we have made two important changes in the original modular invariant combination. First, the sum over l projects onto charge singlet states. Second, the sum over k yields the twisted sector.

From this, we can read off the explicit expression for the new modular matrix M corresponding to the new invariant. We find [23]:

$$M_{J^\alpha a, J^\beta b} = \frac{1}{N} \sum_{l=1}^{N} \delta_{ab}\delta_{\beta,\alpha+l} \sum_{p=1}^{N} \exp\left[2\pi i p\left(Q(a) + \frac{2\alpha+l}{2N}r\right)\right] \tag{4.7.12}$$

where the label $J^\alpha a$ belongs to the orbit of the field Φ_a when we act on it by the current α times. The sum over p has the effect that the expression equals one if the argument within the parenthesis equals 0 mod N and equals zero otherwise. Notice that the argument in the parenthesis is equal to $Q(J^\alpha a) + Q(J^\beta b)$.

The previous equation is the desired expression. It yields new modular invariant combinations based on the method of simple currents. By applying the S and T matrices corresponding to modular transformations, we find that the previous expression is indeed modular invariant. For example, by explicit calculation one can show that $SMS^* = M$, where S_{AB} generates modular transformations.

These results, in turn, can be generalized to include Kac-Moody algebras. For example, let us analyze the affine $SU(N)_k$. Primary fields, as in the classical case, can be characterized by Young tableaux, which is a sequence of boxes representing how we symmetrize or anti-symmetrize the various indices. Let the symbol $[m_1, m_2, ..., m_{N-1}]$ with $m_1 \leq k$ represent the Young tableaux of a primary field where m_i is the length of the ith row in the tableaux.

Consider the primary field $Y_1 = [k, 0, ..., 0]$. One can show that this primary field is a simple current. If we multiply Y_1 with another primary field, then we simply increase the length of the first k columns by one, yielding another Young tableaux. Y_1 is a simple current, and so are the elements of the orbit given by $Y_n = [k, k, k, ...0]$, which has n rows of length k. Furthermore, with a little work, one can show that the fields given by Y_n are the only simple currents. The group generated by the orbit is isomorphic to $\mathbf{Z}_N$, which corresponds to the center of $SU(N)_k$, which is also $\mathbf{Z}_N$. Thus, the center generated by the simple currents corresponds to the center of the corresponding Kac-Moody algebra.

Given the explicit representation of the simple currents of $SU(N)$, we can now calculate the conformal weights of the operators:

$$h_n = \frac{kn(N-n)}{2N} \tag{4.7.13}$$

which agrees with our previous expression for conformal weights if we identify the monodromy parameter r with the level k.

Continuing in this way, we can derive all known results for the modular invariants of the various groups, such as the complete classification for $SU(2)_k$, as well as partial results for $SU(N)_k$. This method also generates new modular invariants.

4.8. Diagonalizing the Fusion Rules

At this point, it appears as if the discussion on characters and modular properties seems divorced from the fusion rules discussed earlier. However, because of the highly restrictive nature of conformal symmetry and because a great deal of information is encoded within the S-matrix, there is a rather remarkable formula found by Verlinde [24] and proven by Moore and Seiberg [25] concerning the relationship between these two concepts. In fact, given the modular S matrix, which governs the modular properties of the characters, we can actually calculate the fusion rules! Specifically, it turns out that the fusion rules, which are determined by the matrix N_{ij}^k, can be written in terms of the S matrix in the following fashion.

First, we will show that S diagonalizes the fusion rules, that is,

$$N_{ij}^k = \sum_n S_j^n \lambda_i^{(n)} S_n^k \qquad (4.8.1)$$

where the $\lambda_i^{(n)}$'s are the eigenvalues of the N matrix. Using this, we can now present the full statement:

$$N_{ij}^k = \sum_n \frac{S_j^n S_i^n S_n^k}{S_0^n} \qquad (4.8.2)$$

This allows us calculate the fusion rules by inserting the modular S matrix into this equation, giving us an independent check on previous results and also giving us new fusion relations. For example, for $SU(2)_k$, we can use the fact that the S matrix is [see Eq. (4.5.21)]:

$$S_{\lambda\lambda'} = \left(\frac{2}{k+2}\right)^{1/2} \sin \frac{\lambda\lambda'}{k+2}\pi \qquad (4.8.3)$$

Inserting this expression into our formula for the N matrix via Eq. (4.8.2), we find that the fusion rules are given by:

$$SU(2)_k : \qquad [\phi_j] \times [\phi_{j'}] = \sum_{j''=|j-j'|}^{min(j+j',k-j-j')} [\phi_{j''}] \qquad (4.8.4)$$

[In passing, we remark that the fusion rules for the minimal model, found in Eq. (2.6.12), can be seen to be related to the $SU(2)_k$ fusion rules if we make the substitution $p_i = 2j_i + 1$ and $q_i = 2j_i' + 1$.]

To prove this remarkable result Eq. (4.8.2), which shows that the S matrix diagonalizes the fusion rules, we first remind ourselves that the character is obtained by tracing over the Verma module associated with the jth primary field $[\phi_j]$:

$$\chi_j(\tau) = \text{Tr}_{[\phi_j]}\left(q^{L_0+\epsilon}\right) \qquad (4.8.5)$$

for $\epsilon = -c/24$ and $q = e^{2\pi i\tau}$. Notice that the S operation changes the ith character into a sum over the jth characters:

$$S: \quad \chi_i\left(-\frac{1}{\tau}\right) = \sum_j S_i^j \chi_j(\tau) \qquad (4.8.6)$$

Our goal is to change the summation over the ith module to the jth module. To do this, we are going to manipulate this expression by changing the basis of the summation in the trace.

Within the trace, let us insert the number one, which does nothing. The trick, however, is to rewrite the number one as the operator product expansion of a primary field ϕ_i and its conjugate field. In the trace operation, let the a cycle represent the line of equal τ, and the b cycle the line of equal σ.

Now, move the ϕ_i field along the τ direction, until it hits the summation over the ϕ_j primary field. When ϕ_i and ϕ_j come close to each other, we must use the fusion rules. Notice that the effect of using the fusion rules is to change the bases of the summation in the trace.

Let us define this operation as $\phi_i(b)$, which is an operator, not a field. Thus, we have:

$$\phi_i(b)\chi_j = N_{ij}^k \chi_k, \qquad \phi_i(a)\chi_j = \lambda_i^{(j)}\chi_j \qquad (4.8.7)$$

In the second equation, we have inserted ϕ_i into the trace and then moved it along an a cycle, that is, a line of equal τ. This does not change the basis of the states at all, but simply inserts a constant matrix into the trace.

There is a big difference between these two expressions. The transformation along the b cycle turns the jth character into a sum over the kth characters, while a transformation along the a cycle simply maps the jth character back into itself.

The important step is now to perform the S operation on both of the above equations. The a cycle and b cycle are interchanged, while the characters change via the S matrix. Thus, the S operation changes these two relations into the following:

$$\phi_i(a)S_j^k \chi_k = N_{ij}^k S_k^l \chi_l, \qquad \phi_i(b)S_j^k \chi_k = \lambda_i^{(j)} S_j^l \chi_l \qquad (4.8.8)$$

By simple manipulations of these two equations, we now have:

$$N_{ij}^k = \sum_n S_j^n \lambda_i^{(n)} S_n^k \qquad (4.8.9)$$

We also know that $N_{i0}^k = \delta_i^k$ (because the fusion of the ith Verma module with the identity again yields the ith Verma module). Putting $j = 0$ into the previous expression, we get:

$$\lambda_i^{(n)} = S_i^{(n)}/S_0^n \qquad (4.8.10)$$

Inserting this expression into the previous one, we now have the desired expression. [We also note that Eq. (4.8.2) can be proven more rigorously using the pentagon and hexagon formula [25] of rational conformal field theory.] It will be helpful to illustrate this with some examples.

Let us take the case when there are no primary fields at all, except for the identity. This may seem trivial, but it is actually quite illustrative of several important principles. In this case, the only primary field is the identity, and we have a simple expression for the action of T and S:

$$T: \quad \chi \to e^{2\pi i(-c/24)}\chi, \qquad S: \quad \chi \to \chi \qquad (4.8.11)$$

so $S = 1$.

We must also satisfy the relation $(ST)^3 = 1$; therefore, we have the constraint $\left[e^{2\pi i(-c/24)}\right]^3 = 1$ or:

$$c = 0 \bmod 8 \qquad (4.8.12)$$

For example, for $k = 1$, this can be satisfied for the affine groups E_8 (where $c = 8$) and for $SO(16)$ (where $c = 16$).

Let us now examine the case when there is only one nontrivial primary field. In this case, the only possible fusion rule is:

$$[\phi] \times [\phi] = 1 + n[\phi] \qquad (4.8.13)$$

Then, a representation of the S and T matrices are as follows:

$$S = \begin{pmatrix} \cos\theta & \sin\theta \\ \sin\theta & -\cos\theta \end{pmatrix}$$
$$T = \begin{pmatrix} e^{2\pi i(-c/24)} & 0 \\ 0 & e^{2\pi i(h-c/24)} \end{pmatrix} \qquad (4.8.14)$$

Now let us impose two constraints. The first is that the S matrix diagonalizes the fusion rules. This easily leads us to $\tan\theta = \lambda$. Then, the second constraint $(ST)^3 = 1$ reduces to:

$$12h - c = 2 \,(\bmod 8), \qquad \cos 2\pi h = -\frac{1}{2}n\lambda \qquad (4.8.15)$$

These values are only defined modulo factors of 8. (This is because we can always tensor any conformal field theory with an independent affine E_8 theory, which has $c = 8$, which does not change the value of h or the fusion rules.)

For $n = 0$, one example is the level 1 $SU(2)$ model. For $n = 1$, some examples include the level 1 G_2 and F_2 WZW models. (For higher n, we find that there are no consistent solutions to the various modular constraints.)

4.9. RCFT: Finite Number of Primary Fields

In this section, we will briefly review the arguments used to show that if the number of primary fields is finite, then the values of h and c must be rational numbers. We call these theories rational conformal field theories (RCFT). Let us begin with a sphere with four punctures, with primary fields ϕ_i located at each of the punctures and then analyze the Dehn twists that one can make on them [26].

Let τ_i equal a Dehn twist where we twist around a circle that wraps around the ith external puncture. Let τ_{ij} equal a Dehn twist where we twist around a circle that wraps around both the ith and jth puncture. By explicitly performing the Dehn twists on a sphere, we can show that the following relation holds:

$$\tau_1 \tau_2 \tau_3 \tau_4 = \tau_{12} \tau_{13} \tau_{23} \qquad (4.9.1)$$

Our strategy is now simple: we will perform the Dehn twists on a tensor defined on the product space of four primary fields. By equating the action of the left-hand side with the action of the right-hand side, we will have an enormously powerful restriction on both the values of h and c.

The action of each τ_i is trivial. The operator $e^{2\pi i (L_0 - \tilde{L}_0)}$ generates the twist. We fix the eigenvalue of L_0 to be equal the eigenvalue of $\tilde{L}_0$ modulo integers. Then, we define the phase:

$$\alpha_i = e^{2\pi i h_i} \qquad (4.9.2)$$

The action of the Dehn twist τ_i on the product space of the primary fields is then just a multiplication by α_i, since the eigenvalue of L_0 (appearing within the twist operator) is h_i.

Let the dimensionality of the product space for the product of four primary fields be labeled N_{ijkl}. Then, the action of the Dehn twists is:

$$\tau_1 \tau_2 \tau_3 \tau_4 \rightarrow \alpha_1 \alpha_2 \alpha_3 \alpha_4 \mathbf{I} \qquad (4.9.3)$$

where $\mathbf{I}$ is the identity operator, which is an $N_{ijkl} \times N_{ijkl}$ matrix. If we take the determinant of this matrix, the answer is:

$$(\alpha_1 \alpha_2 \alpha_3 \alpha_4)^{N_{ijkl}} \qquad (4.9.4)$$

The action of the other Dehn twists on the right-hand side of Eq. (4.9.1), however, is more complicated because the action of τ_{ij} is not diagonal. In general, the action of this Dehn twist mixes up the representations, so that the resulting matrix cannot be simultaneously diagonalized for all such Dehn twists.

The answer is to diagonalize each Dehn twist, one at a time. Take τ_{12}. Now, slice the sphere in half, so that the pairs of punctures 12 and 34 appear on opposite sides of the slice. Place a primary field ϕ_r at the slice.

Notice that we now have two smaller spheres, each with three punctures, with primary fields $1, 2, r$ on one sphere and $3, 4, r$ on the other sphere. The Dehn twist τ_{12} is then represented by $\exp(2\pi i L_0)$, where L_0 acts on the rth primary field. The contribution of this Dehn twist (in the basis where the rth space is diagonal) is given by:

$$\alpha_r^{N_{12r} N_{34r}} \tag{4.9.5}$$

where N_{ijk} are the usual fusion coefficients.

If we take any other Dehn twist and slice along any other channel, we can still use the rth Hilbert space if we use a matrix U that changes basis. Thus,

$$\tau_{23} \rightarrow U\tau_{23}U^{-1} \tag{4.9.6}$$

The action of all three Dehn twists, over all slices, will contain many U matrices. However, if we take the determinant of the resulting product, all of them will conveniently drop out. Thus, taking the determinant of the right-hand side and setting it to the determinant of the left-hand side and equating, we arrive at the final formula:

$$(\alpha_i \alpha_j \alpha_k \alpha_l)^{N_{ijkl}} = \prod_r \alpha_r^{N_{ijkl,r}} \tag{4.9.7}$$

where:

$$N_{ijkl,r} = N_{ijr}N_{klr} + N_{jkr}N_{ilr} + N_{ikr}N_{jlr} \tag{4.9.8}$$

(Each term in the expansion on the right-hand side corresponds to a slice that bisects the sphere in half, such that we place the rth space of primary fields at the slice.)

Now, we make the crucial assumption: the number of primary fields is finite. Assume that there are N primary fields. Then, the previous equation is highly overconstrained. There are only N unknowns (corresponding to the α_r), but there are:

$$(N+1)(N+2)(N+3)(N+4)/4! \tag{4.9.9}$$

relations among them. Our goal is to show that, unless h and c are rational, there is no solution for finite N.

Let us first show that h must be rational. Let us set $i = j = k = l$. Then, there are N equations in N unknowns.

The ith equation reduces to:

$$(\alpha_i)^{4N_{iiii} - N_{iiii,i}} \prod_{r \neq i} \alpha_r^{-N_{iiii,r}} = 1 \tag{4.9.10}$$

If we define the matrix:

$$M_{ir} = \delta_{ri}(4N_{iiii} - N_{iiii,i}) + (1 - \delta_{ri})(-N_{iiii,r}) \tag{4.9.11}$$

then our constraint equation reads:

$$Mh = 0 \bmod 1 \tag{4.9.12}$$

where h is now a column matrix with entries given by h_i. It is not difficult to show that M is invertible and has a nonzero determinant. Then,

$$kh = 0 \bmod 1 \tag{4.9.13}$$

where $k = \det M$. This implies that the h_i are multiples of the inverse of k, so that h_i are rational, as desired. Next, let us show that c is also rational.

We recall that S generates modular transformations that interchange the a and b cycles, while T twists the a cycle. They satisfy:

$$(ST)^3 = 1 \tag{4.9.14}$$

As before, let us take the determinant. One complication is that the representations may or may not be self-conjugate. If they are self-conjugate, then $S^2 = 1$. If they are not, then $S^4 = 1$.

Thus, let us raise the previous equation to the 2nd or 4th power and then take the determinant. We find, in these two distinct cases,

$$\det(T)^6 = 1, \qquad \det(T)^{12} = 1 \tag{4.9.15}$$

Now, treat this equation as an operator equation, operating on the product space of primary fields of the sphere with punctures. Since the eigenvalues of the T matrix are given by $\exp[2\pi i(h_i - c/24)]$, we know that:

$$\det T = \prod_{r=0}^{N} \alpha_r \exp\left(\frac{-2\pi i c}{24}\right) \tag{4.9.16}$$

Raising this to the 6th or 12th power, we then find:

$$\det(T)^6 = \exp\left[-\frac{(N+1)\pi i c}{2}\right] \prod_{r=0}^{N} (\alpha_r)^6 = 1$$
$$\det(T)^{12} = \exp\left[-(N+1)\pi i c\right] \prod_{r=0}^{N} (\alpha_r)^{12} = 1 \tag{4.9.17}$$

Solving for c, we find that it must be rational if h is rational, as desired [26].

Last, before ending this chapter, let us remark on the completeness of our analysis. For tree graphs, it can be shown that the B and F operators and the hexagon and pentagon identities they obey enable us to completely determine the rational conformal field theory. We have replaced the infinite set of Virasoro generators and infinite set of elements within Verma modules to a finite set of equations given by the $6j$ formalism. In some sense, they can be treated as the defining relations for a rational conformal field theory.

At the higher loop level, we see that this is not enough. We must also define the S and T operators. However, it can be shown that if the one-loop graph for a rational conformal field theory is modular invariant, then all higher loop graphs must also be modular invariant using the combined set of operators [27]. This is a gratifying result, because it shows that the braiding and modular operators that we have so patiently constructed in the past few chapters are actually enough to define the entire modular invariant multiloop series. No new operators are necessary. We have, in some sense, finished the program of defining the perturbation series for the rational conformal field theories.

4.10. Summary

Because of the large number of conformal field theories that one can write, we wish to impose physical conditions on them. One of the most important is modular invariance, that is, we wish to subtract those global conformal transformations that cannot be continuously deformed to the identity. Perturbatively, when we sum over inequivalent Riemann surfaces, we must divide by the modular group, or else we will have infinite overcounting.

A torus, for example, can be represented by a parallelogram whose opposite sides are identified. If this parallelogram is placed on the x axis, with one corner at the origin, then the complex parameter τ uniquely specifies the torus. However, if τ is mapped into

$$
\begin{aligned}
T: \quad & \tau \to \tau + 1 \\
S: \quad & \tau \to -1/\tau
\end{aligned}
\tag{4.10.1}
$$

then the torus is mapped into itself. These two transformations, in turn, generate the modular group $SL(2, \mathbf{Z})$, defined by:

$$
\tau \to \frac{a\tau + b}{c\tau + d}, \qquad (a, b, c, d \in \mathbf{Z}, \qquad ad - bc = 1)
\tag{4.10.2}
$$

We wish to calculate the effect of the modular group on a representation of the conformal group. Specifically, we wish to calculate how the characters, which count how many states there are at each level, transform under the modular group.

The character is defined as:

$$
\mathrm{Tr}\, x^{L_0} = \sum_{n=0}^{\infty} x^n \dim V_n = \sum_{n=0}^{\infty} x^n p(n)
\tag{4.10.3}
$$

In general, we will find that modular invariant characters are constructed out of the characters as follows:

$$Z = \sum_{a,b} \chi_a^* N_{ab} \chi_b \qquad (4.10.4)$$

for some constant matrix N_{ab}.

The characters for a free fermion field, $c = \frac{1}{2}$, can be calculated by taking into account the periodic (or antiperiodic boundary conditions). There are four ways in which to specify the boundary conditions, depending on whether the boundary conditions on opposite sides of the parallelogram are periodic (Ramond) or antiperiodic (Neveu–Schwarz).

We can, however, also rearrange them according to the conformal weights. We know that a $c = \frac{1}{2}$ system has three primary fields, with weights 0, $\frac{1}{2}$, $\frac{1}{16}$. By rearranging the above characters, we can write the characters for each conformal weight:

$$\chi_0 = \frac{1}{2}\left[\chi(A,A) + \chi(P,A)\right] = \frac{1}{2}\left(\sqrt{\frac{\vartheta_3}{\eta}} + \sqrt{\frac{\vartheta_4}{\eta}}\right)$$

$$\chi_{\frac{1}{2}} = \frac{1}{2}\left[\chi(A,A) - \chi(P,A)\right] = \frac{1}{2}\left(\sqrt{\frac{\vartheta_3}{\eta}} - \sqrt{\frac{\vartheta_4}{\eta}}\right) \qquad (4.10.5)$$

$$\chi_{\frac{1}{16}} = \frac{1}{\sqrt{2}}\left[\chi(A,P) \pm \chi(P,P)\right] = \frac{1}{\sqrt{2}}\sqrt{\frac{\vartheta_2}{\eta}}$$

The object of this exercise is to write a modular invariant combination of characters. Since a modular transformation reverses the a and b cycles of a torus, it is easy to calculate how the characters transform. Thus, a modular invariant combination is given by:

$$Z_{\text{Ising}} = \frac{1}{2}\left(|\vartheta_3/\eta| + |\vartheta_4/\eta| + |\vartheta_2/\eta| \pm |\vartheta_1/\eta|\right)$$

$$= \chi_0 \bar{\chi}_0 + \chi_{\frac{1}{2}} \bar{\chi}_{\frac{1}{2}} + \chi_{\frac{1}{16}} \bar{\chi}_{\frac{1}{16}} \qquad (4.10.6)$$

Similarly, we can calculate the characters for the minimal models. This can be done by calculating the character for the irreducible Verma module and then successively subtracting out the states given by null modules. The answer, after an infinite series of subtractions, is:

$$\chi_{rs} = q^{-(c-1)/24}\eta(q)^{-1} \sum_{k=-\infty}^{\infty} \left(q^{h_{2mk+r,s}} - q^{h_{2mk+r,-s}}\right) \qquad (4.10.7)$$

Likewise, we can calculate the properties of the minimal model's characters by making modular transformations on them:

$$T: \quad \chi_h(q) \rightarrow e^{2\pi i(h - c/24)}\chi_h(q) \qquad (4.10.8)$$

A bit more difficult is the calculation of the character under an S transformation:

$$\chi_{rs}(\tilde{q}) = \sum_{p,q} S_{rs}^{pq} \chi_{pq}(q) \tag{4.10.9}$$

where:

$$S_{rs}^{pq} = \left[\frac{8}{m(m+1)} \right]^{1/2} (-1)^{(r+s)(p+q)} \sin \frac{\pi r p}{m} \sin \frac{\pi s q}{m+1} \tag{4.10.10}$$

It is now easy to calculate some modular invariant combinations of characters. For example, taking a diagonal combination of characters (as in the free fermion case) yields a modular invariant. Less obvious are combinations such as:

$$Z = \frac{1}{2} \sum_r \sum_{s \text{ odd}} |\chi_{rs} + \chi_{s,m+1-s}|^2 \tag{4.10.11}$$

To find more general modular invariant combinations, it is useful to use the characters of the Kac–Moody algebras, which can be defined as:

$$\chi_{(\lambda)}^k(\theta^i, \tau) = q^{-c_G/24} \, \text{Tr}_{(\lambda),k} \left(q^{L_0} e^{i\theta^i H_0^i} \right) \tag{4.10.12}$$

Notice that it has more parameters because we can trace over both L_0 and the Cartan subalgebra.

The key to constructing these characters is the Weyl–Kac formula, which is a generalization of the classical Weyl formula, which allows us to calculate the dimension of any representation of any Lie group.

The Weyl–Kac character formula is:

$$\text{ch} \, L(\Lambda) = \frac{\sum_{w \in W} \epsilon(w) e^{w(\Lambda+\rho)-\rho}}{\sum_{w \in W} \epsilon(w) e^{w(\rho)-\rho}} \tag{4.10.13}$$

where we sum over Weyl reflections of the root vectors. These reflections are defined as:

$$w_\alpha(\lambda) = \lambda - \alpha \langle \lambda, \alpha \rangle \tag{4.10.14}$$

where the Weyl rotation operator w_α within W is associated with the root α.

When applied to $SU(2)_k$, the Weyl group reduces trivially to reflections around the L_3 axis and translations along this axis. Thus, we find an explicit form for the characters:

$$\chi_\lambda(\tau) = \eta^{-3}(\tau) \sum_{n=-\infty}^{\infty} \left[2n(k+2) + (2j+1) \right] e^{i\pi\tau[2n(k+2)+2j+1]^2/2(k+2)} \tag{4.10.15}$$

where $\lambda = 2j + 1$.

It is now a straightforward exercise to make modular transformations on this character and obtain a representation for the T and S matrices:

$$S_{jj'}^{(k)} = \left(\frac{2}{k+2}\right)^{1/2} \sin \frac{\pi(2j+1)(2j'+1)}{k+2}$$

$$T_{jj'}^{(k)} = \exp\left\{2i\pi\left[\frac{(2j+1)^2}{4(k+2)} - \frac{1}{8}\right]\right\}\delta_{j,j'}$$

(4.10.16)

with $j, j' = 0, ..., k/2$.

The modular invariants that we can construct for the characters for the minimal model and $SU(2)_k$ have a mysterious regularity. In particular, we can place them in one-to-one correspondence with the simply laced A–D–E Lie groups. The origin of correspondence is not well understood.

Because the S matrices contain a large amount of information concerning the conformal field theory, we suspect that many of the properties of the field theory can be rewritten in terms of the S matrix. Specifically, we find that the coefficients found in the fusion rules can be written in terms of the S matrix:

$$N_{ij}^k = \sum \frac{S_j^n S_i^n S_n^k}{S_0^n}$$

(4.10.17)

We see that the S matrix diagonalizes the fusion rules. Because of this relationship, we can independently check many of our results previously obtained for the fusion rules by inserting the S matrix into the above equation. For example, we can derive the fusion rules for $SU(2)_k$:

$$[\phi_j] \times [\phi_{j'}] = \sum_{j''=|j-j'|}^{\min(j+j',k-j-j')} [\phi_{j''}]$$

(4.10.18)

References

1. J. Shapiro, *Phys. Rev.* **D5**, 1945 (1972).
2. K. Kikkawa, B. Sakita, and M. A. Virasoro, *Phys. Rev.* **184**, 1701 (1969).
3. C. S. Hsue, B. Sakita, and M. A. Virasoro, *Phys. Rev.* **D2**, 2857 (1970).

For reviews, see Refs. 4 to 6.

4. P. Ginsparg in *Fields, Strings, and Critical Phenomena*, Elsevier, Amsterdam (1989).
5. J. L. Cardy in *Fields, Strings, and Critical Phenomena*, Elsevier, Amsterdam (1989).
6. J.-B. Zuber, *Fields, Strings, and Critical Phenomena*, Elsevier, Amsterdam (1989).
7. A. Rocha-Caridi, in *Vertex Operators in Mathematics and Physics*, J. Lepowsky, S. Mandelstam, and I.M. Singer, eds., Springer-Verlag, Berlin and New York (1984).
8. A. Rocha-Caridi and N. R. Wallach, *Math. Zeitschr.* **185**, 1 (1984).
9. J. L. Cardy, *J. Phys.* **A17**, L385 (1984).
10. J. L. Cardy, *Nucl. Phys.* **B270 [FS16]**, 186 (1986); **B275**, 200 (1986).
11. D. Friedan and S. Shenker, *Nucl. Phys.* **B281**, 509 (1987).
12. D. Kastor, *Nucl. Phys.* **B280**, 304 (1987).
13. Y. Matsuo and S. Yahikozawa, *Phys. Lett.* **178B**, 211 (1986).

14. A. Erdelyi, *Higher Transcendental Functions*, McGraw Hill, New York (1953).
15. F. Gliozzi, J. Scherk, and D. Olive, *Nucl. Phys.* **B122**, 443 (1983).
16. V. G. Kac and D. Peterson, *Adv. Math.* **53**, 125 (1984).
17. D. Gepner and E. Witten, *Nucl. Phys.* **B278**, 493 (1986).
18. A. Cappelli, C. Itzykson, and J.-B. Zuber, *Nucl. Phys.* **B280**, 445 (1987).
19. A. Cappelli, C. Itzykson, and J.-B. Zuber, *Comm. Math. Phys.* **113**, 1 (1987).
20. D. Gepner, *Nucl. Phys.* **B280 [FS18]**, 445 (1987).
21. C. Itzykson, *Nucl. Phys. Suppl.* **5B**, 150 (1988).
22. A. Kato, *Mod. Phys. Lett.* **B3**, 3918.
23. A.N. Schellekens and S. Yankielowicz, *Phys. Lett.* **227B**, 387 (1989).
24. E. Verlinde, *Nucl. Phys.* **B300**, 493 (1988); R. Dijkgraaf and E. Verlinde, *Nucl. Phys. Suppl.* **5B**, 110 (1988).
25. G. Moore and N. Seiberg, *Phys. Lett.* **212B**, 451 (1988).
26. C. Vafa, *Phys. Lett.* **300B**, 360 (1988); G. Anderson and G. Moore, *Comm. Math. Phys.* **117**, 441 (1988).
27. G. Moore and N. Seiberg, *Lectures on RCFT*, 1989 Trieste Summer School.

Chapter 5

N=2 SUSY and Parafermions

5.1. Calabi–Yau Manifolds

As we have seen, the critical dimension for the bosonic (super)string is 26 (10); therefore, we must compactify the extra dimensions so that we have an acceptable four-dimensional phenomenology. Because, to any order in perturbation theory, the dimension of space–time seems perfectly stable, we must necessarily resort to nonperturbative methods to compactify the unwanted dimensions. However, our techniques for analyzing nonperturbative phenomena are notoriously primitive, and at present there is no way in which nonperturbative phenomena can be systematically analyzed for the string.

Historically, this undesirable situation may be compared with the development of gauge theory itself. Gauge theory was first formulated (using Kaluza–Klein methods) by O. Klein in the 1930s. Its present-day incarnation is due to Yang and Mills, who reformulated the theory in 1954. However, because local gauge invariance was unbroken, it meant that the vector particles were massless and hence unacceptable for any weak interaction phenomenology. The mathematical mechanisms necessary to convert the theory into a useful phenomenological tool were unavailable in the 1950s.

It was not until 20 years later, with the development of spontaneous symmetry breaking, the Higgs mechanism, and the renormalization group, that an acceptable phenomenology became possible. We realize now that gauge theory is the fundamental theory of all particle interactions. The method of spontaneous symmetry breaking made it possible to use $SU(2) \otimes U(1)$ to describe the electro–weak interactions, and the method of the renormalization group made it possible to use $SU(3)$ color to describe the strong interactions.

Today, we may be in a similar situation, where the mathematical tools to break string theory down to four dimensions are simply not available. We will, therefore, have to make certain simple and natural assumptions, without any justification whatsoever. What is remarkable is that, with a few simplifying assumptions, a rich phenomenology that comes remarkably close to describing the real, low-energy world is possible.

At present, compactification schemes [1] come in a bewildering variety of forms, which have given us a rich source of phenomenology:

1. free fermions and free bosons [2–5]
2. orbifolds [6], and
3. Calabi–Yau manifolds [7, 8].

In this chapter, we will concentrate on the last scheme, the Calabi–Yau manifolds.

To see how these manifolds enter into our discussion, our first assumption is that 10-dimensional space–time compactifies to some maximally symmetric manifold M_4, which satisfies:

$$R_{\mu\nu\alpha\beta} = (R/12)(g_{\mu\alpha}g_{\nu\beta} - g_{\mu\beta}g_{\nu\alpha}) \qquad (5.1.1)$$

and some six-dimensional compact manifold, called K, so that:

$$M_{10} \to M_4 \otimes K \qquad (5.1.2)$$

The second assumption is that $N = 1$ space–time supersymmetry survives the compactification process. At present, there is absolutely no physical evidence for supersymmetry. The bosonic partners of the quarks, neutrino, or electron have never been seen. However, supersymmetry is a highly desirable phenomenologically because it solves the "hierarchy," which plagues any unified theory of strong, weak, and electromagnetic interactions. (Briefly, we wish to preserve two different energy scales in any grand unified theory: the GUT energy where unification takes place, which is just short of the Planck energy, and the energy found in our own low-energy world. However, higher order Feynman diagrams will mix these two scales, so the masses of the particles will be unacceptably renormalized. We need a new symmetry, containing the scalar Higgs particles, to keep these two energy scales from mixing. Only supersymmetry, which can put scalar particles and fermions in the same multiplet, can perform this feat.)

For our third assumption, we will postulate about the vanishing of certain fields, which we shall not need in our discussion.

Given these three natural assumptions, we find some powerful results:

1. the manifold K is, in fact, a Calabi–Yau manifold, and M_4 is a Minkowski space. It is possible to find Calabi–Yau manifolds that reproduce the necessary $SU(3) \otimes SU(2) \otimes U(1)$ low-energy symmetry group.
2. there must be a hidden, global, $N = 2$ superconformal symmetry. This $N = 2$ symmetry, in turn, is the key to giving us concrete examples of conformal field theories compactified on Calabi–Yau manifolds.

At first, the assumption that $N = 1$ space–time supersymmetry in four dimensions survives the compactification process seems to be an inconsequential one, without much physical content. However, we will shortly see that it is extremely powerful and gives us enormous restraints on the nature of the four- and six-dimensional manifolds.

Let us begin by analyzing the transformation properties of the massless sector of superstring theory [7]. The spin $\frac{3}{2}$ particle, the gravitino, which is the supersymmetric partner of the graviton, transforms as follows under supersymmetry:

$$\delta\psi_i = [\bar{\epsilon}Q, \psi_i] \tag{5.1.3}$$

where $i = 1, 2, ..., 6$. Because the supersymmetric generator Q annihilates the vacuum, the vacuum expectation value of $\delta\psi_i$ must vanish:

$$\langle 0|\delta\psi|0\rangle = 0 \tag{5.1.4}$$

In the classical limit, the variation of the fermionic field and its vacuum expectation value are the same, so we now have:

$$\delta\psi_i = \kappa^{-1}D_i\epsilon + \cdots = 0 \tag{5.1.5}$$

Therefore, ϵ is a covariantly constant spinor. Usually, in flat space, to say that a scalar field is covariantly constant means that it is a constant. However, for spinors in curved space, this is not so; being covariantly constant places restrictions on the spin connection.

Let us now take the derivative of this equation, and antisymmetrize. We find:

$$D_{[i}D_{j]}\epsilon \sim R_{ijkl}(\Gamma^{kl})\epsilon = 0 \tag{5.1.6}$$

where the Γ are Dirac matrices in six dimensions. By contracting indices, we can, in turn, show that:

$$R_{ij}\Gamma^j\epsilon = 0 \tag{5.1.7}$$

that is, the manifold K has a Ricci flat curvature $R_{ij} = 0$.

Furthermore, the fact that ϵ is covariantly constant means that there is a preferred direction in the six-dimensional tangent space. For example, if we take an unconstrained spinor and move it by parallel displacement, we pick up a factor $D_i\epsilon$. If we take this spinor and move it completely around in a circle, we pick up $D_{[i}D_{j]}\epsilon$. Thus, by being parallel displaced around a circle, the spinor has rotated by a certain angle. If we make repeated circular paths, each time coming back to our starting point, then we will generate a group of displacements, which is nothing but $SO(6)$, which is isomorphic to $SU(4)$. The group $SO(6)$ is called the *holonomy group* of the manifold.

Normally, a spinor with $SO(6)$ symmetry has eight components. However, this can be decomposed as $\mathbf{8} = \mathbf{4} \oplus \mathbf{4}$. Under $SU(4)$, these two quartets $\mathbf{4}$ transform with opposite chirality, so we will only take one of them.

The next question is what is the largest subgroup of $SU(4)$ that leaves invariant the **4** of $SU(4)$? By an $SU(4)$ transformation, we can always put ϵ into the following form:

$$\epsilon = \begin{pmatrix} 0 \\ 0 \\ 0 \\ \epsilon_0 \end{pmatrix} \tag{5.1.8}$$

It is now obvious that the largest subgroup of $SU(4)$ that leaves ϵ invariant is the 3×3 subgroup of $SU(4)$, that is, $SU(3)$. Furthermore, out of the covariantly constant tensor, we can show that the following object is a true tensor:

$$J^i_j = -ig^{ik}\bar{\epsilon}\Gamma_{kj}\epsilon \tag{5.1.9}$$

This tensor plays a key role in the analysis of manifolds, because it has interesting properties:

$$J^j_i J^k_j = -\delta^k_i, \qquad D^i J^k_i = 0 \tag{5.1.10}$$

The first property of the tensor J^k_i is analogous to the number i found in ordinary complex variable theory, which squares to -1. In fact, whenever one can write the tensor J^k_i on a manifold, whose square is -1, it means that the manifold is *almost complex*. (To be fully complex, one has to show that all its transition functions are holomorphic).

The second property of this tensor, that its covariant derivative is zero, means that the manifold is *Kähler*, that is, its metric (in complex coordinates) can always be written as the derivative of a single potential function:

$$g_{i\bar{j}} = \frac{\partial^2 \phi(z_k, \bar{z}_k)}{\partial z_i \, \partial \bar{z}_j} \tag{5.1.11}$$

Thus, from the rather simple assumptions we mentioned earlier, we conclude that the manifold K possesses a large set of stringent properties, which collectively identifies it as a *Calabi–Yau manifold* [7, 8].

In general, these Calabi–Yau manifolds are quite complicated. However, it is possible to write examples of such manifolds and give their topological invariants. One way of constructing such manifolds with $SU(3)$ holonomy is to consider the complex projective space CP_N, which is a complex $N + 1$-dimensional space, where the points Z_i are identified with λZ_i for some nonzero complex number. The simplest six-dimensional Calabi–Yau is then CP_4 (which is eight dimensional) with the following complex constraint:

$$\sum_{i=1}^{5} z_i^5 = 0 \tag{5.1.12}$$

In this way, by taking CP_N and properly placing enough constraints, one may obtain a series of Calabi–Yau manifolds.

Although Calabi-Yau manifolds are in general exceedingly complicated, without explicit expressions for their metric tensor, what is remarkable is that one is able to compute many of their important phenomenological properties. We will conclude this section with a short discussion of how to compute one of the most important phenomenological properties of these manifolds, their Yukawa couplings [9–10]. From this, we can extract a vast number of phenomenological predictions from very general arguments.

We will study one of the most interesting Calabi-Yau manifolds, due to Tian and Yau [11], which has precisely three generations of fermions.

This manifold K is described by two sets of four complex co-ordinates x_i and y_i defined on $CP_3 \otimes CP_3$ (that is, the point x_i and y_i is identified with λx_i and $\lambda' y_i$ for complex λ and λ'). This space is subject to the constraints:

$$p_1 = \sum_{i=0}^{3} x_i^3 = 0; \quad p_2 = \sum_{i=0}^{3} y_i^3 = 0; \quad p_3 = \sum_{i=0}^{3} x_i y_i = 0 \qquad (5.1.13)$$

This space has Euler number -18. However, this number can be reduced by considering the $\mathbf{Z}_3$ symmetry (for $\alpha = e^{2\pi i/3}$):

$$(x_0, x_1, x_2, x_3) \to (x_0, \alpha^2 x_1, \alpha x_2, \alpha x_3)$$
$$(y_0, y_1, y_2, y_3) \to (y_0, \alpha y_1, \alpha^2 y_2, \alpha y_3) \qquad (5.1.14)$$

Then the reduced manifold $K/\mathbf{Z}_3$ has Euler number $-18/3 = -6$, which gives us 3 fermion generations.

Since it is difficult to calculate the Yukawa couplings without a knowledge of the explicit form of the metric tensor, we use a trick, exploiting the fact that one can write everything in terms of the three constraint equations p_i.

To extract the Yukawa couplings, we examine the low-energy limit of the heterotic superstring theory, which yields ordinary supergravity coupled to matter fields. We will find that the Yukawa couplings are contained within the point particle supergravity fermion-boson couplings $\bar{\psi}\gamma \cdot D\psi$. Written out explicitly, this coupling is given by:

$$L = \int d^{10}w \sqrt{-g}\, \bar{\psi}_A \gamma^m \psi_B A_{mC} f^{ABC} \qquad (5.1.15)$$

where A, B, C are $E_8 \otimes E_8$ indices and m is a 10 dimensional Lorentz index.

A vast number of simplifications occurs when we power expand this expression in terms of harmonics defined on the product manifold $K \otimes M_4$. The original 10 dimensional space, labeled by w, splits into the 4 dimensional x space and 6 dimensional y space. Therefore, the zero modes of a field $A(w)$ can be power expanded as:

$$A(w) = \sum_i A^i(x) \otimes A^i(y) \qquad (5.1.16)$$

where we sum over harmonics. Furthermore, because of supersymmetry, we can, to lowest order, write ψ in terms of the vector field via: $\psi^a = A_{\bar{m}}^a \gamma^{\bar{m}} \zeta_+$ where $\gamma \zeta_\pm = \pm \zeta_\pm$ and $\gamma \equiv i \gamma_5 ... \gamma_{10}$.

Splitting the x and y integration, we have:

$$L = g_{ijk} \int d^4x \, \sqrt{-g} \, \bar{\psi}_i^{\bar{A}} \gamma^{\bar{m}} A_{\bar{m}j}^{\bar{B}} \psi_k^{\bar{C}} d_{\bar{A}\bar{B}\bar{C}}$$

$$g_{ijk} = \int d^6y \, \sqrt{g} \, \omega^{\bar{m}\bar{n}\bar{p}} A_{\bar{m}ai} A_{\bar{n}bj} A_{\bar{p}ck} \epsilon^{abc} = \int_K \omega \wedge A_{ai} \wedge A_{bj} \wedge A_{ck} \epsilon^{abc}$$

$$(5.1.17)$$

where $A_{ai} = A_{\bar{m}ai} d\bar{z}^{\bar{m}}$ and where $\omega_{mnp} = \zeta_+^T \gamma_m \gamma_n \gamma_p \zeta_+$. $\bar{A}, \bar{B}, \bar{C}$ are now E_6 indices in the **27** representation, $d_{\bar{A}\bar{B}\bar{C}}$ is the symmetric cubic invariant in the **27** of E_6. a, b, c are $SU(3)$ tangent space indices. m and $\bar{m}$ represent the space indices of the six dimensional manifold.

Notice that all dependence on the fermion field ψ has been collected into the term ω, so that the dependence of the Yukawa coupling g_{ijk} rests entirely on the gauge field A_{ai}.

Although the problem of calculating the Yukawa coupling g_{ijk} at first seems intractable for an arbitrary Calabi-Yau manifold, we can use several more tricks. First, the gauge field A^a is a closed one-form, modulo exact forms, and hence spans the cohomology space H^1. Our task is to re-write these gauge fields in terms of the topological properties of the manifold. To do this, we will use a result from deformation theory [10].

The key observation from deformation theory is that there is a one-to-one correspondence between the elements of H^1 and linearly independent polynomials or deformations q^a that one can add to the defining polynomials p_i. (This correspondence can be seen by noting that different choices for the defining polynomials give rise to physically distinct but topologically equivalent vacua.)

Remarkably, we can re-write the gauge fields A^a in terms of these polynomial:

$$A^a = \chi_{bc}^a q^c dx^{\bar{b}} \tag{5.1.18}$$

where χ_{bc}^a is the extrinsic curvature (which is a known function of the constraint polynomials p^a). One can show that the right hand side is a closed one-form and also spans H^1. We have now made the crucial transition, expressing the gauge field A^a in terms of purely geometric quantities, that is, the polynomials q^a.

There is some arbitrariness, of course, in how we choose the q^a. Since the physics remains the same when we make a gauge transformation on the gauge field, the physics must also remain the same if we make a certain change in the polynomial q^a. More precisely, we can always maintain the properties of H^1 by adding to q^a any linear combination of the original constraints p^a and its derivatives $p_{,A}^a$ with respect to x_i and y_i:

$$q^a \sim q^a + X^A p^a_{,A} + c^{ab} p^b \tag{5.1.19}$$

where X^A and c^{ab} are constant coefficients.

For the case in question, we find that there are nine linearly independent q^a.

Now let us insert the expression for the three gauge fields appearing in Eq. (5.1.18) in terms of three polynomials q^a, r^a, and s^a into the definition of the Yukawa couplings appearing in Eq. (5.1.17). Since we are only interested in the ratio between between different Yukawa couplings, we can factor out unessential terms, so that Eq. (5.1.17) becomes:

$$g_{ijk} \sim \int q^{(a} r^b s^{c)} e^{abc}_{ijk} .... \tag{5.1.20}$$

All Yukawa couplings are now defined in terms of the symmetrized product of three polynomials $q^{(a} r^b s^{c)}$. Because of the degree of freedom in choosing q^a, we can always add combinations of the original constraints p^a and their derivatives $p^a_{,A}$ to this symmetrized polynomial and still preserve the desired properties:

$$q^{(a} r^b s^{c)} \sim q^{(a} r^b s^{c)} + X^{A(ab} \left(p^{c)}_{,A} \right) + c^{abc}_d p^d \tag{5.1.21}$$

In general, different choices of the polynomials $q^{(a} r^b s^{c)}$ yield different Yukawa potentials. However, by repeatedly using these equivalence relations, we find that they can all be set equal to the same polynomial, given by $\prod_{i=1}^{3} x_i y_i$. In other words, no matter which combination of symmetric polynomials we started with, we can use the equivalence relation to reduce them to the same polynomial, times a constant κ:

$$q^{(a} r^b s^{c)} \sim \kappa(q, r, s) \prod_{i=0}^{3} x_i y_i \tag{5.1.22}$$

Notice that the Yukawa coefficients are now just encoded within the $\kappa(q, r, s)$. Different choices for the symmetric polynomials correspond to different $\kappa(q, r, s)$.

Our strategy to calculate numerical values for the Yukawa coefficients is now a follows. A particular choice of lepton and meson fields yields a particular choice of polynomials q^a, r^a, s^a. We then take the symmetrized product of these three polynomials, use equivalence relations to reduce it down to the monomial $\prod_i x_i y_i$, and then calculate the coefficient $\kappa(q, r, s)$. In this way, we can calculate the numerical ratio between any two Yukawa couplings, which was our goal. (There is one technical point: we must also calculate the kinetic terms for each of these fields and diagonalize and normalize them properly. This is easily done by a simple normalization of the fields.)

In summary, we have now given a geometrical derivation of the Yukawa couplings without using an explicit form for the metric tensor of the Calabi-Yau space. The Yukawa couplings $g_{ijk} \sim \kappa(q, r, s)$ depend only on the constraint polynomials p_i, which define the space, and the particular choice of polynomials q^a, which define the isospin of the field we are analyzing. From these couplings, of course, we can extract a wealth of phenomenological information about the theory.

As one can see, the Calabi–Yau manifolds are quite complicated and difficult to construct because of their high degree of nonlinearity. Originally, it was thought that this complication would prevent any simple analysis of their conformal properties. Thus, it was quite remarkable when Gepner used a naive tensoring of the $N = 2$ minimal series to generate conformal field theories that had all the properties of the Calabi–Yau compactification. A seemingly intractable, nonlinear problem was reduced to simple representations of $N = 2$ superconformal models. In this chapter, we will discuss the ramifications of this result. The calculation of Kazama and Suzuki, for example, shows that Gepner's original construction could be generalized to yield $N = 2$ models with the gauge group $E_8 \otimes E_6$, which is phenomenologically desirable.

5.2. N=2 Superconformal Symmetry

Although the $N = 2$ superconformal symmetry is not a symmetry of physical states, the study of such models has proven to be a key aspect of realistic phenomenology. The $N = 2$ superconformal algebra differs from the usual $N = 1$ algebra in that there are now two distinct fermionic components to the energy-momentum tensor, $G_{\pm}(z)$, and that there is also another $U(1)$ current $J(z)$. The operator product expansion is given by:

$$T(w)T(z) \sim \frac{c/2}{(w-z)^4} + \frac{2T(z)}{(w-z)^2} + \frac{\partial T(z)}{w-z}$$

$$T(w)J(z) \sim \frac{J(z)}{(w-z)^2} + \frac{\partial J(z)}{w-z}$$

$$T(w)G_{\pm}(z) \sim \frac{3G_{\pm}(z)}{2(w-z)^2} + \frac{\partial G_{\pm}}{w-z}$$

$$J(w)J(z) \sim \frac{c/3}{(w-z)^2} \tag{5.2.1}$$

$$J(w)G_{\pm}(z) \sim \pm \frac{G_{\pm}(z)}{w-z}$$

$$G_{+}(w)G_{-}(z) \sim \frac{2c/3}{(w-z)^3} + \frac{2J(z)}{(w-z)^2} + \frac{2T(z) + \partial J(z)}{w-z}$$

Written in components, this algebra can be written as:

$$[L_n, L_m] = (n-m)L_{n+m} + (1/4)\tilde{c}(n^3 - n)\delta_{n+m,0}$$
$$[L_n, G_r^i] = (n/2 - r)G_{n+r}^i$$
$$[L_m, J_n] = -nJ_{m+n}$$
$$[J_m, J_n] = \tilde{c}m\delta_{m,-n} \tag{5.2.2}$$
$$[J_m, G_n^i] = i\epsilon^{ij}G_{m+n}^j$$
$$\{G_r^i, G_s^j\} = 2\delta^{ij}L_{r+s} + i\epsilon^{ij}(r-s)J_{r+s} + \tilde{c}\left[r^2 - (1/4)\right]\delta^{ij}\delta_{r,-s}$$

where $\tilde{c} = c/3$ and $G_\pm \sim G^1 \pm iG^2$.

Normally, we consider $N = 2$ superconformal symmetry to be unphysical. However, we now come to the interesting but unexpected observation that $N = 1$ (space–time) supersymmetry can give rise to a global $N = 2$ (world sheet) superconformal symmetry [12–15]. Since $N = 1$ space–time supersymmetry is the key to modular invariance (via the GSO projection), we see that $N = 2$ superconformal symmetry may play a key role in constructing physically realistic conformal field theories. Not only does the $N = 2$ theory arise from $N = 1$ space–time supersymmetry, but it is also useful for compactifying on Calabi–Yau manifolds.

We recall that the NS–R superstring has local $N = 1$ superconformal symmetry and $N = 1$ space–time supersymmetry. However, after compactification, we can show that there is a hidden $N = 2$ superconformal symmetry that emerges. We can see this is several ways, at the level of operators [14] or at the level of the sigma model [15].

First, let us show operatorially how this hidden $N = 2$ superconformal symmetry emerges. This operator approach is rather interesting, because this hidden $N = 2$ symmetry emerges rather unexpectedly when one starts with operators transforming only under the smaller $N = 1$ superconformal symmetry.

Let us write the operator expression for the $N = 1$ space–time supersymmetry operators after compactification:

$$Q_{-1/2;\alpha}(z) = e^{-\phi/2}S_\alpha\Sigma(z)$$
$$Q_{1/2;\dot{\alpha}}(z) = e^{-\phi/2}S_{\dot{\alpha}}\Sigma^\dagger(z) \tag{5.2.3}$$

where:

$$S_\alpha = e^{i\alpha H}, \qquad S_{\dot{\alpha}} = e^{i\dot{\alpha}H} \tag{5.2.4}$$

where:

$$\alpha = \left(\pm\frac{1}{2}, \pm\frac{1}{2}\right), \qquad \dot{\alpha} = \left(\pm\frac{1}{2}, \mp\frac{1}{2}\right) \tag{5.2.5}$$

The four-dimensional S_α only uses two bosonized H_i fields, and hence it is a reduction of the spinor that we originally used to establish the full 10-dimensional space–time supersymmetry, which required five bosonized fields in Eq. (3.2.13). Thus, these reduced spinors have conformal weight $\frac{1}{4}$. The other three bosons H_i contained in the original spinor make up the new field Σ. Since the weight of $e^{q\phi}$ is $-\frac{1}{2}q(q+2)$ [see Eq. (3.2.15)], the Σ fields have weight $\frac{3}{8}$.

Let us now insert these expressions for the four-dimensional spinor Q into the proper relations for the superalgebra. To preserve supersymmetry, we find [14]:

$$\Sigma(z)\Sigma^\dagger(w) \sim (z-w)^{-3/4}I + (z-w)^{1/4}\frac{1}{2}J(w) \qquad (5.2.6)$$

where $J(z)$ is a new field, with weight 1. This new field is the candidate for a $U(1)$ field, which is necessary to build up the $N=2$ superconformal symmetry. Thus, although we started with $N=1$ superconformal fields, the larger set of $N=2$ superconformal fields arose from the product of $N=1$ superconformal fields, if we demand $N=1$ space–time supersymmetry.

The operator product of this new field $J(z)$ with the supersymmetric current gives us a new field T'_F with weight $\frac{3}{2}$:

$$J(z)T_F(w) \sim T'_F/(z-w) \qquad (5.2.7)$$

Now that we have two fields with weight $\frac{3}{2}$, let us define two fields $T_F^\pm$ as:

$$\begin{aligned} T_F &= \frac{1}{\sqrt{2}}\left(T_F^+ + T_F^-\right) \\ T'_F &= \frac{1}{\sqrt{2}}\left(T_F^+ - T_F^-\right) \end{aligned} \qquad (5.2.8)$$

By consistency of the algebra, we can then show:

$$T_F^+(z)T_F^-(w) \sim \frac{\hat{c}/4}{(z-w)^3} + \frac{J(w)/2}{(z-w)^2} + \frac{T(w)/2 + \partial_w J(w)/4}{z-w} \qquad (5.2.9)$$

By calculating the other commutators of the algebra, we find that we have the full $N=2$ commutation relations.

It may seem surprising that we could start with the usual $N=1$ fields T and T_F and the space–time supersymmetry operator, which contained Σ, and then construct the $N=2$ fields and their commutators. The key is that these other operators arose from operator products of the known fields. The $U(1)$ field J arose from the operator product of Σ with $\Sigma^\dagger$, and a new field T'_F arose from the product of $\Sigma\Sigma^\dagger T_F$.

There is a second way [15] of seeing how this $N=2$ superconformal symmetry emerges from $N=1$ space–time symmetry, and this is through the sigma model. We begin by noticing that the supersymmetric generators, like the fields, may be left moving or right moving. Thus, we may have p positive chirality supersymmetries and q negative chirality supersymmetries in the same theory. We refer to this as (p,q) supersymmetry.

We use, for example, $(1,0)$ supersymmetry (sometimes called $N=\frac{1}{2}$ supersymmetry) to describe the right-handed supersymmetry of the heterotic string. In this section, we will be concerned with generating a hidden $(2,0)$ supersymmetry starting from an $N=1$ space–time supersymmetry.

Let us begin with a real scalar superfield $\Phi(x, \theta)$, where the bosonic coordinate is $x^\pm$ but there is only one positive-chirality Fermi coordinate, $\theta^+ = \theta$. The generator of $(1, 0)$ supersymmetry is then given by:

$$Q_+ = i\frac{\partial}{\partial\theta} + \theta\frac{\partial}{\partial x^+} \tag{5.2.10}$$

Let us now construct a sigma model with this symmetry based on the superfield:

$$\Phi(x, \theta) = \phi(x) + \theta\lambda(x) \tag{5.2.11}$$

Let us assume that we have a $N = 1$ superconformal action given by:

$$S = -i \int d^2x \, d\theta \Big[g_{ij}(\Phi) + b_{ij}(\Phi)\Big] D\Phi^i \left(\frac{\partial}{\partial x^-}\right)\Phi^j \tag{5.2.12}$$

where D is the supercovariant derivative:

$$D = \frac{\partial}{\partial\theta} + i\frac{\partial}{\partial x^+} \tag{5.2.13}$$

and g_{ij} is symmetric and b_{ij} is antisymmetric.

If we perform the θ integration, then we arrive at:

$$\begin{aligned} S = \int d^2x \, \Big\{ &\Big[g_{ij}(\phi) + b_{ij}(\phi)\Big]\big(\partial_+\phi^i \, \partial_-\phi^j\big) \\ &+ ig_{ij}(\phi)\lambda^i\big(\partial_-\lambda^j + \Gamma^j_{kl}\partial_-\partial^k\lambda^l\big)\Big\} \end{aligned} \tag{5.2.14}$$

where

$$\Gamma^i_{jk} = \left\{ {i \atop j\,k} \right\} + g^{il}T_{jkl} \tag{5.2.15}$$

$$T_{ijk} = \frac{1}{2}(b_{ij,k} + b_{jk,i} + b_{ki,j})$$

where $\left\{ {i \atop j\,k} \right\}$ is the usual Christoffel connection, and T_{ijk} is a totally anti-symmetric torsion.

The action is invariant under the usual $N = 1$ supersymmetry:

$$\delta\phi^i = \epsilon\lambda^i, \qquad \delta\lambda^i = -i\epsilon\, \partial_+\phi^i \tag{5.2.16}$$

Assume, for the moment, that the action is also invariant under a chiral symmetry:

$$\delta\lambda^i = J^i_j(\phi)\lambda^j \tag{5.2.17}$$

Then, the action is also invariant under a second, hidden superconformal symmetry, given by:

$$\begin{aligned} \delta\phi^i &= \epsilon J^i_j(\phi)\lambda^j \\ \delta(J^i_j\lambda^j) &= -i\epsilon\, \partial_+\phi^i \end{aligned} \tag{5.2.18}$$

But, this is precisely the other supersymmetry transformation necessary to obtain $N = 2$ supersymmetry. Therefore, the question is what conditions do we have to place on the manifold in order that we have chiral symmetry and that this new symmetry anticommutes with the first?

In order for the two supersymmetries to anticommute, we must satisfy:

$$J_j^i J_k^j = -\delta_k^i$$
$$N_{ij}^k \equiv J_i^l J_{[j,l]}^k - J_j^l J_{[i,l]}^k = 0 \tag{5.2.19}$$

This tensor N_{ij}^k is called the *Nijenhuis tensor*, and its vanishing means that the manifold is complex. Also, for the action to be invariant under the chiral transformation, we must impose:

$$g_{kl} J_i^k J_j^i = g_{lj} \tag{5.2.20}$$

that is, the metric is hermitian, and the complex structure is covariantly constant:

$$\nabla_i J_k^j \equiv J_{k,i}^j + \Gamma_{il}^j J_k^l - \Gamma_{ik}^l J_l^j = 0 \tag{5.2.21}$$

If $b_{ij} = 0$, then K is Kähler. If $b_{ij} \neq 0$, then K is an hermitian manifold with torsion.

In conclusion, we see that there are now two supersymmetry generators that anticommute if the background metric satisfies a few mild constraints. Thus, $N = 2$ world sheet supersymmetry has emerged from $N = 1$ supersymmetry at the level of the σ model.

5.3. N=2 Minimal Series

As before, we will briefly repeat the same steps used earlier to calculate the minimal series of the conformal and $N = 1$ superconformal series. Recall that we first write the Verma modules of the theory, consisting of all ladder operators hitting a vacuum state. Then, we take the matrix element of these Verma modules and form the Kac determinant.

By analyzing the Kac determinant, we can determine for which values of c and h we have irreducible and reducible representations of the $N = 2$ algebra. (We will use the convention that the $\tilde{c}$ appearing in the commutation relations of the $N = 2$ algebra are related to c and $\hat{c}$ of the conformal and $N = 1$ algebras by the following: $\tilde{c} = c/3 = \hat{c}/2$.)

The irreducible representations usually have an infinite number of primary fields. However, we will be interested in the reducible series, where we have sequences of null states. By analyzing these representations, we can truncate to a finite number of primary fields.

To construct the Verma modules of the algebra [16, 17], let us first observe that there are three possible modings of the various oscillator states. In the NS sector, the L_n and J_n are integer moded, while the G_m^i are half-integral moded. We will call this the A sector, for anti-periodic boundary conditions. For the R sector, the L_n, J_n, and G_m^i are all integral moded.

We will call this the P sector, for periodic boundary conditions. There is, however, a third possibility, and that is a twisted case when only L_n and G_m^1 are integral moded, and J_m and G_m^2 are half-integral moded. We will call this the T sector.

(We should mention that it is also possible to generalize the boundary conditions so that we interpolate between the R and NS sectors. For example, we can choose L_n and J_n to be periodic, but $m \in \mathbf{Z} + 1/2 + \eta$ for G_m^i, where η is between 0 and 1. Then $G^i(e^{2\pi i}z) \sim e^{2\pi i \eta}G^i(z)$. This boundary condition is useful when studying the spectral flow between theories, since by varying η we can go between the R and NS sectors. We will not, however, study this generalization.)

This means that we will have three sets of coefficients appearing in the Kac determinant. Let us carefully define each set. Although the details of the construction are quite messy, the final answer for the conformal weights of the primary fields is quite simple. (The reader may skip to the end of this section, where the main results are summarized in Eq. (5.3.16-8).)

A Sector

Let us define the A sector partition function p_A, which generalizes the usual partition function for the bosonic or $N = 1$ string. Because we have two supergenerators G_{-n}^i acting on the highest weight vacuum state, the partition function for an irreducible Verma module is more complicated than that of the usual $N = 1$ theory. The partition function is defined by taking the coefficients $p_A(n, m)$ of a double power expansion of the following power series:

$$\sum_{n,m} p_A(n,m)x^n y^m = \prod_{k=1}^{\infty} \frac{(1+x^{k-1/2}y)(1+x^{k-1/2}y^{-1})}{(1-x^k)^2} \qquad (5.3.1)$$

We also introduce the coefficients $\tilde{p}_A$ and $\tilde{p}_T$:

$$\sum_{n,m} \tilde{p}_X(n,m;k)x^n y^m = \left[1 + x^{|k|}y^{\text{sign}(k)}\right]^{-1}\sum_{n,m} p_X(n,m)x^n y^m \qquad (5.3.2)$$

where X = A, P and:

$$\begin{aligned}
\text{sign}(k) &= +1, & k &> 0 \\
\text{sign}(k) &= -1, & k &< 0 \\
\text{sign}(0) &= \pm 1, & \text{for } P_{\pm}
\end{aligned} \qquad (5.3.3)$$

For the NS sector, the Verma modules are given by the standard set of ladder operators multiplying a highest weight vacuum state.

Because the zero modes are given by L_0 and J_0, the vacuums are labeled by two indices corresponding to their eigenvalues: $|h, q\rangle$. Then the Kac determinant for the super-Verma module generalizes the usual Kac determinant of Eqs. (2.5.6) and (2.7.17):

$$\det M_{n,m}^{A}(\tilde{c}, h, q) = \prod_{1 \le rs \le 2n} (f_{rs}^{A})^{p_A(n-rs/2,m)}$$

$$\times \prod_{k \in \mathbf{Z}+1/2} (g_k)^{\tilde{p}_A[n-|k|,m-\text{sign}(k);k]} \tag{5.3.4}$$

for s even, where:

$$f_{r,s}^{A}(\tilde{c}, h, q) = 2(\tilde{c}-1)h - q^2 - \frac{1}{4}(\tilde{c}-1)^2 + \frac{1}{4}[(\tilde{c}-1)r+s]^2$$

$$g_k^{A}(\tilde{c}, h, q) = 2h - 2kq + (\tilde{c}-1)\left(k^2 - \frac{1}{4}\right) \tag{5.3.5}$$

and where $k \in \mathbf{Z} + \frac{1}{2}$.

Repeating the same steps that we used for the minimal model, we now look for the zeros of the Kac determinant in order to find null states. By eliminating these null states, we find the minimal series.

For the A sector, the condition for the unitary minimal series is given by:

$$\tilde{c} = 1 - \frac{2}{\tilde{m}} < 1$$

$$h = \frac{[jk - (1/4)]}{\tilde{m}} \tag{5.3.6}$$

$$q = \frac{(j-k)}{\tilde{m}}$$

for $\tilde{m} \ge 2$ and $j, k \in \mathbf{Z} + \frac{1}{2}$, and $0 < j, k, j+k \le \tilde{m} - 1$.

P Sector

We can repeat the same steps for the P sector, with a few modifications. The partition function for the P sector is given by:

$$\sum_{n,m} p_P(n,m) x^n y^m = \left(\sqrt{y} + \frac{1}{\sqrt{y}}\right) \prod_{k=1}^{\infty} \frac{(1+x^k y)(1+x^k y^{-1})}{(1-x^k)^2} \tag{5.3.7}$$

For the P sector, the zero modes are given by L_0, T_0, and G_0^i. There are two types of highest weight vacuums, given by $|h, q \mp \frac{1}{2}\rangle_{\pm}$, which in turn satisfy $(G_0^1 \mp iG_0^2)|h, q \mp \frac{1}{2}\rangle_{\pm} = 0$.

The Kac determinant is given by:

$$\det M_{n,m}^{P}(\tilde{c}, h, q) = \prod_{1 \le rs \le 2n} (f_{r,s}^{P})^{p_P(n-rs/2,m)} \prod_{k \in \mathbf{Z}} (g_k^{P})^{\tilde{p}_P[n-|k|,m-\text{sign}(k);k]}$$

$$\tag{5.3.8}$$

where:

$$f_{r,s}^{P}(\tilde{c}, h, q) = 2(\tilde{c}-1)\left(h - \frac{\tilde{c}}{8}\right) - q^2 + \frac{1}{4}[(\tilde{c}-1)r+s]^2$$

$$g_k^{P}(\tilde{c}, h, q) = 2h - 2kq + (\tilde{c}-1)\left(k^2 - \frac{1}{4}\right) - \frac{1}{4} \tag{5.3.9}$$

where $k \in \mathbf{Z}$.

As before, by looking for the zeros of the Kac determinant, we can find the unitary minimal series, which is given by:

$$\tilde{c} = 1 - (2/\tilde{m})$$
$$h = \tilde{c}/8 + (jk/\tilde{m}) \qquad (5.3.10)$$
$$q = \text{sign}(0)(j-k)/\tilde{m}$$

for $\tilde{m} \geq 2$, $j, k \in \mathbf{Z}$, and $0 \leq j - 1, k, j + k \leq \tilde{m} - 1$.

T Sector

Last, we analyze the T sector. Its partition function is given by:

$$\sum_n p_T(n)x^n = \prod_{k=1}^{\infty} \frac{(1+x^k)(1+x^{k-1/2})}{(1-x^k)(1-x^{k-1/2})} \qquad (5.3.11)$$

For the twisted T sector, the zero modes are given by L_0 and G_0^1. The highest weight state $|h\rangle$ actually splits into two states of fermion parity $(-1)^F = \pm 1$.

Then, we have:

$$\det M_{\pm,n}^T(\tilde{c}, h) = (h - \tilde{c}/8)^{p_T(n)/2} \prod_{1 \leq rs \leq 2n} (f_{r,s}^T)^{p_T(n-rs/2)} \qquad (5.3.12)$$

where:

$$f_{r,s}^T(\tilde{c}, h) = 2(\tilde{c} - 2)(h - \tilde{c}/8) + (1/4)\left[(\tilde{c}-1)r + s\right]^2 \qquad (5.3.13)$$

where s is odd.

For the T sector, we have:

$$\tilde{c} = 1 - (2/\tilde{m}), \qquad h = (\tilde{c}/8) + (\tilde{m} - 2r)^2/16\tilde{m} \qquad (5.3.14)$$

for integer $\tilde{m}$ and r such that $2 \leq \tilde{m}$ and $1 \leq r \leq \tilde{m}/2$.

Now, let us summarize what we have learned by this exercise. We find that for all three sectors, we have the important condition for the unitary series:

$$\tilde{c} = 1 - (2/\tilde{m}) \qquad (5.3.15)$$

which can be rewritten as:

$$N = 2 \text{ minimal series}: \qquad c = \frac{3k}{k+2}, \qquad k = 1, 2, \ldots \qquad (5.3.16)$$

so that $c < 3$. As in the bosonic case, we find that the primary fields corresponding to the minimal models are labeled by two integers.

For the A sector, the conformal weight of the primary field was found to be $h = (jk - \frac{1}{4})/\tilde{m}$. Let us make the following substitution of variables:

$\tilde{m} \to k+2$, $j-k \to q$, and $j+k-1 \to l$. Then, the conformal weight $\Delta_{l,q}$ and $U(1)$ charge of the minimal primary fields can be rewritten as:

$$\Delta_{l,q} = \frac{l(l+2)}{4(k+2)} - \frac{q^2}{4(k+2)}, \qquad Q = \frac{q}{k+2} \qquad (5.3.17)$$

for $l = 0, ..., k$ and $q = -l, -l+2, ..., l$. In this case, we have l, the principle quantum number, which ranges from 0 to k. Also, we have the integer q, which labels the $U(1)$ charge and is defined modulo $2(k+2)$.

Similarly, in the P sector, the conformal weight of a primary minimal field was given by $h = (\tilde{c}/8) + jk/\tilde{m}$. By making a similar change of variables, we find that the conformal weight and the $U(1)$ charge is given by:

$$\Delta_{l,q} = \frac{l(l+2)}{4(k+2)} - \frac{(q \pm 1)^2}{4(k+2)} + \frac{1}{8}, \qquad Q = \frac{q \pm 1}{k+2} \mp \frac{1}{2} \qquad (5.3.18)$$

These last two equations for the conformal weights and $U(1)$ charges for the primary fields will be crucial for our later discussion.

5.4. N=2 Minimal Models and Calabi–Yau Manifolds

One of the remarkable surprises coming from the $N = 2$ superconformal field theory is the one-to-one relationship between certain Calabi–Yau manifolds and certain minimal theories [18, 19]. Normally, Calabi–Yau manifolds are so nonlinear and complicated that any simple representation of them seems out of the question. However, we will see that by simply tensoring certain minimal theories that we found earlier, we will find a one-to-one correspondence with certain Calabi–Yau manifolds. (In hindsight, perhaps this is not that surprising. Calabi–Yau manifolds, in some sense, make the minimal assumptions that one can make on a manifold and retain $N = 1$ space–time supersymmetry, while $N = 2$ superconformal field theories, we saw earlier, are also associated with manifolds with $N = 1$ space–time supersymmetry. However, it is still remarkable that this nontrivial correspondence appears so early, at the level of tensoring minimal models together.) In Chapter 7, we will give a heuristic argument which helps to explain the origin of this interesting result.

If we take the naive case of simply tensoring several independent minimal models together, then the central term of this product space is just the sum of the central term of each minimal model. This sum, in turn, must equal $\frac{3}{2}$ times the number of dimensions remaining after compactifying the 10-dimensional NS–R space. Thus, the condition we wish to satisfy is given by [18, 19]:

$$c = \sum_i \frac{3k_i}{k_i + 2} = \frac{3}{2}(10 - D) \qquad (5.4.1)$$

Symbolically, we will denote the tensoring of l conformal field theories, each with k_i, by the notation $(k_1 k_2 \cdots k_l)$. We will now show the equivalence of these minimal models and the Calabi–Yau manifolds by showing that they have the same topological properties, that is, they have the same discrete symmetries, the same fermion generation numbers, etc.

We will first calculate the group of symmetries that will leave this conformal field theory invariant. For each k_i, there is a symmetry that leaves the theory invariant given by Z_{k_i+2}. To see this, let us construct the character associated with the primary fields of the theory, which is indexed by the integers l, q, and s (which indicates if the sector is NS or R). Let us call the character associated with this primary field $\chi_q^{l(s)}$, where s is even (odd) for the NS (R) sector.

Then, construct the following partition function:

$$Z(x,y) = \frac{1}{2} \exp\left[\frac{2\pi i x y}{(k+2)}\right] \times \sum_{l,q,s} \exp\left[\frac{2\pi i x q}{(k+2)}\right] \chi_{q+2y}^{l(s)} \chi_q^{l(s)*} \qquad (5.4.2)$$

where x, y are complex numbers. Then, under the standard modular transformation on τ in Eq. (4.1.13), we find:

$$Z(x,y) \rightarrow Z(ax + by, cx + dy) \qquad (5.4.3)$$

which holds if x, y are both members of Z_{k+2}. (In the next section, we will give an explicit form for these characters and partition functions, using the parafermionic representation, from which this identity can be shown.) Thus, the theory has a discrete Z_{k+2} symmetry due to the fact that x, y can be any of the elements in Z_{k+2}.

For the general case of tensoring many minimal models together, each labeled by k_i, the discrete symmetry of such a product is given by the product of the individual discrete symmetries:

$$G = Z_{k_1+2} \otimes Z_{k_2+2} \otimes \cdots \otimes Z_{k_l+2} \qquad (5.4.4)$$

If all the k_i are the same, then the group G has an additional symmetry group given by S_l, the permutation group of l identical objects. Thus, for l identical conformal field theories and $k + 2 = l$, the group of symmetries is:

$$G = S_l \tilde{\otimes} Z_l^l / Z_l \qquad (5.4.5)$$

where the tilde represents the semidirect product.

Example: (3^5)

An example will help to illustrate this construction. Equation (5.4.1) tells us that if we wish to have physical $D = 4$ theories, then we must set $c = 9$. The simplest $c = 9$ theory can be obtained by tensoring 5 copies of the $k = 3$ theory, which we denote by (3^5).

From the above arguments, this manifold has partition functions that can be labeled by elements of Z_5. We have 5 copies of Z_5, and therefore, we also have the symmetry S_5 due to interchanges of the various Z_5 factors. It's symmetry group is therefore given by $S_5 \tilde\otimes Z_5^5/Z_5$. This group has 75,000 elements.

Now, let us compute the generation number (that is, the number of identical fermion multiplets) for the theory. We recall that the $U(1)$ charge for the ith theory is given by Eq. (5.3.18):

$$Q_i = -\frac{k_i - 2q_i}{2(k_i + 2)} \tag{5.4.6}$$

The ground state Ramond state is the tensor product of the ground state of each of the individual theories. However, the sum of these $U(1)$ charges must equal $\pm\frac{1}{2}$. Assuming $-\frac{1}{2}$, we have $\sum Q_i = -\frac{1}{2}$, or:

$$\sum_{i=1}^{l} \frac{q_i}{k_i + 2} = 1; \qquad 0 \le q_i \le k_i \tag{5.4.7}$$

where we have used the fact that $c = 9 = \sum(3k_i)/(k_i + 2)$.

It is now a simple matter to compute the generation number for various theories. One simply counts the number of ways in which we can choose the integers q_i. Each set of integers $\{q_i\}$ that satisfies Eq. (5.4.7) yields an equivalent fermion generation. For example, the number of ways we can choose 5 integers, each between 0 and 3, such that they sum to 5 is equal to 101. Thus, the generation number for (3^5) is 101. Likewise, the generation number of (6^4) is 149.

Now that we have calculated the discrete symmetries and the number of fermion generations associated with some of these tensored minimal models, let us do the same for the Calabi–Yau manifolds. First, let us compare them to the Calabi–Yau manifold CP_4 constrained by:

$$\sum_{i=1}^{5} z_i^5 = 0 \tag{5.4.8}$$

This surface enjoys the global symmetry:

$$G = S_5 \tilde\otimes Z_5^5/Z_5 \tag{5.4.9}$$

This symmetry group arises because S_5 permutes the different variables z_i within Eq. (5.4.8). Also, the manifold is invariant if we make the following transformation in the constraint equation:

$$z_i \rightarrow e^{2\pi i/5} z_i \tag{5.4.10}$$

(The last factor of Z_5 is due to the fact that the overall phase in CP_4 is irrelevant.) This symmetry group is precisely the same as that found for (3^5). Likewise, the Calabi–Yau manifold specified by:

$$\sum_{i=1}^{n} z_i^n = 0 \qquad (5.4.11)$$

has the symmetry group:

$$S_n \tilde{\otimes} Z_n^n / Z_n \qquad (5.4.12)$$

which is precisely the same as the one found for the minimal model in Eq. (5.4.5).

In addition, the generation number for a manifold can also be calculated topologically. Because the number of fermions that can propagate on a manifold is topologically fixed, the generation number is directly related to the Dirac index, Euler number, or Hodge numbers. We find that the generation number is 101. This agrees exactly with the fermion generation number found for (3^5).

Similarly, we can proceed to establish a one-to-one relationship between certain tensor products of minimal models and other Calabi–Yau manifolds by examining their discrete symmetries, their generation number, their fermion content, etc. [18, 19]. Although this does not constitute a rigorous proof of the correspondence between these theories, it gives overwhelming evidence of their equivalence. In Chapter 7, we will give some arguments which help to reveal the origin of this fascinating result.

5.5. Parafermions

As in bosonic theory, we wish to find a specific representation of the $N = 2$ minimal model by which we can calculate the structure constants of the fusion rules and the correlation functions. In particular, we wish to calculate the character $\chi_q^{l(s)}$ for the primary fields of the minimal model and also the partition function. In this way, we can confirm that the partition function has the Z_{k+2} discrete symmetry in Eq. (5.4.2).

There is a representation of this algebra in terms of *parafermions* [20, 21] that allows us to calculate the main features of this algebra. This representation of the $N = 2$ algebra can be given in terms of the parafermion field ψ_k, used to build up representations of the $\mathbf{Z}_k$ spin models found in statistical mechanics. (The $\mathbf{Z}_k$ symmetry refers to the discrete values that the spins in a lattice can assume. For example, in an ordinary Ising model, the spins can take on ± 1, so the spin assumes values in $\mathbf{Z}_2$. The generalization to the $\mathbf{Z}_k$ model is thus straightforward, depending on the roots of unity.)

Let us postulate the existence of a parafermion field ψ_1 and a free boson field ϕ. Combine them in the following fashion:

$$G_+(z) = \sqrt{\frac{2k}{k+2}}\, \psi_1 e^{i\alpha\phi}$$

$$G_-(z) = \sqrt{\frac{2k}{k+2}}\, \psi_1^\dagger e^{-i\alpha\phi} \qquad (5.5.1)$$

$$J(z) = i\sqrt{\frac{k}{k+2}}\, \partial\phi$$

where $\alpha = \sqrt{(k+2)/k}$.

If we plug this representation back into the $N = 2$ operator product expansion, then we find that the $N = 2$ algebra is satisfied as long as we fix:

$$\psi_1(w)\psi_1^\dagger(z) \sim (w-z)^{-2(k-1)/k} + \cdots \qquad (5.5.2)$$

For $k = 2$, this expression reduces to the usual one for ordinary fermions, as in Eq. (2.2.15). For other values of k, however, this expression shows that the fields are parafermionic. From Eq. (5.5.2), we see by scale arguments that the field ψ_1 has conformal weight $(k-1)/k$. From Eqs. (3.1.28) and (3.1.29), we see that the exponential field $e^{i\alpha\phi}$ has weight $\alpha^2/2 = (k+2)/2k$. If we add them together, we find that the operators $G_\pm$ in Eq. (5.5.1) have weight $\frac{3}{2}$, as desired.

If we calculate the operator product expansion of 2 energy-momentum tensors $T(z)$, then the central charge of the full algebra is the sum of the central charge of the free boson field ϕ, which has $c_\phi = 1$, and the parafermion field, which contributes $c_\psi = 2(k-1)/(k+2)$. (To see this, notice that the parafermionic theory can be obtained from a GKO construction with $SU(2)_k/U(1)$, so the central charge is given by $c = [3k/(k+2)] - 1 = 2(k-1)/(k+2)$.) Thus, the full central charge of the $N = 2$ algebra is given by:

$$c = c_\phi + c_\psi = 1 + \frac{2(k-1)}{(k+2)} = \frac{3k}{k+2} \qquad (5.5.3)$$

which confirms that we have a representation of the minimal series as in Eq. (5.3.16).

Now that we have a representation of the minimal $N = 2$ algebra with the correct central charge in terms of parafermions, we would like to construct the primary fields and the characters of the $N = 2$ algebra out of these parafermions and solve for the entire theory. In particular, we will write a parafermionic representation of the primary field V_m^l. To do this, we find it necessary to introduce the complete set of parafermionic operators, of which ψ_1 is only one member.

In the $\mathbf{Z}_k$ model, we have a set of currents ψ_l, $l = 1, 2, ..., k-1$, which have an operator product expansion given by:

$$\psi_l(w)\psi_{l'}(z) \sim c_{l,l'}(w-z)^{-2ll'/k}\big[\psi_{l+l'}(z) + O(w-z)\big]; \qquad l+l' < k$$

$$\psi_l(w)\psi_{l'}^\dagger(z) \sim c_{l,k-l'}(w-z)^{-2l(k-l')/k}\big[\psi_{l-l'}(z) + O(w-z)\big]; \qquad l' < l$$

$$\psi_l(w)\psi_l^\dagger(z) \sim (w-z)^{-2\Delta_l}\big[I + (2\Delta_l/c)(w-z)^2 T(w) + O(w-z)^3\big] \qquad (5.5.4)$$

where $\psi_l^\dagger = \psi_{k-l}$. In addition, we have the operator product expansion with regard to the energy-momentum tensor:

$$T(w)\psi_l(z) \sim \frac{\Delta_l \psi_l(z)}{(z-w)^2} + \frac{\partial_z \psi_l(z)}{(w-z)} + O(1) \tag{5.5.5}$$

where:

$$\Delta_l = \frac{l(k-l)}{k} \tag{5.5.6}$$

In addition to the current ψ_l, we must also introduce the field ψ_m^l, where:

$$\psi_1(w)\psi_m^l(z) \sim \sum_{j=-\infty}^{\infty} (w-z)^{-m/k+j-1} A_{(m+1)/k-j}\psi_m^l(z)$$

$$\psi_1^\dagger(w)\psi_m^l(z) \sim \sum_{j=-\infty}^{\infty} (w-z)^{m/k+j-1} A_{(1-m)/k-j}^\dagger \psi_m^l(z) \tag{5.5.7}$$

where the A's represent certain operators, whose precise nature is not important to our discussion, and where the dimension of ψ_m^l is given by:

$$\Delta_m^l = \frac{l(l+2)}{4(k+2)} - \frac{m^2}{4k} \tag{5.5.8}$$

The point of introducing these extra fields ψ_m^l associated with the $\mathbf{Z}_k$ model is that we can now write an explicit representation of the primary fields of the $N = 2$ theory, which is given by:

$$V_m^l = \psi_m^l : e^{i\alpha_m \phi(z)} : \tag{5.5.9}$$

We demand that the above expression transform as a primary field under the $N = 2$ algebra, which places a constraint on the constant α_m. Since all operator product expansions for a primary field are completely given, we can calculate the operator product of V_m^l, with the generators $G^\pm$, and the value of α_m. The calculation yields:

$$\alpha_m = \frac{m - [(1/2)\mathrm{sign}(0) + a]k}{\sqrt{k(k+2)}}, \qquad m = \dots, l-2, l$$

$$\alpha_m = \frac{m - [(1/2)\mathrm{sign}(0) + 1 + a]k}{\sqrt{k(k+2)}}, \qquad m = l, l+2, \dots, \tag{5.5.10}$$

where, in the NS sector, we have $a = \frac{1}{2}$ and $\mathrm{sign}(0) = -1$, while in the R sector, we have $a = 0$ and $\mathrm{sign}(0) = \pm 1$.

The final step in the proof that V_m^l are primary fields is to calculate their conformal dimension. Since the conformal dimension of V_m^l is the sum of the dimension of the ψ_m^l and the exponential of the free boson field ϕ, we find that the final dimension is given by the following sum:

$$\Delta(V_m^l) = \Delta_m^l + \frac{1}{2}\alpha_m^2 = \frac{l(l+2)}{4(k+2)} - \frac{m^2}{4(k+2)} \qquad (5.5.11)$$

for the NS sector, with $l = 0, 1, ..., k$ and $m = -l, -l+2, ..., l$. For the Ramond sector, we have:

$$\Delta(V_m^l) = \Delta_m^l + \frac{1}{2}\alpha_m^2 = \frac{l(l+2)}{4(k+2)} - \frac{(m \pm 1)^2}{4(k+2)} + \frac{1}{8} \qquad (5.5.12)$$

Comparing these weights with the weights given by Eqs. (5.3.17) and (5.3.18), we find that we have an exact correspondence, as expected. Thus, we now have an explicit representation of the primary fields of the minimal series in terms of parafermions via Eq. (5.5.9).

Our next goal is to calculate the characters and partition functions associated with each primary field. Although this may seem an arduous task, there are many simplifications because of the relationship between parafermions and the WZW model, whose characters and partition functions we calculated in Chapter 4.

We have seen that the $\mathbf{Z}_k$ currents are sufficient to give us a representation of the minimal model of $N = 2$ superconformal symmetry. However, what is also interesting is that these same parafermion fields can give us a representation of the WZW model. The quantum numbers l, m found in the parafermion model, in fact, have a direct counterpart in the quantum numbers found in the $SU(2)$ model. For example, let us construct the following operators from the parafermions:

$$J^+(z) = \sqrt{k}\,\psi_1(z) : e^{i\phi(z)/\sqrt{k}} :$$
$$J^-(z) = \sqrt{k}\,\psi_1^\dagger(z) : e^{-i\phi(z)/\sqrt{k}} : \qquad (5.5.13)$$
$$J^3(z) = \sqrt{k}\,\partial_z\phi(z)$$

Because we know all the operator products, we can then check that the affine $SU(2)$ operator product relations are satisfied:

$$J^a(w)J^b(z) \sim \frac{kq^{ab}}{(w-z)^2} + \frac{f^{abc}J^c(z)}{(w-z)} + \cdots \qquad (5.5.14)$$

as long as we fix:

$$f^{3++} = -f^{+3+} = -f^{-3-} = \frac{1}{2}f^{+-3} = 1$$
$$q^{33} = \frac{1}{2}q^{+-} = \frac{1}{2}q^{-+} = 1 \qquad (5.5.15)$$

Now that we have established a relationship between the $\mathbf{Z}_k$ parafermion fields and the algebra of the WZW $SU(2)$ model, let us establish a relationship between their primary fields. Let us denote the primary fields of the WZW model by $G_{m,\bar{m}}^{l,\bar{l}}$, obeying the usual relations for a primary field:

$$J_n^a G_{m,\bar{m}}^{l,\bar{l}} = \bar{J}_n^a G_{m\bar{m}}^{l,\bar{l}} = 0$$

$$(J_0^3 - m)G_{m,\bar{m}}^{l\bar{l}} = (\bar{J}_0^3 - \bar{m})G_{m,\bar{m}}^{l,\bar{l}} = 0 \tag{5.5.16}$$

What is remarkable is that we can establish a direct relationship between the primary fields found in the $\mathbf{Z}_k$ parafermion model and the WZW $SU(2)$ model:

$$G_{m,\bar{m}}^{l,\bar{l}} = V_{m,\bar{m}}^{l,\bar{l}} : \exp\left[im\phi(z)/2\sqrt{k} + i\bar{m}\bar{\phi}(\bar{z})/2\sqrt{k}\right] : \tag{5.5.17}$$

where $0 \leq l \leq k$ and $-l \leq m \leq l$.

Let us now calculate the final goal, which is to calculate the characters and partition functions of the superconformal minimal model. At first, the thought of calculating the characters of the $N = 2$ superconformal theory may seem prohibitive, until we realize that we can use some tricks. We will calculate the partition function of the $N = 2$ theory in terms of the partition function of the parafermion theory.

Let us define:

$$Z^{l,m} = \mathrm{Tr}\, e^{2\pi i \tau (L_0 - c/24)} \tag{5.5.18}$$

where we only trace over the states in the Verma module given by specific values of $\{l, m\}$. The key is to realize that, because the primary field V_m^l in Eq. (5.5.9) is a product of a bosonic and parafermionic piece, the operator L_0 is just the sum of the L_0 over the bosonic and parafermionic pieces.

The trace then splits up into the product of these two factors. The product over the bosonic piece yields the usual Dedekind η function, and the other piece yields the trace over the parafermionic field:

$$Z^{l,m} = q^{m^2/(4k)} \eta(q)^{-1} Z_p^{l,m} \tag{5.5.19}$$

where $Z_p^{l,m}$ is the contribution over the parafermionic piece. Using Eqs. (5.5.11) and (5.5.12), we find:

$$Z_p^{l,m} = \mathrm{Tr}\, \exp\left\{2\pi i \tau \left[\frac{l(l+2)}{4(k+2)} + N_c - c/24\right]\right\} = \exp\left(\frac{2\pi i \tau m^2}{4k}\right) c_m^l \tag{5.5.20}$$

where N_c is the number operator, where:

$$L_0 = N_c + \frac{l(l+2)}{4(k+2)} \tag{5.5.21}$$

and:

$$c_m^l(\tau) = \exp\left\{2\pi i \tau \left(\frac{l(l+2)}{4(k+2)} - \frac{m^2}{4k} - \frac{c}{24}\right)\right\} \sum_{n=0}^{\infty} p_n q^n \tag{5.5.22}$$

where $c = 3k/(k+2)$, p_n is the number of states in the irreducible representation with highest weight l, and m is the eigenvalue of J_0^3.

Given the partition function over the parafermion theory, we can now construct the partition function over the $N = 2$ theory. It is given by:

$$\chi_m^{l(s)} = \sum_{j \bmod k} c_{m+4j-s}^l(\tau)\Theta_{2m+(4j-s)(k+2),2k(k+2)}(\tau, 2kz, u) \qquad (5.5.23)$$

where:

$$\Theta_{n,m}(\tau, z, u) = e^{-2\pi i u} \sum_{j \in Z+n/2m} e^{2\pi i \tau m j^2 + 2\pi i j z} \qquad (5.5.24)$$

which is the same Θ function that appeared in Eq. (4.5.18) for the WZW model.

We have now succeeded in our goal: an explicit formula for the character of the primary fields of the superconformal minimal model, obtained using the parafermionic representation. From this, we can, in principle, calculate the modular properties of the $N = 2$ minimal series.

5.6. Supersymmetric Coset Construction

One defect of the previous construction is the presence of extra $U(1)$ factors in the gauge group; these factors are known to cause complications at the loop level [22–23]. However, these extra $U(1)$ factors are difficult to remove in the previous construction.

We will now review a more ambitious construction [24] based on conformal field theories with $c > 3$, which generalizes the previous construction. We recall that, with minimal models, this is impossible to implement without taking tensor products of several minimal models, each with $c < 3$. However, we learned from previous chapters that the GKO coset construction allows for representations with $c > 3$, so we will now investigate supersymmetric coset constructions to search for nonminimal models that are not bound by this restriction.

In order to carry out this ambitious plan, we must generalize our previous discussion by repeating the steps used for the bosonic case.

1. First, we will give the operator product relations for the super-Kac–Moody algebra.

2. Second, we will give the operator product expansion of the coset construction based on these super-Kac–Moody generators.

3. Third, we will construct a $U(1)$ operator out of the fields of the super-Kac–Moody algebra.

4. Last, we will postulate a very general *ansatz* for the $N = 2$ superconformal operators in terms of the super-Kac–Moody superalgebra containing a large number of unspecified coefficients. We then force these operators to obey the correct operator product expansion for

the $N = 2$ algebra. This will determine all the coefficients appearing in the *ansatz*, giving us what are called hermitian symmetric spaces. Among these spaces, we will find phenomenologically interesting coset constructions based on the gauge group $E_8 \otimes E_6$ without the unwanted $U(1)$ factors.

We begin this construction by introducing, in addition to the usual Kac–Moody generator $J^A(z)$, its supersymmetric partner $j^A(z)$. To couple these two currents, we also introduce a Grassmann variable θ, such that:

$$J^A(z, \theta) = j^A(z) + \theta J^A(z) \tag{5.6.1}$$

Let us now write the operator product expansion of these currents:

$$J^A(z_1, \theta_1) J^B(z_2, \theta_2) \sim \frac{k}{2(z_{12})} \delta_{AB} \mathbf{1} + \frac{\theta_{12}}{z_{12}} i f_{ABC} J^C(z_2, \theta_2) \tag{5.6.2}$$

where, as usual, we define:

$$\theta_{12} = \theta_1 - \theta_2, \qquad z_{12} = z_1 - z_2 - \theta_1 \theta_2 \tag{5.6.3}$$

The previous expression is written in shorthand. If we expand the previous product relation in detail, we find the following three relations:

$$j^A(w) j^B(w) \sim \frac{(k/2)\delta_{AB}}{w - z}$$

$$J^A(w) j^B(z) \sim j^A(w) J^B(z) \sim \frac{i f_{ABC} j^C(z)}{w - z} \tag{5.6.4}$$

$$J^A(w) J^B(z) \sim \frac{k/2 \delta_{AB}}{(w - z)^2} + \frac{i f_{ABC} J^C(z)}{w - z}$$

Let us now express the usual $N = 1$ superconformal algebra in terms of the super-Kac–Moody algebra via the Sugawara construction, as we did for the bosonic theory. If we define:

$$T(z, \theta) = \frac{1}{2} G(z) + \theta T(z) \tag{5.6.5}$$

then we have the usual operator product expansion for the $N = 1$ SUSY:

$$T(z_1, \theta_1) T(z_2, \theta_2) \sim \frac{c/6}{z_{12}^3} + \left(\frac{3\theta_{12}}{2z_{12}^2} + \frac{1}{2z_{12}} D_2 + \frac{\theta_{12}}{z_{12}} \partial_2 \right) T(z_2, \theta_2) \tag{5.6.6}$$

where $D_2 = \partial/\partial\theta_2 + \theta_2 \partial/\partial z_2$.

The important step is to use the Sugawara construction to find a representation of the $N = 1$ SUSY in terms of the super-Kac–Moody generators via:

$$T(z, \theta) = \frac{1}{k} : DJ^A(z, \theta) J^A(z, \theta) : + \frac{2i}{3k^2} f_{ABC} : J^A(z, \theta) J^B(z, \theta) J^C(z, \theta) : \tag{5.6.7}$$

With this Sugawara construction, we can compute the product between the $N = 1$ SUSY generators and the super-Kac–Moody generators:

$$T(z_1, \theta_1) J^A(z_2, \theta_2) \sim \frac{\theta_{12}}{2z_{12}^2} J^A(z_2, \theta_2) + \frac{1}{2z_{12}} D_2 J^A(z_2, \theta_2)$$
$$+ \frac{\theta_{12}}{z_{12}} \partial_2 J^A(z_2, \theta_2) \tag{5.6.8}$$

The previous relations show that we do not yet have an acceptable Kac–Moody algebra. Notice that the second relation shows that the two currents, J^A and its supersymmetric partner j^B, are not independent. To make them truly independent, we introduce the modified field $\hat{J}^A(z)$ such that:

$$\hat{J}^A(z) = J^A + (i/k) f_{ABC} : j^B(z) j^C(z) : \tag{5.6.9}$$

so that $\hat{J}^A$ and j^B are independent, that is, $\hat{J}^A(z) j^B(w) \sim 0$.

Then, we can rewrite:

$$T(z) = \frac{1}{k} \big[: \hat{J}^A(z) \hat{J}^A(z) : - : j^A(z) \partial j^A(z) : \big]$$
$$G(z) = \frac{2}{k} \big[j^A(z) \hat{J}^A(z) - \frac{i}{3k} : f_{ABC} J^A(z) J^B(z) J^C(z) : \big] \tag{5.6.10}$$

The central charge emerging from this construction is given by:

$$c_G = \frac{1}{2} \dim G + \frac{\hat{k} \dim G}{\hat{k} + g} \tag{5.6.11}$$

where

$$g = c_2(G); \qquad f_{ACD} f_{BCD} = c_2(G) \delta_{AB} \tag{5.6.12}$$

and $\hat{k} = k - g$. We have now successfully constructed the Sugawara representation of the $N = 1$ SUSY in terms of the super-Kac–Moody algebra. However, our goal is to implement the GKO coset construction so that we may have $c > 3$ (which is impossible for any minimal model).

Let G have a subgroup H. Let the indices of the generators of $G(H)$ be represented by $A(a)$. Also let the generators of the coset G/H be represented by the indices $\bar{a}$. Then, as usual, we have the important relationship:

$$c_{G/H} = c_G - c_H \tag{5.6.13}$$

To be specific, we wish to split the supergenerators as follows:

$$T_G(z) = T_H + T_{G/H}$$
$$G_G(z) = G_H(z) + G_{G/H} \tag{5.6.14}$$

Let us now make one more redefinition:

$$\tilde{J}^a(z) = J^a(z) + (i/k) f_{abc} j^b(z) j^c(z) \tag{5.6.15}$$

Let us redefine:

$$T_H(z) = (1/k)\left[\tilde{J}^a(z)\tilde{J}^a(z) - j^a(z)\partial j^a(z)\right]$$
$$G_H(z) = (2/k)\left[j^a\,\tilde{J}^a(z) - (i/3k)f_{abc}j^a(z)j^b(z)j^c(z)\right] \qquad (5.6.16)$$

With this redefinition, we have:

$$T_{G/H} = \frac{1}{k}\left(\hat{J}^{\bar{a}}\hat{J}^{\bar{a}} - \frac{\hat{k}}{k}j^{\bar{a}}\,\partial J^{\bar{a}} + \frac{2i}{k}\hat{J}^{a}f_{a\bar{b}\bar{c}}j^{\bar{b}}j^{\bar{c}} \right.$$

$$\left. - \frac{2}{k}f_{\bar{a}\bar{p}\bar{q}}f_{\bar{b}\bar{p}\bar{q}}j^{\bar{a}}\,\partial j^{\bar{b}} - \frac{1}{k^2}f_{\bar{a}\bar{b}\bar{c}}f_{\bar{a}\bar{d}\bar{e}}j^{\bar{b}}j^{\bar{c}}j^{\bar{d}}j^{\bar{e}} \right) \qquad (5.6.17)$$

$$G_{G/H} = \frac{2}{k}\left[j^{\bar{a}}(z)\hat{J}^{\bar{a}}(z) - \frac{i}{3k}f_{\bar{a}\bar{b}\bar{c}}j^{\bar{a}}(z)j^{\bar{b}}(z)j^{\bar{c}}(z)\right]$$

We find that the Virasoro generators now close correctly, and that:

$$c_G = \frac{1}{2}\dim G + \frac{(k-g)\dim G}{k}$$
$$c_H = \frac{1}{2}\dim H + \frac{(k-h)\dim H}{k} \qquad (5.6.18)$$

5.7. Hermitian Spaces

So far, we have done almost nothing new. All we have shown is how to write the operator product expansion of the super-Kac–Moody representation and its coset construction. In the next step, we add the real physics: out of the $N = 1$ operators, we construct the most general *ansatz* for the $N = 2$ superconformal generators. By demanding that these generators have the correct operator product expansion, we fix the values of the coefficients appearing in the *ansatz*, which then fixes the structure of the group manifold.

We will now show how to generalize the $N = 1$ SUSY generators, given by $\{G(z), T(z)\}$, to form the $N = 2$ SUSY generators, given by $\{G^i(z), T(z), J(z)\}$, where $J(z)$ is the $U(1)$ generator.

First, we define $G^0 \equiv G_{G/H}$. Second, we define G^1 by writing the most general dimension $\frac{3}{2}$ field, which can be constructed from $J^{\bar{a}}(z)$ and $\hat{J}^{\bar{a}}(z)$:

$$G^1(z) = \frac{2}{k}\left[h_{\bar{a}\bar{b}}j^{\bar{a}}(z)\bar{j}^{\bar{b}}(z) - \frac{i}{3k}S_{\bar{a}\bar{b}\bar{c}}j^{\bar{a}}(z)j^{\bar{b}}(z)j^{\bar{c}}(c)\right] \tag{5.7.1}$$

where $h_{\bar{a}\bar{b}}$ and $S_{\bar{a}\bar{b}\bar{c}}$ are arbitrary constants, subject only to the condition that G^i generate the usual $N = 2$ product expansion.

We can also construct the $U(1)$ current, given by:

$$J(z) = \frac{i}{k}h_{\bar{a}\bar{b}}j^{\bar{a}}(z)j^{\bar{b}}(z) + \frac{1}{k}f_{\bar{c}\bar{d}}f_{\bar{c}dE}\left[\hat{j}^E(z) - \frac{i}{k}f_{E\bar{a}\bar{b}}j^{\bar{a}}(z)j^{\bar{b}}(z)\right] \tag{5.7.2}$$

The key step is to force these operators to reproduce the correct commutation relations of the $N = 2$ algebra. We find the following constraints for the coefficients [24]:

$$\begin{aligned}
h_{\bar{a}\bar{b}} &= -h_{\bar{b}\bar{a}}, \qquad h_{\bar{a}\bar{p}}h_{\bar{p}\bar{b}} = -\delta_{\bar{a}\bar{b}} \\
h_{\bar{a}\bar{d}}f_{\bar{d}\bar{b}e} &= f_{\bar{a}\bar{d}e}h_{\bar{d}\bar{b}} \\
f_{\bar{a}\bar{b}\bar{c}} &= h_{\bar{a}\bar{p}}h_{\bar{b}\bar{q}}f_{\bar{p}\bar{q}\bar{c}} + h_{\bar{b}\bar{p}}h_{\bar{c}\bar{q}}f_{\bar{p}\bar{q}\bar{a}} + h_{\bar{c}\bar{p}}h_{\bar{a}\bar{q}}f_{\bar{p}\bar{q}\bar{b}} \\
S_{\bar{a}\bar{b}\bar{c}} &= h_{\bar{a}\bar{p}}h_{\bar{b}\bar{q}}h_{\bar{c}\bar{r}}f_{\bar{p}\bar{q}\bar{r}}
\end{aligned} \tag{5.7.3}$$

The first two conditions state that $h_{\bar{a}\bar{b}}$ defines an almost complex structure. The third and fourth conditions are satisfied if we set:

$$f_{\bar{a}\bar{b}\bar{c}} = S_{\bar{a}\bar{b}\bar{c}} = 0 \tag{5.7.4}$$

Given this last condition, we can solve for these constraints. Notice that these coefficients, because they determine the way that vectors contract to give other vectors and scalars in Eqs. (5.7.1) and (5.7.2), also determine the structure of the group manifold itself. Thus, forcing these operators to obey an $N = 2$ superconformal algebra places constraints on the group structure of the theory. In particular, a close examination of these constraints shows that they give us what are called "hermitian symmetric spaces" [24].

Fortunately, mathematicians have given us a complete classification of these spaces [25]. It is then a simple matter to calculate the c for each of these spaces and show that solutions exist for $c = 9$ without trivially tensoring minimal models together, as we did in Eq. (5.4.1). Thus, we have successfully found an "irreducible" $c = 9$ generalization to the "reducible" $c = 9$ theory discussed in section 5.4.

The hermitian symmetric spaces corresponding to G/H that we have found, with their central charges, are given by the following [24]:

$$\frac{SU(n+m)}{SU(m) \otimes SU(n) \otimes U(1)} : \quad c_{G/H} = \frac{3\hat{k}mn}{(\hat{k}+m+n)}$$

$$\frac{SO(n+2)}{SO(n) \otimes SO(2)} : \quad c_{G/H} = \frac{3\hat{k}n}{(\hat{k}+n)}, \quad n \geq 2$$

$$\frac{SO(3)}{SO(2)} : \quad c_{G/H} = \frac{3\hat{k}}{(\hat{k}+2)}$$

$$\frac{SO(2n)}{SU(n) \otimes U(1)} : \quad c_{G/H} = \frac{3\hat{k}n(n-1)}{2(\hat{k}+2n-2)} \qquad (5.7.5)$$

$$\frac{Sp(n)}{SU(n) \otimes U(1)} : \quad c_{G/H} = \frac{3\hat{k}n(n+1)}{2(\hat{k}+n+1)}$$

$$\frac{E_6}{SO(10) \otimes U(1)} : \quad c_{G/H} = \frac{48\hat{k}}{(\hat{k}+12)}$$

$$\frac{E_7}{E_6 \otimes U(1)} : \quad c_{G/H} = \frac{81\hat{k}}{(\hat{k}+18)}$$

for $\hat{k} = 1, 2, 3, \ldots$ Notice that we have now broken past the $c = 3$ barrier found in the minimal case.

Given this complete classification, we can now begin to look for phenomenologically acceptable solutions to the string equations of motion. Unfortunately, after compactification to 4 dimensions, the usual 10-dimensional type II string cannot generate gauge groups large enough to include the minimal $SU(3) \otimes SU(2) \otimes U(1)$. A careful examination of the central charges of the type II theory reveals that, after compactification, the complete set of possible gauge groups does not include the minimal gauge group [26].

The alternative is to examine the heterotic string, which has a much larger gauge group, and use a trick exploited in Refs. 18 and 19. This allows us to convert a superstring theory into a heterotic one.

Normally, modular invariant partition functions for the heterotic superstring are extremely difficult to construct. In fact, for the heterotic string in 10 dimensions, the only modular invariant combinations come from the groups $E_8 \otimes E_8$ and $Spin(32)/Z_2$. The difficulty arises because the left and right movers are treated differently in the heterotic string, while modular invariance tends to mix both sectors. Thus, it is a highly nontrivial result that the heterotic string has 2 possible modular invariant isospin groups. We may therefore suspect that the set of modular invariant partition functions for compactified 4-dimensional heterotic strings is extremely restrictive.

Actually, there is a trick to create modular invariant partition functions for 4-dimensional heterotic strings starting with modular invariant partition functions for the ordinary superstring in 10 dimensions. The trick is based upon the fact that the modular invariant partition function for the superstring is invariant if we make a subtle interchange between fermions and bosons. Specifically, by direct calculation, we can show that the modular invariant partition function for the superstring is invariant if we make the following series of transformations:

1. replace the character of $SO(d)$ with $SO(24 + d)$,
2. exchange the singlets and vectors appearing in the sum in the partition function, and
3. reverse the sign of the spinors appearing in the sum.

We also can show that the partition function remains the same if we interchange $SO(d)$ with $E_8 \otimes SO(8+d)$ in this fashion. We are particularly interested in the physical situation where the dimension of the transverse states is $d = 2$, which leads to a 4-dimensional theory. (In the language of Calabi–Yau manifolds, this complicated series of transformations is identical to setting the spin connection and gauge connection equal to each other.)

To use this trick, let us start with an ordinary type II superstring in 10 dimensions, and then compactify it to four dimensions. Because the left- and right-moving sectors are symmetrical, there is no problem in finding modular invariant partition functions in two transverse dimensions in the light cone gauge, that is, $d = 2$. Now let us use the trick of exchanging fermion and boson sectors for the left-moving sector only. The transverse group $SO(2)$ becomes $E_8 \otimes SO(10)$ in the left-moving sector, while the right-moving sector remains the same.

Let us take l copies of the $N = 2$ conformal field theory, so the basic gauge group is actually $E_8 \otimes SO(10) \otimes U(1)^l$, where this last $U(1)^l$ factor poses problems at the higher loop level. However, it will turn out that, upon closer analysis, one of the $U(1)$ factors can combine with $SO(d + 8)$ to produce $E_{5+d/2}$. In other words, the basic gauge group of the theory is:

$$E_8 \otimes E_{5+d/2} \otimes U(1)^{l-1} \qquad (5.7.6)$$

For $d = 2$, we find the desirable gauge group $E_8 \otimes E_6$, multiplied by $U(1)^{l-1}$. Since the whole point of this discussion is to get rid of the extra $U(1)$ factors, the obvious choice is to choose just one $N = 2$ representation. In section 5.4, when we considered tensoring minimal models together, the choice $l = 1$ was forbidden, since one minimal model by itself could not produce a $c = 9$ theory. However, this restriction no longer applies for the hermitian symmetric case, where $c = 9$ is easy to obtain. Thus, this more general construction based on hermitian symmetric spaces gives us a phenomenologically desirable compactification with $E_8 \otimes E_6$, which is large enough to include the minimal group $SU(3) \otimes SU(2) \otimes U(1)$. Thus, this new method enjoys considerable advantages over the earlier one based on tensoring minimal models together.

In summary, although modular invariant heterotic partition functions are notoriously difficult to construct because modular transformations mix up left- and right-moving sectors, we have constructed a phenomenologically desirable compactification to four dimensions. Beginning with the ordinary type II superstring compactified to four dimensions with transverse symmetry $SO(2)$, we have flipped fermion and boson sectors in the left movers to obtain $E_8 \otimes SO(10)$. With l $N = 2$ copies, we obtained the gauge group $E_8 \otimes SO(10) \otimes U(1)^l$. We then absorbed one of the $U(1)$ factors into $SO(10)$ to obtain E_6, leaving a factor of E_6^{l-1} behind. However, for the hermitian symmetric spaces, we have the freedom to choose $l = 1$ and still maintain $c = 9$ (which is impossible if we tensor minimal models together). Thus, we have arrived at a phenomenologically acceptable compactification. This sequence of changes can be summarized as:

$$SO(2) \rightarrow E_8 \otimes SO(10) \otimes U(1)^l \rightarrow E_8 \otimes E_6 \otimes U(1)^{l-1} \rightarrow E_8 \otimes E_6 \quad (5.7.7)$$

(There is an important physical difference between the three steps that we have outlined here. The first and third step, in some sense, were implemented by hand. However, the second step was a consequence of implementing $N = 1$ SUSY.)

Finally, we remark that once all the constraints are in place in the theory, we find that the coset construction produces groups that fall into the following 10 coset categories:

$$
\begin{aligned}
SO(N+2)/SO(n) \otimes SO(2): &\quad (n,k) = (6,6),(12,4) \\
SO(2n)/SU(2) \otimes U(1): &\quad (n,k) = (7,2) \\
Sp(n)/SU(n) \otimes U(1): &\quad (n,k) = (3,4) \\
U(n+m)/U(n) \otimes U(m): &\quad (n,m,k) = (1,4,15),(1,5,9),(1,6,7), \\
&\quad (2,2,12),(2,3,5),(3,3,3)
\end{aligned}
$$
$$(5.7.8)$$

All of these coset constructions have $c = 9$ and produce $E_8 \otimes E_6$ as the fundamental gauge group. Notice that all these models are free of the unwanted $U(1)$ factors, as desired, which make them phenomenologically attractive.

5.8. Summary

In the search for the perturbative vacuum, the Calabi–Yau manifold has emerged as one of the most attractive choices. Although a Calabi–Yau manifold is highly nonlinear, what is surprising is that a naive extension of $N = 2$ superconformal models can generate such complex manifolds.

We begin by postulating that, after compactification, the 10-manifold has compactified to the direct product of a 4-dimensional manifold M_4 and a 6-dimensional manifold K:

$$M_{10} \rightarrow M_4 \otimes K \quad (5.8.1)$$

where M_4 satisfies:

$$R_{\mu\nu\alpha\beta} = (R/12)(g_{\mu\alpha}g_{\nu\beta} - g_{\mu\beta}g_{\nu\alpha}) \qquad (5.8.2)$$

Additionally, we assume that $N = 1$ space–time supersymmetry is preserved after compactification.

The key statement is that the survival of $N = 1$ supersymmetry leads to the vanishing of the variation of the gravitino field:

$$\delta\psi_i = \kappa^{-1} D_i\epsilon + \cdots = 0 \qquad (5.8.3)$$

Thus, ϵ is a constant covariant spinor. This statement is highly nontrivial, for it places enormous constraints on the structure of the manifold. In particular, by taking two covariant derivatives:

$$D_{[i}D_{j]}\epsilon \sim R_{ijkl}(\Gamma^{kl})\epsilon = 0 \qquad (5.8.4)$$

we see that the manifold is Ricci flat.

Furthermore, the constant covariant spinor, which transforms under $SO(6)$, can be written as a **4** of $SU(4)$. By an $SU(4)$ transformation, the spinor can be written as:

$$\epsilon = \begin{pmatrix} 0 \\ 0 \\ 0 \\ \epsilon_0 \end{pmatrix} \qquad (5.8.5)$$

which shows that it has $SU(3)$ holonomy. Thus, K can be shown to be a Calabi–Yau manifold. In summary, we see that one of the key ingredients of this procedure was the assumption that $N = 1$ space–time supersymmetry survives the compactification to four dimensions.

A second consequence of this assumption is that $N = 1$ space–time supersymmetry naturally leads to a global $N = 2$ superconformal symmetry on the world sheet. To see this, remember that the $N = 1$ space–time supersymmetry generators, after compactification, can be written as:

$$Q_{-1/2;\alpha}(z) = e^{-\phi/2}S_\alpha\Sigma(z)$$
$$Q_{1/2;\dot\alpha}(z) = e^{-\phi/2}S_{\dot\alpha}\Sigma^\dagger(z) \qquad (5.8.6)$$

where an additional field Σ is required to get the counting correct.

However, one can show that the operator product expansion of two Σ's, in turn, generates a new field J:

$$\Sigma(z)\Sigma^\dagger(w) \sim (z-w)^{-3/4}I + (z-w)^{1/4}\frac{1}{2}J(z) \qquad (5.8.7)$$

This new field J is precisely the $U(1)$ field of the $N = 2$ superconformal theory. Furthermore, all the generators of the $N = 2$ theory can be successively generated in this way. Thus, quite miraculously, a hidden $N = 2$

superconformal symmetry emerges by the assumption of $N = 1$ space–time supersymmetry.

The $N = 2$ algebra is written as:

$$[L_n, L_m] = (n - m)L_{n+m} + \frac{1}{4}\tilde{c}(n^3 - n)\delta_{n+m,0}$$

$$[L_n, G_r^i] = \left(\frac{n}{2} - r\right)G_{n+r}^i$$

$$[L_m, J_n] = -nJ_{m+n}$$

$$[J_m, J_n] = \tilde{c}m\delta_{m,-n} \tag{5.8.8}$$

$$[J_m, G_n^i] = \epsilon^{ij}G_{m+n}^j$$

$$\{G_r^i, G_s^j\} = 2\delta^{ij}L_{r+s} + i\epsilon^{ij}(r - s)J_{r+s} + \tilde{c}\left(r^2 - \frac{1}{4}\right)\delta^{ij}\delta_{r,-s}$$

where the supercurrent G^i is doubled by the index i.

Just as in the bosonic case, we now construct the Verma modules of the $N = 2$ theory and look for zeros of the Kac determinant, which reads (for the NS sector):

$$\det M_{n,m}^A(\tilde{c}, h, q) = \prod_{1 \leq rs \leq 2n} (f_{rs}^A)^{p_A(n - rs/2, m)}$$

$$\times \prod_{k \in \mathbf{Z}+1/2} (g_k)^{\tilde{p}_A[n - |k|, m - \text{sign}(k); k]} \tag{5.8.9}$$

for s even, where:

$$f_{r,s}^A(\tilde{c}, h, q) = 2(\tilde{c} - 1)h - q^2 - \frac{1}{4}(\tilde{c} - 1)^2 + \frac{1}{4}[(\tilde{c} - 1)r + s]^2$$

$$g_k^A(\tilde{c}, h, q) = 2h - 2kq + (\tilde{c} - 1)\left(k^2 - \frac{1}{4}\right) \tag{5.8.10}$$

and where $k \in \mathbf{Z} + \frac{1}{2}$.

As in the case with ordinary minimal models, we set the determinant to zero in order to find the unitary minimal series:

$$c = \frac{3k}{k+2}, \qquad k + 1, 2, \ldots \tag{5.8.11}$$

so that $c < 3$. By examining the values for h, we find that the conformal weight $\Delta_{l,q}$ and the $U(1)$ charge Q of a primary field for the minimal series are labeled by the integers l and q:

$$\Delta_{l,q} = \frac{l(l+2)}{4(k+2)} - \frac{q^2}{4(k+2)}, \qquad Q = \frac{q}{k+2} \tag{5.8.12}$$

for $l = 0, \ldots, k$ and $q = -l, -l + 2, \ldots, l$.

The most naive superconformal theory is the tensoring of several of these minimal theories, each labeled by k_i. The central charge is the sum of the individual central charges, which in turn must equal $\frac{3}{2}$ times the dimension of the manifold K, giving us:

$$c = \sum_i \frac{3k_i}{k_i + 2} = \frac{3}{2}(10 - D) \tag{5.8.13}$$

Remarkably, this naive tensoring gives us an explicit representation of the Calabi–Yau manifold. If we tensor l identical superconformal field theories, for example, the resulting theory has a discrete symmetry. Each individual theory has a discrete symmetry Z_{k+2}, and multiplying l identical ones creates the permutation symmetry among them, which is S_l, so the final symmetry group is:

$$G = S_l \tilde{\otimes} Z_l^l / Z_l \tag{5.8.14}$$

where the tilde represents the semidirect product.

The number of identical fermion generations can also be calculated for this theory. The $U(1)$ charge for the product theory equals the sum of the $U(1)$ charges of the individual theories, which in turn must be $-\frac{1}{2}$. The total number of fermion generations is thus the total number of set of integers $\{q_i\}$, which are solutions to:

$$\sum_{i=1}^{l} \frac{q_i}{k_i + 2} = 1; \qquad 0 \le q_i \le k_i \tag{5.8.15}$$

One of the simplest theories is (3^5), which has 101 generations.

Now, compare these superconformal theories to Calabi–Yau manifolds. One of the simplest is CP_4, subject to:

$$\sum_{i=1}^{5} z_i^5 = 0 \tag{5.8.16}$$

This surface enjoys the global symmetry:

$$G = S_5 \tilde{\otimes} Z_5^5 / Z_5 \tag{5.8.17}$$

and has 101 fermion generations, which strongly suggests that the CP_4 and the (3^5) theories are the same. Similarly, a large number of one-to-one correspondences can be made between minimal superconformal theories and Calabi–Yau manifolds.

As surprising as this construction is, its main problem is the presence of unwanted $U(1)$ factors in the low-energy symmetry. To verify many of the steps in this construction, such as constructing the characters $\chi_q^{l(s)}$ and the partition functions of the theory, it is important to have a specific representation of this $N = 2$ algebra.

One of the most powerful is the parafermion representation. We assume the existence of a ϕ_l parafermion used in constructing $\mathbf{Z}_k$ models in statistical mechanics ($\mathbf{Z}_2$ is the Ising model). Then, the superconformal generators can be written in terms of the parafermion field and a free boson:

$$G_+(z) = \sqrt{\frac{2k}{k+2}} \, \psi_1 e^{i\alpha\phi}$$

$$G_-(z) = \sqrt{\frac{2k}{k+2}} \, \psi_1^\dagger e^{-i\alpha\phi} \qquad (5.8.18)$$

$$J(z) = i\sqrt{\frac{k}{k+2}} \, \partial\phi$$

where $\alpha = \sqrt{(k+2)/k}$. If we calculate the central charge of this representation, we find:

$$c = c_\phi + c_\psi = 1 + \frac{2(k-1)}{(k+2)} = \frac{3k}{k+2} \qquad (5.8.19)$$

which confirms that we have a representation of the minimal series. Then, by introducing another parafermion field ψ_l^1, we can explicitly construct a representation of the primary fields:

$$V_m^l = \psi_m^l : e^{i\alpha_m \phi(z)} : \qquad (5.8.20)$$

Last, with more work, we can construct the characters of this representation:

$$\chi_m^{l(s)} = \sum_{j \bmod k} c_{m+4j-s}^l(\tau) \Theta_{2m+(4j-s)(k+2),2k(k+2)}(\tau, 2kz, u) \qquad (5.8.21)$$

where:

$$\Theta_{n,m}(\tau, z, u) = e^{-2\pi i u} \sum_{j \in Z + n/2m} e^{2\pi i \tau m j^2 + 2\pi i j z} \qquad (5.8.22)$$

Yet another compactification scheme is given by relaxing the condition of tensoring minimal models with $c < 3$. The advantage of this new scheme is that we avoid unnecessary factors of $U(1)$ symmetry that persist in the naive tensoring of minimal models. This new scheme begins with an $N = 1$ superconformal coset theory, and then it postulates that the $N = 2$ generators can be constructed from the $N = 1$ currents as follows:

$$G^1(z) = \frac{2}{k}\left[h_{\bar{a}\bar{b}} j^{\bar{a}}(z) \hat{J}^{\bar{b}}(z) - \frac{i}{3k} S_{\bar{a}\bar{b}\bar{c}} j^{\bar{a}}(z) j^{\bar{b}}(z) j^{\bar{c}}(c) \right] \qquad (5.8.23)$$

where $h_{\bar{a}\bar{b}}$ and $S_{\bar{a}\bar{b}\bar{c}}$ are arbitrary constants, subject to the condition that they generate the usual $N = 2$ product expansion operators. We also have the U(1) current given by:

$$J(z) = \frac{i}{k} h_{\bar{a}\bar{b}} j^{\bar{a}}(z) j^{\bar{b}}(z) + \frac{1}{k} f_{\bar{c}\bar{d}} f_{\bar{c}\bar{d}E}\left[\hat{J}^E(z) - \frac{i}{k} f_{E\bar{a}\bar{b}} j^{\bar{a}}(z) j^{\bar{b}}(z) \right] \qquad (5.8.24)$$

By forcing the above operators to generate the $N = 2$ superconformal series, we find that the constants define an almost complex structure and that:

$$f_{\bar{a}\bar{b}\bar{c}} = S_{\bar{a}\bar{b}\bar{c}} = 0 \qquad (5.8.25)$$

The solution to these constraints gives us hermitian symmetric spaces, which can produce $c = 9$ with just one irreducible representation, without tensoring several of them together.

For the heterotic string, we can tensor l of these representations and find the gauge group $E_8 \otimes SO(10) \otimes U^{l-1}$, where the last $U(1)^{l-1}$ factors are unwanted because they cause problems at the loop level. If $l = 1$, then we can eliminate the last $U(1)$ by combining this factor with $SO(10)$, giving us E_6. For the minimal models considered earlier, this choice is not possible since one minimal cannot give us a $c = 9$ theory. However, for hermitian symmetric spaces, there is no problem in getting $c = 9$ with just one copy, that is, $l = 1$. In summary, by this procedure we have now produced the gauge group $E_8 \otimes E_6$, which is large enough to contain the minimal group $SU(3) \otimes SU(2) \otimes U(1)$.

Last, we mention that the groups appearing among the hermitian symmetric spaces, once all the constraints are taken into consideration, are given by:

$$
\begin{aligned}
SO(N+2)/SO(n) \otimes SO(2): &\quad (n,k) = (6,6), (12,4) \\
SO(2n)/SU(2) \otimes U(1): &\quad (n,k) = (7,2) \\
Sp(n)/SU(n) \otimes U(1): &\quad (n,k) = (3,4) \\
U(n+m)/U(n) \otimes U(m): &\quad (n,m,k) = (1,4,15), (1,5,9), (1,6,7), \\
&\quad (2,2,12), (2,3,5), (3,3,3)
\end{aligned}
$$

$$(5.8.26)$$

References

1. E. Cremmer and J. Scherk, *Nucl. Phys.* **B108**, 409 (1976); **B118**, 61 (1977).
2. K. S. Narain, *Phys. Lett.* **169B**, 41 (1986).
3. W. Lerche, D. Lust, and A. N. Schellekens, *Nucl. Phys.* **B287**, 477 (1987).
4. H. Kawai, D. C. Lewellen, and S.-H. Tye, *Phys. Rev. Lett.* **57** 1832 (1986); *Phys. Rev.* **D34**, 3794 (1986); *Nucl. Phys.* **B288**, 1 (1987).
5. I. Antoniadis, C. P. Bachas, and C. Kounnas, *Nucl. Phys.* **B289**, 87 (1987).
6. L. Dixon, J. Harvey, C. Vafa, and E. Witten, *Nucl. Phys.* **B261** 651, (1985); **B274**, 285 (1986).
7. P. Candelas, G. Horowitz. A. Strominger, and E. Witten, *Nucl. Phys.* **B285**, 56 (1985).
8. S. T. Yau, *Proc. Natl. Acad. Sci.* **74**, 1978 (1977).
9. A. Strominger and E. Witten, *Comm. Math. Phys.* **101**, 341 (1985); A. Strominger, *Phys. Rev. Lett.* **55**, 2547 (1985).
10. P. Candelas and S. Kalara, *Nucl. Phys.* **B298**, 357 (1988); P. Candelas, *Nucl. Phys.* **B298**, 458 (1988).
11. S. T. Yau in *Proc. Argonne Symposium on Anomalies, Geometry and Topology*, World Scientific (1985).
12. D. Friedan, A. Kent, S. Shenker, and E. Witten, unpublished.
13. C. M. Hull and E. Witten, *Phys. Lett.* **160B**, 398 (1985).
14. T. Banks, L. J. Dixon, D. Friedan, and E. Martinec, *Nucl. Phys.* **B299**, 613 (1988).

15. A. Sen, *Nucl. Phys.* **B278**, 289 (1986); *Nucl. Phys.* **B284**, 423 (1987).
16. W. Boucher, D. Friedan, and A. Kent, *Phys. Lett.* **B172**, 316 (1986).
17. S. Nam, *Phys. Lett.* **172B**, 323 (1986).
18. D. Gepner, *Phys. Lett.* **199B**, 380 (1987); *Nucl. Phys.* **B296**, 380, 757 (1987); *Nucl. Phys.* **B311**, 191 (1988–1989).
19. D. Gepner, in Proceedings of the Spring School on Superstrings, Trieste, Italy, 1989.
20. A. B. Zamolodchikov and V. A. Fateev, *Sov. Phys. JETP* **62**, 215 (1985); *Sov. Phys. JETP* **63**, 912 (1986).
21. D. Gepner and Z. Qiu, *Nucl. Phys.* **B285**, 423 (1987).
22. J. J. Atick, L. J. Dixon, and A. Sen, *Nucl. Phys.* **B292**, 109 (1987).
23. M. Dine, I. Ichinose, and N. Seiberg, *Nucl. Phys.* **B292**, 253 (1987).
24. Y. Kazama and H. Suzuki, *Phys. Lett.* **216B**, 112 (1989); *Nucl. Phys.* **B321**, 232 (1989).
25. J. A. Wolf in *Symmetric Space,* Dekker, New York (1972); *Spaces of Constant Curvature,* Publish or Perish Press, Berkeley (1984).
26. L. Dixon, V. Kaplunovsky, and C. Vafa, *Nucl. Phys.* **B294**, 43 (1987).

Chapter 6

Yang–Baxter Relation

6.1. Statistical Mechanics and Critical Exponents

Throughout the previous chapters, we have seen the close relationship between conformal field theory and two-dimensional statistical mechanics. In fact, at criticality, the detailed behavior of a statistical mechanical system gets washed out, and universality sets in. Since we have a complete classification of certain classes of conformal field theories, we should be able to catalog the models of statistical mechanics at criticality according to known representations of conformal field theories.

Before we proceed with a discussion of classifying conformal field theories, it thus becomes important to analyze this remarkable relationship in more detail. In this chapter, we make this relationship between conformal field theory and critical systems explicit, and we also point out the origin of why these statistical mechanical models are exactly solvable, that is, the Yang–Baxter relationship [1, 2]. Surprisingly, we will see the Yang–Baxter relation crop up in numerous other ways in later chapters, such as in our discussion on knot theory and Chern–Simons Yang–Mills theory.

Let us begin our discussion of statistical mechanics by making a few basic definitions [3–6]. When analyzing the properties of a solid, liquid, or gas, the starting point of our discussion will be the Boltzmann partition function:

$$Z = \sum_n \exp\left[-\frac{E(n)}{kT}\right] \tag{6.1.1}$$

where $E(n)$ represents the energy of the nth state, k represents the Boltzmann constant, and T represents the temperature.

Notice the similarity between this partition function and the generating functional found in relativistic quantum field theory:

$$Z = \int D\phi \, \exp i \int L(\phi) \, d^4 x \tag{6.1.2}$$

Notice that there is a correspondence between the two widely divergent formalisms if we take the Euclidian version of quantum theory (so that the factor of i in the exponent becomes -1). In particular, the high-(low-)temperature limit found in statistical mechanics corresponds to the weak (strong) coupling limit of quantum field theory.

The fundamental quantity we wish to calculate for any statistical mechanical system is called the *free energy*, and it is defined by:

$$F = -kT \ln Z \tag{6.1.3}$$

In addition, the statistical average of any observable X is given by:

$$\langle X \rangle = Z^{-1} \sum_n X(n) \exp\left[-\frac{E(n)}{kT}\right] \tag{6.1.4}$$

We say that a two-dimensional statistical model is *exactly solvable* if we can solve for an explicit expression for the free energy. There are remarkably few exactly solvable two-dimensional models, such as the Ising model, ferroelectric six-vertex model, eight-vertex model, three-spin model, and hard hexagon model.

Let us say that we have a collection of spins σ_i arranged in some regular two-dimensional lattice. Then, define the correlation between the ith and jth spin as:

$$g_{ij} = \langle \sigma_i \sigma_j \rangle - \langle \sigma_i \rangle \langle \sigma_j \rangle \tag{6.1.5}$$

In general, we find that the function g_{ij} will depend on the distance x separating the states, and at large distances, it will behave like some decreasing power of x multiplied by some exponential:

$$g_{ij} \sim x^{-\tau} e^{-x/\xi} \tag{6.1.6}$$

where ξ is called the *correlation length*.

At the critical temperature, we find that the correlation length becomes infinite, that is, the system looses all dependence on any fundamental length scale, so the correlation function exhibits a power behavior:

$$g_{ij} \sim x^{-d+2-\eta} \sim x^{-2\Delta} \tag{6.1.7}$$

where η is called a *critical exponent* and Δ is the conformal weight of the field. Likewise, one can also define the "energy operator" as a product of two fields, $\epsilon_n = \sigma_n \sigma_{n+1}$, whose critical behavior is governed by another critical exponent ν:

$$\langle \epsilon_n \epsilon_0 \rangle \sim x^{-2(d-1/\nu)} \tag{6.1.8}$$

For the Ising model, we can actually compute these critical exponents for the spin field and the energy field, and we find:

$$\eta = 1/4, \qquad \nu = 1 \tag{6.1.9}$$

Because Δ, the conformal weight of the field, can be written as $h + \bar{h}$ for the minimal model, we can write the correspondence between Ising fields and the minimal model for $m = 3$:

$$\sigma \leftrightarrow \phi_{1/16,1/16}, \qquad \epsilon \leftrightarrow \phi_{1/2,1/2} \qquad (6.1.10)$$

Because the correlation length becomes infinite at criticality, the properties of the system can be roughly described by the critical exponents for various physical quantities.

For systems with a magnetic field, for example, the *magnetization M* is defined to be the average of the magnetic moment per site:

$$M(H,T) = N^{-1}\langle\sigma_1 + \cdots + \sigma_N\rangle \qquad (6.1.11)$$

where T is the temperature of the system, H is proportional to the magnetic field, and the energy is given by:

$$E = E_0 + H\sum_n \sigma_n \qquad (6.1.12)$$

where E_0 is the energy in the free field limit $H = 0$.

In the limit that $N \to \infty$, we can describe the magnetization as:

$$M(H,T) = -\frac{\partial}{\partial H}f(H,T) \qquad (6.1.13)$$

because taking the derivative with respect to H simply brings down σ_i into the sum. The *susceptibility* of a magnet is then defined as:

$$\chi(H,T) = \frac{\partial M(H,T)}{\partial H} \qquad (6.1.14)$$

Using this formalism, we will make contact with the conformal field theories described in previous chapters.

6.2. One-Dimensional Ising Model

Let us first solve the simplest system, the one-dimensional Ising model. This theory is not very physical because it exhibits no phase transition at all, but it has many of the mathematical ingredients useful for more complicated systems.

We begin by placing a series of spins σ_i along a line, which can take the values of ± 1. The energy of the system can be described as:

$$E(\sigma) = -J\sum_{j=1}^{N}\sigma_j\sigma_{j+1} - H\sum_{j=1}^{N}\sigma_j \qquad (6.2.1)$$

where we assume that the jth spin only interacts with its nearest neighbors at the $j - 1$ and $j + 1$ sites. Then, the partition function can be written as:

$$Z_N = \sum_\sigma \exp\left(K\sum_{j=1}^{N}\sigma_j\sigma_{j+1} + h\sum_{j=1}^{N}\sigma_j\right) \qquad (6.2.2)$$

where we have rescaled the parameters via $K = J/kT$ and $h = H/kT$.

Let us make a most important observation about this system, by rewriting the partition function as a sum over a series of matrices:

$$Z_N = \sum_{\sigma} V(\sigma_1, \sigma_2) V(\sigma_2, \sigma_3) \cdots V(\sigma_{N-1}, \sigma_N) V(\sigma_N, \sigma_1) \qquad (6.2.3)$$

where:

$$V(\sigma, \sigma') = \exp \left[K\sigma\sigma' + \frac{h}{2}(\sigma + \sigma') \right] \qquad (6.2.4)$$

Now, let us regard the elements of V as a two-by-two dimensional matrix $\mathbf{V}$, which is called the *transfer matrix*, which depends on whether the spins are $+1$ or -1, that is,

$$\mathbf{V} = \begin{pmatrix} V(+,+) & V(+,-) \\ V(-,+) & V(-,-) \end{pmatrix} = \begin{pmatrix} e^{K+h} & e^{-K} \\ e^{-K} & e^{K-h} \end{pmatrix} \qquad (6.2.5)$$

Therefore, the partition function can now be succinctly rewritten as:

$$Z_N = \operatorname{Tr} \mathbf{V}^N \qquad (6.2.6)$$

On one hand, we have done nothing. We have merely reshuffled the summation within Z_N by rewriting it as a sum over the two-by-two transfer matrix $\mathbf{V}$. On the other hand, we have made an enormous conceptual difference, because we can now diagonalize the transfer matrix in terms of its eigenvalues, that is, there exists a matrix $\mathbf{P}$ that diagonalizes $\mathbf{V}$:

$$\mathbf{V} = \mathbf{P} \begin{pmatrix} \lambda_1 & 0 \\ 0 & \lambda_2 \end{pmatrix} \mathbf{P}^{-1} \qquad (6.2.7)$$

Substituting this into our original expression for the partition function, we now find:

$$Z_N = \operatorname{Tr} \begin{pmatrix} \lambda_1 & 0 \\ 0 & \lambda_2 \end{pmatrix}^N = \lambda_1^N + \lambda_2^N \qquad (6.2.8)$$

Let λ_1 be the larger of the two eigenvalues, which will then dominate the sum in the limit as $N \to \infty$. We then have:

$$\begin{aligned}
f(H, T) &= -kT \lim_{N \to \infty} N^{-1} \ln Z_N = -kT \ln \lambda_1 \\
&= -kT \ln \left[e^k \cosh h + \sqrt{e^{2K} \sinh^2 h + e^{-2K}} \right]
\end{aligned} \qquad (6.2.9)$$

In addition to having an exact expression for the free energy, we also have an exact expression for the magnetization:

$$M(H, T) = \frac{e^K \sinh h}{\sqrt{e^{2K} \sinh^2 h + e^{-2K}}} \qquad (6.2.10)$$

This is a truly remarkable result, first found by Ising, who also proposed the model in 1925 [7]. Because we have an analytic expression for the free energy in terms of K and h, it shows that the model is exactly solvable. Moreover, it also shows, unfortunately, that the system does not exhibit any phase transition for any positive temperature.

Because we have an exact expression for the transfer matrix, we can now solve for the correlation length and show that it goes to infinity when $H = T = 0$, although this latter point is not a critical point. To do this, we need to calculate the averages $\langle \sigma_i \rangle$ and $\langle \sigma_i \sigma_j \rangle$. We begin by defining the matrix $\mathbf{S}$ in spin space as:

$$\mathbf{S} = \begin{pmatrix} 1 & 0 \\ 0 & -1 \end{pmatrix} \tag{6.2.11}$$

which has elements:

$$S(\sigma, \sigma') = \sigma \delta(\sigma, \sigma') \tag{6.2.12}$$

Therefore, the average can be written as:

$$\langle \sigma_1 \sigma_3 \rangle = Z_N^{-1} \sum_\sigma \sigma_1 V(\sigma_1, \sigma_2) V(\sigma_2, \sigma_3) \sigma_3 \cdots = Z_N^{-1} \operatorname{Tr} \mathbf{S} \mathbf{V}^2 \mathbf{S} \mathbf{V}^{N-2} \tag{6.2.13}$$

So,

$$\langle \sigma_i \sigma_j \rangle = Z_N^{-1} \operatorname{Tr} \mathbf{S} \mathbf{V}^{j-i} \mathbf{S} \mathbf{V}^{N+i-j}$$
$$\langle \sigma_i \rangle = Z_N^{-1} \operatorname{Tr} \mathbf{S} \mathbf{V}^{N} \tag{6.2.14}$$

Now let the matrix $\mathbf{P}$, which diagonalizes the transfer matrix, be parametrized by an angle ϕ:

$$\mathbf{P} = \begin{pmatrix} \cos\phi & -\sin\phi \\ \sin\phi & \cos\phi \end{pmatrix} \tag{6.2.15}$$

Then, we have:

$$\begin{aligned} g_{ij} = \langle \sigma_i \sigma_j \rangle - \langle \sigma_i \rangle \langle \sigma_j \rangle &= \cos^2\phi + \sin^2 2\phi \left(\lambda_2/\lambda_1\right)^{j-i} - \cos 2\phi \\ &= \sin^2 2\phi \left(\lambda_2/\lambda_1\right)^{j-i} \end{aligned} \tag{6.2.16}$$

So, we have the desired result:

$$\xi = \left[\ln(\lambda_1/\lambda_2) \right]^{-1} \tag{6.2.17}$$

which tends to ∞ as $H, T \to 0$. Thus, all reference to a mass scale has disappeared.

6.3. Two-Dimensional Ising Model

Now that we have some experience using the transfer matrix technique, let us tackle a nontrivial problem, the two-dimensional Ising model [8].

We place the spins σ_i on a two-dimensional lattice, except that the lattice sites are placed diagonally. Let W and V represent the lattice sites that are arranged along a horizontal line. The sites along W and V alternate as we descend the lattice. Our strategy is to rewrite the partition function once again in terms of transfer matrices, except that we will perform the sums over spins in a particular fashion.

First, we will sum the spin horizontally, which will give us expressions for W and V. Then, we will sum the lattice vertically, which will give us sums over the product $WVWVW, \ldots$, etc.

To sum the lattice sites horizontally, let $\phi = \{\sigma_1, \sigma_2, \ldots \sigma_n\}$, that is, the lattice sites arranged horizontally along the top of Fig. 6.1, and let ϕ' be the lattice sites arranged below them. Then, define W and V as follows:

$$V_{\phi,\phi'} = \exp\left[\sum_{j=1}^{n}(K\sigma_{j+1}\sigma_j' + L\sigma_j\sigma_j')\right]$$

$$W_{\phi,\phi'} = \exp\left[\sum_{j=1}^{n}(K\sigma_j\sigma_j' + L\sigma_j\sigma_{j+1}')\right] \tag{6.3.1}$$

where W and V are now $2^n \times 2^n$ matrices. As before, we can perform the sum over the two transfer matrices by summing vertically over the lattice:

$$Z_n = \text{Tr}(\mathbf{VW})^{m/2} = \sum_{i=1}^{2^n} \Lambda_i^m \tag{6.3.2}$$

In the thermodynamic limit, as we let the number of points $n, m \to \infty$, the partition function is once again dominated by the largest eigenvalue of the transfer matrix $\mathbf{VW}$:

$$\lim_{n,m\to\infty} Z \sim (\Lambda_{\max})^m$$

We see that the one-dimensional and two-dimensional Ising models are therefore closely related to each other and that the calculation of the free energy reduces to calculating the largest eigenvalue of the transfer matrix.

The actual solution for the free energy in the continuum limit, however, is quite involved for the two-dimensional Ising model, so we will just present the final result. Define a function:

$$F(\theta) = \ln\left[2\big(\cosh 2K \cosh 2L + k^{-1}\sqrt{1 + k^2 - 2K\cos 2\theta}\big)\right] \tag{6.3.4}$$

Then, the largest eigenvalue can be written as:

$$\ln \Lambda_{\max} \sim \frac{1}{2}\sum_{j=1}^{2p} F\left(\frac{\pi}{2p}\left(j - \frac{1}{2}\right)\right) \tag{6.3.5}$$

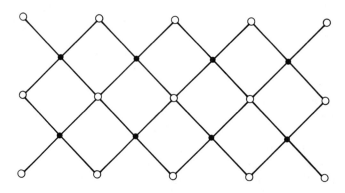

Fig. 6.1.

In the thermodynamic limit, the summation over the evenly spaced θ_i becomes an integral, so we can write [6]:

$$f = -\frac{kT}{2\pi} \int_0^\pi F(\theta)d\theta \qquad (6.3.6)$$

This is our final result for the free energy of the two-dimensional Ising model.

6.4. RSOS and Other Models

There are a number of models [2–6] that generalize the behavior of the Ising model and are exactly solvable. More important, there are a number of models that, although they may not be exactly solvable, exhibit critical behavior that can be described by the known conformal field theories. Let us list some of these models and their properties:

Spherical model

One defect of the Ising model is that it is only solvable in the zero external field limit, which is a feature of many ferromagnetic models. However, one model that can be solved exactly, even in the presence of a field, is the spherical model.

This model is similar to the Ising model, except for several important differences. The spin σ_i can take on real values, not just +1 and -1, and it can interact with all spins in the lattice, subject to the constraint:

$$\sum_{j=1}^{N} \sigma_j^2 = N \qquad (6.4.1)$$

The partition function is now replaced by an integral (not a sum) of the Ising model's partition function, with a delta function insertion that guarantees the constraint:

$$Z = \int_{-\infty}^{\infty} \cdots \int_{-\infty}^{\infty} d\sigma_1 \cdots d\sigma_N$$
$$\times \exp\left(K \sum_{jl} \sigma_j \sigma_l + h \sum_j \sigma_j \right) \delta\left(N - \sum_j \sigma_j^2 \right) \tag{6.4.2}$$

This model may be criticized because it is unphysical, that is, it implies a coupling between all spins on the lattice, no matter how far apart they are. However, this model is exactly solvable and exhibits normal phase transitions, despite its unphysical appearance. This puzzling result has been explained: the spherical model has been shown to be a special limiting case of the n-vertex model with only nearest neighboring interactions.

Ice-type, six-vertex model

The ice-type model was introduced to model the behavior of ferroelectrics. It differs from the usual Ising model, whose spins are located at the lattice sites, because the energy of the ice-type model is defined on the links or edges connecting the sites, not on the sites themselves.

Ice-type models may describe the behavior of crystals with hydrogen bonds, such as ice, with oxygen and hydrogen atoms within the water molecule. We assume that the molecule is placed at each lattice site and that the edges represent the electric dipole field, and hence, the model represents a ferroelectric.

In general, since the electric dipole can assume two directions along the edges, there are $2^4 = 16$ ways in which arrows can be placed along the edges, arranged around a lattice site. To be more concrete, we assume the ice rule, which states that there must be two arrows going into and two arrows going out of each site. Thus, the 16 possible configurations of arrows surrounding each site is reduced to six (see Fig. 6.2).

Let ϵ_i represent the energy associated with each of the six possible configurations of arrows surrounding each site. Let n_i be the number of times the ith configuration is repeated throughout the lattice. Then, the partition function is represented by:

$$Z = \sum_{i=1}^{6} \exp\left(-\frac{[n_i \epsilon_i]}{kT} \right) \tag{6.4.3}$$

For different values of ϵ_i, we have different physical structures. For example, it is thought that the potassium dihydrogen phosphate crystal, KH_2PO_4, can be represented by the following choice:

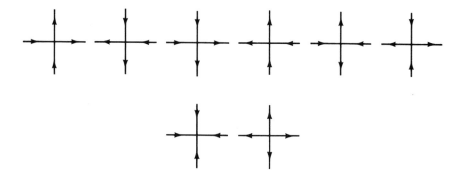

Fig. 6.2.

$$\epsilon_1 = \epsilon_2 = 0; \qquad \epsilon_3 = \epsilon_4 = \epsilon_5 = \epsilon_6 > 0 \tag{6.4.4}$$

On the other hand, it is thought that an antiferroelectric can be modeled by the choice:

$$\epsilon_1 = \epsilon_2 = \epsilon_3 = \epsilon_4 > 0; \qquad \epsilon_5 = \epsilon_6 = 0 \tag{6.4.5}$$

Because the ice-type, six-vertex model is exactly solvable, we can find analytic expressions for its free energy and solve for its critical exponents. There are, however, some defects with the six-vertex model. It turns out that the model has a ferroelectric ordered state that is frozen, that is, the ordering is complete even at nonzero temperatures, and that the antiferroelectric properties do not diverge or vanish at criticality as simple powers of the critical temperature.

Eight-vertex model

Because of the oversimplification present in the six-vertex model, leading to nonphysical results, it was generalized to the eight-vertex model. This model places constraints on the 16 possible configurations of arrows surrounding a lattice site by assuming that there are an even number of arrows going into and out of each site. Thus, we will sum over 8 possible configurations, each with its own energy ϵ_i.

The partition function is:

$$Z = \sum_{i=1}^{8} \exp\left(\frac{-[n_i \epsilon_i]}{kT}\right) \tag{6.4.6}$$

Like the six-vertex model, the eight-vertex model is also exactly solvable. However, the eight-vertex model is considerably more sophisticated than the six-vertex model. In fact, the eight-vertex model, for various choices of the physical parameters, can describe both ferromagnets as well as ferroelectrics. In addition, it can contain the six-vertex model and the Ising model as special cases.

The first statement is easy to understand, because we can always set two of the ϵ_i to zero to obtain the six-vertex model. However, the second statement is surprising, because the eight-vertex model and the Ising model have very different physical structures. The vertex models have their energy based on the edges connecting the sites, while the Ising model has its energy based on the sites themselves.

The fact that the eight-vertex model contains the Ising model, however, can be seen by a change of parameters. Let σ_{ij} be associated with the edge that links the ith and jth lattice site, and let it assume values of $+1$ or -1. Now re-write the energy as follows:

$$
\begin{aligned}
E = - \sum_{i=1}^{M} \sum_{j=1}^{N} & \left(J_v \sigma_{ij}\sigma_{i,j+1} + J_h \sigma_{ij}\sigma_{i+1,j} + J\sigma_{i,j+1}\sigma_{i+1,j} \right. \\
& \left. + J'\sigma_{ij}\sigma_{i+1,j+1} + J''\sigma_{ij}\sigma_{i,j+1}\sigma_{i+1,j}\sigma_{i+1,j+1} \right)
\end{aligned}
\tag{6.4.7}
$$

Notice that we have done nothing; we have merely rewritten the eight-vertex model in a way such that its dependence on Ising-type spins is more apparent.

Now define:

$$
\alpha_{ij} = \sigma_{ij}\sigma_{i,j+1}, \qquad \mu_{ij} = \sigma_{ij}\sigma_{i+1,j}
\tag{6.4.8}
$$

Then, the partition function can be written as:

$$
\begin{aligned}
E = - \sum_{i=1}^{M} \sum_{j=1}^{N} & \left(J_v \alpha_{ij} + J_h \mu_{ij} + J\alpha_{ij}\mu_{ij} \right. \\
& \left. + J'\alpha_{i+1,j}\mu_{ij} + J''\alpha_{ij}\alpha_{i+1,j} \right)
\end{aligned}
\tag{6.4.9}
$$

with the condition:

$$
\mu_{ij}\alpha_{ij}\alpha_{i+1,j}\mu_{i,j+1} = 1
\tag{6.4.10}
$$

Notice that we have now split the partition function into two pieces, each representing a distinct Ising model.

In fact, the explicit relations between the eight-vertex ϵ_i and the $J's$ of the Ising model are given by:

$$
\begin{aligned}
\epsilon_1 &= -J_h - J_v - J - J' - J'', & \epsilon_2 &= J_h + J_v - J - J' - J'' \\
\epsilon_3 &= -J_h + J_v + J + J' + J'', & \epsilon_4 &= J_h - J_v + J + J' - J'' \\
\epsilon_5 &= \epsilon_6 = J - J' + J'', & \epsilon_7 &= \epsilon_8 = -J + J' + J''
\end{aligned}
\tag{6.4.11}
$$

With this choice, we can show that the partition function of the Ising model is just twice the partition function of the eight-vertex model:

$$Z_{\text{Ising}} = 2Z_{\text{eight-vertex}} \tag{6.4.12}$$

Z_N model

Potts originally introduced two different types of models, the $\mathbf{Z}_N$ model and what is usually called the Potts model. The $\mathbf{Z}_N$ model is a straightforward generalization of the usual Ising model. The spins in the Ising model assume only the values of $+1$ or -1. However, we can easily generalize this to the case where the spin σ_i points in N equally spaced directions. Then, the energy associated with the model is the scalar product between nearest neighbors, that is,

$$E = \sum_{i,j} \left\{ \sigma_{ij} \cdot \sigma_{i,j+1} + \sigma_{i,j} \cdot \sigma_{i+1,j} \right\} \tag{6.4.13}$$

Obviously, the Ising model corresponds to the case of $N = 2$. A parafermionic representation of the primary fields of this model was studied in Chapter 5.

Potts model

The Potts model is defined by letting the spin σ_i at the ith lattice site take on values from 1 to q. Two nearest neighbor spins interact via the delta function and are defined as:

$$\begin{aligned} \delta(\sigma, \sigma') &= 1 \quad \text{if} \quad \sigma = \sigma' \\ \delta(\sigma, \sigma') &= 0 \quad \text{if} \quad \sigma \neq \sigma' \end{aligned} \tag{6.4.14}$$

The energy is then defined to be:

$$E = -J \sum_{i,j} \delta(\sigma_i, \sigma_j) \tag{6.4.15}$$

This model can be solved at criticality. The case $q = 1$ is trivial. The case $q = 2$ is the Ising model, which is equivalent to a minimal model with $m = 3$. It can be shown to have critical exponents given by:

$$\alpha = 0, \qquad \beta = \frac{1}{8}, \qquad \delta = 15 \tag{6.4.16}$$

which can then be compared with the minimal model. The case $q = 3$ is also equivalent to a minimal model. Its critical exponents are given by:

$$\alpha = \frac{1}{3}, \qquad \beta = \frac{1}{9}, \qquad \delta = 14 \tag{6.4.17}$$

XYZ Heisenberg model

Closely related to the Ising and the eight-vertex models is the XYZ Heisenberg model. Here, we replace the spin σ_i with a real Pauli spinor. The Hamiltonian is given by:

$$H = -\frac{1}{2} \sum_{j=1}^{N} \left\{ J_x \sigma_j^x \sigma_{j+1}^x + J_y \sigma_j^y \sigma_{j+1}^y + J_z \sigma_j^z \sigma_{j+1}^z + \cdots \right\} \qquad (6.4.18)$$

where the ellipses represent the interactions in the vertical direction.

Not only is σ_i a Pauli spinor, it is also carries the indices of all the spins in the system, that is,

$$\sigma_j^x = 1 \otimes \cdots \otimes \sigma^x \otimes \cdots \otimes 1 \qquad (6.4.19)$$

where 1 is a two-by-two unit matrix, and the only nontrivial entry in the tensor product is at the jth site.

If $J_x = J_y = J_z$, then this is the usual Heisenberg model.

If $J_x = J_y = 0$, then only J_z survives, and hence, we obtain the usual Ising model.

If $J_z = 0$, then we have the XY model.

If $J_x = J_y$, then we have the Heisenberg–Ising model.

It can be shown that the Hamiltonian, for any value of the J's, can be written as the logarithmic derivative of an eight-vertex transfer matrix.

Ashkin–Teller model

The Ashkin–Teller model, like the previous models, was based on a generalization of the Ising model. In this model, there are four types of atoms, called A,B,C, and D. There are three values of the energy ϵ_i, given the different possible nearest neighbor pairings of these atoms. The following energies correspond to the given pairings:

$$
\begin{array}{llll}
\epsilon_0 : & AA, BB, CC, DD; & \epsilon_1 : & AB, CD \\
\epsilon_2 : & AC, BD; & \epsilon_3 : & AD, BC
\end{array}
\qquad (6.4.20)
$$

It is possible however, to rewrite this model in terms of the usual Ising-type spins. Let us introduce two types of spins, s_i and σ_i. Let the pair $\{s_i, \sigma_i\}$ equal $(+, +)$ if there is an A atom at any site i; $(+, -)$ if there is a B atom; $(-, +)$ if there is a C atom; and $(-, -)$ if there is a D atom. Then, the energy can be written as:

$$E(ij) = -J s_i s_j - J' \sigma_i \sigma_j - J_4 s_i \sigma_i s_j \sigma_j - J_0 \qquad (6.4.21)$$

where:

$$-J = (\epsilon_0 + \epsilon_1 - \epsilon_2 - \epsilon_3)/4$$
$$-J' = (\epsilon_0 + \epsilon_2 - \epsilon_3 - \epsilon_1)/4$$
$$-J_4 = (\epsilon_0 + \epsilon_3 - \epsilon_1 - \epsilon_2)/4 \qquad (6.4.22)$$
$$-J_0 = (\epsilon_0 + \epsilon_1 + \epsilon_2 + \epsilon_3)/4$$

(There is yet another representation of the Ashkin–Teller model, as a staggered eight-vertex model.)

The Ashkin–Teller model is not solvable, but its properties at criticality are known. It is known that its phase structure is surprisingly rich. It has five phases, including phases that are ferromagnetically and antiferromagnetically ordered, and that one corresponds to the $\hat{m} = 4$ superconformal unitary minimal series at criticality.

Hard hexagon model

The hard hexagon model is exactly solvable, and it represents a two-dimensional lattice model of a gas of hard, that is, nonoverlapping, molecules. For example, it can be compared to a two-dimensional helium monolayer adsorbed onto a graphite surface.

Imagine that our lattice consists of an infinite series of hexagons, each adjacent to each other and without any spaces in between. The only rule is that a particle may occupy the center of each hexagon. The model is hard, that is, the hexagons do not overlap. The partition function is then defined to be:

$$Z = \sum_{n=0}^{N/3} z^n g(n, N) \qquad (6.4.23)$$

where $g(n, N)$ is the number of ways in which n particles can be placed in each of the various hexagons. There are N sites, and hence, at maximum, only $N/3$ sites can be occupied.

RSOS

One of the most general solvable models is the RSOS (restricted solid-on-solid) model, which is directly related at criticality to the infinite sequence of minimal models found in Chapter 2, as we saw in Eq. (2.5.24). The RSOS model is defined by the plaquettes (squares). At each site i in a square lattice, define an integer l_i, which represents the "height" of that point. The height is restricted to the interval:

$$1 \le l_i \le (r - 1) \qquad (6.4.24)$$

for a fixed integer r $(r \ge 4)$. (For the unrestricted RSOS, the value of l_i has no restrictions, that is, $-\infty < l_i < \infty$.) The relative heights of nearest neighbor sites can only differ by unity, that is,

$$|l_i - l_j| = 1 \qquad (6.4.25)$$

if i and j are nearest neighbors.

To each plaquette, assign a Boltzmann weight W, which has the symmetries:

$$\begin{aligned}
W(l_1, l_2, l_3, l_4) &= W(l_3, l_2, l_1, l_4) \\
&= W(l_1, l_4, l_3, l_2) \\
&= W(r - l_1, r - l_2, r - l_3, r - l_4)
\end{aligned} \qquad (6.4.26)$$

The partition function is then given by the product of the weights:

$$Z = \sum \prod_{i,j,n,m} W(l_i, l_j, l_m, l_n) \qquad (6.4.27)$$

where the sum is over all allowed arrangements of heights on the lattice, and the product is over all faces of the lattice.

The RSOS model exhibits several phases, and contains a wide variety of known models as specific examples. For example, if r is even, we can translate it into an Ising-type model by changing to spin variables:

$$s_i = (r - 2l_i)/4 \qquad (6.4.28)$$

which produces spin$(r-2)/r$ Ising spins on an odd lattice and spin$(r-4)/4$ on an even one.

By comparing the critical exponents of the RSOS model with those found for the minimal model, we can show that the correspondence between the two models is established for

$$r = m + 1 \qquad (6.4.29)$$

that is, the $m = 3$ minimal model, the $r = 4$ RSOS model, and the Ising model at criticality are all the same. For higher values of r, the critical exponents of the RSOS model can be shown to include those of the Ashkin–Teller model and the hard hexagon model. In particular, we have the series in Eq. (2.5.24).

6.5. Yang–Baxter Relation

It is possible to bring some order to the rapid proliferation of models. We find that most of these models fall into two types:

1. vertex models and
2. IRF (interaction around a face) models.

The vertex models are like the ones we have studied, where the energy is defined by the arrows on the edges that surround a given site. The IRF models, like the Ising models, have spins located at each lattice site, with nearest neighbor interactions. If we take one plaquette of a lattice, we can place the spins around the corners of the plaquette, hence the name.

It will turn out that the reason for the exact solvability of these models is that the transfer matrices, which define the partition function and free energy, commute. When expressed mathematically, this relationship becomes the celebrated *Yang–Baxter* relation. In fact, mutually commuting transfer matrices, or equivalently the Yang–Baxter relation, are sufficient conditions for the solvability of any two-dimensional model.

To get a better understanding of the Yang–Baxter relation, let us study the ice-type six-vertex model. For example, in ice, we have the molecules of water held together by electric dipole moments. Let us place water molecules on a square two-dimensional lattice, such that the line segments forming the lattice correspond to the electric fields, represented by arrows.

These arrows only have two directions on any given line segment. Thus, from any lattice site, there are six different possible orientations of the arrows. Each of these six different orientations will have an energy associated with it, called ϵ_i, for $i = 1, 2, \ldots, 6$. Thus, if ϕ represents the lattice sites along a horizontal line, then we have:

$$Z = \sum_{\phi_1} \sum_{\phi_2} \cdots \sum_{\phi_M} V(\phi_1, \phi_2) V(\phi_2, \phi_3) \cdots V(\phi_M, \phi_1) = \text{Tr } V^M \quad (6.5.1)$$

where:

$$V(\phi, \phi') = \sum \exp \left[-\frac{(m_1 \epsilon_1 + m_2 \epsilon_2 + \cdots + m_6 \epsilon_6)}{kT} \right] \quad (6.5.2)$$

One can proceed to solve the system in this fashion. However, for our purposes, let us propose an alternative method in which we see the origin of the Yang–Baxter equation.

The partition function can be totally rewritten in terms of:

$$w(i, j | k, l) \equiv \exp \left[- \epsilon(i, j, k, l)/kT \right] \quad (6.5.3)$$

Different values of $\epsilon(i, j, k, l)$ correspond to different models.

Now, let $w(\mu_i, \alpha_i | \beta_i, \mu_{i+1})$ represent the contribution to the sum from the ith site. Each Greek index, in turn, can have values of ± 1, such that there are only six possible orientations. Let us perform the sum horizontally, as before:

$$\mathbf{V}_{\alpha, \beta} = \sum_{\mu_1} \cdots \sum_{\mu_N} w(\mu_1, \alpha_1 | \beta_1, \mu_2) w(\mu_2, \alpha_2 | \beta_2, \mu_3) \quad (6.5.4)$$

Let V' represent another transfer matrix, except that

$$(VV')_{\alpha, \beta} = \sum_{\gamma} V_{\alpha\gamma} V'_{\gamma, \beta} = \sum_{\mu_1 \cdots \mu_N} \sum_{\nu_1 \cdots \nu_N} \prod_{i=1}^{N} S(\mu_i, \nu_i | \nu_{i+1}, \nu_{i+1} | \alpha_i, \beta_i)$$

$$(6.5.5)$$

where:

$$S(\mu, \nu | \mu', \nu | \alpha, \beta) = \sum_{\gamma} w(\mu, \alpha | \gamma, \mu') w'(\nu, \gamma | \beta, \nu') \qquad (6.5.6)$$

We can therefore write:

$$(VV')_{\alpha,\beta} = \text{Tr}\, \mathbf{S}(\alpha_1, \beta_1) \mathbf{S}(\alpha_2, \beta_2) \cdots \mathbf{S}(\alpha_N, \beta_N)$$
$$(V'V)_{\alpha,\beta} = \text{Tr}\, \mathbf{S}'(\alpha_1, \beta_1) \mathbf{S}'(\alpha_2, \beta_2) \cdots \mathbf{S}'(\alpha_N, \beta_N) \qquad (6.5.7)$$

We wish to show that V and V' commute, so that the two previous expressions are identical. This is obviously possible if there exists a four-by-four matrix $\mathbf{M}$ such that:

$$\mathbf{S}(\alpha, \beta) = \mathbf{MS}'(\alpha, \beta)\mathbf{M}^{-1} \qquad (6.5.8)$$

Let us multiply the previous relation from the right by $\mathbf{M}$, which we can also represent as a matrix called w''. Then, we have the relationship:

$$\sum_{\gamma, \mu'' \nu''} w(\mu, \alpha | \gamma, \mu'') w'(\nu, \gamma | \beta, \nu'') w''(\nu'', \mu'' | \nu', \mu')$$
$$= \sum_{\gamma, \mu'' \nu''} w''(\nu, \mu | \nu'', \mu'') w'(\mu'', \alpha | \gamma, \mu') w(\nu'', \gamma | \beta, \nu') \qquad (6.5.9)$$

If we redefine:

$$w(\mu, \alpha, \mu'', \gamma) = S_{\alpha\gamma}^{\mu\mu''}(u), \qquad w'(\nu, \gamma, \nu'', \beta) = S_{\gamma\beta}^{\nu\nu''}(u+v)$$
$$w''(\nu'', \mu'', \mu', \nu') = S_{\mu''\nu}^{\nu''\mu'}(v') \qquad (6.5.10)$$

then we can write the Yang–Baxter relationship in the form:

$$\sum_{\alpha\beta\gamma} S_{j\beta}^{i\alpha}(u) S_{k\gamma}^{\alpha p}(u+v) S_{\gamma r}^{\beta q}(v) = \sum_{\alpha\beta\gamma} S_{k\gamma}^{j\beta}(v) S_{\gamma r}^{i\alpha}(u+v) S_{\beta q}^{\alpha p}(u) \qquad (6.5.11)$$

If we graphically represent this relationship, then we find the pattern expressed in Fig. 6.3, which pictorially displays the Yang–Baxter relation.

The second type of exactly solvable model is the IRF, which includes the Ising model and many of the other exactly solvable models. (However, we should stress that, as in the eight-vertex model, there are ways in which certain models can be formulated in both languages.) If we place four spins a,b,c, and d (which can equal $+1$ or 0) around the four corners of a plaquette, the energy associated with the plaquette will be $\epsilon(a, b, c, d)$, so we define the Boltzmann weight of the plaquette as:

$$w(a, b, c, d) = \exp[-\epsilon(a, b, c, d)/kT] \qquad (6.5.12)$$

For different choices of $\epsilon(a, b, c, d)$, we can represent a wide variety of models. For example, the Ising model can be represented as:

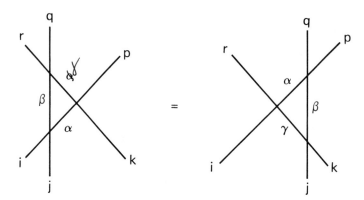

Fig. 6.3.

$$\epsilon(a, b, c, d) = -\frac{1}{2}J[(2a-1)(2b-1) + (2c-1)(2d-1)]$$
$$-\frac{1}{2}J[(2c-1)(2b-1) + (2d-1)(2a-1)] \tag{6.5.13}$$

and the eight-vertex model can be written as:

$$\epsilon(a, b, c, d) = -J(2a-1)(2c-1) - J'(2b-1)(2d-1)$$
$$- J_4(2a-1)(2b-1)(2c-1)(2d-1) \tag{6.5.14}$$

for $a, b, c, d = 0, 1$.

The Hamiltonian is then represented as:

$$H = \sum_{\text{faces}} \epsilon(\sigma_i, \sigma_j, \sigma_k, \sigma_l) \tag{6.5.15}$$

and the partition function for the IRF model is given by:

$$Z = \sum_{\sigma_1} \cdots \sum_{\sigma_N} \prod_{i,j,k,l} w(\sigma_i, \sigma_j, \sigma_k, \sigma_l) \tag{6.5.16}$$

We will now repeat the same steps that we used in studying the Ising model. We wish to express the partition function as a trace over the transfer matrix and then isolate the condition for commuting transfer matrices. Let us define the partial sum:

$$V_{\sigma\sigma'} = \prod_{j=1}^{n} w(\sigma_j, \sigma_{j+1}, \sigma_{j+1}', \sigma_j') \tag{6.5.17}$$

where the sum over σ is shorthand for:

$$\sigma = \{\sigma_1, \sigma_2, \ldots, \sigma_n\}$$
$$\sigma' = \{\sigma'_1, \sigma'_2, \ldots, \sigma'_n\} \tag{6.5.18}$$

and $\sigma_{n+1} = \sigma_1$ and $\sigma'_{n+1} = \sigma'_1$.

We similarly define V' by replacing w with w':

$$V'_{\sigma\sigma'} = \prod_{j=1}^{n} w'(\sigma_j, \sigma_{j+1}, \sigma'_{j+1}, \sigma'_j) \tag{6.5.19}$$

This allows us to form the product VV' defined by:

$$(VV')_{\sigma\sigma'} = \sum_{\sigma''} V_{\sigma\sigma''} V'_{\sigma''\sigma'} = \sum_{\sigma''} \prod_{j=1}^{n} X(\sigma_j, \sigma''_j, \sigma'_j | \sigma_{j+1}, \sigma''_{j+1}, \sigma'_{j+1})$$
$$\tag{6.5.20}$$

where we introduce the quantity:

$$X(a, b, c | a', b', c') = w(a, a', b', b) w'(b, b', c', c) \tag{6.5.21}$$

The whole point of performing this decomposition is to be able to write the sums as traces over transfer matrices:

$$(VV')_{\sigma\sigma'} = \operatorname{Tr} X(\sigma_1, \sigma'_1 | \sigma_2, \sigma'_2) X(\sigma_2, \sigma'_2 | \sigma_3, \sigma'_3) \cdots X(\sigma_n, \sigma'_n | \sigma_1, \sigma'_1) \tag{6.5.22}$$

Similarly, we now define X' with w and w' interchanged:

$$(V'V)_{\sigma\sigma'} = \operatorname{Tr} X'(\sigma_1, \sigma'_1 | \sigma_2, \sigma'_2) X'(\sigma_2, \sigma'_2 | \sigma_3, \sigma'_3) \cdots X'(\sigma_n, \sigma'_n | \sigma_1, \sigma'_1) \tag{6.5.23}$$

As usual, we find that, for the transfer matrices V and V' to commute, we need to postulate the existence of an **M** matrix, such that:

$$X(a, a' | b, b') = M(a, a') X'(a, a' | b, b') M(b, b')^{-1} \tag{6.5.24}$$

Multiplying by **M** from the right, we now find that the condition for commuting transfer matrices is:

$$\sum_{c} w(b, d, c, a) w'(a, c, f, g) w''(c, d, e, f)$$
$$= \sum_{c} w''(a, b, c, g) w'(b, d, e, c) w(c, e, f, g) \tag{6.5.25}$$

where we have defined $M(a, a')$ as $w''(c, a, d, a')$.

This is now the Yang–Baxter relation for the IRF model. In Fig. 6.4, we have graphically displayed the structure of the Yang–Baxter relation, which differs only in form from the Yang–Baxter relation obtained from

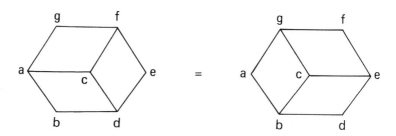

Fig. 6.4.

the vertex models in Eq. (6.5.10). (Because of the shape of this graph, this equation also goes by the name "star–triangle" relation.)

Now that we have derived the Yang–Baxter relationship for both the vertex models and the IRF models, we have an alternative method of solving these models. Instead of trying to maximize the eigenvalues of the transfer matrices, which is how the Ising model was historically solved, we solve the Yang–Baxter relation directly.

This second approach to solving statistical mechanical models is much more elegant than the brute force, hit-or-miss methods employed over the past decades. In fact, the method is so powerful that we can even see how new infinite classes of models might be solved by looking for solutions of the Yang–Baxter relation.

The trick behind solving the Yang–Baxter relation is to reduce the Boltzmann weight function $w(a, b, c, d)$ to a few independent parameters, and reexpress the Yang–Baxter relation in terms of this set. Then, we notice that these relations are identical to the addition formulas found in ordinary trigonometry or the theory of theta functions. The solution to the Yang–Baxter equation can be given in terms of known analytic functions satisfying these addition formulas. Once this analytic solution to the Yang–Baxter relation is found, we can insert this into the partition function and calculate the free energy.

Example: Ising Model

Let us illustrate this procedure for the Ising model. The partition function in Eq. (6.5.12) can be written in terms of four independent functions ω_i:

$$\begin{aligned}
\omega_1(u) &= w(2,3,2,1;u) = w(2,1,2,3;u) \\
\omega_2(u) &= w(2,1,2,1;u) = w(2,3,2,3;u) \\
\omega_3(u) &= w(1,2,3,2;u) = w(3,2,1,2;u) \\
\omega_4(u) &= w(1,2,1,2;u) = w(3,2,3,2;u)
\end{aligned} \qquad (6.5.26)$$

Inserting this into Eq. (6.5.25), we find the Yang–Baxter equation simplifies and reduces to the following equations for the Boltzmann functions:

$$\omega_4(u)\omega_2(u+v)\omega_4(v) + \omega_3(u)\omega_1(u+v)\omega_3(v) = \omega_2(v)\omega_4(u+v)\omega_2(u)$$
$$\omega_4(u)\omega_1(u+v)\omega_4(v) + \omega_3(u)\omega_2(u+v)\omega_3(v) = \omega_1(v)\omega_4(u+v)\omega_1(u)$$
$$\omega_4(u)\omega_2(u+v)\omega_3(v) + \omega_3(u)\omega_1(u+v)\omega_4(v) = \omega_2(u)\omega_3(u+v)\omega_1(u)$$
$$(6.5.27)$$

The key step is to notice the similarity between these equations and the addition formulas found in the classical theory of theta functions. This is how we will find a solution to these reduced Yang–Baxter equations.

The addition formulas for the theta functions are:

$$
\begin{aligned}
&\vartheta_1(u+x)\vartheta_1(u-x)\vartheta_1(v+y)\vartheta(v-y) \\
&\quad - \vartheta_1(u+r)\vartheta_1(u-y)\vartheta_1(v+x)\vartheta_1(v-x) \\
&= \vartheta_1(u+v)\vartheta_1(u-v)\vartheta_1(x+y)\vartheta_1(x-y) \\
&\vartheta_4(u+x)\vartheta_4(u-x)\vartheta_4(v+y)\vartheta_4(v-y) \\
&\quad - \vartheta_4(u+y)\vartheta_4(u-y)\vartheta_4(v+x)\vartheta_4(v-x) \qquad (6.5.28) \\
&= -\vartheta_1(u+v)\vartheta_1(u-v)\vartheta_1(x+y)\vartheta_1(x-y) \\
&\vartheta_4(u+x)\vartheta_4(u-x)\vartheta_1(v+y)\vartheta_1(v-y) \\
&\quad - \vartheta_4(u+y)\vartheta_4(u-y)\vartheta_1(v+x)\vartheta_1(v-x) \\
&= \vartheta_4(u+v)\vartheta_4(u-v)\vartheta_1(x+y)\vartheta_1(x-y)
\end{aligned}
$$

By comparing the reduced Yang–Baxter relation and the addition formulas for theta functions, we can find the solution:

$$
\omega_1(u) = \frac{\vartheta_1(u+\lambda, p)}{\vartheta_1(\lambda, p)}, \qquad \omega_2(u) = \frac{\vartheta_1(\lambda - u, p)}{\vartheta_1(\lambda, p)}
$$
$$
\omega_3(u) = \epsilon \frac{\vartheta_1(u, p)}{\vartheta_1(2\lambda, p)}, \qquad \omega_4(u) = \frac{\vartheta_1(2\lambda - u, p)}{\vartheta_1(2\lambda, p)}
$$
$$(6.5.29)$$

where $\epsilon = \pm 1$ and $\lambda = \pi/4$. Following in this fashion, we can use the Yang–Baxter relations to find exact solutions for the various statistical mechanical models.

Example: Hard Hexagon Model

For example, for the hard hexagon model, there are five independent Boltzmann weights:

$$
\begin{aligned}
&\omega_1 = w(0,0,0,0; u), \qquad \omega_2 = w(0,1,0,0; u) = w(0,0,0,1; u) \\
&\omega_3 = w(1,0,0,0; u) = w(0,0,1,0; u) \qquad\qquad\qquad (6.5.30) \\
&\omega_4 = w(0,1,0; u), \qquad \omega_5 = w(1,0,1,0; u)
\end{aligned}
$$

Inserting these weights into the Yang–Baxter relation, we find that the equations reduce to a set of five equations:

$$\omega_1\omega_2'\omega_1'' + \omega_3\omega_4'\omega_3''' = \omega_2\omega_1'\omega_2'', \qquad \omega_3\omega_1'\omega_1'' + \omega_5\omega_2'\omega_3'' = \omega_1\omega_3'\omega_3''$$
$$\omega_1\omega_2'\omega_3'' + \omega_3\omega_4'\omega_5'' = \omega_4\omega_3'\omega_2'', \qquad \omega_3\omega_1'\omega_3'' + \omega_5\omega_2'\omega_5'' = \omega_2\omega_5'\omega_2'' \quad (6.5.31)$$
$$\omega_3\omega_2'\omega_3'' + \omega_5\omega_4'\omega_5'' = \omega_4\omega_5'\omega_4''$$

Once again, by comparing these equations with the classical theta addition formulas, we find that the solution can be written as:

$$\omega_1 = \frac{\vartheta_1(3\lambda - u)}{\vartheta_1(3\lambda)}, \qquad \omega_2 = \frac{\vartheta_1(\lambda - u)}{\vartheta_1(\lambda)}, \qquad \omega_3 = \frac{\vartheta_1(u)}{\sqrt{\vartheta_1(\lambda)\vartheta(2\lambda)}}$$

$$\omega_4 = \frac{\vartheta_1(4\lambda - u)}{\vartheta_1(4\lambda)}, \qquad \omega_5 = \frac{\vartheta_1(2\lambda - u)}{\vartheta_1(2\lambda)} \qquad\qquad (6.5.32)$$

where $\lambda = \pi/5$.

Example: Eight-Vertex Model

Last, we can also compare the addition formulas with the Yang–Baxter relations coming from the eight-vertex model. We find the exact solution:

$$w(l, l+1, l, l-1; u) = w(l, l-1, l, l+1; u) = \frac{\vartheta_1(\lambda - u)}{\vartheta_1(\lambda)}$$

$$w(l+1, l, l-1, l; u) = w(l-1, l, l+1, l; u) = \pm \frac{\sqrt{\psi(l-1)\psi(l+1)}}{\psi(l)}\frac{\vartheta_1(u)}{\vartheta_1(\lambda)}$$

$$w(l, l+1, l, l+1, ; u) = \frac{\vartheta_1\big[(l+1)\lambda + \omega_0 - u\big]}{\vartheta_1(l\lambda + \omega_0)}$$

$$(6.5.33)$$

where:

$$\psi(a) = \vartheta_1(a\lambda + \omega_0) \qquad\qquad (6.5.34)$$

and λ and ω_0 are arbitrary constants.

6.6. Solitons and the Yang–Baxter Equation

Before leaving this chapter, let us briefly sketch another exactly solvable two-dimensional theory, that of solitons. They will become important for two reasons. First, the heart of their integrability condition is once again the Yang–Baxter relation and the commuting of the transfer matrices. Thus, the language of conformal field theory can be used to describe solitons. Second, as we will see in Chapter 13, the Korteweg–de Vries (KdV) soliton equations become important when we solve matrix models, which give us the first nonperturbative information concerning strings and two-dimensional gravity (albeit in the unphysical dimension $D < 1$). In particular, we will see that the reason why the matrix models are solvable in this domain is the existence of an infinite soliton hierarchy, called the KdV hierarchy.

This points to one of the intriguing mysteries of two-dimensional physics, the wealth of exactly soluble but highly nonlinear field theories. This must, in some sense, have some common origin. Systems as vastly separated as ice crystals of hydrogen, Korteweg–de Vries or sine-Gordon soliton descriptions of water waves, and the string vacuum of the universe are strangely linked together by two-dimensional conformal systems.

Solitons (for solitary wave) have two distinct qualities. They are two-dimensional solutions of nonlinear equations that

1. are localized waves that propagate without changing their properties, such as shape, energy, or velocity, and
2. are stable against mutual collisions; in multiple soliton scattering, the solitons maintain their shape, although they are phase shifted.

What is remarkable is that these models are exactly soluble in two dimensions, that is, they possess an infinite number of conserved quantities I_i. By Liouville's theorem, the model is exactly soluble if there are an infinite number of conserved quantities I_i that are in involution (their Poisson brackets among themselves are all zero).

In solving these nonlinear equations, ingenious methods, such as the *inverse scattering method*, have been devised. However, over the decades, it has become increasingly obvious that the essence of why this inverse scattering method works so well is because of the Yang–Baxter equation. Let us list some of the more well-known soluble models.

Korteweg–de Vries Equation

The first and best-known integrable model is the Korteweg–de Vries equation, formulated by J. Scott Russel to explain the behavior of water waves along the Edinburgh Glasgow canal, which would travel long distances without dispersing. The equation has the form:

$$\frac{\partial u}{\partial t} + au\frac{\partial u}{\partial x} + b\frac{\partial^3 u}{\partial x^3} = 0 \qquad (6.6.1)$$

where $u(x,t)$ is the height of the water wave.

Sine-Gordon Equation

The sine-Gordon equation is given by:

$$\frac{\partial^2 u}{\partial t^2} - \frac{\partial^2 u}{\partial x^2} = \sin u \qquad (6.6.2)$$

It's soliton solution can be written as:

$$u = 4\tan^{-1}\left\{ C\exp\left[\frac{-2(t+vx)}{\sqrt{1-v^2}}\right]\right\} \qquad (6.6.3)$$

Non-Linear Schrödinger Equation

This generalizes the usual linear Schrödinger equation in two dimensions by adding an explicit cubic term to the equations of motion:

$$i\frac{\partial u}{\partial t} = -\frac{\partial^2 u}{\partial x^2} + 2\kappa(u^*u)u \qquad (6.6.4)$$

To analyze the solutions to these systems, we will use the method of Lax pairs and the inverse scattering method, perhaps the most powerful method devised to solve these models. Although the original equations themselves are quite nonlinear, the trick is to invent an auxiliary set of linear equations whose solution is well understood.

Let $u(x,t)$ be a solution of one of the above equations. Then, let $\psi(x,t)$ be a solution of the ordinary linear Schrödinger equation, moving in a potential that is precisely $u(x,t)$. Usually, when solving the linear Schrödinger equation, we begin with a potential and then solve $\psi(x,t)$ moving in that potential. However, the key observation is that this process also works backward: given the scattering data for $\psi(x,t)$, we can reconstruct the potential $u(x,t)$, which is a solution of the original nonlinear equation.

Therefore, we will start with the linear Schrödinger equation with a field $\psi(x,t)$ moving in potential $u(x,t)$, governed by the equation:

$$-\frac{\partial^2 \psi}{\partial x^2} + u(x,t)\psi(x,t) = \lambda\psi(x,t) \qquad (6.6.5)$$

where λ is independent of the time and $u(x,t)$ is a solution of the original nonlinear equation of motion.

The asymptotic form of $\psi(x,t)$ for discrete eigenvalues $\lambda_n = -\kappa_n^2$ is:

$$\psi_n(x,t) = c_n(t)e^{\kappa x}, \qquad x \to -\infty$$
$$c_n(t) = c_n(0)e^{-4\kappa_n^3 t} \qquad (6.6.6)$$

and for continuous eigenvalues $\lambda = \kappa^2$, we also have:

$$\psi = \begin{cases} T(k,t)\exp(ikx + 4ik^3t), & x \to \infty \\ \\ \exp(ikx + 4ik^3t) + R(k,t)\exp(-ikx + 4ik^3t), & x \to -\infty \end{cases} \qquad (6.6.7)$$

where:

$$T(k,t) = T(k,0)$$
$$R(k,t) = R(k,0)\exp(-8ik^3t) \qquad (6.6.8)$$

where $T(k,t)$ is the transmission coefficient and $R(k,t)$ is the reflection coefficient.

Given the scattering data, it a straightforward problem to reconstruct the potential function $u(x,t)$, which in turn is a solution of the original nonlinear equation that we seek to solve. A more systematic way of solving the inverse scattering problem is to set up the equations in the Heisenberg picture.

We recall that in the Heisenberg picture, the operators of the theory are functions of time. Once again, we will set up an auxiliary problem, except this time we will introduce two new operators, called L_n and M_n, which are the *Lax pairs*. They are $M \times M$ matrices, which determine the evolution of the wave function ψ_n via the equations:

$$\psi_{m+1} = L_m(\lambda)\psi_m, \qquad d\psi_m/dt = M_m\psi_m \qquad (6.6.9)$$

Let λ be a constant in time. Now, differentiate the first equation with respect to time and insert the second equation into the first. Then, it is easy to show:

$$dL_m/dt = M_{m+1}L_m - L_m M_m \qquad (6.6.10)$$

which is an operator expression that acts on the wave function $\psi(x,t)$.

We now claim that each of the previous nonlinear equations can be recast in the above form, with suitable choices of the Lax pair. In fact, the model is completely integrable if a Lax pair can be found that is equivalent to the above consistency condition. For example, let us take the simplest example of the KdV equation. If we choose:

$$L(t) = D^2 + \frac{1}{6}u$$
$$M(t) = 4D^3 + \frac{1}{2}(Du + uD) \qquad (6.6.11)$$

then the equation:

$$dL/dt = [M, L] \qquad (6.6.12)$$

is equivalent to the KdV equation.

Let us now construct the transfer matrix by taking a product of N matrices and taking their trace:

$$T_N(\lambda) = \mathrm{Tr}\, t_N$$
$$t_N \equiv [L_N(\lambda)L_{N-1}(\lambda)\cdots L_1(\lambda)] \qquad (6.6.13)$$

Differentiate this equation by t. Because of the Lax equation, it is easy to see that $T_N(\lambda)$ is a constant in time:

$$dT_N(\lambda)/dt = 0 \qquad (6.6.14)$$

Now assume, for the moment, that two transfer matrices with different spectral parameters commute:

$$[T_N(\lambda), T_N(\mu)] = 0 \qquad (6.6.15)$$

If we power expand the transfer matrix $T_N(\lambda)$ as a function of the spectral parameter λ, then we will have an infinite number of conserved quantities, I_i, such that they are in involution:

$$[I_i, I_j] = 0 \qquad (6.6.16)$$

for all i and j. Then, by Liouville's theorem, the model is solvable. For our purposes, however, we recognize the commuting of the transfer matrix $T_N(\lambda)$ as a signal that there is a Yang–Baxter relation at work.

The point of this discussion is to reveal that the essential ingredient for solubility of two-dimensional models is the Yang–Baxter relationship, which is equivalent to commuting transfer matrices. The Yang–Baxter relation, in fact, is so powerful that we can write down analytic solutions to infinite classes of models based on its solutions. The key to understanding two-dimensional quantum systems, and in turn classes of conformal field theory, seems to lie in understanding better the meaning behind the Yang–Baxter relationship.

Last, we mention that we will be meeting the KdV equations from an entirely new point of view when we encounter matrix models in Chapter 13. We will find that the KdV equations are the key to the exact nonperturbative solvability of the string theory in low dimensions.

6.7. Summary

One of the principle uses for conformal field theory, in addition to searching for string vacuums, is to analyze two-dimensional statistical mechanical models at criticality, where the details of the models are washed out and universal characteristics remain that typify a conformal theory. What is interesting is that so many of these models are exactly solvable. The origin of this remarkable property is the Yang–Baxter relation.

We begin with the partition function for a one- or two-dimensional discrete lattice:

$$Z = \sum_n \exp\left[-\frac{E(n)}{kT}\right] \qquad (6.7.1)$$

where $E(n)$ represents the energy of the nth state, k represents the Boltzmann constant, and T represents the temperature. The object is to calculate an exact expression for the free energy, defined as:

$$F = -kT \ln Z \qquad (6.7.2)$$

Correlation functions between spins σ_i and σ_j at criticality exhibit scaling behavior, which can be parameterized by critical exponents. Specifically, the correlation function:

$$g_{ij} = \langle \sigma_i \sigma_j \rangle - \langle \sigma_i \rangle \langle \sigma_j \rangle \qquad (6.7.3)$$

will depend on the distance x separating the states, and at large distances, it will behave like some decreasing power of x multiplied by some exponential:

$$g_{ij} \sim x^{-\tau} e^{-x/\xi} \qquad (6.7.4)$$

where ξ is called the *correlation length*. When the correlation length becomes infinite, we have a phase transition.

For the Ising model, we can introduce the "energy operator" $\epsilon_n = \sigma_n \sigma_{n+1}$. At criticality, when the theory becomes conformally invariant, we have the one-to-one association between the familiar minimal primary fields of conformal field theory and the fields σ and ϵ of the Ising model:

$$\sigma \leftrightarrow \phi_{1/16,1/16}, \qquad \epsilon \leftrightarrow \phi_{1/2,1/2} \qquad (6.7.5)$$

What is surprising is that so many two-dimensional models are exactly solvable. The simplest example is the one-dimensional Ising model, whose partition function describing spins (which take on values of ± 1) can be arranged on a line:

$$Z_N = \sum_\sigma \exp\left(K \sum_{j=1}^{N} \sigma_j \sigma_{j+1} + h \sum_{j=1}^{N} \sigma_j \right) \qquad (6.7.6)$$

We can rewrite this in matrix form as:

$$Z_N = \sum_\sigma V(\sigma_1, \sigma_2) V(\sigma_2, \sigma_3) \cdots V(\sigma_{N-1}, \sigma_N) V(\sigma_N, \sigma_1) \qquad (6.7.7)$$

where we introduce the 2×2 dimensional transfer matrix:

$$V(\sigma, \sigma') = \exp\left[K\sigma\sigma' + \frac{h}{2}(\sigma + \sigma') \right] \qquad (6.7.8)$$

The key is that we can rewrite the partition function as a matrix product over the transfer matrix:

$$Z_N = \operatorname{Tr} \mathbf{V}^N \qquad (6.7.9)$$

This means that we can solve the entire system by diagonalizing the transfer matrix, as follows:

$$Z_N = \operatorname{Tr} \begin{bmatrix} \lambda_1 & 0 \\ 0 & \lambda_2 \end{bmatrix}^N = \lambda_1^N + \lambda_2^N \qquad (6.7.10)$$

Let λ_1 be the larger of the two eigenvalues, which will then dominate the sum in the limit as $N \to \infty$. We then have:

$$f(H,T) = -kT \lim_{N \to \infty} N^{-1} \ln Z_N = -kT \ln \lambda_1$$
$$= -kT \ln \left[e^k \cosh h + \sqrt{e^{2K} \sinh^2 h + e^{-2K}} \right] \qquad (6.7.11)$$

so the system is exactly solvable. (Unfortunately, the system does not exhibit a phase transition at finite temperature.)

More complicated is the two-dimensional Ising model, which can be exactly solved only in the zero magnetic field limit. The two-dimensional Ising model, like its simpler one-dimensional cousin, can be expressed in terms of transfer matrices:

$$V_{\phi,\phi'} = \exp\left[\sum_{j=1}^{n}(K\sigma_{j+1}\sigma'_j + L\sigma_j\sigma'_j)\right]$$

$$W_{\phi,\phi'} = \exp\left[\sum_{j=1}^{n}(K\sigma_j\sigma'_j + L\sigma_j\sigma'_{j+1})\right]$$

(6.7.12)

where W and V are now $2^n \times 2^n$ matrices. We can now perform the sum over the two transfer matrices by summing vertically over the lattice:

$$Z_n = \text{Tr}(\mathbf{VW})^{m/2} = \sum_{i=1}^{2^n} \Lambda_i^m \qquad (6.7.13)$$

As before, we now diagonalize the transfer matrices and then look for the largest eigenvalue. (The details, however, are rather involved.) A careful examination of its critical exponents at the phase transition shows that the model becomes the well-known $m = 3$ minimal model discussed in Chapter 2.

After the original one-dimensional Ising model was proposed in 1925, a wide variety of exactly solvable models were studied. Some of these models can be solved exactly in the continuum limit for all values of the temperature. Some can only be solved exactly at the phase transition. However, all of these models exhibit universality at the critical temperature, so they can be compared to conformal field theories. Since we have exhausted all classifications of the simple bosonic conformal field theories, we should be able to group together certain statistical mechanical models based on their critical exponents and fusion rules according to conformal field theory. Thus, we have a simple classification scheme for statistical mechanical models at criticality. Let us just briefly sketch some of these models.

The *spherical model* has the same partition function as the Ising model, except that the spins obey the constraint:

$$\sum_{j=1}^{N} \sigma_j^2 = N \qquad (6.7.14)$$

(Although this constraint seems unphysical, linking together spins no matter how far apart they are, it can be shown to be a special limiting case of the n-vertex model.) The advantage of this model is that it can be solved exactly in the presence of a magnetic field, which is an advantage over the Ising model.

The *ice-type, six-vertex model* differs from the Ising model qualitatively, because the energy is contained not at the sites but at the links or edges between the sites (that is, the energy is concentrated in the chemical bonds between atoms, and hence, this model can describe systems with hydrogen, such as ice or dihydrogen phosphate crystals).

Its partition function is given by:

$$Z = \sum_{i=1}^{6} \exp \left(-\frac{[n_i \epsilon_i]}{kT} \right) \qquad (6.7.15)$$

where ϵ_i is the energy associated with each link, and n_i is the number of times the ith site is repeated in the lattice. Different values of ϵ_i and n_i describe different types of crystals. The model is exactly solvable, but it exhibits some nonphysical properties, such as complete ordering even at nonzero temperatures. Because of these nonphysical properties, the model was generalized to the *eight-vertex model*, which places constraints on the 16 possible ways in which arrows can be placed going into and out of each site.

The eight-vertex model, for various values of ϵ_i and n_i, is so general it can model both ferromagnetics and ferroelectrics, and in fact, it includes the six-vertex and Ising models as special cases. In fact, for a special case, we can show:

$$Z_{\text{Ising}} = 2 Z_{\text{eight-vertex}} \qquad (6.7.16)$$

(which shows that a model with energy associated with sites, like the Ising model, can be rewritten as a model with energy associated with links, such as the vertex model).

Also, Potts introduced two new models. The first is the $\mathbf{Z}_k$ *model*, where the spins can assume values in $\mathbf{Z}_k$ instead of just ± 1. Spins can now be represented as vectors in this space, so the energy becomes:

$$E = \sum_{i,j} \left\{ \sigma_{ij} \cdot \sigma_{i,j+1} + \sigma_{i,j} \cdot \sigma_{i+1,j} \right\} \qquad (6.7.17)$$

(obviously, this is the familiar Ising model for $N = 2$).

However, the *Potts model* is defined by letting the spin σ_i at the ith lattice site take on values from 1 to q. Two nearest neighbor spins interact via the delta function and are defined as:

$$\begin{aligned} \delta(\sigma, \sigma') &= 1 \quad \text{if} \quad \sigma = \sigma' \\ \delta(\sigma, \sigma') &= 0 \quad \text{if} \quad \sigma \neq \sigma' \end{aligned} \qquad (6.7.18)$$

where δ is 1 if the two spins are the same and 0 if they differ. This model can be solved exactly. For the $q = 2$ case, we have the Ising model. For the case of $q = 3$, we also have a minimal model.

The well-known *XYZ Heisenberg model* has a partition function given by:

$$H = -\frac{1}{2} \sum_{j+1}^{N} \left\{ J_x \sigma_j^x \sigma_{j+1}^x + J_y \sigma_j^y \sigma_{j+1}^y + J_z \sigma_j^z \sigma_{j+1}^z + \cdots \right\} \qquad (6.7.19)$$

where the ellipses represent the interactions in the vertical direction.

If $J_x = J_y = J_z$, then this is the usual Heisenberg model.

If $J_x = J_y = 0$, then only J_z survives, and hence, we obtain the usual Ising model.

If $J_z = 0$, then we have the XY model.

If $J_x = J_y$, then we have the Heisenberg–Ising model.

It can be shown that the Hamiltonian, for any value of the J's, can be written as the logarithmic derivative of an eight-vertex transfer matrix.

The *Ashkin–Teller model* is another generalization of the Ising model, except now we have different species of atoms. Its partition function is roughly the same as the Ising model, except we have different types of spinors to sum over. (The model is not solvable, but its properties at criticality are known.)

The *hard hexagon model* is exactly solvable and represents a gas of hard (that is, non-overlapping) molecules. The partition function is:

$$Z = \sum_{n=0}^{N/3} z^n g(n, N) \qquad (6.7.20)$$

where $g(n, N)$ is the number of ways in which n particles can be placed in each of the various hexagons. There are N sites, and hence, at maximum, only $N/3$ sites can be occupied.

Now, we turn our attention to the main problem, which is the origin of why these models are solvable. A close examination of the steps used to solve these models shows that the key ingredient is that partition functions can be expressed entirely in terms of transfer matrices and that these transfer matrices commute. The mathematical expression of commuting transfer matrices, in turn, is expressed by the Yang–Baxter relationship, one of the deepest results of two-dimensional statistical mechanics.

The Yang–Baxter relation can be expressed graphically in two ways, depending on whether we are studying vertex-type models or IRF (interaction around a face) models.

Let us begin with a vertex model. Let $w(\mu_i, \alpha_i | \beta_i, \mu_{i+1})$ represent the contribution to the sum from the ith site. Each Greek index, in turn, can have values of ± 1, such that there are only six possible orientations. Now, let us introduce the transfer matrix:

$$\mathbf{V}_{\alpha,\beta} = \sum_{\mu_1} \cdots \sum_{\mu_N} w(\mu_1, \alpha_1 | \beta_1, \mu_2) w(\mu_2, \alpha_2 | \beta_2, \mu_3)$$

$$(6.7.21)$$

The partition function can be represented totally in terms of transfer matrices. By demanding that transfer matrices commute, we have a nontrivial relation, the Yang–Baxter equation:

$$\sum_{\alpha\beta\gamma} S^{i\alpha}_{j\beta}(u)S^{\alpha p}_{k\gamma}(u+v)S^{\beta q}_{\gamma r}(v) = \sum_{\alpha\beta\gamma} S^{j\beta}_{k\gamma}(v)S^{i\alpha}_{\gamma r}(u+v)S^{\alpha p}_{\beta q}(u) \qquad (6.7.22)$$

where the S are defined in terms of the w's. Similarly, the IRF models can also be solved via the Yang–Baxter equation, except that the topology of the relation resembles a star and a triangle, hence the name "star–triangle relation."

Last, we conjecture that perhaps all two-dimensional soluble systems have the same origin in the Yang–Baxter relation. One example of this is soliton theory (which we will meet again in Chapter 13). Solitons can be described by $M \times M$ matrices, which determine the evolution of the wave function ψ_n via the equations:

$$\psi_{m+1} = L_m(\lambda)\psi_m, \qquad d\psi_m/dt = M_m\psi_m \qquad (6.7.23)$$

Let λ be a constant in time. Now, differentiate the first equation with respect to time and insert the second equation into the first. Then, it is easy to show:

$$dL_m/dt = M_{m+1}L_m - L_mM_m \qquad (6.7.24)$$

which is an operator expression that acts on the wave function $\psi(x,t)$.

We now claim that each of the previous nonlinear equations can be recast in the above form, with suitable choices of the Lax pair. In fact, the model is completely integrable if a Lax pair can be found that is equivalent to the above consistency condition.

Let us now construct the transfer matrix by taking a product of N L matrices and taking their trace:

$$\begin{aligned} T_N(\lambda) &= \operatorname{Tr} t_N \\ t_N &\equiv \left[L_N(\lambda)L_{N-1}(\lambda)\cdots L_1(\lambda)\right] \end{aligned} \qquad (6.7.25)$$

Differentiate this equation by t. Because of the Lax equation, it is easy to see that $T_N(\lambda)$ is a constant in time:

$$dT_N(\lambda)/dt = 0 \qquad (6.7.26)$$

Now assume, for the moment, that two transfer matrices with different spectral parameters commute:

$$[T_N(\lambda), T_N(\mu)] = 0 \qquad (6.7.27)$$

If we power expand the transfer matrix $T_N(\lambda)$ as a function of the spectral parameter λ, then we will have an infinite number of conserved quantities, I_i, such that they are in involution:

$$[I_i, I_j] = 0 \qquad\qquad (6.7.28)$$

for all i and j. Then, by Liouville's theorem, the model is exactly solvable. Thus, we see once again the strong relationship between exactly solvable two-dimensional systems.

References

1. C. N. Yang, *Phys. Rev.* **85**, 808 (1952); *Phys. Rev. Lett.* **19**, 1312 (1967).
2. R. J. Baxter, *Exactly Solved Models in Statistical Mechanics*, Academic Press, San Diego, 1982; *Ann. Phys.* **70**, 193 (1972).

    For reviews, see Refs. 3–6.

3. E. H. Lieb and F. Y. Wu, in *Phase Transitions and Critical Phenomena*, C. Domb and M. S. Green, eds., Academic Press, San Diego (1972).
4. M. N. Barber in *Phase Transitions and Critical Phenomena*, C. Domb and J. L. Lebowitz, eds., Academic Press, San Diego (1983).
5. H. W. Diehl in *Phase Transitions and Critical Phenomena*, **10**, C. Domb and J. L. Lebowitz, eds., Academic Press, San Diego (1986).
6. B. M. McCoy and T. T. Wu, *The Two-Dimensional Ising Model*, Harvard Univ. Press, Cambridge, Massachusetts (1973).
7. E. Ising, *Z. Physik.* **31**, 253 (1925).
8. L. Onsager, *Phys. Rev.* **65**, 117 (1944).

Chapter 7

Towards a Classification of Conformal Field Theories

7.1. Feigin–Fuchs Free Fields

In order to make some sense out of the jungle of conformal field theories that have been discovered from string theory, physicists have tried to classify these vacuums using various techniques, with varying degrees of success. At present, no comprehensive classification scheme exists that gives us insight into the structure of these vacuums. In fact, it is still largely a mystery why conformal field theories behave as they do. There has been some progress in understanding conformal field theories with finite numbers of primary fields, but there is almost no real understanding of conformal field theories with infinite numbers of primary fields. If a convenient and powerful classification scheme could be devised, then it may be possible to see nontrivial relationships between different conformal field theories, which in turn may help us to understand which, if any, of these conformal field theories have a physical application.

Although a satisfactory classification scheme does not yet exist, in the last few years much progress has been made toward developing different schemes that can partially catalog the multitude of conformal field theories. Let us list some of the major formalisms that have been proposed, mentioning their strong and weak points.

1. *Coset Construction [1]*: The GKO coset construction, reviewed in Chapter 2, was one of the earliest to be discovered and is still one of the most powerful techniques for categorizing conformal field theories. The minimal conformal field theories can be easily constructed via this method. In fact, all known rational conformal field theories can be constructed using the coset construction.

   However, one drawback to this construction is that it reduces the problem of finding representations of the Virasoro algebra to an equally difficult problem, finding representations of the Kac–Moody algebras. In particular, the method has somewhat limited usefulness because, in order to construct the correlation functions for G/H, one must know them for G and H. Although this procedure provides a remarkably versatile method by which to construct conformal field theories, it is sometimes a rather clumsy way in which to actually calculate the characters, primary fields, etc., of the model. More important, however, it give us no deeper understanding into the reason why the multitude of conformal field theories exists, nor does it help us understand the relationships between these theories.

2. *Feigin–Fuchs [2–9]*: The Feigin–Fuchs free field method, in contrast to the coset method, gives us information about computing correlation functions by reducing them to a series of line integrals over the complex plane. By adding fields at infinity, it gives the ability to reduce complicated conformal field theory correlation functions to correlation functions of free fields. Its power is that it gives us a very practical way in which to calculate with a conformal field theory. In fact, it can be shown that all GKO coset constructions can be derived via Feigin–Fuchs free fields. Its disadvantage is that, like the GKO construction, it gives us no deeper understanding of the relationships between conformal field theories.

3. *Landau–Ginzburg and Catastrophe Theory [10–16]*: The Landau–Ginzburg method gives us a new way of looking at the relationship between conformal field theories. The Landau–Ginzburg potential, coupled with renormalization group methods, gives us a way in which certain conformal field theories may flow into each other via renormalization group flows. The Zamolodchikov c theorem, especially, gives us a powerful way in which to see how certain conformal field theories can flow into other ones. For the $N = 2$ superconformal field theories, catastrophe theory may be applied to these potentials. Because the mathematicians have already made great strides in the classification of catastrophe theory, perhaps one can use this classification scheme to classify the Landau–Ginzburg potentials of the $N = 2$ theory.

    Although this formalism is quite beautiful, it is not as general as the other methods. Many conformal field theories cannot be written in terms of Landau–Ginzburg potentials, and they lie outside the classification scheme of catastrophe theory.

4. *Knots and Chern–Simons Theory [17]*: Perhaps the most original attempt in which to approach the classification problem is Witten's use of knot theory. Because a three-dimensional Chern–Simons gauge theory is purely topological (that is, it is generally covariant without any metric tensor), its correlation functions are also purely topological. The correlation functions over Wilson lines give us invariant knot polynomials, generalizing the knot polynomials found independently by the mathematicians.

    If we quantize the system and take a time slice, one dimension is lost, and the theory becomes a two-dimensional conformal field theory. If we apply the Dirac constraints directly onto the Hilbert space, then the physical space is equivalent to the conformal blocks found in Chapter 2. The three-dimensional action, in the Coulomb gauge, becomes a version of the two-dimensional WZW model.

Like the coset and free field construction, knot theory can also describe all known rational conformal field theories. The drawback with this approach is that the beautiful geometry behind knot theory does not, at the moment, give us any insight into the classification of conformal field theory. Like the other schemes, it is still obscure how this approach can reveal to us the relationships between conformal field theories.

In this chapter, we will review the free field and the Landau–Ginzburg approaches. (We will save our discussion of knot theory until Chapter 8, where we will use knot theory to unify many of the features of conformal field theory and statistical mechanics.)

The Feigin–Fuchs method begins with the observation that the N-point correlation functions of the minimal model, which in general are difficult to compute, have a representation entirely in terms of free boson fields. All correlation functions, as well as all structure constants, can be explicitly calculated.

We start by making some elementary observations about the N-point function of N interacting tachyons, which were discussed in Chapter 1:

$$\left\langle V_{\alpha_1}(z_1)V_{\alpha_2}(z_2)\ldots V_{\alpha_N}(z_N)\right\rangle = \prod_{i<j}^{N}(z_i - z_j)^{\alpha_i \alpha_j} \tag{7.1.1}$$

which vanishes unless $\sum_i \alpha_i = 0$. In general, if we wish to take the correlation function of several vertex functions of the same type V_α, then, since the α are all positive and can never sum to zero, we find that the correlation function is trivially zero. Feigin–Fuchs, however, discovered a trick by which general correlation functions can be constructed, even when the α_i do not sum to zero.

Let us recall the discussion of the energy-momentum tensor and vertex operators given in Eqs. (3.1.28)–(3.1.33). Let us choose:

$$T(z) = -\frac{1}{2} : \partial_z\phi\,\partial_z\phi : +i\alpha_0\,\partial_z^2\phi \tag{7.1.2}$$

where we have set $\epsilon = -1$ and $Q = -2i\alpha_0$. Then, the central term is given by $c = 1 - 12\alpha_0^2$, and the conformal weight of the vertex function $: e^{i\alpha\phi(z)} :$ is equal to $\frac{1}{2}\alpha(\alpha - 2\alpha_0)$, that is,

$$T(w) : e^{i\alpha\phi(z)} :\sim \frac{\alpha^2 - 2\alpha\alpha_0}{2(w-z)^2} : e^{i\alpha\phi} : +\frac{1}{w-z}\partial : e^{i\alpha\phi} : \tag{7.1.3}$$

Now, consider the operator [2–5]:

$$Q = \oint_C dz\, J(z), \qquad J(z) =: e^{i\alpha\phi(z)} : \tag{7.1.4}$$

If we choose α_0 such that the integrand $J(z)$ has weight 1, then we have:

$$\alpha^2 - 2\alpha\alpha_0 = 2 \qquad (7.1.5)$$

which has two solutions for α_0:

$$\alpha_\pm = \alpha_0 \pm \sqrt{\alpha_0^2 + 2} \qquad (7.1.6)$$

The whole point of the Feigin–Fuchs construction is that Q is conformally invariant (and hence may be inserted into a correlation function without affecting its conformal properties) but carries a nontrivial "momentum" α, which can be adjusted so that the total "momentum" of an N-point correlation function vanishes. For example, the correlator:

$$\langle V_\alpha V_\alpha V_\alpha V_{2\alpha_0 - \alpha} \rangle \qquad (7.1.7)$$

is usually equal to zero because the momenta do not sum to zero.

However, we will alter this correlator in several ways. First, we will insert a new vertex into the correlator, located at $z = \infty$ with momentum $-2\alpha_0$. This will represent the "screening charge." It will also partially cancel the momentum due to the last vertex function. Second, we can insert as many $Q_\pm$ with momentum $\alpha_\pm$ into the correlator as we want, so let us insert $n - 1$ operators Q_- and $m - 1$ operators Q_+. Third, set all α_i equal to α.

To have a nonzero matrix element, we must have the sum of all momenta, coming from both V_α and $Q_\pm$, equal $2\alpha_0$. Depending on n and m, this fixes the value of α to be α_{nm} via:

$$(3 - 1)\alpha_{n,m} + (n - 1)\alpha_- + (m - 1)\alpha_+ + 2\alpha_0 - 2\alpha_0 = 0 \qquad (7.1.8)$$

Solving for $\alpha_{n,m}$, we find that the conformal weight Δ_{nm} of the vertex $V_{\alpha_{nm}}$ must satisfy:

$$2\Delta_{nm} = \alpha_{n,m}^2 - 2\alpha_{n,m}\alpha_0 = \frac{1}{4}\left[(\alpha_- n + \alpha_+ m)^2 - (\alpha_+ + \alpha_-)^2\right] \qquad (7.1.9)$$

But this, however, is precisely the form of the Kac formula [see Eq. (2.5.7)]. The correspondence becomes complete if we set $\alpha_0^2 = \left[2m(m+1)\right]^{-1}$ and $p = m$ and $q = n$.

This is a rather unexpected, but fortunate, result. If we set the conformal weight of V_α to be the conformal weight found in the minimal model, then we can obtain a nonzero correlation function by inserting a certain number of Q's into Eq. (7.1.7), which do not change the conformal structure of the correlation function but do change the momentum-conservation equation so that nonzero correlation functions are found.

This means that we have found a representation of the correlation functions of the minimal model in terms of free boson fields. Since the transformation properties of these free boson vertex functions agree precisely with the transformation properties of the minimal model's fields $\phi_{n,m}$, then we now have a convenient way in which to calculate the N-point functions and structure constants of the minimal model. For example, we can find an explicit expression of the four-point function over minimal fields $\phi_{n,m}$ if we insert $n-1$ currents with weight α_- and $m-1$ currents with weight α_+ as follows:

$$
\left\langle \phi_{n,m}(z_1)\phi_{n,m}(z_2)\phi_{n,m}(z_3)\phi_{n,m}(z_4) \right\rangle
$$
$$
= \oint_{C_1} du_1 \ldots \oint_{C_{n-1}} du_{n-1} \oint_{S_1} dv_1 \ldots \oint_{S_{m-1}} dv_{m-1}
$$
$$
\times \left\langle V_{\alpha_{n,m}}(z_1) \ldots V_{\alpha_{n,m}}(z_4) J_-(u_1) \ldots J_-(u_{n-1}) \right.
$$
$$
\left. \times J_+(v_1) \ldots J_+(v_{m-1}) \right\rangle
$$

(7.1.10)

where the contour integrals are over circles $\{C_1, \ldots, S_{m-1}\}$, which enclose the points z_i so that they cannot be shrunk down to a point.

We have now replaced an abstract field from the minimal model $\phi_{n,m}$ with a specific representation given by free fields whose matrix elements are all known. We can do this because the left- and right-hand sides of the previous equation have the same conformal properties. The final contraction over the vertices and currents is now trivial, since all fields are written in terms of free bosons. Thus, we have reduced a potentially difficult problem, the calculation of N-point functions over minimal fields, to a much simpler, almost trivial one: performing line integrals over complex-valued vertices constructed from free fields.

Example: Four-Point Function

For example, let us use this deceptively simple method to calculate the explicit value of a correlation function of minimal fields, such that one of the fields is equal to $\phi_{1,2}$ (we now rescale $\alpha \to \sqrt{2}\alpha$ to agree with the literature):

$$
\left\langle \phi_{n_1,m_1}(0)\phi_{1,2}(z)\phi_{n_3,m_3}(1)\phi_{n_4,m_4}(\infty) \right\rangle
$$
$$
= \oint_C dt \left\langle V_{\alpha_1}(0)V_{\alpha_2}(z)V_{\alpha_3}(1)V_{\alpha_4}(\infty)J_+(t) \right\rangle
$$

(7.1.11)

$$
= z^{2\alpha_1\alpha_2}(1-z)^{2\alpha_2\alpha_3} \oint_C dt\, t^a(t-1)^b(t-z)^c
$$

where $a = 2\alpha_1\alpha_+$, $b = 2\alpha_3\alpha_+$, $c = 2\alpha_2\alpha_+$, $\alpha_2 = -\frac{1}{2}\alpha_+$, and $\alpha_4 = 2\alpha_0 - \alpha_1 - \alpha_2\alpha_3 - \alpha_+$.

Although the final answer is unique, there is some arbitrariness in defining the line integrals (which is eliminated once we fix the monodromy properties of the integral). An incorrect choice of the line integral, for example, could lead to a vanishing result.

There are two line integrals in this calculation, corresponding to the two linearly independent solutions to the hypergeometric equation. It is straightforward to write these line integrals, in turn, as hypergeometric functions. Let us define the following:

$$I_1(a, b, c; z) = \int_1^\infty dv\, v^a (v-1)^b (v-z)^c = \frac{\Gamma(-a-b-c-1)\Gamma(b+1)}{\Gamma(-a-c)}$$
$$\times F(-c, -a-b-c-1, -a-c; z)$$

$$I_2(a, b, c; z) = \int_0^z dv\, v^a (1-v)^b (z-v)^c = z^{1+a+c} \frac{\Gamma(a+1)\Gamma(c+1)}{\Gamma(a+c+2)}$$
$$\times F(-b, a+1, a+c+2; z)$$

$$(7.1.12)$$

where F is the standard hypergeometric function. The final result for the correlation function is a function of both z and $\bar{z}$ and is constructed out of I_1 and I_2. Thus, the correlation function must be a function of:

$$G(z, \bar{z}) = \sum X_{ij} I_i \bar{I}_j(\bar{z}) \tag{7.1.13}$$

where the X_{ij} can be determined by making changes in the contour integrations. The final answer is [3–5]:

$$\left\langle \phi_{n_1, m_1}(z_1) \phi_{n_2, m_2}(z_2) \phi_{n_3, m_3}(z_3) \phi_{n_4, m_4}(z_4) \right\rangle$$
$$\sim \frac{|z_{13}|^{\beta_{13}} |z_{24}|^{\beta_{24}}}{|z_{12}|^{\beta_{12}} |z_{23}|^{\beta_{23}} |z_{34}|^{\beta_{34}} |z_{14}|^{\beta_{14}}} G(z, \bar{z}) \tag{7.1.14}$$

where:

$$G(\eta) = \sin \pi(a+b+c) \sin \pi(b) |I_1(a, b, c; \eta)|^2$$
$$+ \sin \pi(a) \sin \pi(c) |I_2(a, b, c; \eta)|^2 \tag{7.1.15}$$

and $\eta = z_{12} z_{34} / z_{13} z_{24}$ and:

$$\beta_{13} = 2\big[\Delta(\alpha_1 + \alpha_3 + \alpha_+) - \Delta_1 - \Delta_3 + 2\alpha_+\alpha_2\big]$$
$$\beta_{24} = 2\big[\Delta(\alpha_2 + \alpha_4) - \Delta_2 - \Delta_4 + 2\alpha_+\alpha_2\big]$$
$$\beta_{12} = -2\big[\Delta(\alpha_1 + \alpha_2) - \Delta_1 - \Delta_2\big]$$
$$\beta_{23} = -2\big[\Delta(\alpha_2 + \alpha_3) - \Delta_2 - \Delta_3\big] \tag{7.1.16}$$
$$\beta_{34} = -2\big[\Delta(\alpha_3 + \alpha_4 + \alpha_+) - \Delta_3 - \Delta_4\big]$$
$$\beta_{14} = -2\big[\Delta(\alpha_1 + \alpha_4 + \alpha_+) - \Delta_1 - \Delta_4\big]$$

where Δ are the conformal weights.

Example: General Case

We can now present the general case, which appears formidable but is actually a straightforward application of the ideas presented previously.

First, we start with the correlation function of a product of a series of $\phi_{n,m}$. Then, we replace each minimal field with a vertex function:

$$\phi_{n,m} \to V_{\alpha_{n,m}} \tag{7.1.17}$$

To prevent the correlation function from vanishing, we have to insert the requisite number of charges $Q_\pm$ (with weight one) within the correlation function, such that the total "momentum" still vanishes. The contraction over the free boson field can now be trivially performed, and we are left with a series of line integrals. The last step is to write the final correlation function as:

$$\sum_{ij} X_{ij} I_i \bar{I}_j \tag{7.1.18}$$

where I_i are various line integrals, and then use the monodromy properties of the correlation function to fix X_{ij}.

This procedure can be performed for any N-point function, but let us present only the final result for the four-point function with four totally arbitrary minimal fields. As before, the four-point function can be written as a function of a set of contour integrals $I_{lk}^{(nm)}$ multiplied by normalization factors $X_{lk}^{(nm)}$ [3–5]:

$$\left\langle \phi_{n_1,m_1}(0)\phi_{n_2,m_2}(z)\phi_{n_3,m_3}(1)\phi_{n_4,m_4}(\infty) \right\rangle = \sum_{l,k} X_{lk}^{(nm)} |I_{lk}^{(nm)}|^2 \tag{7.1.19}$$

These contour integrals, in turn, can be written in terms of the following factors:

$$I_{lk}^{(nm)} = N_{lk}^{(nm)} F_{lk}^{(nm)} \tag{7.1.20}$$

where:

$$N_{lk}^{(nm)} = J_{n-l,m-k}\left[-a-b-c-2\rho(m-2)+2(n-2),b;\rho\right] J_{l-1,k-1}(a,c;\rho) \tag{7.1.21}$$

and:

$$\begin{aligned}
J_{nm}(\alpha,\beta;\rho) = (\rho')^{2mn} \prod_{i=1}^{n} \frac{\Gamma(i\rho'-m)}{\Gamma(\rho')} \prod_{i=1}^{m} \frac{\Gamma(i\rho)}{\Gamma(\rho)} \\
\times \prod_{i=1}^{n} \frac{\Gamma(1-m+\alpha'+i\rho')\Gamma(1-m+\beta'+i\rho')}{\Gamma[2-m+\alpha'+\beta'+(n-1+i)\rho']} \\
\times \prod_{i=1}^{m} \frac{\Gamma(1+\alpha+i\rho)\Gamma(1+\beta+i\rho)}{\Gamma[2-2n+\alpha+\beta+(m-1+i)\rho]}
\end{aligned} \tag{7.1.22}$$

and the normalization factors X are given by:

$$X_{lk}^{(nm)}(a,b,c;\rho) = X_l^{(n)}(a',b',c';\rho')X_k^{(m)}(a,b,c;\rho) \tag{7.1.23}$$

where:

$$X_k^{(m)} = \prod_{i=1}^{k-1} s(i\rho) \prod_{i=1}^{m-k} s(i\rho) \prod_{i=0}^{k-2} \frac{s(1+a+i\rho)s(1+c+i\rho)}{s[2+a+c+(k-2+i)\rho]}$$

$$\times \prod_{i=0}^{m=k-1} \frac{s(1+b+i\rho)s[-1-a-b-c-2\rho(m-2)+i\rho]}{s[-a-b-2\rho(m-2)+(m-k-1+i)\rho]}$$

$$(7.1.24)$$

where $s(a) = \sin \pi a$ and:

$$F_{lk}^{(nm)}(z) \equiv f_{lk}^{(nm)}(z)(z)^q$$
$$q \equiv (l-1)[1+a'+c'+\rho'(l-2)] \qquad (7.1.25)$$
$$+ (k-1)(1+a+c+\rho(k-2)) - 2(l-1)(k-1)$$

where $f_{lk}^{(nm)}(z)$ is regular at $z = 0$ and $f_{lk}^{(nm)}(0) = 1$. Also: $\rho = \alpha_+^2$, $a' = -\rho^{-1}a$, $b' = -\rho^{-1}b$, $c' = -\rho^{-1}c$.

7.2. Free Field Realizations of Coset Theories

So far, we have only used the free field method to analyze simple models, like the minimal model. However, the free field method is much more powerful than that. Now, we wish to analyze generalized free field constructions in order to realize Kac–Moody algebras, coset constructions, and $N = 2$ superconformal theories [8, 9]. Thus, the free field construction is one of the most general schemes proposed.

Free Field Construction of Kac–Moody Algebras

First, let us analyze the free field Kac–Moody representation by introducing three sets of free fields $\beta_{-\alpha}, \gamma_\beta, \phi^i$, where α and β represent the positive roots of some Lie algebra and i is the index for the Cartan subalgebra. Our goal is to construct the Kac–Moody generators out of these free fields. We postulate the following operator product expansions:

$$\beta_{-\alpha}(z)\gamma_\beta(w) \sim \frac{\delta_{\alpha\beta}}{z-w} \qquad (7.2.1)$$
$$\phi^i(z)\phi^j(w) \sim -\delta^{ij} \ln(z-w)$$

First, the generators of the Cartan subalgebra can be written as:

$$H^i(z) = -i\alpha_+ \partial \phi^i + \sum_{\alpha \in G} \alpha^i \beta_{-\alpha} \gamma_\alpha(z) \qquad (7.2.2)$$

where the sum runs over positive roots and where $\alpha_+ = \sqrt{k+g}$ is the second-order Casimir of the group G, that is, $g = n$ for A_n, $g = n - 2$ for

$SO(n)$ $(n \geq 5)$, and $g = n + 1$ for C_n. Likewise, the currents for negative roots can be written as:

$$E_{-\alpha}(z) = \beta_{-\alpha}(z) + \sum_{\rho - \sigma = -\alpha} N_{\rho\sigma}\gamma_\rho(z)\beta_{-\sigma}(z) \qquad (7.2.3)$$

where the N matrices can always be chosen so that the algebra formed by the E's and H's agrees with the usual definition of the Lie algebra, as in Eq. (3.2.6).

Last, the energy-momentum tensor can be written as:

$$T(z) = \sum_{\alpha \in G} \beta_{-\alpha} \partial\gamma_\alpha - \frac{1}{2} \sum_{i=1}^{\text{rank } G} [\partial\phi^i(z)]^2 - \frac{i}{\alpha_+} \sum_{i=1}^{\text{rank } G} \rho^i \partial^2 \phi^i(z) \qquad (7.2.4)$$

where ρ^i is half the sum of the positive roots. (Notice the last term in the expression for the energy-momentum tensor. Because it is linear in the fields, it shows the presence of the screening charges that are typically found in the Feigin–Fuchs construction.)

It is now a simple matter to calculate the operator product expansion of the various operators. By multiplying two energy-momentum tensors, we can calculate the central charge of the algebra, which yields [8, 9]:

$$c_G = \dim G - 12\rho^2/\alpha_+^2 \qquad (7.2.5)$$

If we use the "strange formula" of Freudenthal and de Vries,

$$\rho^2 = \frac{1}{12} g^2 \dim G \qquad (7.2.6)$$

we find the correct central charge of the Kac–Moody algebra:

$$c_G = \frac{k \dim G}{k + g} \qquad (7.2.7)$$

Equations (7.2.2), (7.2.3), and (7.2.4) then define the complete generators of the Kac–Moody and Virasoro algebras in terms of Feigin–Fuchs free fields.

Free Field Coset Construction

Next, we will use the free field construction to give us the GKO coset construction. As before, we begin with the same set of fields and the same T_G. However, there is a small complication in constructing the current J^a for the subgroup H. We demand that the current have the operator product expansion:

$$T_G(z)J'^a(w) \sim \frac{J'^a}{(z - w)^2} + \frac{\partial J'^a}{z - w} \qquad (7.2.8)$$

We demand that J'^a also have the same operator product expansion with respect to T_H. Thus, the difference $T_{G/H} = T_G - T_H$ has the following operator product expansion:

$$T_{G/H}(z)J'^a(w) \sim 0 \tag{7.2.9}$$

The trick is to find the representation of J'^a in terms of free fields. If we naively take the construction used previously for Kac–Moody operators, we find that free fields will not work.

We will, therefore, have to modify some of our operators in order to construct J'^a. We will construct the generators of the subgroup H out of modified fields denoted by a prime. We construct the generators of the subgroup H by identifying the fields $\beta'_{-\alpha}$ and γ'_α with those of the full group G. To calculate ϕ', we equate the generators H^i and H'^i to be the same. Equating the two, we find all the terms are the same except:

$$-i\sqrt{k+h}\,\partial\phi'^i(z) = -i\sqrt{k+h}\,\partial\phi^i(z) + \sum_{\alpha \in G/H} \alpha^i \beta_{-\alpha}\gamma_\alpha \tag{7.2.10}$$

Then, we easily find the expression for J'^a, as well as [8, 9]:

$$T_{G/H}(z) = \sum_{\alpha \in G/H} \beta_{-\alpha}(z)\,\partial\gamma_\alpha(z) - \frac{1}{2}[\partial\rho(z)]^2 - \frac{i}{\sqrt{k+g}}\,\rho_G\,\partial^2\phi(z)$$
$$- \left\{ -\frac{1}{2}[\partial\phi'(z)]^2 - \frac{i}{\sqrt{k+h}}\,\rho_H\,\partial\phi'(z) \right\} \tag{7.2.11}$$

Free Fields and Supercoset Models

Next, we can use the free field construction to give us the superconformal theory as well. The generalization to the superconformal case is straightforward. We simply double all the fields by including a Grassmann variable, so $Z = (z, \theta)$. The operator product expansion of the free fields now generalize to:

$$B_{-\alpha}(Z)C_\beta(Z') \sim \frac{\delta_{\alpha\beta}(\theta - \theta')}{Z - Z'}$$
$$\Phi^i(Z)\Phi^j(Z') \sim -\delta^{ij}\,\ln(Z - Z') \tag{7.2.12}$$

The generators of the Cartan subalgebra become:

$$H^i(Z) = -i\sqrt{k}\,D\Phi^i(Z) + \sum_{\alpha \in G} \alpha^i B_{-\alpha}(Z)C_\alpha(Z) \tag{7.2.13}$$

In this way, we find that the energy-momentum tensor for the coset is given by:

$$T_G(Z) \sim \frac{1}{2} \sum_{\alpha \in G} \left[B_{-\alpha}(Z)\,\partial C_\alpha(Z) + DB_{-\alpha}DC_\alpha(Z) \right]$$
$$- \frac{1}{2}\,D\Phi(Z)\,D^2\Phi(Z) - \frac{i}{\sqrt{k}}\,\rho_G\,D^3\Phi(Z) \tag{7.2.14}$$

As expected, we find that the central charge is given by [8, 9]:

$$\frac{c}{3} = \frac{1}{2} \dim G - \frac{4\rho_G^2}{k} = \left(\frac{1}{2} - \frac{g}{3k}\right) \dim G \qquad (7.2.15)$$

As before, we find that the naive construction of the Kac–Moody current fails for free fields. Once again, we construct the generators of the subgroup H by identifying the fields $B'_{-\alpha}$ and C'_α with those of the group G, and the fields Φ' are also chosen by setting H^i equal to H'^i, giving us:

$$D\Phi'^i(Z) = D\Phi^i(Z) + \frac{i}{\sqrt{k}} \sum_{\alpha \in G/H} \alpha^i B_{-\alpha} C_\alpha(Z) \qquad (7.2.16)$$

The energy-momentum tensor of the coset can now be represented by:

$$\begin{aligned}
T_{G/H}(Z) = \frac{1}{2} \sum_{\alpha \in G/H} &[B_{-\alpha}(Z)\,\partial C_\alpha(Z) + DB_\alpha(Z)\,DC_\alpha(Z)] \\
&- \frac{1}{2} D\Phi(Z)\,D^2\Phi(Z) - \frac{i}{\sqrt{k}} \rho_G\, D^3\Phi(Z) \\
&+ \frac{1}{2} D\Phi'(Z)\,D^2\Phi'(Z) + \frac{i}{\sqrt{k}} \rho_H\, D^3\Phi'(Z)
\end{aligned} \qquad (7.2.17)$$

Free Fields and $N = 2$ Superconformal Algebra

Last, we show that the $N = 2$ superconformal field theories can also be written via free fields. The problem, as we saw in the previous chapter, is to find a scalar current $J(Z)$ that will generalize the $N = 1$ algebra into an $N = 2$ algebra.

In terms of free fields, the current is:

$$\begin{aligned}
J(Z) = \sum_{\alpha \in G/H} \Big\{ &B_{-\alpha}DC_\alpha \\
&+ \frac{2}{k}(\rho_G \cdot \alpha)D(B_{-\alpha}C_\alpha) + \frac{i}{\sqrt{k}}\left[B_{-\alpha}C_\alpha(\alpha \cdot D\Phi) + \alpha \cdot D^2\Phi\right] \Big\} \\
&+ \sum_{\alpha,\beta \in G/H} A_{\alpha,\beta}\left[2D(B_{-\alpha}C_\alpha) + B_{-\alpha}C_\alpha B_{-\beta}C_\beta\right]
\end{aligned}$$
$$(7.2.18)$$

where $A_{\alpha\beta}$ is an antisymmetric matrix that does not contribute to the energy-momentum tensor. This field, in turn, has the correct operator product expansion [8, 9]:

$$J(Z)J(Z') \sim \frac{c}{3(Z-Z')} + \frac{2(\theta - \theta')}{Z-Z'}T(Z')$$
$$T(Z)J(Z') \sim \frac{(\theta - \theta')}{(Z-Z')^2}J(Z') + \frac{1}{2(Z-Z')^2}D'J(Z') + \frac{(\theta - \theta')}{Z-Z'}\partial'J(Z')$$
$$(7.2.19)$$

In this way, we can construct free field representations for the $N = 2$ theories of Gepner, Kazama, and Suzuki.

7.3. Landau–Ginzburg Potentials

The Landau–Ginzburg method [10] approaches conformal field theory from a different point of view, using renormalization group arguments with an initially nonconformally invariant theory. We start with a scalar theory with the following interaction term:

$$L \sim g \int d^2x \ \Phi^{2(p-1)} \tag{7.3.1}$$

for fixed p. We notice immediately that the theory is not conformally invariant by dimensional arguments. However, it is possible to calculate the β function of the theory and find where it vanishes, that is, its fixed points. When the β function vanishes, then the theory becomes conformally invariant.

In this way, it is possible to find the relationship between the fixed points of the Landau–Ginzburg action, where the theory becomes conformally invariant, and known conformal field theories. In particular, we will compare the composite operators for fixed p that one can construct from the above interaction and then compare them with the operators appearing in the minimal series, and we shall argue that there seems to be a correspondence between them. We find that the theory defined by the potential $\Phi^{2(p-1)}$ at criticality corresponds to the familiar unitary minimal model with $c = 1 - 6/p(p+1)$.

To show the relation, we will argue that $: \Phi^k :$ has the same conformal expansion as one of the minimal fields $\phi_{k+1,k+1}$, and hence we can establish a one-to-one correspondence between Landau–Ginzburg composite operators and minimal primary fields.

At the fixed point, one finds a series of composite operators that are equal to powers of the scalar field Φ^n for $n = 1, 2, \ldots, 2p-4$, as well as derivatives of these fields. (For the sake of argument, we will assume that we have averaged over all two-dimensional directions so that derivative terms will be dropped.) Fields with powers higher than $2p-4$ are discarded. (This is because the equations of motion for the theory:

$$\partial_{\bar{z}}\partial_z\Phi \sim \Phi^{2p-3} \tag{7.3.2}$$

show that Φ^{2p-3} must have dimension greater than 2. However, according to renormalization group theory, the addition of operators with dimension greater than 2 does not change the point to which the renormalization group flows.)

These composite fields, of course, have no meaning until we define what they mean by proper normal ordering. However, for higher powers of the field Φ^n, the definition of normal ordering is ambiguous (because we must subtract divergent terms that are now operators, not just ordinary numbers).

To provide a self-consistent definition of normal ordering for higher order operators, let us first reexamine the operator product expansion for two fields:

$$\Phi(z)\Phi(0) - \langle \Phi(z)\Phi(0)\rangle \sim |z|^{d_2 - 2d_1}\Phi_2(0) + \cdots \qquad (7.3.3)$$

which serves to define the composite field $:\Phi^2: \equiv \Phi_2$ (where d_i is the anomalous dimension of the ith field). We can use the above equation to successively define what we mean by normal ordering for higher powers.

Thus, assuming that the kth composite field is well defined, we can make sense out of expressions like Φ^{k+1} via:

$$:\Phi^{k+1}:(0) \equiv \lim_{z \to 0} |z|^{d_k + d_1 - d_{k+1}} \left[:\Phi(z): \Phi^k(0) \right.$$
$$\left. - \sum_{q=1}^{k/2} A_q |z|^{d_{k-2q} - d_1 - d_k} :\Phi^{k-2q}(0): \right] \qquad (7.3.4)$$

where the coefficients A_k are chosen so that the series is well defined.

Now that the operator product expansion for all composite fields is well defined, let us compare this with the operator product expansion found for minimal models, for example,

$$\phi_{2,2}\phi_{n,m} \sim \sum_{k,l} X_{n,m}^{(k,l)} \left[\phi_{n+k,m+l} + \cdots \right] \qquad (7.3.5)$$

If we make the correspondence $\Phi \leftrightarrow \phi_{2,2}$ and compare the two sets of operator product expansions, then we find that we can make the correspondence between the two sets of fields. In the previous chapters, we computed the value of the structure constants, so it is now a simple matter to compare the two operator product expansions for Φ^k and for the minimal primary fields $\phi_{n,m}$ and find the correspondence between the two sets of fields. We find that we can make the correspondence [10]:

$$\begin{aligned} :\Phi^k: &\leftrightarrow \phi_{k+1,k+1}, & k &= 0,1,\ldots,p-2 \\ :\Phi^k: &\leftrightarrow \phi_{k-p+2,k-p+3}, & k &= p-1,p,p+1,\ldots,2p-4 \end{aligned} \qquad (7.3.6)$$

Now, let us analyze the correspondence between the superconformal minimal series and a superfield Landau–Ginzburg theory at criticality. Let us start with the Landau–Ginzburg action:

$$L = \int d^2z\, d^2\theta \left[\frac{1}{2} D\Phi\, \bar{D}\Phi + g\Phi^p \right] \qquad (7.3.7)$$

where we introduce the following derivatives:

$$D = \partial_\theta - \theta\partial_z, \qquad \bar{D} = \partial_{\bar\theta} - \bar\theta\partial_{\bar z} \qquad (7.3.8)$$

and:

$$\Phi = \phi + \theta\psi + \bar\theta\bar\psi + \theta\bar\theta\chi \qquad (7.3.7)$$

By contrast, the superconformal minimal series is defined by the series [see Eq. (2.7.10)]:

$$c = \frac{3}{2} - \frac{12}{p(p+2)}, \qquad p = 2, 3, \dots \qquad (7.3.10)$$

The primary fields $\phi_{n,m}$ obey:

$$\phi_{n,m} = \phi_{p+2-n,p-m}, \qquad n = 1, 2, \dots, p+1; \qquad m = 1, 2, \dots, p-1 \qquad (7.3.11)$$

for $m + n$ odd. This field has dimension:

$$h_{nm} = \frac{[(np - m(p+2))]^2 - 4}{4p(p+2)} \qquad (7.3.12)$$

Then, by once again making the correspondence between $\Phi = \phi_{2,2}$ and by checking the one-to-one correspondence between the operator product expansion of Φ^k and $\phi_{n,m}$, we can make the correspondence:

$$: \Phi^k : \leftrightarrow \phi_{k+1,k+1}, \qquad k = 0, 1, 2, \dots, p-2 \qquad (7.3.13)$$

The same correspondence can be established for the $N = 2$ superconformal series. The Landau–Ginzburg action for this theory is given by:

$$L = \int d^2z \, d^4\theta \, D(\Phi, \bar\Phi) + g \int d^2z \, d^2\theta \, F(\Phi) \qquad (7.3.14)$$

where the first term is called the D term and is integrated over all four values of θ, while the chiral F term is integrated over only two of them. We choose $F = \Phi^n$.

Let us now compare this with the superconformal minimal series, which is given by:

$$\tilde{c} = 1 - (2/n) \qquad (7.3.15)$$

The fields are given by ϕ_{jm}, which have conformal weights h_{jm} and $U(1)$ charges q_{jm} given by:

$$h_{jm} = \frac{j(j+2) - m^2}{4n}, \qquad q_{jm} = \frac{m}{n} \qquad (7.3.16)$$

Finally, we make the crucial identification of

$$\Phi^p \leftrightarrow \phi_{p,p} \qquad (7.3.17)$$

7.4. N=2 Chiral Rings

The most physically interesting case is studying the renormalization flows of the Landau–Ginzburg potentials of $N = 2$ superconformal symmetry. As we emphasized earlier, the only consistent theories of interacting superstrings have at least $N = 1$ supersymmetry once we restrain the one-loop amplitudes to be modular invariant. Thus, $N = 1$ space–time supersymmetry (or $N = 2$ superconformal symmetry) seems to be the minimum symmetry required in model building. (If $N = 2$ space–time supersymmetry survived after compactification, then left and right multiplets would appear in the same supersymmetric representation, which is phenomenologically undesirable.)

Several new features emerge when we discuss renormalization flows and Landau–Ginzburg potentials for $N = 2$ superconformal models. First, there is an interesting rotation one can perform on the generators of the $N = 2$ algebra, which turns the NS sector into the R sector and vice versa. Let us define an operator U_θ that has the following properties:

$$U_\theta^{-1} L_n U_\theta = L_n + \theta J_n + \frac{c}{6}\theta^2 \delta_{n,0}$$
$$U_\theta^{-1} J_n U_\theta = J_n + \frac{c}{3}\theta \delta_{n,0}$$
$$U_\theta^{-1} G_r^+ U_\theta = G_{r+\theta}^+$$
$$U_\theta^{-1} G_r^- U_\theta = G_{r-\theta}^+$$

$$(7.4.1)$$

It is easy to check that the deformed generators still satisfy the same commutation relations as the original algebra. Thus, the operator U_θ maps the original Hilbert space into a rotated Hilbert space parameterized by θ.

Under this rotation, the $U(1)$ charge and dimension of a state shift by the following amount:

$$q \to q + (c\theta/3)$$
$$h \to h + \theta q + (c\theta^2/6)$$

$$(7.4.2)$$

If θ is half-integral, then the rotation maps the integer (half-integer)-valued G_r operators into half-integer (integer)-valued operators. Thus, we have the most remarkable fact that, for $\theta = Z + \frac{1}{2}$, the NS algebra rotates into an R algebra and vice versa. This deformation of the algebra is called the *spectral flow* connecting the NS and R algebras [18]. To show that this spectral flow is not a fluke, an explicit representation of U_θ can be written if we introduce a scalar ϕ field, which bosonizes the J current:

$$J(z) = \sqrt{\frac{c}{3}}\,\partial\phi(z), \qquad U_\theta = e^{i\theta\sqrt{c/3}\phi}$$

$$(7.4.3)$$

Next, we wish to construct representations of this $N = 2$ algebra, in order to construct the Landau–Ginzburg potentials. We make a few definitions. A left *chiral* NS field is one that satisfies:

$$G^+_{-1/2}|\phi\rangle = 0 \qquad (7.4.4)$$

This notation comes from the theory of supersymmetry, where a chiral superfield $\phi(x,\theta)$ is one that satisfies $D\phi(x,\theta) = 0$. (Since Q and D anticommute, placing this restriction on ϕ does not affect the fact that it is still a representation of supersymmetry.)

Second, we define a *primary field* for the $N = 2$ theory as one that satisfies both:

$$G^-_{n+1/2}|\phi\rangle = G^+_{n+1/2}|\phi\rangle = 0; \qquad n \geq 0 \qquad (7.4.5)$$

in analogy with the usual definition of primary fields for bosonic fields. A primary chiral field is one which satisfies both conditions.

Let us take the commutator of these conditions:

$$\{G^-_{1/2}, G^+_{-1/2}\}|\phi\rangle = (2L_0 - J_0)|\phi\rangle = 0 \qquad (7.4.6)$$

Therefore, a primary chiral field satisfies:

$$h = q/2 \qquad (7.4.7)$$

(If the condition $h = -q/2$ is satisfied, then we call it an *antichiral field*.) Similarly, if we take the commutator:

$$\{G^-_{3/2}, G^+_{-3/2}\} = 2L_0 - 3L_0 + 2c/3 \qquad (7.4.8)$$

then we have:

$$h \leq c/6 \qquad (7.4.9)$$

for any primary chiral field.

This has two very interesting consequences. First, it shows that there are only a finite number of primary chiral operators, which is unexpected. (This is because the dimension of each primary chiral operator is less and or equal to $c/6$, but the spectrum of L_0 is discrete, which can only be satisfied if we have a finite number of primary chiral operators.) Second, it shows that the algebra formed by taking operator product expansions of products of primary chiral fields produces a *finite chiral ring* R_{chiral} of operators [11–16].

If we take the operator product expansion of two primary chiral fields ϕ_1 and ϕ_2, then we will produce a composite operator ϕ_{12} with the following dimension:

$$h_{12} \geq \frac{1}{2}(q_1 + q_2) = h_1 + h_2 \qquad (7.4.10)$$

The product of primary chiral fields also produces primary chiral fields. Since there are only a finite number of them, we obtain a finite chiral ring R_{chiral} of such operators whose products form a closed system.

Since many of the properties of an $N = 2$ superconformal field theory are determined once R_{chiral} is fixed, our goal in the next section is to find some way in which to mathematically categorize the various possible R_{chiral}. This is where catastrophe theory enters our discussion of string theory.

7.5. N=2 Landau–Ginzburg and Catastrophe Theory

We now make contact with the $N = 2$ superconformal chiral rings and the Landau–Ginzburg formalism, which we began earlier. Previously, in Eq. (7.3.14), we constructed the most general superpotential involving the chiral superfields Φ_i and $\bar{\Phi}_i$. The first contained an integration over all four θ's, and is called the D term, while the second contained an integration over only two θ's, meaning that $F(\Phi)$ is a chiral superfield.

Let us introduce a new F term, called W. Let us scale the superfields x_i contained within W according to $x_i \rightarrow \lambda^{n_i} x_i$. Then, we define W to have the following scaling property:

$$W(\lambda^{n_i} x_i) = \lambda^d W(x_i) \qquad (7.5.1)$$

If W has $U(1)$ charge $(1,1)$, then this means that X_i must have charge $q_i = n_i/d$. We see that each superfield scales differently according to n_i, but that the overall function W scales by the same amount, regardless of how it depends on the various x_i.

Our next task is to construct the ring formed by forming all products of the superfields x_i, modulo terms that contain factors of $\partial_j W(x_i)$. This ring has a finite number of terms, and can be written symbolically as:

$$R_{\mathrm{LG}} = \frac{\prod_i x_i}{[\partial_j W]} \qquad (7.5.2)$$

(This means that whenever factors of the derivatives of W appear in the monomials formed by x_i, we set them to zero). The interesting conclusion that we will draw is that, for a wide variety of models, the two rings are the same [11–16]:

$$R_{\mathrm{chiral}} = R_{\mathrm{LG}} \qquad (7.5.3)$$

This is a powerful result that will significantly help us in the task of categorizing the possible $N = 2$ superconformal field theories via Landau–Ginzburg potentials.

Second, we can further the identification of superconformal theories by calculating their central charge. This will help to identify the various possible conformal field theories. Let us rescale the two-dimensional metric on the world sheet by an overall factor λ. In this case, the partition function Z for a conformal field theory defined on a sphere also gets rescaled. The effect of this rescaling has already been computed using functional methods [11, 19]:

$$g_{ab} \rightarrow \lambda^2 g_{ab}$$

$$Z \rightarrow \left(\exp \frac{c}{48\pi} \ln \lambda \int R \right) Z = \lambda^{c/6} Z \qquad (7.5.4)$$

where R is the curvature on the world sheet. The last integral can be evaluated since the integral of the curvature tensor yields 8π for the sphere.

The previous result was independent of the specific model we are analyzing. Now, let us take a specific model and perform this rescaling. We have to take the product of several different contributions.

First, we have the rescaling of the functional measure. The measure gets rescaled by $\lambda^{c/6}$, so we must calculate the c for each superfield. Each boson contributes $c = 1$, and each fermion $c = \frac{1}{2}$. Because a superfield has a complex boson and a complex fermion, the value of c is 3, so the measure scales as $\lambda^{1/2}$. Next, we have the rescaling of the W term, because the fields rescale as:

$$\Phi_i \to \lambda^{-d_i}\Phi_i \qquad (7.5.5)$$

where d_i is the $U(1)$ charge of the field.

Now, let us calculate the contribution to the functional measure due to rescaling, using the fact that the potential is quasi-homogeneous. The contributions are:

$$\text{bosons}: \quad \lambda^{-2d_i\,\text{Tr}(1)}; \qquad \text{Tr}(1) = \frac{1}{24\pi}\int R = \frac{1}{3}$$

$$\text{fermions}: \quad \lambda^{+2d_i\,\text{Tr}(1)}; \qquad \text{Tr}(1) = -\frac{1}{48\pi}\int R = -\frac{1}{6} \qquad (7.5.6)$$

so that the Jacobian contributes a total factor of λ^{-d_i}.

Putting all factors together, we find that the partition function scales as:

$$Z \to \lambda^{\Delta}Z; \qquad \Delta \equiv \sum_i \left(\frac{1}{2} - d_i\right) \qquad (7.5.7)$$

Since this scale factor must equal $\lambda^{c/6}$, we have:

$$\frac{c}{6} = \sum_i \left(\frac{1}{2} - d_i\right) \qquad (7.5.8)$$

so c is simply defined via the $U(1)$ charges d_i of the various independent superfields within the potential.

Example: Free Boson on a Circle

To illustrate these ideas, let us consider several examples. First, let us consider the simplest possible $N = 2$ superconformal model, the theory of a free boson ϕ (with $c = 1$) defined on a circle with fixed radius. One finds that an explicit representation of the $N = 2$ algebra is given by:

$$G^{\pm} = e^{\pm i\sqrt{3}\phi_L}; \qquad \bar{G}^{\pm} = e^{\pm i\sqrt{3}\phi_R}$$

$$J(z) = (i/\sqrt{3})\,\partial\phi; \qquad \bar{H}(z) = -(i/\sqrt{3})\,\bar{\partial}\phi \qquad (7.5.9)$$

subject to the condition that the allowed winding (momentum) modes are of the form:

$$\exp[i(n_L\phi_L - n_R\phi_R)/\sqrt{12}] \qquad (7.5.10)$$

with $n_L - n_R = 0 \bmod 6$ (before a GSO projection).

The only primary chiral states are the vacuum and the state that has $h_L = q_L/2 = h_R = q_R/2 = 1/6$, which we denote by X. Since the product of primaries is either another primary or zero, we find that X^2 is not a primary and hence must be zero. The chiral ring of this superconformal model is simple:

$$R_{\text{chiral}} = \{1, X\}; \qquad XX \equiv 0 \qquad (7.5.11)$$

Now, compare this chiral ring with the ring formed by starting with the Landau–Ginzburg potential:

$$W = x^3 \qquad (7.5.12)$$

where x is a chiral superfield (not the X of the previous discussion). The Landau–Ginzburg ring is formed by taking all possible products of 1 and x, modulo all possible derivatives of W, that is, modulo x^2. But, this leaves only two elements in the ring, 1 and x; so,

$$R_{\text{LG}} = \{1, x\}; \qquad x \cdot x \equiv 0 \qquad (7.5.13)$$

Comparing Eqs. (7.5.11) and (7.5.13), we find that we are back to the same ring structure as the chiral ring, that is, $R_{\text{chiral}} = R_{\text{LG}}$.

The final link between these two rings is their central charge. We know that $c = 1$ for the chiral ring. If we scale by $x \to \lambda^{n_1} x$, then, from Eq. (7.5.1),

$$(\lambda^{n_1} x)^3 = \lambda^d (x^3) \qquad (7.5.14)$$

so that $3n_1 = d$, or that $q_1 = n_1/d = 1/3$. Since the central charge c in Eq. (7.5.8) equals $3 - 6q$, we find that $c = 1$, as expected.

Example: Catastrophe Theory

Fortunately, it is now possible to use the mathematical theory of catastrophes (singularity theory) [16] in order to classify the various types of supersymmetric Landau–Ginzburg potentials. Catastrophe theory, like superconformal field theory, is interested in the behavior of functions such as W under a rescaling. Let us introduce a few simple definitions from catastrophe theory. Let the dimension of the ring R_{LG} equal μ, which is called the criticality type. For example, $\mu = 2$ in the previous discussion.

The question that we will address is: if we add a deformation δW to W, will the deformation change the criticality type? We will therefore introduce the "modality" of a singularity, that is, the number of parameters in δW that one can add to W without changing μ and that cannot be eliminated by a coordinate transformation.

It can be shown that the list of potentials with zero modality can be arranged according to an A–D–E classification, that is, there is a one-to-one correspondence between a zero modality potential and a self-dual Lie group. If N is the Coxeter number of the Lie group, then it can be shown that the zero modality potentials have the central charge:

$$c = 3 - \frac{6}{N} \qquad (7.5.15)$$

This means that the zero modality potentials can be represented as:

$$A_k : \quad x^{k+1}, \qquad c = 3 - \frac{6}{k+1}; \qquad (k \geq 1)$$

$$D_k : \quad x^{k-1} + xy^2, \qquad c = 3 - \frac{6}{2(k-1)}; \qquad (k > 2)$$

$$E_6 : \quad x^3 + y^4, \qquad c = 3 - \frac{6}{12} \qquad\qquad (7.5.16)$$

$$E_7 : \quad x^3 + xy^3, \qquad c = 3 - \frac{6}{18}$$

$$E_8 : \quad x^3 + y^5, \qquad c = 3 - \frac{6}{30}$$

Let us analyze some of these examples in more detail.

Example: A_k

Using the definition of a quasi-homogeneous function, we can calculate the $U(1)$ charge for each of the variables in the potential, and then we can calculate the central charge. For the first example, A_k, the charge of x can be calculated using the quasi-homogeneous equation Eq. (7.5.1):

$$W(\lambda^{n_x} x) = \lambda^{(k+1)n_x} x^{k+1} = \lambda^d W(x) \qquad (7.5.17)$$

which gives us the $U(1)$ charge:

$$\frac{n_x}{d} = \frac{1}{k+1} \qquad (7.5.18)$$

We can then plug this expression for the charge into the equation for the central charge, Eq. (7.5.8), giving us:

$$\frac{c}{3} = 1 - 2\frac{1}{k+1} \qquad (7.5.19)$$

which is the expression in Eq. (7.5.16).

Next, we can calculate the ring associated with this potential by taking all possible monomials (x^n) modulo the derivative:

$$dW/dx = (k+1)x^k \qquad (7.5.20)$$

This, in turn, means that $x^k \sim 0$, so the elements in the ring stop at x^{k-1}:

$$R_{\text{LG}} = \{1, x, x^2, \ldots, x^{k-1}\} \qquad (7.5.21)$$

The criticality type is $\mu = k$. The modality is also zero. [This is because we cannot add δW to W without changing μ. If we add x^m ($m < k+1$), then, since the derivatives of W are changed, we find that μ also changes. If we add x^m ($m \geq k+1$), then this can be absorbed by a general coordinate transformation.]

Example: D_k

For the second example, D_k, we calculate the charges for x and y by solving the quasi-homogeneous equations:

$$\lambda^{nn_x} = \lambda^d$$
$$\lambda^{n_x+2n_y} = \lambda^d \tag{7.5.22}$$

where $n = k - 1$. This gives us $n_x = d/n$ and $n_y = d(n-1)/2n$. Now, let us insert these values for the $U(1)$ charges into the equation for the central charge in Eq. (7.5.8), and we find:

$$\frac{c}{3} = 2 - \frac{2}{n} - \frac{n-1}{n} \tag{7.5.23}$$

giving us the expression in Eq. (7.5.16).

Now, let us calculate the Landau–Ginzburg ring for this potential, which is constructed out of all possible monomials $x^i y^j$ modulo the derivatives of W, that is, we set equal to zero the following:

$$\partial W/\partial x = n x^{n-1} + y^2 \sim 0, \qquad \partial W/\partial y = 2xy \sim 0 \tag{7.5.24}$$

It is now easy to show that the complete set of monomials is equal to:

$$R_{\mathrm{LG}} = \{1, x, x^2, \ldots, x^{n-1}, y\} \tag{7.5.25}$$

Then, we have $\mu = n + 1$, and the modality is zero.

Example: E_6

The third example, E_6, can also be analyzed the same way. The rescaling of W gives us:

$$\lambda^{3n_x} x^3 + \lambda^{4n_y} y^4 = \lambda^d (x^3 + y^3) \tag{7.5.26}$$

which gives us $n_x = d/3$ and $n_y = d/4$. Examining the central charge with these values for d_i, we find the value in Eq. (7.5.16):

$$c = \left(3 - \frac{6}{3}\right) + \left(3 - \frac{6}{4}\right) = 3 - \frac{6}{12} \tag{7.5.27}$$

Now, let us calculate the derivatives of the potential:

$$\partial W/\partial x = 3x^2 \sim 0, \qquad \partial W/\partial y = 4y^3 \sim 0 \tag{7.5.28}$$

The elements of the ring are, therefore,

$$R_{\mathrm{LG}} = \{1, x, y, xy, y^2, xy^2\} \tag{7.5.26}$$

The criticality type $\mu = 8$ and the modality is zero.

Example: E_8

The last example, E_8, has charges for x and y given by $\frac{1}{3}$ and $\frac{1}{4}$, respectively. The value of the central charge is given by $\left(3 - \frac{6}{3}\right) + \left(3 - \frac{6}{4}\right) = 3 - \frac{6}{30}$, which agrees with the value in Eq. (7.5.16). Since the derivatives of the potential are x^2 and y^4, the elements of the ring must be given by:

$$R_{LG} = \{1, x, y, xy, y^2, xy^2, y^3, xy^3\} \tag{7.5.30}$$

Then, $\mu = 8$ and the modality is zero.

Armed with this new, powerful formulation, let us now re-investigate the miraculous relationship between $N = 2$ superconformal field theories and Calabi-Yau manifolds which we studied in Chapter 5. Each formulation is based on a different series of assumptions and mathematics, yet both seem to be equivalent. We end this section by presenting a heuristic argument which attempts to explain the deeper, underlying reason why tensoring $N = 2$ superconformal field theories yields Calabi-Yau manifolds [20].

Let us begin by studying $N = 2$ superconformal field theory by considering a Landau-Ginzburg superpotential given by the F-term $W(\varPhi) = \varPhi^{P+2}$. We know that, at criticality, this simple superpotential yields a $N = 2$ minimal theory of level P with central charge given by Eq. (7.3.15): $c = 3P/(P + 2)$. In this language, describing the tensor product of several $N = 2$ superconformal field theories is rather simple: we just add several superpotentials together:

$$W(\varPhi_i) = \varPhi_1^{P_1+2} + \cdots + \varPhi_N^{P_N+2} \tag{7.5.31}$$

For example, the (3^5) model discussed earlier, formed by tensoring five copies of the $P = 3$ discrete series, corresponds to a superpotential with $W(\varPhi) = \sum_{i=1}^{5} \varPhi_i^5$. Our goal is to show the relationship between this superpotential and the Calabi-Yau manifold given by $Y_{4;5}$, which is given by CP^4 constrained by $z_1^5 + z_2^5 + z_3^5 + z_4^5 + z_5^5 = 0$.

To reveal the relationship between the tensoring of $N = 2$ minimal models and Calabi-Yau manifolds, let us analyze the path integral defined over this superpotential given by:

$$\int D\varPhi_1 \cdots D\varPhi_N \, \exp\left[i \int d^2z d^2\theta \left(\varPhi_1^{l_1} + \cdots + \varPhi_N^{l_N}\right)\right] \tag{7.5.32}$$

where l_i are integers.

Let us now define the variables:

$$\xi_1 = \varPhi_1^{l_1}, \qquad \xi_i^{l_i} = \varPhi_i^{l_i}/\varPhi_1^{l_1} \tag{7.5.33}$$

By factoring out ξ_1, the original path integral can be written as:

$$\int D\xi_1 \cdots D\xi_N \, \Omega \exp\left[i \int d^2z d^2\theta \, \xi_1(1 + \xi_2^{l_2} + \cdots + \xi_N^{l_N})\right] \qquad (7.5.34)$$

where Ω is the Jacobian for this co-ordinate transformation from $\Phi_i^{l_i}$ to ξ_i. It is easy to show that this Jacobian is proportional to Φ_1^j where $j = 1 - l_1 + l_1(\sum_{i=2}^4 (1/l_i))$. Therefore, the Jacobian drops out if we set $j = 0$, or:

$$\sum_{i=1}^N \frac{1}{l_i} = 1 \qquad (7.5.35)$$

If this condition is met, then $\Omega = 1$ and the integration over ξ_1 can be performed, yielding the following delta function:

$$\delta\left(1 + \xi_2^{l_2} + \cdots + \xi_N^{l_N}\right) \qquad (7.5.36)$$

This complex constraint is identical to the constraint found in what is called weighted CP_N manifolds. The manifold described by this delta function is identical to the manifold arising from the constraint:

$$\sum_{i=1}^N z_i^{l_i} = 0 \qquad (7.5.37)$$

Assuming that z_1 is regular at the origin, we see that we can divide by z_1 and arrive at the same condition as the delta function. We can make the identification $\xi_i = z_i/z_1$.

This weighted CP_{N-1} manifold is defined as a complex space where we identify the point $[z_1, z_2, \cdots, z_N]$ with another point given by:

$$[z_1, z_2, \cdots, z_N] = \left[\lambda^{k_1} z_1, \cdots, \lambda^{k_N} z_N\right] \qquad (7.5.38)$$

for some complex λ. We call this manifold $WCP_{k_1, \cdots, k_N}^{N-1}$. If we apply this transformation on the constraint defined by the delta function, we see that the constraint remains invariant. (Notice that if all the integers k_i are identical, then we have an ordinary CP_{N-1} manifold.)

To make the identification more precise, let d be the least common multiple of the integers l_i. Then the superpotential $\sum_{i=1}^N \Phi_i^{l_i}$, via the delta function, corresponds to the space:

$$WCP_{d/l_1, \cdots, d/l_N}^{N-1} \qquad (7.5.39)$$

Lastly, we remark that the condition $\Omega = 1$ also has a counterpart from the point of view of complex z space. It turns out that this condition is identical to the vanishing of the first Chern class c_1 for the manifold. Since a Calabi-Yau manifold is a complex Kähler manifold with vanishing first Chern class, we have now shown that the superpotential $\sum_{i=1}^N \Phi_i^{l_i}$ corresponds to a Calabi-Yau manifold if the Jacobian $\Omega = 1$.

In summary, the integration over the superpotential has become the defining relation for the weighted CP_{N-1} manifold, and the vanishing of the Jacobian has become the condition for the vanishing of the first Chern class, yielding a Calabi-Yau manifold:

$$\sum_{i=1}^{N} \Phi_i^{l_i} \quad \rightarrow \quad WCP_{d/l_1,\cdots,d/d_N}^{N-1}$$
$$\Omega = 1 \quad \rightarrow \quad c_1 = 0 \tag{7.5.40}$$

(We caution, however, that there are some loose ends in this heuristic derivation. For example, we assumed that we could ignore the D term appearing in the super path integral. This is a reasonable, though not rigorous, assumption, because we expect the D term to contribute only small perturbations to the theory.)

As a check on our results, let us investigate the $c = 9$ theories found by Gepner. Since $p_i = l_i - 2$, we can write the central charge corresponding to tensoring superconformal field theories:

$$c = \sum_{i=1}^{5} 3(l_i - 2)/l_i = 9 \tag{7.5.41}$$

This, in turn, can be reduced back to $\sum_{i=1}^{5}(1/l_i) = 0$, which is precisely the condition for the Jacobian being equal to one. Once again, this method reveals the origin of the relationship between the two different formalisms for the case $c = 9$.

In this fashion, it is now easy to write down the Calabi-Yau manifold which corresponds to the tensoring of various superconformal field theories [20]:

$$
\begin{aligned}
(3^5): & \quad z_1^5 + z_2^5 + z_3^5 + z_4^5 + z_5^5 = 0 \in CP_4 \\
(4^41): & \quad z_1^6 + z_2^6 + z_3^6 + z_4^6 + z_5^3 = 0 \in WCP_{1,1,1,1,2}^4 \\
(6^4): & \quad z_1^8 + z_2^8 + z_3^8 + z_4^8 + z_5^2 = 0 \in WCP_{1,1,1,2,6}^4 \\
(8^33): & \quad z_1^{10} + z_2^{10} + z_3 10 + z_4^5 + z_4^2 + z_5^2 = 0 \in WCP_{1,1,1,2,5}^4 \\
(7^31^2): & \quad z_1^9 + z_2^9 + z_3^9 + z_4^3 + z_5^3 = 0 \in WCP_{1,1,1,3,3}^4 \\
(10^22^21): & \quad z_1^{12} + z_2^{12} + z_3^4 + z_4^4 + z_5^3 = 0 \in WCP_{1,1,3,3,4}^4 \\
(5^21^219): & \quad z_1^{21} + z_2^7 + z_3^7 + z_4^3 + z_5^3 = 0 \in WCP_{1,3,3,7,7}^4 \\
(16^31): & \quad z_1^{18} + z_2^{18} + z_3^{18} + z_4^3 + z_5^2 = 0 \in WCP_{1,1,1,6,9}^4
\end{aligned} \tag{7.5.42}
$$

We should also mention that the class of potentials that we have been considering, of the type Φ^{P+2}, only corresponds to one type of manifold. Earlier, we saw that there is an A-D-E classification of zero modality singular functions. The A_P singular functions are of the form z^{P+1} which we have considered so far.

For the D_{P+2} series, we need to use monomials of the form $z^{P+1}+zw^2$, while monomials of the form $z^3 + w^4$, $z^3 + zw^3$, and $z^3 + w^5$ are in the E_6, E_7, and E_8 series, respectively. In this way, it is also straightforward to construct the correspondence between tensoring superconformal field theories and Calabi-Yau manifolds for the D-E series, as well.

7.6. Zamolodchikov's c Theorem

One major defect in the previous presentation has been the fact that our discussion has focused on systems that were exactly conformally invariant at criticality. From the string point of view, this means that we have been studying vacuums that are on shell. However, this does not tell us which vacuums the theory prefers and how it tunnels between possible vacuums. Some insight can be gained by studying conformal theories that are allowed to go off criticality. For example, in solid-state physics, two-dimensional systems, such as the Ising model, are exactly solvable both at criticality as well as off criticality. Thus, we should reexamine our approach to conformal systems by analyzing our equations off criticality.

The c theorem [10] provides a powerful way in which to analyze systems off criticality. In short, the c theorem states that there is a function C with two properties. First, at criticality, the function C reduces to the usual central term c for some conformal field theory. Second, the value of C along renormalization group flows decreases.

The proof of this theorems is quite general and deceptively simple. It is based on analyzing the full energy-momentum tensor T_{ab} for systems that are not critical.

For example, T_{ab} can be broken down into four pieces:

1.   the antisymmetric piece $T_{[a,b]}$,
2.   the trace, given by Θ, and
3.   the symmetric parts $T = T_{zz}$ and $\bar{T} = T_{\bar{z}\bar{z}}$.

The energy-momentum is conserved, which means $\partial_a T^{ab} = 0$, or:

$$\partial_{\bar{z}}T + \frac{1}{4}\partial_z\Theta = 0, \qquad \partial_z\bar{T} + \frac{1}{4}\partial_{\bar{z}}\Theta = 0 \qquad (7.6.1)$$

The antisymmetric piece can be set equal to zero if the system is rotationally invariant, which we will assume. Normally, for conformally invariant systems, we also set the trace Θ equal to zero. This, in turn, means that T ($\bar{T}$) is function of $z(\bar{z})$. However, we will now keep the trace arbitrary for a noncritical system.

Because T, Θ, and $\bar{T}$ have conformal spins equal to 2 , 0 , -2, we can write the new operator product expansion as:

$$\langle T(z, \bar{z})T(0,0)\rangle = F(z\bar{z})/z^4$$
$$\langle T(z, \bar{z})\Theta(0,0)\rangle = G(z\bar{z})/z^3\bar{z} \qquad (7.6.2)$$
$$\langle \Theta(z, \bar{z})\Theta(0,0)\rangle = H(z\bar{z})/z^2\bar{z}^2$$

Now, let us take the correlation function between the equation of motion and $T(0,0)$ or $\Theta(0,0)$, that is,

$$\left\langle \left(\partial_{\bar{z}}T + \frac{1}{4}\partial_z\Theta\right)T(0,0)\right\rangle = 0 \qquad (7.6.3)$$

We then find two equations:

$$\dot{F} + \frac{1}{4}(\dot{G} - 3G) = 0$$
$$\dot{G} - G + \frac{1}{4}(\dot{H} - 2H) = 0 \qquad (7.6.4)$$

where we have defined $\dot{F} = z\bar{z}F'(z\bar{z})$.

Now, we can define the function C, which reduces to the central term c at criticality:

$$C \equiv 2F - G - \frac{3}{8}H \qquad (7.6.5)$$

which obeys the equation:

$$\dot{C} = -\frac{3}{4}H \qquad (7.6.6)$$

By reflection positivity, we know that $H \geq 0$, so C is a decreasing function of $R \equiv \sqrt{z\bar{z}}$.

In a theory with coupling constants $\{g_i\}$, we can write a renormalization group equation for $C(r, \{g\})$ as follows:

$$\left[R\frac{\partial}{\partial R} + \sum_i \beta_i(\{g\}) \frac{\partial}{\partial g_i}\right]C(\{g_i\}, R) = 0 \qquad (7.6.7)$$

Notice that, at a fixed point where $\beta_i = 0$, we have $G = H = 0$ and $F = \frac{1}{2}c$, so that $C = c$ at the fixed point.

In summary, we have now shown that, if renormalization flows connect different conformal field theories, then C decreases along the flows and that $C = c$ at criticality.

7.7. A-D-E Classification of c=1 Theories

So far, we have reviewed the major methods that have been devised which can give us, for $c < 1$, a complete classification of the unitary representations of the conformal group, including its modular invariants, in terms of finite numbers of primary fields. For $c > 1$, there are an infinite number of primary fields and much less is known about their representations.

Questions remain, however, about the case $c = 1$. Will it behave more like $c < 1$ and give us exactly solvable representations, or will it behave more like $c > 1$ and be, at least with present methods, intractable?

The answer is rather unexpected. We find that at $c = 1$ we can find the complete set of unitary representations, as in the $c < 1$ case [21].

A careful analysis shows that there are only three classes of solutions, corresponding to a boson propagating on

1) a torus of radius r

2) an orbifold parametrized by radius r

3) three discrete orbifold spaces defined on $SU(2)/\Gamma_i$, where Γ_i are discrete elements of $SU(2)$.

The first two solutions represent continuous classes of solutions which can be parametrized by r, where $0 \leq r \leq \infty$, while the third solution is discrete.

To understand how to construct these $c = 1$ representations, we begin with the usual action for a spin 0 boson:

$$S = \frac{1}{2\pi} \int d^2z \ \partial_z x \bar\partial_{\bar z} x \qquad (7.7.1)$$

If we compactify this on a circle S^1 with x being identified with $x + 2\pi r$ and trace over $q^{L_0 - 1/24} \bar q^{\bar L_0 - 1/24}$, we find the partition function $Z(r)$ which we calculated earlier in Eq. (4.2.17), which obeys the duality condition:

$$Z(r) = Z(1/r) \qquad (7.7.2)$$

Notice that we have a continuous set of solutions defined on S^1 indexed by the radius r.

We obtain the second set of continuous solutions when the boson propagates on an orbifold. In particular, let us divide out by the discrete symmetry $x \to -x$, so the boson propagates on $S^1/\mathbf{Z}_2$ instead of the circle S^1. This alteration leaves c invariant, but changes the boundary condition for the trace operation. In general, if we perform the functional integral over an orbifold where we have divided out by the action of a discrete group G with elements g_i, then we must sum over all possible boundary conditions in the σ_1 and σ_2 direction, with g_i acting on both sides of the parallelogram. Since the discrete group G has two elements, the identity and the parity operator, we must therefore sum over four possible boundary conditions.

Evaluating the trace over these new boundary conditions, we find:

$$Z_{\text{orb}}(r) = \frac{1}{2}\left\{ Z(r) + \frac{|\vartheta_3\vartheta_4|}{\bar\eta\eta} + \frac{|\vartheta_2\vartheta_3|}{\bar\eta\eta} + \frac{|\vartheta_2\vartheta_4|}{\bar\eta\eta} \right\}$$
$$= \frac{1}{2}[Z(r) + 2Z_2 - Z_2] \qquad (7.7.3)$$

where we define:

$$Z_n \equiv Z(1/n\sqrt{2}) = Z(n/\sqrt{2}) \qquad (7.7.4)$$

and where $\vartheta_i = \vartheta(0, \tau)$.

Then we have the desired result for the partition function over the $\mathbf{Z}_2$ orbifold. By explicit calculation, one can show that both partition functions are modular invariant.

We have now constructed modular invariants for two continuous classes of representations. Before discussing the third solution, we note that within these two continuous classes given by the torus and the orbifold, there are interesting special values for r which yield some insight into the structure of these solutions.

For example, we can re-express the above results in the language of bosons propagating on the orbifold $SU(2)/\Gamma$, where Γ represents the various discrete subgroups of $SU(2)$. This means that the point $g \in SU(2)$ is equivalent to the point hgh^{-1} if $h \in \Gamma$.

If we choose $h = \exp(2\pi J_3/n)$, then the elements J_3 and $J_\pm$ of $SU(2)$ must be identified according to the following:

$$hJ_3h^{-1} = J_3; \quad hJ_\pm h^{-1} = e^{\pm 2\pi i/n}J_\pm \qquad (7.7.5)$$

This simply means that the point x is to be identified with the point $x + 2\pi/(n\sqrt{2})$. In other words, the boson propagates on a circle S^1 with discrete radius $1/(n\sqrt{2})$. This just selects out special radii for the circle.

The element h generates the discrete subgroup of $SU(2)$ called the binary cyclic group $\mathcal{C}_{2n}$. This is a group of order $2n$, whose projection $\mathbf{C}_n \equiv \mathcal{C}_{2n}/\mathbf{Z}_2 \in SO(3)$ is the group which describes rotations about an axis of n-fold symmetry. Thus, there is a one-to-one correspondence between the special radii we have found for the torus and the elements of this particular discrete subgroup of $SU(2)$.

We can also establish:

$$Z[SU(2)/\mathcal{C}_{2n}] = Z_n \qquad (7.7.6)$$

Similarly, we can also choose another discrete subgroup Γ to be generated by the element $\tilde{h} = \exp(i\pi J_1)$. The group action yields:

$$\tilde{h}J_3\tilde{h}^{-1} = -J_3; \quad \tilde{h}J_\pm \tilde{h}^{-1} = J_\mp \qquad (7.7.7)$$

This identification, in turn, can be shown to correspond to identifying $x \to -x$, as before, for the $S^1/\mathbf{Z}_2$ orbifold. If we combine the action of h and $\tilde{h}$, then we can describe propagation on the orbifold $S^1/\mathbf{Z}_2$ with the orbifold radius being $r = 1/(n\sqrt{2})$.

These special radii can be placed in one-to-one correspondence with the elements of the binary dihedral group $\mathcal{D}_n$, which is a discrete subgroup of $SU(2)$ of order $4n$ formed generated by h and $\tilde{h}$. Their projections $\mathbf{D}_n = \mathcal{D}_n/\mathbf{Z}_2 \in SO(3)$ have n axes of two-fold symmetry perpendicular to an n-fold axis. For example, the group $\mathbf{D}_4$ corresponds to the 8 element symmetry group of the square.

We can also establish:

$$Z[SU(2)/\mathcal{D}_2] = \frac{1}{2}(Z_n + 2Z_2 - Z_1) = Z_{\text{orb}}\left(1/(n\sqrt{2})\right) \qquad (7.7.8)$$

This concludes our discussion of the first two continuous classes of solutions and their special points. Now, we describe the third class of solutions to the $c = 1$ theory, which is given by three discrete solutions. In addition to the $\mathcal{C}_n$ and $\mathcal{D}_n$ discrete subgroups of $SU(2)$, there are also the binary tetrahedral, octahedral, and icosahedral groups, labeled by $\mathcal{T}$, $\mathcal{O}$, and $\mathcal{I}$ of order 24, 48, 120, which are related to the symmetry groups of the regular polyhedra found in ordinary solid geometry.

We can calculate the modular invariants associated with each of these three discrete subgroups by breaking them down further into their various elements and expressing them as rotations about various axes of the regular polyhedra. For example, the mutually commuting elements of the discrete subgroup $\mathcal{T}$ lie in 4 $\mathbf{C}_3$'s (acting about axes through the centers of the various faces) and one $\mathbf{D}_2$ (of rotations about axes through the centers of the opposite edges).

We find, therefore, that a modular invariant combination for $SU(2)/\mathcal{T}$ is given by:

$$\begin{aligned} Z[SU(2)/\mathcal{T}] &= \frac{1}{24}\Big\{4(3Z_2 - Z - 1) + 4Z[SU(2)/\mathcal{D}_2]\Big\} \\ &= \frac{1}{2}(2Z_3 + Z_2 - Z_1) \end{aligned} \qquad (7.7.9)$$

For the octahedral group $\mathcal{O}$, its generators lie in 3 $\mathbf{C}_4$'s (acting about axes through the centers of opposite faces of a cube), 4 $\mathbf{C}_3$'s (acting about axes through antipodal vertices), one $\mathbf{D}_2$ (containing the elements of three $\mathbf{C}_4$'s), and 3 $\mathbf{D}_2$'s (each of which contains one of these elements and two others which are associated with orthogonal axes through the centers of opposite edges).

We find, therefore:

$$\begin{aligned} Z[SU(2)/\mathcal{O}] &= \frac{1}{24}\Big\{3(4Z_4 - 2Z_2) + 4(3Z_3 - Z - 1) + \\ &\quad 4Z[SU(2)/\mathcal{D}_2] + 3(4Z[SU(2)/\mathcal{D}_2] - 2Z_2)\Big\} \\ &= \frac{1}{2}(Z_4 + Z_3 + Z_2 - Z_1) \end{aligned} \qquad (7.7.10)$$

Finally, the icosahedral group's elements lie in 6 $\mathbf{C}_5$'s (acting about axes through opposite faces of a dodecahedron), 10 $\mathbf{C}_3$'s (acting about axes through antipodal vertices), and 5 $\mathbf{D}_2$'s (comprised of the rotations about axes through the centers of opposite edges). We have, therefore:

$$Z[SU(2)/\mathcal{I}] = \frac{1}{60}\Big\{(5Z_5 - Z_1) + 10(3Z_3 - Z_1)$$
$$+ 5(4Z[SU(2)/\mathcal{D}_2] - Z_1) + Z_1\Big\} \qquad (7.7.11)$$
$$= \frac{1}{2}(Z_5 + Z_3 + Z_2 - Z_1)$$

We now remark on a curious mathematical fact. When we analyzed the modular invariants of $SU(2)_k$, we noticed that they could be placed in one-to-one correspondence with the A-D-E classification of simply laced groups (which have simple root vectors of the same length). We now show that the special solutions of the $c = 1$ theory for the two continuous classes and the three discrete solutions can also be placed in a one-to-one correspondence with the A-D-E classification. This is because there is a one-to-one correspondence between the simply laced groups and the finite subgroups Γ of $SU(2)$. Since each of the discrete solutions that we have found can be described by propagation on $SU(2)/\Gamma$, we now have a one-to-one correspondence between the simply laced groups and the complete discrete modular invariant solutions of the $c = 1$ series.

In particular, we find that the special values for the two continuous series can be identified as:

$$SU(n) = A_{n-1} \leftrightarrow \mathcal{C}_n; \quad SO(2n) = D_n \leftrightarrow \mathcal{D}_{n-2} \qquad (7.7.12)$$

while the three discrete solutions can be identified as:

$$E_6 \leftrightarrow \mathcal{T}; \quad E_7 \leftrightarrow \mathcal{O}; \quad E_8 \leftrightarrow \mathcal{I} \qquad (7.7.13)$$

Lastly, it can be proven that the three classes of solutions that we have found for the $c = 1$ theory are, in fact, the only solutions [22]. Thus, we have now achieved a rather interesting result, the complete classification of the representations of the $c = 1$ theory.

7.8. Summary

At present, there is no generally accepted classification scheme that reveals the deep relationship between various conformal field theories. Several methods have been devised that can catalog vast numbers of conformal field theories, especially the rational conformal field theories, although none of these methods give us much insight into the relationship between the theories or how to select out the true vacuum of the string. The various methods are summarized here.

1. The GKO coset method is still one of the most powerful methods by which all known rational conformal field theories may be represented. However, one has to know the representations of G and its subgroup H in order to determine the representations of the coset G/H.

2. The Feigin–Fuchs method uses a set of almost trivial free fields in which to generate the primary fields of a wide variety of models. It can include the Kac–Moody algebras, cosets, and $N = 2$ superconformal field theory.

3. The Landau–Ginzburg method and catastrophe theory is not as powerful as the others, since some conformal field theories cannot be represented in this fashion, but it is the most physical and may eventually explain how one conformal field theory may "flow" into another.

4. The Chern–Simons knot theory method can also represent all known conformal field theories. It will be presented in the next chapter.

The Feigin–Fuchs free field method begins with an energy-momentum tensor with a linear term:

$$T(z) = -\frac{1}{2} : \partial_z \phi \, \partial_z \phi : +i\alpha_0 \, \partial_z \phi \qquad (7.8.1)$$

Then the central term can be calculated to be:

$$c = 1 - 12\alpha_0^2 \qquad (7.8.2)$$

and the conformal weight of the vertex function $: e^{i\alpha\phi(z)} :$ is equal to $\alpha^2 - 2\alpha\alpha_0$.

Now, consider the operator:

$$Q = \oint_C dz \, J(z), \qquad J(z) =: e^{i\alpha\phi(z)} : \qquad (7.8.3)$$

If we choose α_0 such that the integrand has weight 1, then we have:

$$\alpha^2 - 2\alpha\alpha_0 = 2 \qquad (7.8.4)$$

which has two solutions for α_0:

$$\alpha_\pm = \alpha_0 \pm \sqrt{\alpha_0^2 + 2} \qquad (7.8.5)$$

The trick behind the Feigin–Fuchs method is that, because $Q_\pm$ is conformally invariant, we can insert as many of them into a correlation function as required until the sums of the momenta add up to zero. For example, the correlator:

$$\langle V_\alpha V_\alpha V_\alpha V_{2\alpha_0 - \alpha} \rangle \qquad (7.8.6)$$

is nonzero if we put momentum $-2\alpha_0$ at infinity and insert $n - 1$ Q_-'s and $m - 1$ Q_+'s. Then, the sum of the momenta must be zero, so:

$$2\alpha = 2\alpha_{n,m} = (1 - n)\alpha_- + (1 - m)\alpha_+ \qquad (7.8.7)$$

The conformal weight of the vertex V_α must therefore be:

$$2\Delta_{nm} = \alpha_{n,m}^2 - 2\alpha_{n,m}\alpha_0 = \frac{1}{4}\left[(\alpha_- n - \alpha_+ m)^2 - (\alpha_+ + \alpha_-)^2\right] \qquad (7.8.8)$$

But, this is just the conformal weight of a minimal field. Since the correlator of vertex functions is trivial to calculate, we have now reduced the problem of finding the correlation function of the minimal model to evaluating line integrals (arising from Q). For example,

$$
\begin{aligned}
\Big\langle \phi_{n_1,m_1}(0)&\phi_{1,2}(z)\phi_{n_3,m_3}(1)\phi_{n_4,m_4}(\infty)\Big\rangle \\
&= \oint_C dt \Big\langle V_{\alpha_1}(0)V_{\alpha_2}(z)V_{\alpha_3}(1)V_{\alpha_4}(\infty)J_+(t)\Big\rangle \\
&= z^{2\alpha_1\alpha_2}(1-z)^{2\alpha_2\alpha_3}\oint_C dt\, t^a(t-1)^b(t-z)^c
\end{aligned}
\tag{7.8.9}
$$

where $a = 2\alpha_1\alpha_+$, $b = 2\alpha_3\alpha_+$, $c = 2\alpha_2\alpha_+$, $\alpha_2 = -\frac{1}{2}\alpha_+$, and $\alpha_4 = 2\alpha_0 - \alpha_1 - \alpha_2\alpha_3 - \alpha_+$.

The Feigin–Fuchs method can be generalized by introducing more free fields $\beta_{-\alpha}, \gamma_\alpha$, and ϕ^i, with an operator product expansion:

$$
\begin{aligned}
\beta_{-\alpha}(z)\gamma_\beta(w) &\sim \frac{\delta_{\alpha\beta}}{z-w} \\
\phi^i(z)\phi^j(w) &\sim -\delta^{ij}\ln(z-w)
\end{aligned}
\tag{7.8.10}
$$

With these free fields, one can represent the generators of a Kac–Moody algebra. For example, the generators of the Cartan subalgebra can be written as:

$$
H^i(z) = -i\alpha_+\,\partial\phi^i + \sum_{\alpha\in G}\alpha^i\beta_{-\alpha}\gamma_\alpha(z)
\tag{7.8.11}
$$

where the sum runs over positive roots and where $\alpha_+ = \sqrt{k+g}$, and the other generators can be written as:

$$
E_{-\alpha}(z) = \beta_{-\alpha}(z) + \sum_{\rho-\sigma=-\alpha} N_{\rho\sigma}\gamma_\rho(z)\beta_{-\sigma}(z)
\tag{7.8.12}
$$

Also, the energy-momentum tensor can be written as:

$$
T(z) = \sum_{\alpha\in G}\beta_{-\alpha}\partial\gamma_\alpha - \frac{1}{2}\sum_{i=1}^{\text{rank }G}[\partial\phi^i(z)]^2 - \frac{i}{\alpha_+}\sum_{i=1}^{\text{rank }G}\rho^i\,\partial^2\phi^i(z)
\tag{7.8.13}
$$

where ρ^i is half the sum of the positive roots.

Furthermore, we can write the free representation of the coset energy-momentum tensor:

$$
\begin{aligned}
T_{G/H}(z) = \sum_{\alpha\in G/H}&\beta_{-\alpha}(z)\,\partial\gamma_\alpha(z) - \frac{1}{2}[\partial\rho(z)]^2 - \frac{i}{\sqrt{k+g}}\rho_G\,\partial^2\phi(z) \\
&- \left\{-\frac{1}{2}[\partial\phi'(z)]^2 - \frac{i}{\sqrt{k+h}}\rho_H\,\partial\phi'(z)\right\}
\end{aligned}
\tag{7.8.14}
$$

In addition, this method easily generalizes to the $N = 1$ and $N = 2$ super-conformal cases.

Next, we study the Landau–Ginzburg approach, which begins with the observation that a theory with the potential:

$$L \sim g \int d^2 x \, \Phi^{2(p-1)} \qquad (7.8.15)$$

is not usually conformally invariant except at criticality. At this point, however, it must equal one of the classes of conformal field theories that has been proposed. Specifically, the potential $\Phi^{2(p-1)}$ at criticality corresponds to the unitary minimal model with $c = 1 - 6/p(p+1)$.

To see the relationship between a Landau–Ginzburg potential at criticality and a standard conformal field theory, take the fusion relations for a minimal model:

$$\phi_{2,2}\phi_{n,m} \sim \sum_{k,l} X_{n,m}^{(k,l)} \Big[\phi_{n+k,m+l} + \cdots \Big] \qquad (7.8.16)$$

and compare them to the operator product expansion of the Landau–Ginzburg theory. By examining these relations, we can show the correspondence between conformal field theories and Landau–Ginzburg models:

$$\begin{aligned}
: \Phi^k : &\leftrightarrow \phi_{k+1,k+1}, \qquad k = 0, 1, \ldots, p-2 \\
: \Phi^k : &\leftrightarrow \phi_{k-p+2,k-p+3}, \qquad k = p-1, p, p+1, \ldots, 2p-4
\end{aligned} \qquad (7.8.17)$$

Similarly, we can analyze the $N = 2$ theory and find the relationship:

$$: \Phi^k : \leftrightarrow \phi_{k+1,k+1}, \qquad k = 0, 1, 2, \ldots, p-2 \qquad (7.8.18)$$

The situation for the $N = 2$ theories, however, is more complicated. First, we have the fact that, by defining a U_θ operator, one can smoothly transform the generators from NS to R:

$$\begin{aligned}
U_\theta^{-1} L_n U_\theta &= L_n + \theta J_n + \frac{c}{6}\theta^2 \delta_{n,0} \\
U_\theta^{-1} J_n U_\theta &= J_n + \frac{c}{3}\theta \delta_{n,0} \\
U_\theta^{-1} G_r^+ U_\theta &= G_{r+\theta}^+ \\
U_\theta^{-1} G_r^- U_\theta &= G_{r-\theta}^+
\end{aligned} \qquad (7.8.19)$$

where the $U(1)$ charge and dimension change as:

$$\begin{aligned}
q &\to q + (c\theta/3) \\
h &\to h + \theta q + (c\theta^2/6)
\end{aligned} \qquad (7.8.20)$$

where θ will determine the spectral flow of the theory.

We define a *primary* field as one that satisfies:

$$G^-_{n+1/2}|\phi\rangle = G^+_{n+1/2}|\phi\rangle = 0 \tag{7.8.21}$$

in analogy with the usual definition of primary fields for bosonic fields. What is interesting is that the algebra formed by these chiral primaries forms a closed ring. The strategy is then to find the equivalence between this chiral ring R_{chiral} and the algebra formed by a Landau–Ginzburg potential.

Given a Landau–Ginzburg potential, we can form the ring that is created by taking all possible monomials of the fields, divided by all possible derivatives of the potential, that is,

$$R_{\text{LG}} = \frac{\prod_i x_i}{[\partial_j W]} \tag{7.8.22}$$

The object is to establish the relationship:

$$R_{\text{chiral}} = R_{\text{LG}} \tag{7.8.23}$$

For example, the simplest representation of the chiral ring is in terms of a free boson defined on a circle. We can represent the $N = 2$ generators as:

$$G^\pm = e^{\pm i\sqrt{3}\phi_L}; \qquad \bar{G} = e^{\pm i\sqrt{3}\phi_R}$$
$$J(z) = (i/\sqrt{3})\partial\phi; \qquad \bar{H}(z) = -(i/\sqrt{3})\,\bar{\partial}\phi \tag{7.8.24}$$

The only primary chiral states are the vacuum and the state that has $h_L = q_L/2 = h_R = q_R/2 = 1/6$, which we denote by X. The chiral ring of this superconformal model is simple:

$$R_{\text{chiral}} = \{1, X\}; \qquad XX \equiv 0 \tag{7.8.25}$$

Now, compare this chiral ring with the Landau–Ginzburg potential:

$$W = x^3 \tag{7.8.26}$$

where x is a chiral superfield. The Landau–Ginzburg ring is formed by taking all possible products of 1 and x, modulo all possible derivatives of x, that is, modulo x^2. But, this leaves only two elements in the ring, 1 and x; so,

$$R_{\text{LG}} = \{1, x\}; \qquad x \cdot x \equiv 0 \tag{7.8.27}$$

Notice we are back to the same ring structure as the chiral ring, $R_{\text{chiral}} = R_{\text{LG}}$.

A more systematic search for the equivalences between R_{chiral} and R_{LG} involves analyzing the scaling property of the potential:

$$W(\lambda^{n_i} x_i) = \lambda^d W(x_i) \tag{7.8.28}$$

However, this equation is also the defining equation for a class of catastrophes that has been studied by mathematicians. Thus, systematic study of catastrophes (which have been cataloged) may give us insight into the classification of these types of superconformal field theories.

Last, we analyze the possibility that various conformal field theories may flow into others via the renormalization group equations. This may give us a clue as to the nature of the true string vacuum. When the transition is made between one conformal field theory and another, the theory goes off criticality, so it is important to reanalyze the energy-momentum tensor when scale invariance is violated and it has a nonzero trace, denoted by T and $\bar{T}$. With a nonzero trace, the conservation of the energy-momentum tensor becomes:

$$\partial_{\bar{z}} T + \frac{1}{4} \partial_z \Theta = 0 \tag{7.8.29}$$

Now, let us take the correlation function between the equation of motion and $T(0,0)$ or $\Theta(0,0)$, that is,

$$\left\langle \left(\partial_{\bar{z}} T + \frac{1}{4} \partial_z \Theta \right) T(0,0) \right\rangle = 0 \tag{7.8.30}$$

We then find two equations:

$$\begin{aligned} \dot{F} + \frac{1}{4}(\dot{G} - 3G) &= 0 \\ \dot{G} - G + \frac{1}{4}(\dot{H} - 2H) &= 0 \end{aligned} \tag{7.8.31}$$

where we have defined $\dot{F} = z\bar{z}F'(z\bar{z})$.

Now, we can define the function C, which reduces to the central term c at criticality:

$$C \equiv 2F - G - \frac{3}{8}H \tag{7.8.32}$$

which obeys the equation:

$$\dot{C} = -\frac{3}{4}H \tag{7.8.33}$$

By reflection positivity, we know that $H \geq 0$, so that C is a nondecreasing function of $R \equiv \sqrt{z\bar{z}}$.

In a theory with coupling constants $\{g_i\}$, we can write a renormalization group equation for $C(r, \{g\})$ as follows:

$$\left[R \frac{\partial}{\partial R} + \sum_i \beta_i(\{g\}) \frac{\partial}{\partial g_i} \right] C(\{g_i\}, R) = 0 \tag{7.8.34}$$

In summary, we have defined a new function C, which equals the usual c when we are sitting at criticality, which generalizes the concept of a central charge for off-critical systems. Then the c theorem tells us that C must decrease monotonically along the path connecting two conformal field theories. Thus, via the c-theorem, we have established an important link between different conformal field theories, that is, the renormalization group flows connecting them have decreasing central charge.

References

1. P. Goddard, A. Kent, and D. Olive, *Comm. Math. Phys.* **103**, 105 (1986).
2. B. L. Feigin and D. B. Fuchs, unpublished.
3. Vl. S. Dotsenko and V. A. Fateev, *Nucl. Phys.* **B240 [FS12]**, 312 (1984).
4. Vl. S. Dotsenko and V. A. Fateev, *Nucl. Phys.* **F251 [FS13]**, 691 (1985).
5. Vl. S. Dotsenko, *Lectures on Conformal Field Theory*, Adv. Studies in Pure Math. **16** (1988).
6. M. A. Bershadsky, V. G. Knizhnik, and M. G. Teitelman, *Phys. Lett.* **151B**, 31 (1984).
7. J. Bagger, D. Nemeschansky, and J. Zuber, *Phys. Lett.* **216B**, 320 (1989).
8. N. Ohta and H. Suzuki, *Nucl. Phys.* **B332**, 146 (1990).
9. M. Kuwahara, N. Ohta, and H. Suzuki, *Phys. Lett.* **235B**, 57 (1989).
10. A. B. Zamolodchikov, *JETP Lett.* **43**, 731 (1986); *Sov. J. Nucl. Phys.* **46**, 1090 (1987); *Sov. J. Nucl. Phys.* **44**, 529 (1987).
11. C. Vafa at the Symposium on Fields, Strings, and Quantum Gravity in Beijing, 1989.
12. C. Vafa and N. P. Warner, *Phys. Lett.* **218B**, 51 (1989).
13. C. Vafa, *Mod. Phys. Lett.* **A4**, 1169, 1615 (1989).
14. W. Lerche, C. Vafa, N. P. Warner, *Nucl. Phys.* **B324**, 427 (1989).
15. E. Martinec in "Criticality, Catastrophe, and Compactifications," in *Physics and Mathematics of Strings,* World Scientific, Singapore (1990).
16. V. I. Arnold, *Singularity Theory,*, London Math. Lec. Notes Series, **53**, Cambridge Univ. Press, London and New York (1981); V. I. Arnold, S. M. Gusein-Zade, and A. N. Varchenko, *Singularities of Differentiable Maps*, **1**, Birkhauser, Basel (1985).
17. E. Witten, *Comm. Math. Phys.* **121**, 351 (1989).
18. A. Schwimmer and N. Seiberg, *Phys. Lett.* **184B**, 191 (1987).
19. O. Alvarez, *Nucl. Phys.* **B216**, 125 (1983).
20. B.R. Greene, C. Vafa, and N.P. Warner, *Nucl. Phys.* **B324**, 317 (1989).
21. P. Ginsparg, *Nucl. Phys.* **B295**, 153 (1988).
22. E.B. Kiritsis, *Phys. Lett.* **217B**, 427 (1988).

Chapter 8

Knot Theory and Quantum Groups

8.1. Chern–Simons Approach to Conformal Field Theory

In the previous chapters, we analyzed the various schemes that have been proposed to catalog large numbers of conformal field theories, especially the rational ones with a finite number of primary fields. In this chapter, we will explore the most ambitious one, which is the use of Chern–Simons gauge theory [1] to classify conformal field theories. In the process, we will uncover a deep but unexpected relationship between conformal field theories and knot theory. Surprisingly, we will be able to use quantum field theory to generate new knot polynomials and analytic expressions for them. Knot theory, in turn, will be a tool by which we study conformal field theories and statistical mechanics, giving us a topological meaning to the Yang–Baxter relation.

In contrast to the previous approaches to classifying conformal field theories, which were all completely defined in two-dimensions, our starting point will be the Yang–Mills theory formulated as a pure Chern–Simons term in three dimensions. Our philosophy will be that the "miracles" that occur in conformal field theory are by-products of simpler structures that exist in three dimensions. Viewed from the perspective of "flatland," many puzzling relationships may have no obvious origin. However, once we "leave flatland," as Atiyah and Witten have suggested, the origin of these relations can be seen transparently. The basic premise, therefore, is to see gauge symmetry and general covariance as the origin of the fortunate "accidents" appearing in conformal field theory.

Our starting point will be the action [1]:

$$L = \frac{k}{8\pi} \int \epsilon^{ijk} \operatorname{Tr} \left[A_i(\partial_j A_k - \partial_k A_j) + \frac{2}{3} A_i[A_j, A_k] \right] \qquad (8.1.1)$$

There are several unusual features to this action. First, the integrand is a total derivative, that is, it is a topological term that can be written as the integral of a derivative. With the usual boundary conditions at infinity, the Chern–Simons action is zero. However, if the fields do not vanish at infinity so rapidly, one can show that the action does not vanish. The action is invariant under gauge transformations that contain the identity, but it is not invariant under gauge transformations that have nonzero "winding numbers." In fact, under such a gauge transformation, the action changes by an integer:

$$L \rightarrow L + \text{constant } m \qquad (8.1.2)$$

The second unusual feature of this action is that it is generally covariant without the presence of a metric tensor. For decades, physicists have viewed this symmetry as a result of integrating over all possible metric tensors in the functional integral.

However, because the constant tensor ϵ^{ijk} transforms as a true density under coordinate transformations, we see that the Chern–Simons action is actually a generally covariant object, even if it lacks any metric tensor. There is no need to insert the determinant of the metric tensor $\sqrt{-g}$ into the action to form scalar densities. This leads to the unusual conclusion that the observables of the theory must be independent of the parameterization, that is, they must be topological objects, with *finite* degrees of freedom.

Topological field theories of this type are very strange when viewed from the point of view of ordinary quantum field theory. For example, even a point particle field theory, the simplest possible quantum field theory, has an infinite number of degrees of freedom. These topological theories, because they only have a finite number of degrees of freedom, describe topological objects and are not physical in the strict sense of the word. In fact, we will see that the space of observables consists of pure numbers, for example, topological invariants associated with knots.

The observables are gauge invariant and independent of the background metric (that is, they are topological) and consist of Wilson loops:

$$W_R(C) = \text{Tr} \, P \exp i \int_C A_i dx^i \qquad (8.1.3)$$

where we take the path-ordered P product of exponentials around a knot loop or knot C and trace over the R representation of the Lie group. The object that we wish to study is the functional average of these Wilson loops defined over knots C_i (which are closed) or a series of "links," which intertwine several closed knots:

$$\left\langle \prod_{i=1} W_{R_i} \right\rangle = \int DA \exp(iL) \prod_{i=1} W_{R_i}(C_i) \qquad (8.1.4)$$

This object must be a topological object, that is, by changing the background metric, the correlation function remains the same. The topological invariants of knot theory are called "knot invariants," so we now have an analytic way in which to generate knot invariants via quantum field theory. (However, it should be emphasized that the string itself moves in 26-dimensional space and does not form knots. In fact, in four and higher dimensions, it can be shown that knots formed by the string can be untied or unraveled. Knots, consisting of one-dimensional lines, only exist in three dimensions. To create higher dimensional knots, one has to twist planes and solids.)

The next step is to quantize the theory. There are two ways in which to perform quantization.

1. One may impose the constraints first and then quantize the theory in the reduced system, as in Coulomb gauge quantization. This has the advantage of working directly with the reduced space, which will turn out to be finite dimensional. If we work in the Coulomb gauge, we define the fields along a slice in three-dimensional space. Along the two-dimensional slice, we can, in turn, rewrite the coordinates in terms of z and $\bar{z}$ coordinates, that is, in terms of holomorphic and antiholomorphic coordinates.

2. One may alternatively quantize the system first and then impose the constraint (Gauss's law) on the Fock space, as in Gupta–Bleuler quantization. In this way, we work with the full infinite dimensional space spanned by A_i^a, and only at the end, do we see the finite dimensional space emerge.

Let us analyze the first case. As in ordinary gauge theory, the constraint is the coefficient of the Lagrange multiplier A_0 in the action. In our case, however, the constraint is qualitatively different from that found in ordinary gauge theory:

$$\epsilon^{ij} F_{ij} = 0 \qquad (8.1.5)$$

Gauss's law tells us that the space M spanned by the connections A_i^a must be reduced to the space of flat (that is, zero curvature) connections, modulo gauge transformations.

Naively, having zero curvature means that the connection is a pure gauge field, and hence, the theory is empty. In our case, the meaning of this strange constraint is that an infinite dimensional system, labeled by A_i^a, has been reduced to a system with only a finite number of degrees of freedom. The constraint destroys an infinite number of degrees of freedom, but it leaves a finite dimensional system intact, spanned by topological objects.

The space of flat connections modulo gauge transformation has already been classified by mathematicians, and it is the space of conformal blocks [1–3]. Thus, conformal field theory enters at the level of the Hilbert space of the theory after quantization.

Flat connections, in turn, are characterized by their holonomies (that is, Wilson loops) around closed paths. These paths, unlike the situation in ordinary gauge theory, are topologically defined; if one distorts them continuously, then the holonomy remains the same.

That the Wilson loops are topologically defined can be seen by pinching off a small circular deformation in a path ordered product. The change in holonomy around this small loop will be proportional to the curvature in that loop. But, since the curvature is zero, there is no change if we make a small deformation in the path. Thus, these paths are defined topologically, unlike the usual case in ordinary gauge theory. They depend only on the homology cycles on the Riemann surface, that is, the a and b cycles. The dimension of the conformal blocks on this genus g Riemann surface for a Lie group G is finite and is given by:

$$2(g-1)\dim G; \qquad g > 1 \qquad\qquad (8.1.6)$$

The situation differs slightly if we quantize in the presence of a knot or Wilson loop defined in this space. When one slices a knot by taking equal time slices, one see charges distributed along the surface, so the constraint equation becomes:

$$\frac{k}{8\pi}\epsilon^{ij}F_{ij}^a = \sum_{s=1}^{r}\delta^2(x - P_s)T_{(s)}^a \qquad\qquad (8.1.7)$$

where the sources are located at points P_s. Then, the physical space of the theory describes the conformal blocks for an r point function for fields defined in various representations of the algebra.

Let us analyze this quantization more carefully. Before gauge fixing, let us split the three-dimensional space into $Y = \Sigma \otimes R$, a Riemann surface Σ times the time direction R. Let us use the language of forms for convenience. Then, the exterior derivative d and the gauge field A split up as:

$$d = dt\,\partial/\partial t + \tilde{d}, \qquad A = A_t + \tilde{A} \qquad\qquad (8.1.8)$$

With this splitting the action becomes:

$$S = -\frac{k}{4\pi}\int_Y \mathrm{Tr}\left(\tilde{A}\frac{\partial}{\partial t}\tilde{A}\,dt\right) + \frac{k}{2\pi}\int_Y \mathrm{Tr}(\tilde{d}\tilde{A} + \tilde{A}^2) \qquad\qquad (8.1.9)$$

Since the last term fixes the curvature to be zero, we can solve this to give:

$$\tilde{A} = -(\tilde{d}U)U^{-1} \qquad\qquad (8.1.10)$$

which has zero curvature. Assuming Σ is bounded, we can plug this into the original action and find

$$S = kS_{WZW} = \frac{k}{4\pi}\int_{\partial Y}\mathrm{Tr}(U^{-1}\partial_\phi U\,U^{-1}\partial_t)d\phi\,dt + \frac{k}{12\pi}\int_Y \mathrm{Tr}(U^{-1}\tilde{d}U)^3 \qquad\qquad (8.1.11)$$

where ϕ is an angular variable defined around the perimeter of Σ.

We see that the Chern–Simons action has become a version of the WZW action, which we know is conformally invariant. Here, we see how the full Kac–Moody algebra emerges from the Chern–Simons theory, as well as conformal field theory. Likewise, we can define the theory on cosets and retrieve all the results of the GKO coset construction.

Another way of viewing these results is to quantize first and then apply the constraints on the wave function later. In this scheme, A_i^a still has an infinite number of degrees of freedom, which become finite only when one applies the constraints. The quantization is carried out by postulating the commutation relations:

$$[A_z^a(z), A_{\bar{z}}^b(w)] = \frac{4\pi}{k}\delta^{ab}\delta^2(z-w) \tag{8.1.12}$$

where we have converted the coordinates on Σ into holomorphic and anti-holomorphic coordinates.

We now postulate the existence of a wave functional $\Psi(A_z)$, where the field A_z has an infinite number of degrees of freedom. We wish to impose Gauss's law constraint directly onto this wave functional, thereby reducing the number of degrees of freedom of the system. This wave functional transforms under a gauge transformation labeled by g as:

$$\left[U(g)\Psi\right](A_z) = e^{ikS(g,A_z)}\Psi(A_z^g) \tag{8.1.13}$$

where S is the WZW action. If we apply the constraint onto $\Psi(A_z)$, it means that the wave function must be gauge invariant. Thus, the physical wave function is given by [2]:

$$\Psi_{\text{phy}}(A_z) = \int Dg\, e^{ikS(g,A_z)}\Psi_0(A_z^g) \tag{8.1.14}$$

Our task is to show that Ψ_{phy} depends only on a finite number of degrees of freedom because it is invariant under the gauge constraint, while the original Ψ depends on an infinite number of degrees of freedom.

In contrast to ordinary gauge theory, in the Chern–Simons theory, we have the freedom of gauging A_z to a constant a in the Cartan subalgebra, that is,

$$A_z = gag^{-1} - \partial g\, g^{-1} \tag{8.1.15}$$

so A_z^g can be replaced by a constant a. Thus, the wave function's infinite degrees of freedom have been reduced. However, the remaining Ψ_{phy} still has a finite number of degrees of freedom. For example, let Σ equal the torus T^2. Then Ψ_{phy} is a function of the modular parameter τ and a. We can power expand the wave function in a complete set of states which have the correct boundary conditions for T^2, which are the characters $\chi_\lambda(\tau, a)$. This is equivalent to writing Ψ_{phy} as a trace over the torus and then inserting a complete set of operators that belong to a Verma module into the trace. We get:

$$\Psi_{\text{phy}}(\tau, a) = \sum_\lambda A_\lambda \chi_\lambda(\tau, a) \tag{8.1.16}$$

Thus, the wave function, defined on a torus, can be expanded in the space of characters defined on the torus, which in turn can be written as a Weyl–Kac character formula.

Likewise, the same reasoning can be applied when Σ equals a disk. Again, we see that a gauge transformation on Ψ can reduce the infinite degrees of freedom contained within A_z, leaving only a function depending on the parameters of the disk and the value of the constant a. We then expand the remaining wave function, which is now power expanded in terms of a complete set of functions defined on the disk, that is, conformal blocks. In this way, conformal blocks enter the theory using the wave functional method.

In summary, we find a large set of correspondences between the Chern–Simons theory and conformal field theory that allow us to classify the latter via the gauge group G and the coupling constant of the Chern–Simons theory. Specifically, we found that

1. the space of flat connections modulo gauge transformations is equivalent to the space of conformal blocks;

2. the space of states of the wave function defined on a torus is equivalent to the space of Weyl–Kac character functions;

3. when we choose the gauge group G with a subgroup H, we can construct the apparatus of the GKO coset method; and

4. a system with an infinite number of degrees of freedom is reduced, after quantization, to a space with a finite number of degrees of freedom, related to the invariant holonomies one can construct on Σ with genus g.

A systematic study shows that all known rational conformal field theories can be constructed in this way. This is remarkable, because we have only assumed general covariance and gauge invariance, that is, we have derived the results of conformal field theory using much simpler structures.

8.2. Elementary Knot Theory

The next step is to calculate the correlation functions of the theory, which are matrix elements of Wilson loops. To understand how knot invariants arise from these correlation function, it will be useful to review the developments in knot theory [4, 5]. Historically, knot theory has, over the decades, been a stagnant area of mathematics. The central problem of knot theory, the complete classification of all possible knots, has consistently eluded mathematicians. The problem is to construct certain expressions, called *knot invariants*, that can be placed in one-to-one correspondence with topologically distinct knots.

The problem is an exceedingly difficult one because knots, even if they have only a small number of loops, become quickly snarled and entangled, making it difficult to decide which ones are really distinct and which ones can be deformed into each other without cutting the knot.

Let us make a few simple definitions. To be more specific, we say a knot is a single strand or line that is closed, that is, if we cut the knot, it unravels into a single strand. Then, we define the *unknot* as a knot which can be topologically deformed into a circle, without cutting. Also, define a *link* as a series of knots that are intertwined and cannot be separated. If we cut N distinct strands that form a link, then it reduces to N independent single strands.

The first major topological invariant in knot theory was found by Gauss; it is called the *linking number* and is defined as:

$$\Phi(C_a, C_b) = \frac{1}{4\pi} \int_{C_a} dx^i \int_{C_b} dx^j \, \epsilon_{ijk} \frac{(x-y)^k}{|x-y|^3} \qquad (8.2.1)$$

where dx^i are defined along the knot.

The Gauss linking number is an analytic expression defined on a link that is topologically invariant. By performing the integration around the stands, one can, in principle, determine the degree to which several closed loops are linked together. Notice that the linking number remains invariant even if one smoothly deforms the contour, as long as one does not move contours past each other or cut them. Thus, this expression must be a topological invariant.

In the 19th century, Tait and Little began the arduous task of beginning a classification of topologically distinct knots. In 1970, Conway pushed the classification to 10 double points [6].

One fundamental advance was made in 1928 by Alexander [7], who showed that it was possible to associate a polynomial, called the *Alexander polynomial*, with each knot. If two knots had different polynomials, then they were topologically distinct. This made it possible to take two complicated knots, calculate their Alexander polynomial, and rapidly decide whether they were topologically distinct or not.

For example, the circle is a trivial knot and has an Alexander polynomial given by the number one. The clover leaf knot, shown in Fig. 8.1a, has the Alexander polynomial given by:

$$\text{clover leaf}: \quad \Delta = t^2 - t + 1 \qquad (8.2.2)$$

Another example is the figure eight knot, shown in Fig. 8.1b, with an Alexander polynomial given by:

$$\text{figure eight}: \quad \Delta = t^2 - 3t + 1 \qquad (8.2.3)$$

A more complicated knot is called the Stevedore's knot, shown in Fig. 8.1c, with an Alexander polynomial given by:

$$\text{Stevedore's knot}: \quad \Delta = 2t^2 - 5t + 2 \qquad (8.2.4)$$

The Alexander polynomials Δ, furthermore, can be shown to obey the following identity. Let a L_+, L_-, and L_0 be three links that are completely

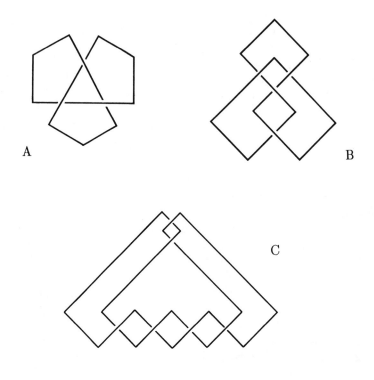

Fig. 8.1.

identical, except that, at one juncture, they have the topology as shown in Fig. 8.2. Then, the Alexander polynomial, for these three different links, satisfies the following relation:

$$\Delta_{L_+} - \Delta_{L_-} = \left(\sqrt{t} - 1/\sqrt{t}\right) \Delta_{L_0} \qquad (8.2.5)$$

which is called the "skein relation."

However, Alexander polynomials have an important defect; they are not powerful enough to distinguish between all topologically distinct knots. In fact, they often fail to distinguish between elementary knots. For example, both the granny knot and the square knot have the same Alexander polynomial, but they are topologically distinct. Also, the Alexander polynomial cannot distinguish between knots that are mirror reflections of each other.

Fig. 8.2.

Another important development in knot theory was Artin's theory of braids [8]. Because knots are such difficult objects to manipulate, Artin introduced a much simpler system by which to analyze knots. A braid begins as a series of parallel strands with equal length arranged in a definite sequence. We are allowed to cross one strand over another by a braiding operator. Let U_i be the braiding operator that moves the ith line across the $(i+1)$th line. Then, Artin showed that these braiding operators form a group:

$$\text{braid group :} \quad \begin{cases} U_i U_j = U_j U_i; & |i-j| \geq 2 \\ \\ U_i U_{i+1} U_i = U_{i+1} U_i U_{i+1} \end{cases} \tag{8.2.6}$$

In Fig. 8.3, we have an example of a braid specified by the U operation.

The advantage of using the braid group is that we can form knots and links out of these braids. If we identify the two sets of endpoints of the braid, then the lines close on themselves and we get a link. For example, n parallel lines forming a braid, when wrapped in this fashion, make n closed loops, or unknots. All knots and links can be generated by wrapping a braid. (However, the power of this technique is limited because two different braids, when wrapped, may yield the same knot or link.)

After years of slow progress in knot theory, two recent breakthroughs in this area came quite suddenly, within the last few years. The first breakthrough, after an interval of almost 60 years, was the development of the *Jones polynomial* in 1985 [9, 10], which associated a new polynomial to every knot, more powerful than the earlier Alexander polynomial. With the Jones polynomial, the classification of all possible knots was suddenly within reach. The origin of the Jones polynomial, however, was quite obscure. Also, the Jones polynomial still was not powerful enough to generate a one-to-one relation between a polynomial and a knot.

The second development, which we will soon discuss, came from an entirely unexpected area, two- and three-dimensional quantum field theories that could be written in terms of knots. In fact, it became possible to rederive the Jones polynomial from physics and to create infinitely many more polynomials. Eventually, with this new generation of knot invariants

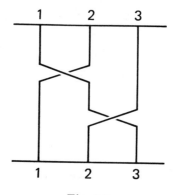

Fig. 8.3.

coming from physics, it may be possible to find a one-to-one description of knots in terms of polynomials, although this is still not certain.

8.3. Jones Polynomial and the Braid Group

Jones made the following observation [9]. In mathematics, there is the *von Neumann algebra* A_n, a finite dimensional algebra whose elements e_i obey the following relations:

$$e_i^2 = e_i, \qquad e_i^* = e_i$$
$$e_i e_{i\pm1} e_i = \left[t/(1+t)^2\right]e_i \qquad (8.3.1)$$
$$e_i e_j = e_j e_i, \qquad |i-j| \geq 2$$

Jones noticed the similarity between the von Neumann algebra A_n and Artin's braid group B_n. Specifically, one can make the following correspondence between the two algebras:

$$U_i \leftrightarrow \sqrt{t}\left[te_i - (1-e_i)\right] \qquad (8.3.2)$$

Although this correspondence between two seemingly unrelated algebras appears to have no consequence, there is an important difference: on the von Neumann algebra (in contrast to the Artin algebra), it is possible to define the trace operation, which sends elements of A_n into a complex number, such that:

$$\text{Tr}(ab) = \text{Tr}(ba)$$
$$\text{Tr}(we_{n+1}) = \left[t/(1+t)^2\right]\text{Tr}\,w \qquad \text{if} \quad w \text{ is in } A_n \qquad (8.3.3)$$
$$\text{Tr}(a'a) > 0 \qquad \text{if} \quad a \neq 0$$

and $\text{Tr}(1) = 1$. Because an invariant trace is defined on the von Neumann algebra and a correspondence can be made between the generators of the von Neumann algebra and the generators of the Artin algebra, we can now define the invariant trace operation on the Artin algebra:

$$V_L(t) = \left[-(t+1)/\sqrt{t}\right]^{n-1} \text{Tr}\left[r_t(b)\right] \tag{8.3.4}$$

where r_t is the operation that sends generators of one algebra into the other.

This invariant polynomial $V_L(t)$ is called the Jones polynomial. Like the Alexander polynomial, it also satisfies a skein relationship:

$$(1/t)V_{L_-} - tV_{L_+} = \left(\sqrt{t} - \frac{1}{\sqrt{t}}\right)V_{L_0} \tag{8.3.5}$$

(Via the skein relation, one can recursively generate the Jones polynomials. One starts with the unknot, which has the knot polynomial 1, and then successively applies the skein relation to get the polynomial of increasingly more complicated knots.)

The advantage of the Jones polynomial is that it reveals elegant relations between knots and other types of algebras, as well as being much more powerful than the Alexander polynomial. Topologically distinct knots, which have the same Alexander polynomial, can be shown to have different Jones polynomials. (However, the Jones polynomial, as we mentioned, fails to establish a one-to-one relationship between a polynomial and a knot, that is, two topologically distinct knots may have the same Jones polynomial.) After the initial discovery of the Jones polynomial, several other, more powerful polynomials were then discovered.

To analyze these higher polynomials, let us make a few definitions. If two links L_1 and L_2 are topologically equivalent, we say that they are *ambient isotopic*. It has been shown that two links are ambient isotopic if and only if there is a finite sequence of moves (called *Reidemeister moves*) that can deform L_1 into L_2. These moves, which come in three types, are shown in Fig. 8.4.

However, there is also a second concept, called *regular isotopy*. Two links are regular isotopic if one can be deformed into the other by using only type II and type III moves.

Regular isotopic links have a simple interpretation. Let us temporarily replace each strand in a knot by a long, flat ribbon, which makes a "framed knot." An ordinary knot and a framed knot are topologically distinct because twisting a ribbon within a framed knot produces a new configuration, while twisting a strand of a knot does nothing.

Notice that type I Reidemeister moves are not allowed for two-dimensional or framed knots. If we execute the Reidemeister type I move for ribbons, we find that they are equivalent to twisted ribbons when they are straightened out.

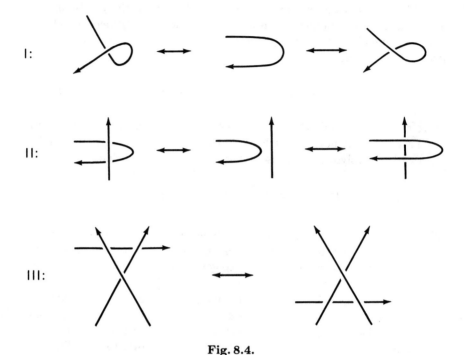

Fig. 8.4.

Let us now introduce the Hoste, Ocneanu, Millet, Freyd, Lickorish, Yetter HOMFLY polynomial $P_L(t, z)$ [11], which is more powerful than the Jones polynomial, which is a finite Laurent polynomial in two variables. If two links are ambient isotopic, then they have the same HOMFLY polynomial. These HOMFLY polynomials are defined via:

$$tP_{L_+} - t^{-1}P_{L_-} = zP_{L_0} \tag{8.3.6}$$

where $L_\pm$ and L_0 were defined in Fig. 8.2.

The HOMFLY polynomials are defined recursively. If we set the unknotted knot as $P_0 = 1$, then, by successively knotting various parts of the line via $L_\pm$ and L_0, we can gradually build up knots and links of arbitrary complexity.

Both the Alexander polynomial and the Jones polynomial can be represented as special cases of the HOMFLY polynomial. From Eqs. (8.2.5) and (8.3.5), we see:

$$P_L(t = 1, z = \sqrt{t} - 1/\sqrt{t}) = \text{Alexander polynomial}$$
$$P_L(t, z = -\sqrt{t} + 1/\sqrt{t}) = \text{Jones polynomial} \tag{8.3.7}$$

The deficiency of the HOMFLY polynomial, however, is that it is not powerful enough to distinguish between all topologically distinct links, that is, like the Jones polynomial, it is possible to construct two that which are ambient isotopically distinct but have the same HOMFLY polynomial.

Notice that the Alexander, Jones, and HOMFLY polynomials share the property of categorizing links that are ambient isotopically the same. However, it is possible to introduce another polynomial, the Kauffman polynomial [12], that is used to analyze links which are only regular isotopically the same (that is, for framed links or ribbons).

The Kauffman polynomial R_L is defined by:

$$R_{L_+} - R_{L_-} = z R_{L_0} \qquad (8.3.8)$$

with the additional condition:

$$R_{\hat{L}_+} = \alpha R_{\hat{L}_0}, \qquad R_{\hat{L}_-} = \alpha^{-1} R_{\hat{L}_0} \qquad (8.3.9)$$

where the $\hat{L}_\pm$ and $\hat{L}_0$ are given in Fig. 8.5. Because the HOMFLY and Kauffman polynomials are related to ambient and regular isotopy, they cannot, of course, be directly related to each other via an ambient isotopic equation. However, it is possible to write the following relationship, which is defined only up to regular isotopy:

$$P_L(t = \alpha, z) = \alpha^{-w(L)} R_L(\alpha, z) \qquad (8.3.10)$$

where $w(L)$, called the *wraith*, is given by:

$$w(L) = \sum_p \epsilon(p) \qquad (8.3.11)$$

and $\epsilon(L_\pm) = \pm 1$. The important point is that the wraith is only a regular isotopic invariant, and hence, there is no contradiction in the above equation. The distinction between the ambient isotopically defined HOMFLY polynomial and the regular isotopically defined Kauffman polynomial will soon become important when we use quantum field theory to generate these knot invariants.

8.4. Quantum Field Theory and Knot Invariants

The remarkable feature about the Chern–Simons field theory is that the correlation functions must be topological (since the metric never appears in the action), and hence, it reproduces the known invariants of topology and generates entirely new classes of invariants as well. For example, take the matrix element between two gauge fields:

$$\left\langle A_\mu^a(x) A_\nu^b(y) \right\rangle = \frac{i}{k} \delta^{ab} \epsilon_{\mu\nu\rho} \frac{(x-y)^\rho}{|x-y|^3} \qquad (8.4.1)$$

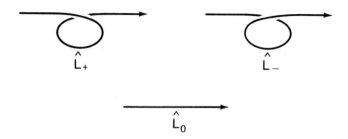

Fig. 8.5.

Notice that the two-point function in a Chern–Simons theory is not the usual propagator, with a well-defined light cone and propagation of states. Also, notice that the Chern–Simons theory is linear in derivatives.

Next, let us power expand the Wilson loop as follows [13, 14]:

$$
W(C) = \text{Tr} \left[1 + i \oint_C dx^\mu A_\mu - \oint_C dx^\mu \int^x dy^\nu A_\nu(y) A_\mu(x) \right.
$$
$$
- i \oint_C dx^\mu \int^x dy^\nu \int^y dz^\rho A_\rho(z) A_\nu(y) A_\mu(x)
$$
$$
\left. + \oint_C dx^\mu \int^x dy^\nu \int^y dz^\rho \int^z dw^\sigma A_\sigma(w) A_\rho(z) A_\nu(y) A_\mu(x) + \cdots \right]
$$
(8.4.2)

The advantage of this power expansion is that we can now take the matrix element of the Wilson loop and calculate the topological invariants of knot theory. However, there are some ambiguities that have to be clarified.

Expanding in powers of $1/k$, we find that the first term in the correlation function is:

$$
\langle W(C) \rangle_0 = \dim R \tag{8.4.3}
$$

where R^a are the generators of the group. To the next order in $1/k$, we have:

$$
\langle W(C) \rangle_1 = -\text{Tr}(R^b R^a) \oint_C dx^\mu \int^y dy^\nu \langle A_\nu^b(y) A_\mu^a(x) \rangle
$$
$$
= -\frac{2\pi}{k} \dim R \, c_2(R) \phi(C) \tag{8.4.4}
$$

where $c_2(R)\mathbf{1} = R^a R^a$ and:

$$\phi = \frac{1}{4\pi} \oint_C dx^\mu \oint_C dy^\nu \epsilon_{\mu\nu\rho} \frac{(x-y)^\rho}{|x-y|^3} \qquad (8.4.5)$$

Notice that the above expression is actually ambiguous when $x \to y$. Oddly enough, the answer is dependent on how we regularize the correlation function. There are many ways to regularize this function when x and y are coincident, and the correlation function changes with each one. (This is actually reasonable, because the regularization dependent terms are phases, which drop out when calculating the squares of amplitudes.)

The most common procedure is to introduce a *framing* for each contour, that is, expand each line into a flat, two-dimensional ribbon. We can frame a knot by introducing a vector n^μ that is orthogonal to C, that is, we can replace the contour C via a framing contour C_f defined as:

$$x^\mu(t) \to y^\mu(t) + \epsilon n^\mu(t) \qquad (8.4.6)$$

for small ϵ. Thus, we have now reproduced a result from classical mathematics, the Gauss winding number for a knot from quantum field theory.

We can do better than just rederive known topological invariants; in fact, we can derive analytical expressions for knot invariants that are only known abstractly to mathematicians.

For example, taking the next order in $1/k$, we find an analytical expression for the second coefficient of the Alexander polynomial [13, 14]:

$$\langle W(C)_2 \rangle = (2\pi/k)^2 \dim R \left[-\frac{1}{2} c_2^2(R) \phi^2(C) + c_v c_2 \rho(C) \right] \qquad (8.4.7)$$

where $\rho(C)$ is $\rho_1 + \rho_2$ and:

$$\rho_1(C) = -\frac{1}{32\pi^3} \oint_C dx^\mu \int^x dy^\nu \int^y dz^\rho \, \epsilon^{\alpha\beta\gamma} \epsilon_{\mu\alpha\sigma} \epsilon_{\nu\beta\lambda} \epsilon_{\rho\gamma\tau} I^{\sigma\lambda\tau}$$

$$I^{\sigma\lambda\tau}(x,y,z) = \int d^3w \frac{(w-x)^\sigma (w-y)^\lambda (w-z)^\tau}{|w-x|^3 |w-y|^3 |w-z|^3}$$

$$\qquad (8.4.8)$$

and:

$$\rho_2(C) = \frac{1}{8\pi^2} \oint_C dx^\mu \int^x dy^\nu \int^y dz^\rho \int^z dw^\sigma$$

$$\times \epsilon_{\sigma\nu\alpha} \epsilon_{\rho\mu\beta} \frac{(w-y)^\alpha (z-x)^\beta}{|w-y|^3 |z-x|^3} \qquad (8.4.9)$$

In this way, we can obviously continue and obtain analytic expressions for knots of arbitrary complexity. However, we now would like to make contact with the Jones polynomial and the Kauffman polynomials.

The essential point is that the matrix elements of Wilson lines are ill defined when two points coincide, and hence we must introduce framed links. Thus, the knot invariants that arise from quantum field theory must be regular isotopically defined links.

Now, let us derive the knot invariants for arbitrary links via quantum field theory. We will find it convenient to introduce yet another invariant, called $S_L(\alpha, \beta, z)$, which is a function of three variables and is defined for regularly isotopic links.

This new knot invariant satisfies [13, 14]:

$$
\begin{aligned}
S_{\hat{L}_+} &= \alpha S_{\hat{L}_0} \\
S_{\hat{L}_-} &= \alpha^{-1} S_{\hat{L}_0} \\
\beta S_{L_+} - \beta^{-1} S_{L_-} &= z S_{L_0}
\end{aligned}
\tag{8.4.10}
$$

We can relate this new knot invariant to the HOMFLY invariant via:

$$
P_L(t = \alpha\beta, z) = \alpha^{-w(L)} S_L(\alpha, \beta, z) \tag{8.4.11}
$$

If we define $S_0 = 1$ for the unknot, then we can, via these generalized skein relations, gradually build up links of arbitrary complexity.

The whole point of introducing yet another knot invariant is that correlation functions of the Wilson loop generate S_L, which in turn can be related to HOMFLY, Kauffman, and Jones polynomials. To prove this, we will show that $\langle W(C) \rangle$ satisfies these skein relations, which will be sufficient to show that the correlation functions equal S_L for a link of arbitrary complexity. We will demonstrate that the correlation functions of Wilson loops satisfy these generalized skein relations by explicit calculation.

Let us analyze the difference between $\hat{L}_\pm$ and $\hat{L}_0$ from the quantum field point of view. We see that the only difference between these three configurations is the insertion of a loop, which can be taken to be infinitesimally small. Our task, therefore, is to see how a quantum correlation function changes when we insert a small loop into a Wilson line.

We begin by defining a Wilson line from x_1 to x_2 and then inserting an infinitesimally small loop at point x, which lies between the endpoints. Since the line integral around a small loop generates the curvature tensor, the Wilson line operator $U(x_1, x_2)$ becomes:

$$
U(x_1, x_2) \rightarrow U(x_1, x) i \Sigma^{\mu\nu} F^a_{\mu\nu} R^a U(x, x_2) \tag{8.4.12}
$$

where:

$$
\Sigma^{\mu\nu} = dx^\mu \, dx^\nu \tag{8.4.13}
$$

is the area tensor characterizing the small loop and R^a is a generator of the algebra.

Fortunately, the insertion of $F_{\mu\nu}$ inside a correlation function is equivalent to taking the functional derivative of e^{iS} with respect to the field. We can then integrate by parts to obtain the following:

$$\langle F_{\mu\nu}^a O_1 O_2 \cdots \rangle = Z^{-1} \int DA_\mu^a \left(\frac{-i4\pi}{k} \right) \epsilon_{\mu\nu\rho} \frac{\delta e^{iS}}{\delta A_\rho^a(x)} O_1 O_2 \cdots$$

$$= Z^{-1} \int DA_\mu^a e^{iS} \left(\frac{i4\pi}{k} \right) \epsilon_{\mu\nu\rho} \frac{\delta}{\delta A_\rho^a} (O_1 O_2 \cdots)$$

(8.4.14)

where we have integrated by parts. The next step is to take the derivative of the various O_i, which pulls down an R^a, so the insertion of $F_{\mu\nu}^a$ generates the insertion of:

$$-i \left(\frac{4\pi}{k} \right) \delta^3(x-y) \epsilon_{\mu\nu\rho} \Sigma^{\mu\nu} \, dy^\rho \times \langle \cdots U(x_1,x) \sum_a R^a R^a U(y,x_2) \cdots \rangle$$

(8.4.15)

Thus, we see that the insertion of a small loop at point x yields the insertion of the curvature tensor, which in turn generates a factor proportional to:

$$\delta^3(x-y) \epsilon_{\mu\nu\rho} \Sigma^{\mu\nu} \, dy^\rho$$

(8.4.16)

which is a volume element oriented along the tangent to the Wilson line. We can normalize v to be 0 or ± 1, depending on the orientation of $\Sigma_{\mu\nu}$. Thus, the change in the correlation function by inserting a small loop along the Wilson loop is:

$$\delta \langle W(L) \rangle = \mp(i4\pi/k) c_2(R) \langle W(L) \rangle$$

(8.4.17)

Now, compare this to the first of the two generalized skein relations. We see that we can now set:

$$\langle W(\hat{L}_+) \rangle = \alpha \langle W(\hat{L}_0) \rangle$$
$$\langle W(\hat{L}_-) \rangle = \alpha^{-1} \langle W(\hat{L}_0) \rangle$$

(8.4.18)

where we now have an explicit quantum theoretic expression for α:

$$\alpha = 1 - i \frac{2\pi}{k} c_2(R) + O \left(\frac{1}{k^2} \right)$$

(8.4.19)

Now, to prove the last of the skein relations, we repeat the same steps as before, except for the configurations $L_\pm$ and L_0.

In order to reduce the skein relations to the insertion of small loops, we must analyze the crossing of two Wilson lines. Notice that the crossing of L_+ can be deformed into the crossing of L_- if we insert a small loop at the precise point of the crossing. For example, if the Wilson line $U(x_1, x_2)$ passes over the line $U(x_3, x_4)$ at point x, then the insertion of a loop into $U(x_1, x_2)$ at the crossing point x changes the topology: now, $U(x_1, x_2)$ passes *under* $U(x_3, x_4)$. Concretely, we see that we can pass from L_- to L_+ by the insertion of this loop:

$$\langle W(L_+) \rangle = \langle W(L_-) \rangle + \langle \cdots U(x_1,x) i \Sigma^{\mu\nu} F_{\mu\nu}^a R^a U(x,x_2) \cdots U(x_3,x_4) \cdots \rangle$$

(8.4.20)

Repeating the same steps as before, we find that the insertion of the curvature tensor is equivalent to taking the derivative with respect to the field, which in turn pulls down an R^a. However, it is important to realize that the R^a factors no longer occur at the same point along the Wilson loop:

$$\langle W(L_+)\rangle = \langle W(L_-)\rangle$$
$$- \frac{i4\pi}{k} \sum_a \langle \cdots U(x_1, x) R^a U(x, x_2) \cdots U(x_3, x) R^a U(x, x_4) \cdots \rangle$$

(8.4.21)

Because the R^a matrices are defined at different points along the path, we must use the Fierz identity on the sum:

$$\sum_a R^a_{ij} R^a_{kl} = \frac{1}{2} \delta_{il} \delta_{jk} - \frac{1}{2N} \delta_{ij} \delta_{kl}$$

(8.4.22)

Notice that this Fierz identity, inserted at the crossing point x, has the effect of changing the $SU(N)$ topology of the graph, so that the Wilson loop defined at L_+ spins off terms related to L_- and L_0 as follows:

$$\langle W(L_+)\rangle = \left(1 + \frac{i2\pi}{kN}\right) \langle W(L_-)\rangle = -i\frac{2\pi}{k} \langle W(L_0)\rangle$$

(8.4.23)

Putting everything together, we find that we can satisfy the last of the generalized skein relations [1, 13, 14]:

$$\beta\langle W(L_+)\rangle - \beta^{-1}\langle W(L_-)\rangle = z\langle W(L_0)\rangle$$

(8.4.24)

where:

$$\beta = 1 - \frac{i2\pi}{kN} + O\left(\frac{1}{k^2}\right)$$
$$z = -\frac{i2\pi}{k} + O\left(\frac{1}{k^2}\right)$$

(8.4.25)

All three of the generalized skein relations [Eq. (8.4.10)] are now satisfied, so that we have a precise relationship (at least to order $1/k^2$) between correlation functions of Wilson loops and knot polynomials.

In summary, we have been able to obtain new knot invariants, unite the old ones, and generate analytic expressions for them by using Chern–Simons theory. Because this theory is generally covariant without the presence of a metric, its states must be topological and gauge invariants, that is, knot polynomials.

8.5. Knots and Conformal Field Theory

It is also possible to use knot theory to analyze conformal field theories directly, without having to use a three-dimensional theory. This reveals yet another application of knot theory to physics and gives us a powerful tool by which to analyze the properties of conformal field theories, such as their correlation functions.

The braid group of Artin emerges quite naturally if one views the *monodromy* properties of N-point functions in conformal field theory, that is, taking points z_i of an N-point function and then moving them around loops or interchanging them. If, for example, we interchange two points z_i and z_j in an N-point function, this is mathematically equivalent to braiding the ith string with the jth string. Thus, the N-point functions of conformal field theory, via monodromy relations, give us a representation of the braid group.

For example, consider the Knizhnik–Zamolodchikov relations derived earlier for the Kac–Moody algebra in Eq. (3.4.8):

$$\kappa \frac{\partial}{\partial z_i} \Psi = \sum_{j \neq i} \frac{\mathbf{t}_i \cdot \mathbf{t}_j}{z_i - z_j} \Psi \qquad (8.5.1)$$

Let us say that there are l solutions to this equation. If we interchange the ith and $(i+1)$th point, then we find a linear combination of these same solutions [15], that is,

$$\Psi_I(z_1, \ldots, z_{i+1}, z_i, \ldots, z_N) = \sum_{J=1}^{l} B_I^J \Psi_J(z_1, \ldots, z_i, z_{i+1}, \ldots, z_N) \qquad (8.5.2)$$

Because the B_I^J operation simply interchanges the location of two points, it can be shown that the matrices B_I^J form representations of the braid group.

Let us now be concrete about our discussion of the properties of correlation functions, by writing everything in terms of conformal blocks. We recall from Chapter 3 that there are two operations that we can perform on these conformal blocks, called B (corresponding to braiding or interchanging two points) and F (corresponding to pinching the graph, that is, making an s-channel graph into a t-channel graph).

On the space of conformal blocks, we wrote a representation of the B-twist operation on a four-point function, representing the interaction of conformal fields labeled by i, j, k, l:

$$\Phi_{ip}^j(z_1)\Phi_{pl}^k(z_2) = \sum_q B_{pq} \begin{bmatrix} j & k \\ i & l \end{bmatrix} \Phi_{ip}^k(z_2)\Phi_{ql}^j(z_1) \qquad (8.5.3)$$

It is easy to see that the effect of the B operation is to interchange z_1 and z_2, that is, it is a braiding operator.

Now we wish to relate the Yang–Baxter relationship and knot theory directly to conformal field theory. First, we note the similarity between Fig. 8.4 and Fig. 6.3, which represents the Yang–Baxter relationship. In fact, we see that they are the same. This means that the Yang–Baxter relationship is a specific realization of the braid group.

Next, we notice that the Yang–Baxter relationship can be rewritten as a series of braiding operations on conformal blocks, that is, conformal field theory gives us a representation of the Yang–Baxter relationship in terms of the braiding matrix of conformal field theory. It is now a simple matter to represent the Yang-Baxter relationship on the space of braiding matrices by labeling all of the legs and singling out each braiding operation. At the end of the calculation, we will set the two equations equal to each other.

By carefully following the strands making up the Yang–Baxter relationship in Fig. 8.6 and isolating the braiding matrices, we easily find [16]:

$$
\sum_p B_{j_6 p} \begin{bmatrix} j_2 & j_3 \\ j_1 & j_7 \end{bmatrix} (\epsilon) B_{j_7 j_9} \begin{bmatrix} j_2 & j_4 \\ p & j_5 \end{bmatrix} (\epsilon) B_{p j_8} \begin{bmatrix} j_3 & j_4 \\ j_1 & j_9 \end{bmatrix} (\epsilon)
$$

$$
= \sum_p B_{j_7 p} \begin{bmatrix} j_3 & j_4 \\ j_6 & j_5 \end{bmatrix} (\epsilon) B_{j_6 j_8} \begin{bmatrix} j_2 & j_4 \\ j_1 & p \end{bmatrix} (\epsilon) B_{p j_9} \begin{bmatrix} j_2 & j_3 \\ j_6 & j_5 \end{bmatrix} (\epsilon)
$$

(8.5.4)

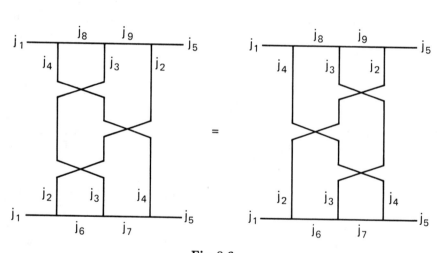

Fig. 8.6.

In summary, we see here the tight relationship between knot theory, the Yang–Baxter relationship, and the braiding matrices of rational conformal field theory. We have used the equivalent topology of all three theories to write an explicit representation of the Yang–Baxter relation in terms of the braiding matrices of rational conformal field theory. Knot theory has thus proven to a powerful tool in which to analyze conformal field theory. It should be no surprise, therefore, that knot theory should also be a useful tool in which to analyze statistical mechanical systems. Specifically, we will show that knot theory gives us a new way in which to view the Yang–Baxter relations, which we saw was the basis for integrability of two-dimensional statistical mechanical systems. To see how knot theory gives us a way in which to reanalyze statistical mechanical systems, recall that in Chapter 6, we found that there were two large classes of models, the vertex models and the IRF models, and that the essence of commuting transfer matrices was the Yang–Baxter relationship. Now, let us probe the topological structure of the Yang–Baxter relationship. Because the Yang–Baxter relationship is a tangle of indices, it helps to introduce operators that are defined in the space of lattice indices, such that the tangle of indices disappears. Let us now introduce a Yang–Baxter operator $X_i(u)$, which we will show is identical to the braid operator. We will define this operator as a generalization of the Boltzmann matrix shown above, operating on the space of the lattice. Thus, it will consist of the Boltzmann matrix multiplied by the unit matrix defined on the lattice. For the vertex model, written in terms of the S matrix, this operator is explicitly:

$$X_i(u) = \sum_{k,l,m,p} S_{lp}^{km}(u)\, I^{(1)} \otimes \cdots \otimes \delta_{pk}^{(i)} \otimes \delta_{ml}^{(i+1)} \otimes I^{(i+2)} \otimes \cdots \otimes^{(n)} \quad (8.5.5)$$

where $I^{(i)}$ is the identity matrix at the ith position, and $(\delta_{nk})_{ab} = \delta_{na}\delta_{kb}$. For the IRF representation, the Yang–Baxter operator is also a series of delta functions multiplied by the Boltzmann weight:

$$\{X_i(u)\}_{l_0 \cdots l_n}^{p_o \cdots p_n} = \prod_{j=0}^{i-1} \delta(p_j, l_j) w(l_i, l_{i+1}, p_i, l_{i-1}; u) \prod_{j=i+1}^{n} \delta(p_j, l_j) \quad (8.5.6)$$

(In other words, X_i consists of a product of delta functions except for the ith entry, which consists of the w matrix.)

Then, the Yang–Baxter operator satisfies the relations:

$$X_i(u)X_j(v) = X_j(v)X_i(u), \qquad |i - j| \geq 2$$
$$X_i(u)X_{i+1}(u+v)X_i(v) = X_{i+1}(v)X_i(u+v)X_{i+1}(u) \qquad (8.5.7)$$

For $u = u + v = v$, we find precisely the braid group relations of Artin. We will, in fact, take the limit $u, v \to \infty$ and set:

$$U_i = \lim_{u \to \infty} X_i(u)/\rho(u), \qquad i = 1, 2, \ldots, n$$
$$U_i^{-1} = \lim_{u \to \infty} X_i(-u)/\rho(-u), \qquad i = 1, 2, \ldots, n \qquad (8.5.8)$$

so, we arrive at:

$$U_i U_j = U_i U_j, \qquad |i - j| \geq 2$$
$$U_i U_{i+1} U_i = U_{i+1} U_i U_{i+1} \qquad (8.5.9)$$

Notice that these relations form Artin's braid group found earlier in Eq. (8.2.6).

Thus, the Yang–Baxter relationship has now been reduced to the relations of braid group theory, which have proven to be one of the essential ingredients in the study of knots.

8.6. New Knot Invariants from Physics

The Jones polynomial, after its discovery, quickly led to other proposals for new knot polynomials. There exists, however, a powerful way of generating new knot polynomials that comes from statistical mechanics. In fact, an infinite number of new knot polynomials can be generated in this fashion [17, 18].

The key to this program is the use of *Markov moves* [19]. If two closed braids represent the same link, then it is possible to deform one link into the other link by a succession of these Markov moves. Let A and B be elements in Artin's braid group B_n. Then equivalent braids expressing the same link can be mutually transformed by successive applications of two types of operations, called type I and type II moves:

$$\text{type I} : \quad AB \to BA$$
$$\text{type II} : \quad A \to AU_n, \qquad A \to U_n^{-1} \qquad (8.6.1)$$

for A, B and products with U_n defined to exist within the braid group.

Given Markov moves on equivalent braids, we now have sufficient information to construct the desired link polynomial. For a link polynomial to be a topological invariant, it must be obey the relationships created by Markov moves. Thus, any link polynomial, to be a topological invariant, must satisfy the following properties:

$$\alpha(AB) = \alpha(BA)$$
$$\alpha(AU_n) = \alpha(AU_n^{-1}) = \alpha(A) \qquad (8.6.2)$$

The important point is to notice that *any* nontrivial function α, which obeys these rules, via Markov moves, is guaranteed to be a topological invariant. If two links have different values of α, then they must, by construction, be topologically invariant. (However, as before, the converse may not be true, that is, this is still not powerful enough to establish that the correspondence between classes of topologically equivalent knots and topological invariants is one-to-one.)

The next step is to notice that Artin's braid relation has precisely the same topology as the Yang–Baxter relationship. Thus, the transfer matrices of statistical mechanics, because they have the same topological structure as the braid group, can be used to define new knot polynomials. The key step is to define, as with the Jones polynomial, a trace operation on the transfer matrices that obey the braid relations. Then, by construction, this trace must be the basis of a new knot polynomial.

Let us define, for the moment, a linear functional ϕ that satisfies:

$$\phi(AB) = \phi(BA)$$
$$\phi(AU_n) = \tau\phi(A) \qquad (8.6.3)$$
$$\phi(AU_n^{-1}) = \bar{\tau}\phi(A)$$

where the constants τ and $\bar{\tau}$ are given by:

$$\tau = \phi(U_i), \qquad \bar{\tau} = \phi(U_i^{-1}) \qquad \text{for all} \quad i \qquad (8.6.4)$$

Then, the desired value of the knot polynomial is given by [17]:

$$\alpha(A) = (\tau\bar{\tau})^{-(n-1)/2}(\tau/\bar{\tau})^{e(A)/2}\phi(A) \qquad (8.6.5)$$

where $e(A)$ is equal to the sum of the exponents appearing in the braid representation of A.

Our goal, therefore, is to find a solution for the trace operation in terms of two dimensional quantum systems. The key to this will be the Yang–Baxter relationship.

We now have an explicit representation of braid operators in terms of Yang–Baxter operators for the N-state vertex model. This, in turn, allows us to introduce the ϕ trace operator, which will allow us to create a polynomial that is invariant under Markov moves.

The last step is filled by noticing that ϕ can be represented by the ordinary trace over the transfer matrix:

$$\phi(A) = \text{Tr}(HA); \qquad A \in B_n \qquad (8.6.6)$$

for:

$$H = h^{(1)} \otimes h^{(2)} \otimes \cdots \otimes h^{(n)} \qquad (8.6.7)$$

where:

$$(h)_{pq} = t^{-p}\delta_{pq}/\left(\sum_{k=-s}^{s} t^{-k}\right) \qquad (8.6.8)$$

where $p, q = (-s, -s+1, \ldots, s)$ and τ and $\bar{\tau}$ can be fixed as:

$$\tau(t) = \text{Tr}(HU_i) = \left(\sum_{j=1}^{N-1} t^j\right)^{-1}$$
$$\bar{\tau}(t) = \text{Tr}(HU_i^{-1}) = t^{N-1}\left(\sum_{j=1}^{N-1} t^j\right)^{-1} \qquad (8.6.9)$$

Thus, the final value of the invariant knot polynomial, in terms of transfer matrices, is given by [17]:

$$\alpha(A) = \left[t^{-(N-1)/2}(1+t+\cdots+t^{N-1})\right]^{n-1}\left[t^{(N-1)/2}\right]^{e(A)}\text{Tr}(HA) \qquad (8.6.10)$$

Our strategy is now simple. The trace operation in Eq. (8.6.10) is guaranteed to generate knot invariants, by construction. Our task is to simply look up the various S matrices that have been computed in the past for various two-dimensional statistical models and insert them into Eq. (8.5.5), giving us the generators $X_i(u)$ of the braid group constructed out of the S matrix. Then, we select a particular knot, reexpress it in terms of braiding operators, and insert them into the trace operation in Eq. (8.6.10). In this way, we are using the S matrices found in statistical mechanics as a way in which to generate braiding operators with a trace operation defined on them.

For example, for the $N = 2$ model, (corresponding to the six-vertex model), we find the following S matrix:

$$S_{1/2\,1/2}^{1/2\,1/2}(u) = \frac{\sinh(\lambda - u)}{\sinh(\lambda)}$$

$$S_{-1/2\,1/2}^{1/2\,-1/2} = 1 \tag{8.6.11}$$

Inserting this S matrix into Eq. (8.5.5) and then inserting the resulting braiding operators into Eq. (8.6.10), we find that we can rederive the old Jones polynomials.

The great advantage of using this statistical mechanical construction, however, is that there are an infinite number of such models, labeled by N, that allow us to generate new knot polynomials beyond the Jones polynomial. For example, for $N = 3$ (corresponding to the 19-vertex model), we find:

$$S_{1\,1}^{1\,1}(u) = \frac{\sinh(\lambda - u)\sinh(2\lambda - u)}{\sinh\lambda\,\sinh(2\lambda)}, \qquad S_{-1\,1}^{1\,-1}(u) = 1$$

$$S_{0\,0}^{1\,1}(u) = \frac{\sinh(u)\sinh(\lambda - u)}{\sinh\lambda\,\sinh 2\lambda}, \qquad S_{0\,1}^{1\,0}(u) = \frac{\sinh(\lambda - u)}{\sinh\lambda} \tag{8.6.12}$$

$$S_{0\,0}^{0\,0}(u) = \frac{\sinh\lambda\,\sinh 2\lambda - \sinh u\,\sinh(\lambda - u)}{\sinh\lambda\,\sinh 2\lambda}$$

The power of this infinite set of knot polynomials is that they can distinguish between knots where the Jones polynomial fails. For example, Birman has shown that the two knots given in Fig. 8.7 by Artin's braid relations have the same Jones polynomial:

$$A = (U_1 U_2 U_1)^4 U_1^{-12} U_2^6$$

$$B = U_1^{-6} U_2^{12} \tag{8.6.13}$$

The Jones polynomial for both these knots is given by:

$$V_A = V_B = t^{-3}(t^{18} - t^{17} + 2t^{16} - 3t^{15} + 4t^{14} - 5t^{13} + 6t^{12}$$

$$- 6t^{11} + 6t^{10} - 6t^9 + 6t^8 - 5t^7 + 6t^6 - 4t^5 + 4t^4 - 3t^3 + 2t^2 - t + 1) \tag{8.6.14}$$

However, for $N = 3$, these two knots have different knot invariants. Specifically [17],

$$\alpha(A) = t^{-10}(1, -1, 0, 2, -2, -1, 4, -2, -2, 4, -3, 0, 5, -7, 1, 11, -11, -4,$$
$$16, -8, -9, 14, -4, -8, 11, -3, -7, 12, -5, -7, 13, -7, -6, 14, -7,$$
$$-9, 16, -3, -10, 10, 1, -5, 4, -1, -1, 4, -4, -1, 5, -2, -2, 3, 0, -1, 1)$$
$$\alpha(B) = t^{-12}(1, -1, 0, 2, -2, -1, 4, -3, -2, 6, -4, -2, 8, -4, -4, 10,$$
$$-6, -5, 12, -7, -5, 12, -7, -5, 13, -7, -5, 12, -7, -5, 12,$$
$$-6, -6, 12, -5, -6, 12, -4, -6, 10, -3, -6,$$
$$8, -2, -5, 6, -1, -3, 4, 0, -2, 2, 0, -1, 1)$$

$$(8.6.15)$$

where we only present the coefficients of each power of t, beginning with t^{54} for knots A and B. In summary, we have succeeded in using the S matrix found in vertex models in statistical mechanics to generate braiding operators and a trace operation, out of which an invariant knot polynomial can be constructed that gives us an infinite family of invariants beyond the Jones polynomial.

8.7. Knots and Quantum Groups

So far, our discussion of the Yang–Baxter relation has been rather ad hoc and not very systematic. In studying the symmetry properties of the Yang–Baxter equation and knots, one feels that there must be a deeper, underlying group theoretical origin to many of the relations miraculous properties. In fact, there is indeed a rich mathematical structure, called *quantum groups* [20–24], that accounts for many of the properties of the Yang–Baxter relation and gives us a systematic way in which to analyze them. It also give us explicit representations of the polynomial $6j$ equations studied in Chapter 3.

The name "quantum group" comes from the very specific way in which the Yang–Baxter relationship is realized. We recall that the Yang–Baxter relationship can be written as:

$$R_{12}(u)R_{13}(u-v)R_{23}(v) = R_{23}(v)R_{13}(u-v)R_{12}(u) \qquad (8.7.1)$$

where R, of course, is a matrix, and we have suppressed indices. Because the Yang–Baxter relation gives solutions to complex two-dimensional quantum mechanical systems, it will be a function of Planck's constant. Therefore, sometimes the previous relationship is called the *quantum Yang–Baxter relation*.

In the limit of small $\hbar$, however, one should be able to obtain the classical limit of the relation. Let us, therefore, power expand the R matrix in terms of Planck's constant:

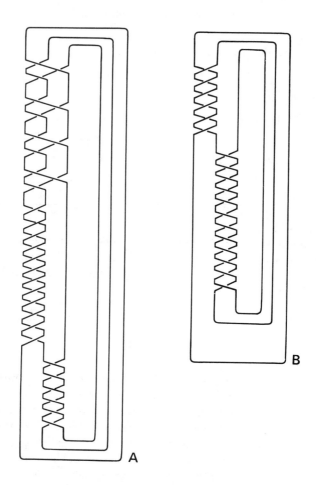

Fig. 8.7.

$$R_{ij}(u) = I - i\hbar r_{ij}(u) \qquad (8.7.2)$$

Then, as a function of the r matrix, the Yang–Baxter relationship reduces to:

$$[r_{12}(u), r_{13}(u - v)] + [r_{12}(u), r_{23}(v)] + [r_{13}(u - v), r_{23}(v)] = 0 \qquad (8.7.3)$$

This latter relationship is called the *classical Yang–Baxter relation*. In the limit that $\hbar$ goes to zero, the quantum and the classical Yang–Baxter relations become equivalent.

The important thing to notice is that the classical Yang–Baxter relationship is expressed in terms of ordinary commutators. In fact, we see that the classical Yang–Baxter relation can be expressed as the Jacobi identity associated with some Lie algebra. Thus, a classification of the classical Yang–Baxter equations in terms of standard Lie algebras is possible.

The full quantum Yang–Baxter relation [Eq. (8.7.11)], however, cannot be written in terms of commutators and is therefore considerably more complicated. Lie algebras cannot express the full quantum Yang–Baxter equation written in terms of R. However, because the limit $\hbar \to 0$ exists, one suspects that a generalization, or deformation, of a Lie algebra-type structure must exist for the full quantum Yang–Baxter relation, which reduces to the usual one in the limit as $\hbar \to 0$.

It turns out that this conjecture is correct, and the generalization of the Lie group is the quantum group. The quantum group is to the quantum Yang–Baxter relation as the classical Lie group is to the classical Yang–Baxter equation:

$$\begin{aligned} \text{Lie group} &\to \text{classical Yang} - \text{Baxter} \\ \text{quantum group} &\to \text{quantum Yang} - \text{Baxter} \end{aligned} \qquad (8.7.4)$$

Notice that the quantum group is necessarily a function of $\hbar$, and hence, possesses a smooth limit in which it reduces back to the usual classical Lie group. We will find it convenient to introduce a parameter q, which is proportional to $\hbar$, so that:

$$\lim_{q \to 1} \text{quantum group} = \text{Lie group} \qquad (8.7.5)$$

A quantum group must share many similarities with an ordinary classical Lie group, but it must also differ from the Lie group in subtle but important ways. There are at least two intuitive ways in which to see these differences.

First, we notice that the conformal field theory formed out of a Kac–Moody algebra is based entirely on its primary fields and its fusion rules. The primary fields, in turn, are labeled by the irreducible representations of the ordinary classical Lie group. Thus, much of the information and detail contained within the Kac–Moody algebra is actually washed out in the process of constructing the conformal field theory. We only see remnants of the full Kac–Moody algebra in the fusion rules. Thus, at the conformal field theory level, the resulting group structure resembles that of an ordinary Lie group in many ways. However, the resemblance cannot be exact because the original Kac–Moody algebra depended on central charges, while ordinary Lie algebras are not compatible with central charges, that is, the Jacobi identity of a classical Lie algebra cannot accomodate the c-number term. Because the conformal field theory associated with a Kac–Moody algebra depends on k and c in important ways, the reduced system cannot, therefore, define a classical Lie group. Although most of the structure of a Kac–Moody algebra is lost in the transition to the fusion rules, the central charges still make their presence felt in many important ways in the conformal field theory.

Second, another way to see the subtle difference between classical and quantum groups is to notice that their braiding operations differ in a small but important way. For a classical Lie group, the braiding operation simply interchanges the representations forming the tensor product of two representations. The braiding operation thus picks up factors of $+1$ or -1, depending on whether we have symmetric or antisymmetric combinations of the two representations. For example, tensor products can be constructed on the basis of the Young tableau, where composite representations are based on symmetrization or antisymmetrization. However, for a quantum group, the braiding operation picks up crucial phase factors. In Eq. (3.5.13), we recall, the braiding operator Ω applied to the tensor product of two primary fields picked up phase factors, which were functions of the conformal weights of the various representations. Because of the presence of these crucial phase factors, we know that the group structure underlying the conformal field theory of a Kac–Moody algebra cannot be an ordinary classical Lie group.

Now that we have established the differences and similarities between classical and quantum groups, let us be specific. Let us take the deformation of the algebra of $SL(2)_q$ and show how it is related to the quantum Yang–Baxter relation and how it may generate solutions to the polynomial equations. The commutators of $SL(2)_q$ are given by the following:

$$[J_3, J_\pm] = \pm 2J_\pm, \qquad [J_+, J_-] = [J_3] \qquad (8.7.6)$$

where, by convention, the brackets in $[J_3]$ mean:

$$[\lambda] = \frac{q^{\lambda/2} - q^{-\lambda/2}}{q^{1/2} - q^{-1/2}} \qquad (8.7.7)$$

The algebra of the quantum group $SL(2)_q$ reduces to the usual algebra in the limit $q \to 1$. This limit corresponds to taking Planck's constant to zero. This also gives us a powerful way to see intuitively how quantum groups differ from classical ones.

The representation of any quantum group necessarily mimics many of the properties of the usual group. In fact, many of the corresponding representations for the quantum group are usually found by taking the usual representation and replacing all c numbers by their quantum analog given by the brackets.

For example, the irreducible representations of $SL(2)_q$ are labeled by integers or half-integers j and have dimensions $2j + 1$. The representation space of the group is spanned by $|j, m\rangle$, just as in the ordinary case, and:

$$J_\pm |j, m\rangle = \sqrt{[j \mp m][j \pm m + 1]} \, |j, m \pm 1\rangle \qquad (8.7.8)$$

In fact, if we label the various representations by $\mathbf{V}^j$, then the usual Clebsch–Gordan tensor product decomposition remains the same:

$$\mathbf{V}^{j_1} \otimes \mathbf{V}^{j_2} = \sum_{j=|j_1-j_2|}^{j_1+j_2} \mathbf{V}^j \qquad (8.7.9)$$

For these representations, $\mathbf{V}^j$, we will now introduce two operations that have their direct analog in the usual Lie group theory. Let the operator $K_j^{j_1 j_2}$ project the jth representation out of the tensor product of the j_1 and j_2 representations. In other words:

$$K_{j_3}^{j_1 j_2} : \quad \mathbf{V}^{j_1} \otimes \mathbf{V}^{j_2} \to \mathbf{V}^{j_3} \qquad (8.7.10)$$

This is the counterpart of the usual Clebsch–Gordan coefficient.

Similarly, let us introduce the operator $R^{j_1 j_2}$ to be the braiding operator that interchanges 1 and 2:

$$R^{j_1 j_2} : \quad \mathbf{V}^{j_1} \otimes \mathbf{V}^{j_2} \to \mathbf{V}^{j_2} \otimes \mathbf{V}^{j_1} \qquad (8.7.11)$$

As we mentioned earlier, the quantum braiding operator differs from the usual braiding operator found in classical Lie group theory because it picks up important phase factors when acting on tensor products.

We now wish to construct identities for the K and R operators. We define:

$$R^{j_1 j_2} R^{j_1 j_3} R^{j_2 j_3} = R^{j_2 j_3} R^{j_1 j_3} R^{j_1 j_2}$$

$$R^{j_1 j_2} K_{j_1}^{j_3 j_4} = K_{j_1}^{j_3 j_4} R^{j_3 j_2} R^{j_4 j_2} \qquad (8.7.12)$$

$$K_j^{j_1 j_2} R^{j_2 j_1} = (-1)^{j_1 j_2 - j} q^{(c_j - c_{j_1} - c_{j_2})/2} K_j^{j_2 j_1}$$

We recognize the first relation as the Yang–Baxter equation. The first two relations are operator expressions acting on the tensor product of three representations:

$$\mathbf{V}^{j_1} \otimes \mathbf{V}^{j_2} \otimes \mathbf{V}^{j_3} \qquad (8.7.13)$$

The last relation can be understood by examining it graphically, that is, it represents the twisting of the two legs of a three-vertex.

The advantage of using quantum groups is that they give us an explicit expression for the "universal R matrix":

$$R = q^{(J_3 \otimes J_3)/4} \sum_{n \geq 0} \frac{(1 - q^{-1})^n}{[n]!} \times q^{-n(n-1)/4} q^{n J_3/4} (J_+)^n \otimes q^{-n J_3/4} (J_-)^n \qquad (8.7.14)$$

The $R^{j_1 j_2}$ matrix can be defined in terms of the universal R matrix by evaluating it on the tensor product $\mathbf{V}^{j_1} \otimes \mathbf{V}^{j_2}$ (and permuting indices).

Let us compare these relations for the K operation with the twist of two legs of a three-point function. In the WZW model $SU(2)_k$, we find that twisting two legs yields a phase factor:

$$(-1)^{j_1 + j_2 - j} \exp[2\pi i(\Delta_j - \Delta_{j_1} - \Delta_{j_2})] \qquad (8.7.15)$$

where the conformal weight of the primary field is given by:

$$\Delta_j = \frac{j(j+1)}{k+2} \tag{8.7.16}$$

Comparing the two expressions for twisting the legs of the three-point function for $SL(2)_q$ in Eq. (8.7.12) and the WZW model in Eq. (8.7.15), we are then led to postulate that they are the same, provided we make the identification:

$$q \leftrightarrow e^{2\pi i/(k+2)} \tag{8.7.17}$$

In fact, by examining other identities as well, we can show that the braiding properties of the WZW model at level k are determined by the representation theory of quantum groups if we make the above correspondence. As we mentioned, the presence of the crucial phase factors in the braiding operation in Eq. (8.7.12) separates a quantum group from an ordinary Lie group.

We can also make a trivial check on the relationship between quantum groups and $SU(2)_k$. . If we take the limit $q \to 1$, the quantum group reduces to an ordinary group. For the WZW model, we find from Eq. (8.7.17) that the corresponding limit is $k \to \infty$, which is also the limit in which a Kac–Moody algebra reduces to an ordinary one.

This relationship is quite remarkable, and once again it reveals the richness of conformal systems. On the left is the parameter labeling the quantum group $SL(2)_q$, and on the right, we have an expression labeling the level number k of $SU(2)_k$. Apparently, there is a deep relationship between these two, which enables us to compute the values of the B and F matrices exactly.

In fact, with a bit of work, we can find the exact numerical value of the B and F matrices for $SL(2)_q$. In this way, we can make a correspondence between the B and F matrices of $SU(2)_k$ and the $3j$ and $6j$ coefficients for $SL(2)_q$. For example, it is possible to show that the B and the F matrices can be represented exactly as follows:

$$F_{jj'}\begin{bmatrix} j_2 & j_3 \\ j_1 & j_4 \end{bmatrix} = \begin{bmatrix} j_1 & j_2 & j \\ j_3 & j_4 & j' \end{bmatrix}$$
$$B_{jj'}\begin{bmatrix} j_2 & j_3 \\ j_1 & j_4 \end{bmatrix} = (-1)^{j+j'-j_1-j_4}q^{(c_{j_1}+c_{j_4}-c_j-c_{j'})/2}\begin{bmatrix} j_2 & j_1 & j \\ j_3 & j_4 & j' \end{bmatrix} \tag{8.7.18}$$

where the matrices on the right are the deformed analog of the usual $3j$ and $6j$ symbols found in ordinary quantum mechanics. Specifically, these deformed $3j$ and $6j$ symbols can be written as [24]:

$$\begin{bmatrix} j_1 & j_2 & j_{12} \\ j_3 & j & j_{23} \end{bmatrix} = \sqrt{[2j_{12}+1][2j_{23}+1]}\,(-1)^{j_1+j_2-j-j_{12}}$$

$$\times\, \Delta(j_1,j_2,j_{12})\Delta(j_3,j,j_{12})\Delta(j_1,j,j_{23})\Delta(j_3,j_2,j_{23})$$

$$\times \sum_{z\geq 0}(-1)^z[z+1]!\Big\{[z-j_1-j_2-j_{12}]![z-j_3-j-j_{12}]! \qquad (8.7.19)$$

$$\times\, [z-j_1-j-j_{23}]![z-j_3-j_2-j_{23}]![j_1+j_2+j_3+j-z]!$$

$$\times\, [j_1+j_3+j_{23}-z]![j_2+j+j_{12}+j_{23}-z]!\Big\}^{-1}$$

where:

$$\Delta(a,b,c) = \sqrt{[-a+b+c]![a-b+c]!/[a+b+c+1]!} \qquad (8.7.20)$$

Thus, it seems near miraculous that the deformation of the $3j$ and $6j$ symbols found in ordinary quantum mechanics can provide us with explicit representations of the B and F matrices found in conformal field theory.

We have tried to present a discussion of quantum groups that stresses the intuitive relationship between them and Kac–Moody algebras. However, before concluding our remarks about quantum groups, let us generalize some of our previous statements.

The algebra of a quantum group is a special case of a larger class of algebras, that is, an associative *Hopf algebra A*, which has three operations that generalize the usual operations of multiplication and of taking the inverse.

1. The relation Δ (comultiplication) maps:

$$\Delta: \quad A \rightarrow A \otimes A; \qquad \Delta(ab) = \Delta(a)\Delta(b) \qquad (8.7.21)$$

Here, Δ is the generalization of the addition of angular momentum. This rule defines the associativity of comultiplication.

2. The relation γ (antipode) maps:

$$\gamma: \quad A \rightarrow A; \qquad \gamma(ab) = \gamma(b)\gamma(a) \qquad (8.7.22)$$

This rule defines the antipode, which is the generalization of the inverse.

3. The relation ϵ (co-unit) maps:

$$\epsilon: \quad A \rightarrow C; \qquad \epsilon(ab) = \epsilon(a)\epsilon(b) \qquad (8.7.23)$$

where $a, b \in A$.

These three relations satisfy the following:

$$\begin{aligned} (\text{id} \otimes \Delta)\Delta(a) &= (\Delta \otimes \text{id})\Delta(a) \\ m(\text{id} \otimes \gamma)\Delta(a) &= m(\gamma \otimes \text{id})\Delta(a) = \epsilon(a)1 \qquad (8.7.24) \\ (\epsilon \otimes \text{id})\Delta(a) &= (\text{id} \otimes \epsilon)\Delta(a) = a \end{aligned}$$

where m is the multiplication in the algebra. This general Hopf algebra, based on these multiplication rules, becomes a "quasi-triangular Yang–Baxter algebra" if we make a few more restrictions on these operations.

Let σ represent the permutation map:

$$\sigma(x \otimes y) = y \otimes x \qquad (8.7.25)$$

Then, we realize that Δ' and $\sigma \cdot \Delta$ are two different comultiplication operators. First, we define the universal R as the operator that establishes the link between Δ' and $\sigma \cdot \Delta$ by conjugation:

$$\sigma \cdot \Delta(a) = R\Delta(a)R^{-1} \qquad (8.7.26)$$

Second, we impose the following conditions:

$$(\mathrm{id} \otimes \Delta)(R) = R_{13}R_{12}; \qquad (\Delta \otimes \mathrm{id})(R) = R_{13}R_{23}$$
$$(\gamma \otimes \mathrm{id})(R) = R^{-1} \qquad (8.7.27)$$

[that is, $(\mathrm{id} \otimes \Delta)(R) \in A \otimes A \otimes A$, so that R_{13} acts as the identity on the second factor and R in the first and third factors.]

These definitions, of course, exist independent of conformal field theory and were introduced because they generalize the usual definitions of multiplication, etc. Their application to conformal field theory arises as follows. In ordinary Lie group theory, when taking the tensor product of a large number of representations, we would like to know when the resulting product yields irreducible representations and how many copies of each irreducible representation appear. In general, this is a difficult question. However, one convenient way in which to analyze this question is to construct the "centralizer."

If we are taking tensor products of representations $\mathbf{V}^i$, then let A and B be two algebras that act on these spaces and map $\mathbf{V} \to \mathbf{V}$. Then, roughly speaking, B is the centralizer of the action of A if it commutes with the action of A [24]. For ordinary Lie groups, this usually means that the centralizer B is the set of braiding operations generated by permutations on the factors appearing in the tensor product $\mathbf{V} \otimes \mathbf{V} \otimes \mathbf{V} \otimes \cdots$. The important point is that the irreducible representations appearing in this large tensor product are labeled by the irreducible representations of B. This gives us a convenient way in which quickly see the irreducible representations arising from this tensor product.

Now, let us reanalyze Eq. (8.7.26) from this perspective. For our purposes, we notice that R is the centralizer of the quantum group because it commutes with the comultiplication Δ. This gives us a general way in which to extract the quantum group associated with a conformal field theory. Given the fusion rules of any conformal field theory, the tensor products obey certain braiding operations. By treating the braiding operator R as the centralizer of an algebra, we can construct the quantum group via Eq. (8.7.26).

The braiding operator R, however, obeys certain complicated relations given by the $6j$ rules. This hexagon and pentagon rules must also be part of the definition of the quantum group. In this light, we can reanalyze Eq. (8.7.27), which we now recognize as containing the information of the hexagon graphs.

In sum, by using the theory of centralizers appearing in ordinary classical Lie group theory, we are led to define the braiding operator R as the centralizer of the quantum group since R commutes with the comultiplication operation defining the quantum group in Eq. (8.7.26). Second, the $6j$ polynomial equations found in Chapter 3 are now reinterpreted as Eq. (8.7.27).

There still remains one last step. We have to show that the Eqs. (8.7.26) and (8.7.27), which restrict the Hopf algebra, generate the Yang–Baxter relation. This can be accomplished in a few steps. Let us define the R operator as:

$$R \equiv \sum_i a_i \otimes b_i \tag{8.7.28}$$

Let the R operator become R_{ij} when it acts on the specific representations i and j. We are interested in its action on the tensor product of three representations $\mathbf{V}^1$, $\mathbf{V}^2$, and $\mathbf{V}^3$. For example, we have the following identity [24]:

$$R_{13}R_{23} = \sum_i \sum_j a_i \otimes a_j \otimes b_i b_j \tag{8.7.29}$$

(Since the index 3 appears twice on the left-hand side, we notice that the third factor in the product contains the ordinary product of two b's.)

Now, let us construct the following sequence of operations, using only the definitions appearing in the Hopf algebra:

$$
\begin{aligned}
(\sigma \cdot \Delta \otimes \mathrm{id})R &= \sum_i \Delta'(a_i) \otimes b_i \\
&= \sum_i R_{12}\Delta(a_i)R_{12}^{-1} \otimes b_i \\
&= R_{12} \sum_i \Delta(a_i) \otimes b_i R_{12}^{-1} \\
&= R_{12}(\Delta \otimes id(R))R_{12}^{-1} \\
&= R_{12}R_{13}R_{23}R_{12}^{-1}
\end{aligned}
\tag{8.7.30}
$$

But, we also know:

$$
\begin{aligned}
(\sigma \cdot \Delta \otimes \mathrm{id})(R) &= \sigma_{12}\big[(\Delta \otimes \mathrm{id})(R)\big] \\
&= \sigma_{12}(R_{13}R_{23}) = R_{23}R_{13}
\end{aligned}
\tag{8.7.31}
$$

By equating these two expressions, we now have:

$$R_{12}R_{13}R_{23} = R_{23}R_{13}R_{12} \tag{8.7.32}$$

which is the Yang–Baxter relation.

In summary, given a conformal field theory, we can always define the braiding operator as the universal R matrix and R_{ij}. Treating the braiding operator as a centralizer, we can then construct the quantum group by the statement that R commutes with the comultiplication defined on the quantum group; R_{ij}, in turn, satisfies the Yang–Baxter relation.

Example: $SU(2)_q$

Some examples will help to clarify these rather arbitrary definitions, which were originally motived by examining integrable systems. For $SU(2)_q$, for example, on the generators of the algebra:

$$[X^+, X^-] = [H], \qquad [H, X^\pm] = \pm 2X^\pm \qquad (8.7.33)$$

we can define the comultiplication, antipode, and co-unit operations:

$$
\begin{aligned}
\Delta(H) &= H \otimes 1 + 1 \otimes H \\
\Delta(X_i^\pm) &= X_i^\pm \otimes q^{H/4} + q^{-H/4} \otimes X_i^\pm \\
\epsilon(H) &= \epsilon(X^\pm) = 0; \qquad \epsilon(1) = 1 \\
\gamma(H) &= -H; \qquad \gamma(X^\pm) = -q^{\pm 1/2} X^\pm
\end{aligned}
\qquad (8.7.34)
$$

We can show that the universal R matrix given earlier satisfies Eqs. (8.7.26) and (8.7.27) if we choose the above rules for the comultiplication, antipode, and co-unit operations.

8.8. Hecke and Temperley-Lieb Algebras

So far, we have seen that the Yang-Baxter relation, in various forms, appears at the very heart of knot theory, the polynomial equations of conformal field theory, quantum groups, etc.

However, it turns out that the generators that we have been studying actually obey additional relations beyond those which define the Yang-Baxter relation.

In this section, we will try to investigate these additional constraints and rigorize some our discussion of the Yang-Baxter relation by introducing the Hecke and the Temperley-Lieb algebras, which allow us to systematically explore the mathematical structure of these various formulations. This will allow us to tie the various loose ends and link up the various themes that we have stressed in the past two chapters.

A Hecke algebra is one in which the generators ρ_i obey the following relation:

$$\rho_i\rho_{i\pm1}\rho_i = \rho_{i\pm1}\rho_i\rho_{i\pm1}$$

Hecke algebra : $$\rho_i\rho_j = \rho_j\rho_i$$ (8.8.1)

$$\rho_i^2 = (1-q)\rho_i + q$$

We immediately recognize that the first relation is equivalent to the Yang-Baxter relation. The first two relations, in fact, are nothing but the braid relations. However, the third relation places a new constraint on the generators beyond the usual braid relations.

We will find it useful to construct yet another algebra out of the Hecke algebra, called the Temperley-Lieb algebra.

We start with the generators ρ_i of the Hecke algebra and then impose one more additional constraint:

$$\rho_i\rho_{i+1}\rho_i - \rho_i\rho_{i+1} - \rho_{i+1}\rho_i + \rho_i + \rho_{i+1} - 1 = 0 \qquad (8.8.2)$$

Let us define the generators e_i of the Temperley-Lieb algebra as follows:

$$e_i = \frac{1-\rho_i}{1+q} \qquad (8.8.3)$$

With this new constraint and definition, we can show that these new generators e_i satisfy:

$$e_ie_j = e_je_i \quad |i-j| \geq 2$$

Temperley $-$ Lieb algebra : $$e_i^2 = e_i$$ (8.8.4)

$$e_ie_{i\pm1}e_i = \beta^{-1}e_i$$

where $\beta = 2 + q + q^{-1}$.

The advantage of introducing the Hecke and Temperley-Lieb algebras is that we can now rigorously isolate the mathematical form that the Yang-Baxter and braid relations take in conformal field theory, quantum groups, and knot theory.

Let us now recast our previous discussion of quantum groups in terms of the Hecke and Temperley-Lieb algebras. If we study $SL(N)_q$, for example, we find that the R matrix commutes with the comultiplication operator Δ, that is, the Yang-Baxter operators are the centralizers of $SL(N)_q$. More precisely, the centralizer of $SL(N)_q$ is given by the Hecke generators ρ_i with parameter q.

To see this, let us find an explicit representation of this algebra. We introduce the symbol e_{ij}, which is a $N \times N$ matrix such that the only non-vanishing element is equal to 1 and is located in the i,j position in the matrix. Because e_{ij} actually has four sets of indices, we will suppress the $N \times N$ indices. Now define the R matrix as:

$$R = \sum_{i\neq j} e_{ij} \otimes e_{ji} + q^{1/2}\sum_i e_{ii} \otimes e_{ii} + (q^{1/2} - q^{-1/2})\sum_{i\leq j} e_{jj} \otimes e_{ii} \quad (8.8.5)$$

(Notice that R is the tensor product of two separate $N \times N$ matrix spaces.) Now define the multiplication operation on R such that the two individual tensor spaces multiply separately, so that R^2 lies in the same space as R.

Then it is an easy matter to show the following:

$$R^2 = (q^{1/2} - q^{-1/2})R + 1 \qquad (8.8.6)$$

Now define the ρ_i matrix as follows:

$$\rho_i = -(q)^{1/2} \, 1 \otimes \cdots \otimes R_{i,i+1} \otimes \cdots \otimes 1 \qquad (8.8.7)$$

where the indices $i, i+1$ indicate the spaces in the tensor product where ρ_i acts non-trivially.

Then, by explicit calculation, we can show that the ρ_i satisfy the defining relations of the Hecke algebra. Thus, the Hecke algebra and quantum groups are related in the most intimate way via the centralizer.

However, there is one additional constraint that we must impose if we wish to study $SU(N)_q$. The irreducible representations of $SU(N)_q$ have the same Young tableaux as the representations of the classical Lie group $SU(N)$. The difference, however, between the two sets of Young tableaux is that they have different operators which symmetrize the indices of the tableaux. For an ordinary Lie group, ordinary transpositions can symmetrize the indices, which form the symmetric group S_n. However, for the quantum group, the symmetrizer is given by the braiding operator which interchanges the indices. In other words, the symmetrizer is given by a generator of the Hecke algebra. However, we know from ordinary classical group theory that the Young tableaux can have at most N rows. In other words the $N + 1$ row antisymmetrizer vanishes. Since manipulations of the Young tableaux are given by the generators of the Hecke algebra, this in turn, gives us an additional constraint on the generators.

Written out explicitly, we find that the extra constraint on the generators is given precisely by Eq. (8.8.2). But this extra constraint allows us to re-write the Hecke generators in terms of the Temperley-Lieb generators. Thus, the algebra of the centralizers that we have been studying in this chapter for $SU(N)_q$ is actually the Temperley-Lieb algebra [24].

Another example where the Hecke and Temperley-Lieb algebras play an important role is in knot theory, where the Hecke relations are equivalent to the skein relations. We recall that the skein relations express a relationship between knot invariants defined for $L_\pm$ and L_0. Let us call the knots $K_\pm$ and K_0 formed by from $L_\pm$ and L_0 by tieing opposite pairs of strings together (such that K_+ consists of a link, K_0 consists of two separate unknots, and K_0 reduces to a single unknot). Then these three knots are related to each other by the following operations: $K_+ = \sigma_1^2 K_-$ and $K_0 = \sigma_1 K_-$. Now let us examine the skein relation, defined in terms of the knot polynomials and the variable t. Therefore we have the relation:

$$K_+ - qK_- = (q - 1)K_0 \qquad (8.8.8)$$

which, in turn, is equivalent to the relation $\sigma_1^2 = (q - 1)\sigma_1 + q$. This relation, in turn, is identical to the constraint found in the Hecke algebra in Eq. (8.8.1). In this way, starting from the skein relation, we recover the constraint which defines the Hecke algebra. This shows the relation between these two formalisms.

In statistical mechanics, we also see the importance of these algebras. Previously, we were able to show that the vertex models could be written in terms of an S-matrix S_{pq}^{rs} and that the IRF models could be written in terms of $w(p, q|r, s)$. Both these operators, in turn, could be expressed in terms of $X_i(u)$, which in turn obeyed the braid relations.

However, these $X_i(u)$ operators can be shown to obey the Temperley-Lieb relations. We recall that these operators obeyed the braid relations:

$$
\begin{aligned}
X_i(u)X_j(v) &= X_j(v)X_i(u); \quad |i - j| \geq 2 \\
X_i(u)X_{i+1}(u + v)X_i(v) &= X_{i+1}(v)X_i(u + v)X_{i+1}(u)
\end{aligned}
\qquad (8.8.9)
$$

Now let us re-define:

$$X_i(u) = I + \beta^{1/2}[\sin u/\sin(\lambda - u)]e_i \qquad (8.8.10)$$

where e_i obey the Temperley-Lieb relation, and where $r = 4\cos^2 \lambda$. Thus, the operators of the IRF models obey the Temperley-Lieb algebra.

Let us now summarize the results of the last few chapters and isolate once again the remarkable relationship between the Yang-Baxter relation (or more specifically, the Hecke and Temperley-Lieb algebras) and a bewildering variety of quantum systems, such as conformal field theory, quantum groups, knot theory, soliton theory, and statistical mechanics.

First, we first encountered these relations when we studied rational conformal field theories for $c < 1$ and then constructed their conformal blocks, which are the building blocks for the theory. These conformal blocks, in turn, obeyed certain identities when we twisted or pinched internal lines. These identities were called the polynomial equations, such as the pentagon and hexagon relation. The algebra obeyed by these twisting and pinching operations, in turn, were equivalent to the Yang-Baxter relation.

Second, we investigated various statistical mechanical models, such as the Ising model or the RSOS model. We found that the partition function for these models could all be expressed in terms of the transfer matrix. The essential feature which allowed us to solve these models exactly was that the transfer matrices commuted. The algebraic statement of commuting transfer matrices, in turn, was the Yang-Baxter relation.

Third, we studied soliton theory, which are exactly solvable, topologically stable non-linear solutions of two dimensional wave equations. Again, we found that the dynamics of their evolution could be governed by a transfer matrix, and that the essential feature which made the system solvable was commuting transfer matrices. Expressed mathematically, this gave us the Yang-Baxter relation.

Fourth, we investigated knot theory. The essential step in knot theory was to cut the knot and form a braid. Then we could systematically deform the topology of the knot by braiding the strands, that is, using the braid relations of Artin. Defining a trace operation on the braid relations, in turn, gave us the various knot polynomials, such as the Jones, HOMFLY, and Kauffman polynomials. These braid relations, in turn, were identical to the Yang-Baxter relation.

Fifth, we investigated quantum groups, which differ from ordinary Lie groups by a continuous parameter q. The relation to the Yang-Baxter relation could be established in a number of ways. Crudely, we could say that the Jacobi relation found in ordinary classical group theory corresponds to the Yang-Baxter relation for quantum groups. We also saw that the Clebsch-Gordan coefficients created by taking tensor products of various representations obeyed certain braiding relations when we twisted lines, which gave us a one-to-one correspondence between these coefficients and conformal blocks. We also found that from the generators, we could define a comultiplication operator Δ. We found that the R matrix commutes with Δ: $R\Delta = \Delta R$. We say that the Temperley-Lieb algebra is the centralizer of the quantum group.

In summary, we find that the essential reason why these various two dimensional systems are exactly solvable is because they are ultimately based on the Yang-Baxter relation, or, more precisely, either the Hecke algebra or the Temperley-Lieb algebra. We summarize this by:

$$\text{conformal field theory} \rightarrow \text{polynomial equations} \rightarrow \text{Yang} - \text{Baxter}$$
$$\text{statistical mechanics} \rightarrow \text{commuting T matrices} \rightarrow \text{Yang} - \text{Baxter}$$
$$\text{soliton theory} \rightarrow \text{commuting T matrices} \rightarrow \text{Yang} - \text{Baxter}$$
$$\text{knot theory} \rightarrow \text{braid relation} \rightarrow \text{Yang} - \text{Baxter}$$
$$\text{quantum groups} \rightarrow \text{centralizer} \rightarrow \text{Yang} - \text{Baxter}$$

$$(8.8.11)$$

8.9. Summary

The latest method by which to categorize conformal field theories is the use of Chern–Simons gauge theory and knot theory. It is the only method that tries to explain conformal field theory by starting with simpler structures in three dimensions to explain the "accidents" of two-dimensional physics.

Our starting point is the action:

$$L = \frac{k}{8\pi} \int \epsilon^{ijk} \mathrm{Tr} \left[A_i(\partial_j A_k - \partial_k A_j) + \frac{2}{3} A_i[A_j, A_k] \right] \tag{8.9.1}$$

which is generally covariant without introducing a metric tensor (because ϵ^{ijk} transforms as a tensor density). The physical states will then be composed of Wilson loops:

$$W_R(C) = \mathrm{Tr}\, P \exp i \int_C A_i \, dx^i \tag{8.9.2}$$

and the correlation functions consist of invariants defined with the topology of knots and links:

$$\left\langle \prod_{i=1} W_{R_i} \right\rangle = \int DA \exp(iL) \prod_{i=1} W_{R_i}(C_i) \tag{8.9.3}$$

To quantize the system, in the Coulomb gauge, we can impose the gauge constraint:

$$\epsilon^{ij} F_{ij} = 0 \tag{8.9.4}$$

which shows that the physical space is spanned by the moduli of flat curvatures modulo gauge transformations. This space is well known to mathematicians, and it is the space of conformal blocks.

If we take a time slice of a three-dimensional space, the resulting surface has a complex structure, and the Hilbert space consists of conformal blocks. Thus, we have made a transition from a theory of infinite degrees of freedom to a topological system of finite degrees of freedom.

We can also make the link to the WZW model by solving the gauge constraint. A solution is given by:

$$\tilde{A} = -(\tilde{d}U)U^{-1} \tag{8.9.5}$$

which has zero curvature. Assuming Σ is bounded, we can plug this into the original action and find

$$S = kS_{WZW} = \frac{k}{4\pi} \int_{\partial Y} \mathrm{Tr}(U^{-1} \partial_\phi U \, U^{-1} \partial_t) d\phi \, dt + \frac{k}{12\pi} \int_Y \mathrm{Tr}(U^{-1} \tilde{d}U)^3 \tag{8.9.6}$$

where ϕ is an angular variable defined around the perimeter of Σ. This is a version of the WZW action. Thus, all the previous results on conformal field theory, including a complete representation of the rational conformal field theories, emerge from Chern–Simons gauge theory.

Because the states are completely generally covariant, the correlation functions must be topological invariants defined on knots and links. The classical Gauss linking number:

$$\Phi(C_a, C_b) = \frac{1}{4\pi} \int_{C_a} dx^i \int_{C_b} dx^j \, \epsilon_{ijk} \frac{(x-y)^k}{|x-y|^3} \tag{8.9.7}$$

(which tells us the degree to which a series of knots is intertwined) emerges when we compute correlation functions.

The goal of knot theory is to find a set of invariants (polynomials) that are in one-to-one correspondence with topologically distinct knots. The classical Alexander polynomial Δ, for example, is defined recursively via the skein relations:

$$\Delta_{L_+} - \Delta_{L_-} = (\sqrt{t} - 1/\sqrt{t})\Delta_{L_0} \tag{8.9.8}$$

A series of knot polynomials has been discovered within the last few years, beginning with the celebrated Jones polynomial, which satisfies

$$(1/t)V_{L_-} - tV_{L_+} = \left(\sqrt{t} - 1/\sqrt{t}\right)V_{L_0} \tag{8.9.9}$$

The HOMFLY polynomial contains both the Jones and Alexander polynomials as subsets and satisfies:

$$tP_{L_+} - t^{-1}P_{L_-} = zP_{L_0} \tag{8.9.10}$$

where $L_\pm$ and L_0 are defined in Fig. 8.2. The Kauffman polynomial R_L is actually defined on knots made of ribbons (framed knots) and satisfies:

$$R_{L_+} - R_{L_-} = zR_{L_0} \tag{8.9.11}$$

with the additional condition:

$$R_{\hat{L}_+} = \alpha R_{\hat{L}_0}, \qquad R_{\hat{L}_-} = \alpha^{-1}R_{\hat{L}_0} \tag{8.9.12}$$

where the $\hat{L}_\pm$ and $\hat{L}_0$ are given in Fig. 8.5. Fortunately, quantum field theory can generate all these knot invariants (and give analytic expressions for all of them) and infinite classes of new ones.

If we quantize the Chern–Simons theory covariantly, we have:

$$\langle A_\mu^a(x)A_\nu^b(y)\rangle = \frac{i}{k}\delta^{ab}\epsilon_{\mu\nu\rho}\frac{(x-y)^\rho}{|x-y|^3} \tag{8.9.13}$$

so we can, by brute force, power expand the Wilson loops and take the matrix elements of the power series. We find, to first order in $1/k$,

$$\langle W(C)\rangle = -\text{Tr}(R^b R^a)\oint_C dx^\mu \int^y dy^\nu \langle A_\nu^b(y)A_\mu^a(x)\rangle$$
$$= -\frac{2\pi}{k}\dim R\,c_2(R)\phi(C) \tag{8.9.14}$$

where $c_2(R)\mathbf{1} = R^a R^a$ and:

$$\phi = \frac{1}{4\pi}\oint_C dx^\mu \oint_C dy^\nu\,\epsilon_{\mu\nu\rho}\frac{(x-y)^\rho}{|x-y|^3} \tag{8.9.15}$$

So, the Gauss linking number comes out of the $1/k$ expansion of the Chern–Simons correlation functions over Wilson loops.

The complete correlation functions over Wilson loops can be shown to satisfy:

$$S_{\hat{L}_+} = \alpha S_{\hat{L}_0}$$
$$S_{\hat{L}_-} = \alpha^{-1} S_{\hat{L}_0} \qquad (8.9.16)$$
$$\beta S_{L_+} - \beta^{-1} S_{L_-} = z S_{L_0}$$

which, in turn, generate a new knot invariant. We can relate this new knot invariant to the HOMFLY invariant via:

$$P_L(t = \alpha, \beta, z) = \alpha^{-w(L)} S_L(\alpha, \beta, z) \qquad (8.9.17)$$

(This new knot invariant is defined only on ribbons, or framed knots, because quantum field theory is ambiguous when two strands are defined at the same point.)

An even larger class of knot invariants can be constructed via conformal field theory. We first note that the correlation functions of conformal field theory obey certain relations when we change the order of the points z_i:

$$\Psi_I(z_1, \ldots, z_{i+1}, z_i, \ldots, z_N) = \sum_{j=1}^{l} B_I^J \Psi_J(z_1, \ldots, z_i, z_{i+1}, \ldots, z_N) \qquad (8.9.18)$$

These B matrices, in turn, form a representation of Artin's braid group. (Braids can be turned into knots by wrapping the ends of the various strands together.) In addition, by examining the $6j$ transformation rules of rational conformal field theory, we can generate a representation of the Yang–Baxter relation. Let the B matrix correspond to twisting two external lines of a four-point correlation function, and let F represent the matrix corresponding to fusing two legs (so an s-channel graph turns into a t-channel graph). Then, these matrices satisfy:

$$\sum_p B_{j_6 p} \begin{bmatrix} j_2 & j_3 \\ j_1 & j_7 \end{bmatrix} B_{j_7 j_9} \begin{bmatrix} j_2 & j_4 \\ p & j_5 \end{bmatrix} B_{p j_8} \begin{bmatrix} j_3 & j_4 \\ j_1 & j_9 \end{bmatrix}$$
$$= \sum_p B_{j_7 p} \begin{bmatrix} j_3 & j_4 \\ j_6 & j_5 \end{bmatrix} B_{j_6 j_8} \begin{bmatrix} j_2 & j_4 \\ j_1 & p \end{bmatrix} B_{p j_9} \begin{bmatrix} j_2 & j_3 \\ j_6 & j_5 \end{bmatrix} \qquad (8.9.19)$$

which is the Yang–Baxter relation.

Notice that the topology of the Yang–Baxter relationship has precisely the topology found in Artin's braid group representation. Thus, we suspect that a representation of the knot invariants should be possible via statistical mechanics.

To show this, we note that if two closed braids represent the same link, then it is possible to deform one link into the other link by a succession of Markov moves. Thus, a knot invariant must necessarily remain invariant under these Markov moves:

$$\alpha(AB) = \alpha(BA)$$
$$\alpha(AU_n) = \alpha(AU_n^{-1}) = \alpha(A) \tag{8.9.20}$$

The problem, therefore, is to find an object in statistical mechanics that is invariant under these Markov moves. However, we notice that the matrices found in statistical mechanics obey precisely the relations of the braid group because they satisfy the Yang–Baxter relations. Thus, given a braid A in this representation, the knot invariant associated with A is given by a trace:

$$\alpha(A) = \left[t^{-(N-1)/2}\right]^{n-1}\left[t^{(N-1)/2}\right]^{e(A)}\mathrm{Tr}(HA) \tag{8.9.21}$$

One advantage of this formalism is that we have a vast number of statistical mechanical models obeying the Yang–Baxter relationship and, hence, forming a representation of the braid group. By forming the appropriate traces over these braid matrices, we can generate all the known link polynomials, as well as infinite classes of new link polynomials.

Last, we note that when constructing modular invariants out of Kac–Moody algebras in terms of primary fields, much of the information concerning the structure of the group is lost in the process. We only need to manipulate the primary fields and their infinite descendants in order to construct these characters. However, we are not reducing the Kac–Moody algebras to ordinary Lie algebras, because the latter is not compatible with the central charge. Thus, the reduction process cannot yield the usual Lie algebras, but something more general. These are the quantum groups, which were originally discovered by examining the Yang–Baxter relation.

The defining relation of the quantum group $SU(2)_q$ is

$$[J_3, J_\pm] = \pm 2J_\pm, \qquad [J_+, J_-] = [J_3] \tag{8.9.22}$$

where the brackets in $[J_3]$ mean:

$$[\lambda] = \frac{q^{\lambda/2} - q^{-\lambda/2}}{q^{1/2} - q^{-1/2}} \tag{8.9.23}$$

Because only the last commutator is changed, much of the representation of the quantum groups is identical to the usual representation of the classical groups. Notice that in the limit of $q \to 1$, we retrieve the usual classical theory.

Last, by examining the "Clebsch–Gordan" coefficients generated by these algebras, one can make an association between the q found in quantum groups and the k found in Kac–Moody algebras:

$$q \leftrightarrow e^{2\pi i/(k+2)} \tag{8.9.24}$$

References

1. E. Witten, *Comm. Math. Phys.* **121**, 351 (1989).
2. S. Elitzur, G. Moore, A. Schwimmer, and N. Seiberg, *Nucl. Phys.* **B326**, 108 (1989).
3. G. Moore and N. Seiberg, *Phys. Lett.* **220B**, 422 (1989).
4. S. Moran, *The Mathematical Theory of Knots and Braids*, North-Holland Publ., Amsterdam (1983).
5. D. Rolfsen, *Knots and Links*, Publish or Perish, Berkeley (1976).
6. J. H. Conway, in *Computational Problems in Abstract Algebra*, Pergamon, Oxford (1970).
7. J. W. Alexander, *Proc. Natl. Acad. Sci.* **9**, 93 (1928); *Trans. Am. Math. Soc.* **20**, 275 (1923).
8. E. Artin, *Ann. Math.* **48**, 101 (1947).
9. V. F. R. Jones, *Inv. Math* **72**, 1 (1983); *Bull. A. M. S.* **12**, 103 (1985); *Ann. Math.* **12**, 239 (1985).
10. J. S. Birman, *Inven. Math.* **81**, 138 (1985).
11. P. Freyd, D. Yetter, J. Hoste, W. B. R. Lickorish, K. Millet, and A. Ocneanu, *Bull. A. M. S.* **12**, 239 (1985).
12. L. Kauffman, *Topology* **26**, 395 (1987); *On Knots*, Princeton Univ. Press, Princeton, New Jersey (1987).
13. E. Guadagnini, M. Martellini, M. Mintchev, *Nucl. Phys.* **B330**, 575 (1990).
14. P. Cotta-Ramusino, E. Guadagnini, M. Marellini, M. Mintchev, *Nucl. Phys.* **B330** 557 (1990).
15. A. Tsuchiya and Y. Kanie, in *Conformal Field Theory and Solvable Lattice Models*, Adv. Studies in Pure Math. **16**, 297 (1988); *Lett. Math. Phys.* **13**, 303 (1987).
16. G. Moore and N. Seiberg, *Lectures on RCFT*, 1986 Summer Trieste Summer School.
17. Y. Akutsku, T. Deguchi, and M. Wadati, *Physics Reports* **180**, 248 (1989); *J. Phys. Soc. Jpn.* **56**, 3039 (1987); **57**, 757 (1988); **57**, 1905 (1988).
18. J. Frohlich, *Nonperturbative Quantum Field Theory*, 1987 Cargese Lectures, Plenum, New York (1987).
19. A. A. Markov, *Recueil Math.* **1**, 73 (1935).
20. V. G. Drinfeld, *Proc. of Int. Cong. of Math.*, Berkeley, California (1986).
21. M. Jimbo, *Lett. Math. Phys.* **10**, 63 (1985); **11**, 247 (1986); *Comm. Math. Phys.* **102**, 537 (1986).
22. L. D. Faddeev, N. Yu. Reshetikhin, and L. A. Takhtajan, LOMI preprint E-14-87 (1987).
23. A. Kirilov and N. Yu. Reshestikhin, LOMI preprint E9-88 (1988).
24. L. Alvarez-Gaumé, C. Gomez, and G. Sierra, *Nucl. Phys.* **B330**, 347 (1990); *Phys. Lett.* **220B**, 142 (1989); "Topics in Conformal Field Theory," in *Physics and Mathematics of Strings*, World Scientific, Singapore (1990).

Part II

Nonperturbative Methods

Chapter 9

Beyond the Planck Length

9.1. Need for a Nonperturbative Approach

In the previous chapters, we have treated string theory as primarily a way in which to analyze perturbative, low-energy phenomena at everyday energies. In this regard, string theory not only gives us a way of renormalizing graviton interactions, but it also gives us constraints on what the low-energy symmetries should be. The success of this perturbative approach is that, with rather mild assumptions, one can effortlessly reproduce certain features of the our low-energy world that are quite difficult to duplicate using standard GUTs.

For example, in Kaluza–Klein theory, it is difficult, if not impossible, to create a theory with left-handed fermions that contains the standard group $SU(3) \otimes SU(2) \otimes U(1)$. However, the heterotic string naturally solves this problem. Furthermore, standard GUTs fail to provide any solutions to the problems of (1) fermion generations, (2) quark masses, (3) hierarchy of energy scales, and (4) and consistent coupling to gravitons. Rather miraculously, superstring theory solves all of these.

However, let us also describe the failings of superstring theory, which lead us to conclude that nonperturbative effects, as in QCD, are decisive. Some of the problems with the perturbative approach include the following.

1. The dimension of space–time is perfectly stable perturbatively, so the process of compactification to four physical dimensions must be dynamical.
2. Supersymmetry is preserved to all orders in perturbation theory, but in our low-energy world, supersymmetry is obviously broken. Thus, supersymmetry must also be broken nonperturbatively.
3. Although many of the general features of our low-energy world are rather easy to deduce as a conformal field theory, getting the precise description of the universe is quite difficult. In particular, a completely satisfactory conformal field theory that has the precise low-energy symmetry, the exact number of generations, and a satisfactory proton lifetime and can eliminate all extraneous factors, such as extra $U(1)$'s,

does not yet exist. Thus, we cannot say that we have constructed a conformal field theory that is phenomenologically perfect.

4. The low-energy mass spectrum is still wrong (for example, the zero mass dilaton field).

5. Most important, the theory cannot select the true vacuum from among the millions or billions of possible conformal field theories.

Thus, although superstring theory has produced a wealth of phenomenological information, the big disappointment is that it has failed to give us any hint as to which vacuum the theory prefers, which probably can only be decided by analyzing the nonperturbative behavior of the theory. Out of the millions or perhaps billions of possible conformal field theories, we do not have the slightest notion of which vacuum the theory really prefers.

The origin of this problem is that, in string theory, the breaking of the original 10-dimensional string down to our 4-dimensional world took place at Planck distances and energies. Thus, the "natural home" for string theory is not everyday energies, but the mysterious region beyond the Planck length itself. By studying conformal field theories near our low-energy world, we are stretching the perturbative approach to its very limits, where we expect it to fail. Even by intensely studying all possible conformal field theories, we are in some sense probing the broken phase of the theory, rather than concentrating on unlocking the secrets of the unbroken theory.

One analogy might be QCD, where its low-energy behavior yields the string-like resonances of hadron physics. However, studying low energy gluon condensation and resonances will never allow us to probe the natural home of QCD, which is the realm of point particle gauge connections.

In a similar way, our emphasis in the future should be to probe the natural home of string theory, which is physics beyond the Planck length, where we find new, unexpected features emerging. In this regard, in this chapter, we will first describe several observations concerning superstring physics beyond the Planck energy.

1. If we assume that conformal field theories are good approximations beyond the Planck scale, then a new phenomenon appears, called "duality," which indicates that the Big Bang may not have been a singular event after all.

2. There is perhaps a phase transition in string theory near the Hagedorn temperature.

3. One can show that new relations arise between different string amplitudes in the limit of $\alpha \to \infty$, indicating the emergence of a powerful new symmetry beyond the Planck scale.

4. One can show that bosonic perturbation theory is not Borel summable, which indicates that one should abandon perturbative string theory altogether and make a concerted assault on the nonperturbative theory.

Assuming the last step is valid, then perhaps we are forced to consider the various nonperturbative proposals seriously. Perhaps Riemann surfaces are unstable near the Planck scale, so that an entirely new, nonconformal approach is warranted.

Last, in this chapter, we will begin an analysis of the various nonperturbative formalisms advanced so far, beginning with the renormalization group approach. In the succeeding chapters, we will discuss string field theory and matrix models.

9.2. Duality at the Planck Scale

For the moment, let us assume that the conformal field theory picture is still valid at energies far beyond the Planck scale. Momentarily ignoring quantum corrections (which are bound to be large at that energy scale), we find a surprising new symmetry emerging, which is sometimes called "duality." (This "duality" should not be confused with the traditional duality associated with the Veneziano model.)

Normally, when we compactify a space, the momentum associated with the compactified direction becomes quantized. If we assume that a field exists on this compactified space, then:

$$\phi(x) \equiv \phi(x + 2\pi R) \qquad (9.2.1)$$

If we expand the field in eigenfunctions:

$$\phi(x) = \sum_n \phi_n e^{ipx} \qquad (9.2.2)$$

then the momentum p must be quantized in order for the field to be single valued:

$$p = n/R \qquad (9.2.3)$$

Now, let us generalize to the open string living on a compactified space. Then, the harmonic oscillator decomposition:

$$X^i(\sigma, \tau) = x^i + 2\alpha' p^i \tau + \sum_{n \neq 0} \frac{1}{n} \alpha_n^i \cos n\sigma e^{-in\tau} \qquad (9.2.4)$$

is altered by the fact that the momentum is quantized:

$$p^i = M_i/R_i \qquad (9.2.5)$$

Then, the masses of the states are also changed by the following:

$$\alpha' m^2 = \frac{\alpha'}{R^2} \sum_{i=1}^{10-D} M_i^2 + N$$

$$N = \sum_{n=1}^{\infty} \sum_{i=1}^{D} \alpha_{-n}^i \alpha_n^i$$

(9.2.6)

where N is the mass operator.

An interesting feature emerges, however, if we have a closed string vibrating in this compactified space. Unlike the open string, the closed string has the possibility of wrapping itself completely around the compactified direction an integer number of times. For the free string, this means that these modes correspond to solitons, because it is impossible to unravel these states without breaking the string.

For the compactified closed string, the masses are shifted by the following:

$$\frac{1}{2}\alpha' m^2 = \frac{1}{2} \sum_{i=1}^{10-D} \left(\frac{\alpha'^2 M_i}{R^2} + \frac{R^2 N_i^2}{\alpha'^2} \right) + N + \tilde{N}$$

(9.2.7)

The important point is to observe that the mass operator is invariant under the transformation:

$$M_i \leftrightarrow N_i, \qquad R/\sqrt{\alpha'} \leftrightarrow \sqrt{\alpha'}/R$$

(9.2.8)

This symmetry is quite unusual. If correct, the emergence of this new symmetry seems to give us a new picture of the Big Bang. Traditionally, one views the Big Bang (or the Big Crunch in an oscillating universe) as a violent explosion or contraction. However, the emergence of this duality seems to indicate that they may have been soft, that is, no singularity was ever reached.

For example, assume that the expansion of our universe is gradually reversed over tens of billions of years, and space–time once again contracts back to the Planck scale. As the size of the universe becomes smaller than the Planck scale, we find that the universe is equivalent to the universe where R is substituted for $1/R$ [1–3]. Thus, there is a smooth "bounce," as a collapsing universe becomes mathematically interchangeable with an expanding universe. The universe never reaches a singularity, but smoothly reverses its contraction and reemerges as an expanding universe.

This curious effect also has philosophical implications. It means that the Planck length, in some sense, is the smallest distance that one can study via high-energy probes, that is, a new Heisenberg uncertainty principle seems to be emerging. Because of the uncertainty principle, increasing the momentum of a probe decreases the uncertainty in distance, that is, high-energy probes can peer into smaller and smaller distances. However, if duality is correct, this means that the Heisenberg uncertainty principle fails at the Planck length and a new principle emerges. This means that a hypothetical beam of high-energy particles, whose energies exceed that of the Planck length, will no longer be able to probe within the Planck length. Increasing the energy only increases the uncertainty in position, making the world within the Planck scale impossible to probe.

Although this picture is attractive from an aesthetic point of view, it is likely that quantum corrections will completely destroy it. The conformal field theories that exhibit duality may be unstable beyond the Planck length, meaning that our equations are quite useless, and it is premature to draw any conclusions about the possibility of duality.

9.3. Possible Phase Transition at the Hagedorn Temperature

Ever since the early days of string theory, when it was called the dual resonance model of hadron physics, it was recognized that the theory set a maximum temperature that could be achieved, no matter how much energy was supplied to a system. The origin of this strange behavior is easy to understand. If we have a box of vibrating strings or resonances and add more energy into it, the extra energy is translated into creating more resonances. The strings break repeatedly, meaning that we have increased the number of particles in the box. Because the number of resonances that can be excited increases exponentially with the energy, it is not surprising that eventually a maximum temperature is reached, such that the addition of more energy does not lead to an increase in temperature but an increase in the number of particles. This temperature is called the *Hagedorn temperature* [4].

To determine this maximum temperature, we perform a simple calculation of the number of resonance states at large energies. The partition function is:

$$f_D(x) = \prod_{m=1}^{\infty} (1 - x^m)^{-D} = \sum_{n=0}^{\infty} P_D(n) x^n \qquad (9.3.1)$$

where $P_D(n)$ is the number of states at level n. The asymptotic behavior of $P_D(n)$ for large n is related to the $z \sim 1$ limit for $f_D(z)$ and can be calculated via the saddle point method. Let us write:

$$P_D(n) = \frac{1}{2\pi i} \oint \frac{f_D(z) dz}{z^{n+1}} \qquad (9.3.2)$$

Near $z \to 1$, we have:

$$f_D(z) \sim A(1 - x)^{D/2} \exp\left[\frac{\pi^2 D}{6(1 - x)}\right] \qquad (9.3.3)$$

and

$$P_D(n) \sim A' n^{-n(D+3)/4} \exp\left(2\pi \sqrt{nD/6}\right) \qquad (9.3.4)$$

For $n \sim \alpha' m^2$, we find the asymptotic number of states per unit mass is:

$$D(m) \sim m^{-(D+1)/2} e^{m/m_0} \tag{9.3.5}$$

where:

$$m_0 = \frac{1}{\pi} \sqrt{\frac{3}{2\alpha' D}} \tag{9.3.6}$$

where m_0 (in suitable units) can be interpreted as the maximum possible temperature of hadronic matter.

Hagedorn, even before the birth of dual models, was then led to calculate the maximum temperature, given an exponential rise in the number of states with temperature, and found that the maximum temperature was 160 MeV for a hadronic system. From gauge theories, we know that the interactions of the Higgs sector, in the limit of high temperature, changes so that the broken symmetries are restored, indicating a phase transition. The same reasoning might apply to the string model, where an unknown symmetry is restored at high temperatures. In this framework, we should no longer view the Hagedorn temperature as the maximum temperature, but instead view it as an indication of a phase transition.

The results are rather surprising. Usually, in point particle theory in D dimensions, the contribution to the free energy, in the limit of high temperature, goes as $T^{(D-1)}$. We therefore expect that in string theory the free energy will go as T^{25}. However, this is not the case. In fact, we find that in the high-temperature limit the free energy goes as T, indicating that the theory in this phase has *fewer* degrees of freedom than any point particle theory! Thus, we may conclude that, beyond the limiting temperature, a new phase of string theory opens up that has fewer degrees of freedom than any known quantum theory [5].

The quantity we wish to calculate is the free energy F, which is given in terms of the partition function Z:

$$F = -kT \ln Z \tag{9.3.7}$$

where:

$$Z = \int DX \exp\left(-\frac{1}{\hbar g^2} \int_0^{\hbar\beta} d\tau \int d^{D-1}x \, L\right) \tag{9.3.8}$$

where $\beta = 1/kT$ (we will take $k = 1$) and the time direction is wrapped into a circle with circumference $\hbar\beta$. Thus, we will consider string motions in the space $R^{25} \times S^1$.

Let us consider the one-loop contribution to the free energy and then the multiloop contribution. We wish to perform the loop integral over the space $R^{25} \times S^1$, since the time direction has been compactified. On the compactified torus, the string variable X_0 becomes:

$$X_0 = x_0 + \beta n \sigma_1 + \beta m \sigma_2 + \cdots \tag{9.3.9}$$

where the ellipses are oscillator terms that will not contribute in the high-temperature limit, and n and m are integers that indicate the winding modes of solitons that wrap around the torus.

Let us insert this contribution into the expression for the free energy. We find:

$$\frac{F}{TV} = -\frac{1}{2}\left(\frac{1}{4\pi^2\alpha'}\right)^{13}\int d^2\tau (\mathrm{Im}\,\tau)^{-14}\,e^{4\pi\,\mathrm{Im}\,\tau}|\eta(e^{2\pi i\tau})|^{-48}G(T) \quad (9.3.10)$$

where the temperature dependence is contained within $G(T)$, which is defined:

$$G(T) = \beta\sum_{n,m}e^{-S(n,m)} \quad (9.3.11)$$

where $S(n,m)$ is the soliton contribution to the action:

$$S(n,m) = \frac{\beta^2}{4\pi\alpha'\,\mathrm{Im}\,\tau}(m^2 + n^2|\tau|^2 - 2(\mathrm{Re}\,\tau)nm) \quad (9.3.12)$$

This expression, in fact, can actually be evaluated to all loop orders. We recall that the period matrix τ generalizes to the $h \times h$ matrix Ω_{ij} for the Riemann surface of genus h.

Let us take the line integral around each of the a and b cycles, which are defined on a torus of genus h:

$$\oint_{a_i} dX^0 = \beta n_i; \qquad \oint_{b_j} dX^0 = \beta m_j \quad (9.3.13)$$

If ω_i are the first Abelian differentials that are defined on the genus g surface, then we can always power expand dX in terms of these differentials:

$$dX^0 = \partial X^0 + \bar{\partial}X^0 \quad (9.3.14)$$

where:

$$\partial X^0 = \sum_{i=1}^{h}\lambda_i\omega_i; \qquad \bar{\partial}X^0 = \sum_{i=1}^{h}\bar{\lambda}_i\bar{\omega}_i \quad (9.3.15)$$

where we have split the differential dX_0 into holomorphic and antiholomorphic pieces. Performing the integrals, we can solve for the coefficients λ_i:

$$\lambda_j = \frac{1}{2}i\beta\left(\Omega_2^{-1}\right)_{jk}\left(\bar{\Omega}_{kl}n_l - m_k\right) \quad (9.3.16)$$

where we have split the period matrix in terms of its real and imaginary parts: $\Omega = \Omega_1 + i\Omega_2$.

Let us now insert all this into the action:

$$S = \frac{1}{4\pi\alpha'}\int d^2\sigma\,\partial_a X^0\,\partial^a X^0 = \frac{i}{2\pi\alpha'}\int \partial X \wedge \bar{\partial}X$$

$$= \frac{\beta^2}{4\pi\alpha'}(\bar{\Omega}_{kl}n_l - m_k)(\Omega_2^{-1})_{kj}(\Omega_{js}n_s - m_s) = \frac{\beta^2}{4\pi\alpha'}N^T A N \quad (9.3.17)$$

where:

$$N = \begin{pmatrix} n_i \\ m_j \end{pmatrix} \tag{9.3.18}$$

and

$$A = \begin{pmatrix} \Omega_1 \Omega_2^{-1} \Omega_1 + \Omega_2 & -\Omega_1 \Omega_2^{-1} \\ -\Omega_2^{-1} \Omega_1 & \Omega_2^{-1} \end{pmatrix} \tag{9.3.19}$$

The result that we want is an expression for $G(T)$ for higher loops. We find:

$$G(\beta, \Omega) = \beta \sum_N e^{-S(N)} = \beta \sum_N \exp\left(-\frac{\beta^2}{4\pi\alpha'} N^T A N\right)$$
$$= \beta \left(\frac{4\pi^2\alpha'}{\beta^2}\right)^h (\det A)^{-12} \tag{9.3.20}$$

where we have summed over N in the first step of the calculation.

By direct calculation, we can show that the determinant of A is equal to one, so the period matrix makes no contribution to the final result. This greatly simplifies the calculation, and yields the answer:

$$\frac{F_h}{VT} = \frac{4\pi^2\alpha'T}{\hbar} \left(\frac{g^2 T^2 4\pi^2\alpha'}{\hbar}\right)^{h-1} \Lambda_h \tag{9.3.21}$$

where Λ_h is the genus h contribution to the cosmological constant.

To take the high-temperature limit of the free energy, we must take into account that the effective coupling constant g_{eff} is also a function of the temperature. We impose:

$$g_{\text{eff}}^2 = \frac{g^2 T^2 4\pi^2\alpha'}{\hbar} \tag{9.3.22}$$

so in the high-temperature limit, we keep $g^2 T^2$ fixed. (In QCD, by analogy, we keep $g^2 T$ fixed at high energy.) Thus, we are left with the final result [5]:

$$F/VT \sim T \tag{9.3.23}$$

as desired.

This calculation, however, made a large number of assumptions that may invalidate the result. We must be cautious in interpreting these results because the very definition of high temperature at Planck times may be ambiguous. (For example, at the Planck length, we do not expect space–time to be flat, and the $R^{25} \times S^1$ space that we have been considering may not be realistic. For that matter, even the definitions of "time" and "temperature" at Planck energies may not exist.)

However, if the main features of the calculation survive, then the interpretation is quite interesting. It means that the Hagedorn temperature may be reinterpreted as indicating a phase transition and that the theory in the high-temperature phase must possess fewer degrees of freedom than even a point particle theory. One example of such a system is topological field theory, which will be fully discussed in Chapter 14.

9.4. New Symmetries at High Energy

Usually in string theory, when we want to make comparisons with point particle theory, we take the zero slope limit $\alpha' \to 0$. In this limit, the Regge trajectories become horizontal, and massive particles begin to have infinite mass (that is, they decouple from the theory). In this limit, we only have massless particles surviving, for example, supergravity theory in 10 dimensions.

However, as we have stressed, the natural home for string theory is in the opposite direction, in the $\alpha' \to \infty$ limit. In this limit, the trajectories become vertical. We will find that, at these enormous energies, a remarkable symmetry among string amplitudes emerges [6, 7], once again indicating that beautiful but mysterious things might have happened during the early universe.

We know that, even at the four-point level, string theory differs markedly from ordinary point particle theory. For example, the high-energy limit of the four-string scattering amplitude is:

$$A_4 \sim (stu)^{-3} e^{-\alpha' S/4}; \qquad S = s \ln s + t \ln t + u \ln u \qquad (9.4.1)$$

which exhibits remarkably soft exponential behavior, unlike ordinary point particle theory.

Let us generalize this discussion to higher loops by considering the string scattering amplitude over the genus h surface:

$$A_h(p_i) = \int \frac{Dg_{ab}}{N} DX^\mu \prod_i V_i(X_i^\mu, P_i)$$
$$\times \exp\left(\frac{-\alpha'}{2\pi} \int d^2\xi \sqrt{g}\, g^{ab}\, \partial_a X^\mu\, \partial_b X_\mu \right) \qquad (9.4.2)$$

where V_i is the vertex for the ith particle carrying momentum P_i:

$$V(P_i) = \int d^2\xi_i \sqrt{g}\, \exp\left[i\alpha' P_i \cdot X(\xi_i) \right] \tilde{V}\left[X^\mu(\xi_i), P_i \right] \qquad (9.4.3)$$

Using saddle point methods, we will integrate this expression and take the high-energy limit. (We will not be concerned about infrared divergences, since these take place in the boundary of moduli space, while the saddle point method picks out points away from the boundary.)

Equation (9.4.2) is dominated by the contribution at a point ξ_i on the world sheet with moduli $\mathbf{m}_i$, generalizing Eq. (1.3.3). We can write:

$$X^\mu(\xi) = i \sum_i P_i^\mu G_{\hat{\mathbf{m}}}(\xi, \hat{\xi}_i) + O\left(\frac{1}{\alpha'} \right) \qquad (9.4.4)$$

where G is the multiloop Green's function defined on the Riemann surface. Consequently, we can approximate Eq. (9.4.2) as:

$$A_h(P_i) = g^{2h+2}\Gamma_h(P_i, \hat{\mathbf{m}}_{\mathbf{i}}, \hat{\xi}_i) \prod_i \tilde{V}_i(X^\mu, \hat{\xi}_j, P_j) + O\left(\frac{1}{\alpha'}\right) \qquad (9.4.5)$$

where Γ_h contains factors arising from the saddle point method, which behaves as:

$$e^{-\left[\alpha'/4(h+1)\right]S} \qquad (9.4.6)$$

The important fact, however, is that all dependence on the external particles is concentrated in the vertex function $\prod_i V_i$. All other dependences on the external particles have been washed out in the high-energy limit. Thus, we can also write the following relationship between any two N-point scattering amplitudes with different external particles a_i or b_j but with the same momenta:

$$A_h^{a_i}(P_i) = \frac{\prod_j V_{a_j}(X_{cl}^\mu, P_j)}{\prod_j \tilde{V}_{b_j}(X_{cl}^\mu)} A_g^{b_i}(P_i)\left[1 + O(\alpha')\right] \qquad (9.4.7)$$

As it stands, this equation is still useless because it only holds at a given loop order h in perturbation theory. However, it is possible to remove this restriction by carefully analyzing the dependence of the vertices on $N = h + 1$.

The problem is that the vertex $\tilde{V}$ is a function of X_{cl}^μ, which in turn behaves at the saddle point as:

$$X_{cl}^\mu \sim \frac{i}{N} \sum_{i=1}^4 P_i^\mu \ln|z - a_i| + O\left(\frac{1}{\alpha'}\right) \qquad (9.4.8)$$

We wish to remove the $1/N$ dependence occurring here. To do this, we define an operator D as:

$$D \equiv \left(\frac{2}{\alpha'S}\right)\sum_i P_i \frac{\partial}{\partial P_i} \qquad (9.4.9)$$

By analyzing the effect that this operator has on factors such as $e^{-\alpha'S/4}$, we see that we can replace $1/N$ by D in the vertex function. The operator D simply pulls down factors of $1/N$. Making this replacement, we find:

$$\prod_i \tilde{V}_{b_i}(X_{cl}^\mu D, P_i)A_h^{a_i}(P_i) = \prod_i \tilde{V}_{a_i}(X_{cl}^\mu D, P_i)A_h^{b_i}(P_i)\left[1 + O(\alpha')\right] \quad (9.4.10)$$

The important point is that this equation is valid independent of the genus h. Since the equation holds, order by order, in perturbation theory, we can argue that it must also hold for the complete amplitude as well (barring unforeseen nonperturbative effects). In this case, we find the following relationship for the entire amplitude [6, 7]:

$$A_{a_i}(P_i) = \prod_i \tilde{V}_{a_i}(X_{cl}^\mu D, P_i)A_{\text{tachyon}}(P_i)[1 + O(\alpha')] \qquad (9.4.11)$$

This is the desired result, which shows the remarkable symmetry that emerges in the high-energy limit, linking different scattering amplitudes in unexpected ways. If we knew, for example, just the dilaton scattering amplitude, then we could in principle determine all scattering amplitudes for all particles. This points to the power of this new symmetry emerging at high energy. One possibility for this new symmetry is exact general covariance, which may indicate that a topological phase exists beyond the Planck length. This possibility will be further discussed in Chapter 14.

9.5. Is String Theory Borel Summable?

Ever since the original proposal of KSV to expand the Veneziano function as the Born term for a perturbation series, there was speculation as to whether the series was summable. Even after the divergence structure of the multiloop graphs was first calculated, it was known that the divergence structure of the higher loop amplitudes would correspond to topological deformations of the Riemann surface.

However, it was recently noticed that, at least for the bosonic string, the string perturbation series appears not to be Borel summable, that is, genus h terms diverge as $h!$ [8]:

$$A(g) \sim \sum_h g^h h! \qquad (9.5.1)$$

which may have a great impact on how we view the whole perturbation theory of strings. If true, it may mean that perturbation theory may be the wrong framework in which to find the true vacuum of the string.

It was also conjectured that the non-Borel summability of the bosonic string is unrelated to the presence of the tachyon. If this is true, then perhaps superstring theory is also non-Borel summable. However, this may actually be a blessing in disguise. If perturbation theory was Borel summable, it would mean that many features found to hold order by order in perturbation theory would also hold to all orders. Thus, the millions of supersymmetric vacuums found in conformal field theory would be perfectly stable. This would mean that there were millions of allowable, self-consistent universes, the overwhelming majority having little resemblance to our physical universe. This, in turn, would be quite an embarrassment for string theory.

Let us begin our discussion of the question of Borel summability by reminding ourselves that, even if a series:

$$A(g) = \sum_h g^h f_h \qquad (9.5.2)$$

formally diverges, we can still define its Borel transform:

$$A_{\text{Borel}}(g) = \sum_h \frac{g^h f_h}{h!} \tag{9.5.3}$$

We could then recover the original function by taking the inverse transform:

$$A(g) = \int_0^\infty A_{\text{Borel}}(gt)e^{-t}dt \tag{9.5.4}$$

However, if the series is not Borel summable, that is, if $f_h \sim h!$, then the Borel transform function A_{Borel} is singular along the real axis, and hence we cannot take the inverse Borel transform without additional (that is, nonperturbative) information concerning the integration near the singularity. We will how show that this actually occurs for bosonic string theory for the simplest case, the partition function.

The partition function for genus h in string theory is given by:

$$\int d\mu_{WP}\, g^h Z(2)Z'(1)^{-13} \tag{9.5.5}$$

where WP stands for the Weil–Petersson measure on the moduli space of genus h graphs, and $Z(s)$ stands for the Selberg Z function,

$$Z'(1) = \det'(\Delta) \tag{9.5.6}$$

where Δ is the Laplacian defined on the two-dimensional Riemann surface. (The prime means that we remove the zero eigenvalue.) The ghost contribution gives us $Z(2)$, while the usual bosonic part gives us $Z'(1)^{-13}$.

The calculation comes in two parts. First, we have to give a lower bound on the contributions coming from the Z functions, showing that the integrand does not behave like $1/h!$ for a large genus h. Second, we have to show that the measure does grow as $h!$. Multiplying the two, we then have the partition function diverging as $h!$, thus spoiling Borel summability.

We will calculate both contributions on a negative curvature metric (that is, $R = -1$). On such a surface, the area of a genus h surface is:

$$A = 2\pi(2h - 2) \tag{9.5.7}$$

We will find it convenient to write the Selberg function via the McKean formula:

$$\frac{Z'(s)}{Z(s)} = (2s - 1)\int_0^\infty dt\, \exp[-s(s - 1)][\theta(t) - Ak(t)] \tag{9.5.8}$$

where

$$k(t) \equiv \frac{\sqrt{2}e^{-t/4}}{(4\pi t)^{3/2}} \int_0^\infty b\, db\, \frac{e^{-b^2/4t}}{(\cosh b - 1)^{1/2}} \sim \frac{1}{4\pi t} - \frac{1}{12\pi} + O(t) \tag{9.5.9}$$

and:

$$\theta(t) = \operatorname{Tr} e^{-t\Delta} \sim \frac{A}{4\pi t} - \frac{A}{12\pi} + O(t) \qquad (9.5.10)$$

for small t.

Let us integrate the expression for Z. We find:

$$\frac{d}{ds} \ln \frac{Z(s)}{s(s-1)} = -\frac{d}{ds} \int_0^\infty e^{-ts(s-1)}[f(t) - f(0)] \frac{dt}{t}$$

$$- f(0) \frac{d}{ds} \left\{ \int_0^\infty \left[e^{-ts(s-1)} - e^{-ta(a-1)} \right] \frac{dt}{t} + C \right\}$$

$$(9.5.11)$$

where $\hat{\theta}(t) = \theta(t) - 1$ and:

$$f(t) \equiv \hat{\theta}(t) - Ak(t) \sim -1 + O(t^n) \qquad (9.5.12)$$

Setting $Z(\infty) = 1$, we can determine the constants a, C.

We finally arrive at:

$$\ln \frac{Z(s)}{s(s-1)} = -\int_0^\infty \frac{dt}{t} \left[e^{-ts(s-1)} f(t) + e^{-t} \right] \qquad (9.5.13)$$

Because the ghost (string) Z function lies in the numerator (denominator), we wish to calculate the lower (upper) bound of these Z functions.

The previous equation implies:

$$Z'(1) = \exp \left\{ -\int_0^\infty \frac{dt}{t} \left[\hat{\theta}(t) - Ak(t) + e^{-t} \right] \right\} \qquad (9.5.14)$$

This, in turn, allows us to calculate a lower bound on $Z(1)$. The key is to observe that the dependence on h does not appear in $\hat{\theta}$ and that:

$$Z'(1) < e^\epsilon \exp \left\{ -\int_\epsilon^\infty \frac{dt}{t} \left[\hat{\theta}(t) - Ak(t) + e^{-t} \right] \right\}$$

$$\sim c^A \exp \left[-\int_\epsilon^\infty \frac{dt}{t} \hat{\theta}(t) \right]$$

$$(9.5.15)$$

Since $\hat{\theta} > 0$, we find that the total contribution of the genus h to the lower bound comes in the form of a constant raised to the h power, which can be absorbed into the coupling constant.

Similarly, we can repeat the analysis for the ghost contribution $Z(2)$. Putting everything together, we find:

$$Z(2) > Z(1) > Z'(1)c^h \qquad (9.5.16)$$

which is the desired result. Thus, the measure terms cannot give us $1/h!$ terms.

Second, we now have to show that the Weil–Petersson metric $d\mu_{WP}$ does, in fact, grow as $h!$.

If we have a closed string that smoothly breaks up into two other closed strings, we have a three-string vertex that has the topology of a pair of pants. Let us introduce coordinates at the base of each leg of the pants. Let l represent the geodesic circumference around each leg and θ a twist angle. Then, the Weil–Petersson measure is given by:

$$\omega = \sum l \, dl \wedge d\theta \qquad (9.5.17)$$

Since there are two coordinates per pants leg, we find 6 coordinates per loop, which is the correct number. This means that the Weil–Petersson volume of a small annulus in moduli space goes as c^{3h-3}, where c is a constant depending only on the length of the shortest geodesic in the annulus and the thickness of the annulus. (Weil–Petersson coordinates do not correctly triangulate moduli space. They yield an infinite over-cover due to the mapping class group or the modular group. However, they correctly express the invariant measure on Teichmüller space. So, keeping in mind that we must eventually divide out by the effect of the mapping class group, we can use the Weil–Petersson measure.)

There is a theorem due to Bollobas, which states that the number of isomorphism classes of trivalent graphs (that is, graphs with three legs in a vertex, as in the Weil–Petersson case) without loops or multiple edges increases factorially with the genus. Putting both contributions to the Weil–Petersson measure together, we find that it grows as [8]:

$$c^{3h-3}h! \qquad (9.5.18)$$

which completes the second half of the proof. Multiplying the first half of the proof with the second, we find that the partition function grows as $h!$, which makes it non-Borel summable.

One essential feature of this proof is that we avoided the region of moduli space that gives rise to the infrared divergence associated with the tachyon. By making the effective cutoff the minimum length of the geodesic, we have introduced a genus-independent regulator to control the infrared contribution of the tachyon. Thus, this makes us suspect that the superstring (which has no tachyon) will also be non-Borel summable (although this is still conjectural). There is, therefore, compelling (but not rigorous) reasons for believing that all string theories are not Borel summable.

This feature of non-Borel summability is actually a familiar one that is found in ordinary gauge theory, which is also non-Borel summable. Normally, this question poses no problem for gauge theories. For example, we can always use the perturbation theory as an asymptotic theory, which approximates the true theory only for low orders.

Also, the failure of Borel summability means that nonperturbative effects become important, such as

1.   the presence of instantons, where instantons give us the mixing of vacuums;

2. the presence of Euclidean bounces, meaning that the naive vacuum decays into the true vacuum; and

3. the presence of renormalons, which are linked to ultraviolet problems.

In summary, we may take the non-Borel summability of the perturbation series as a blessing in disguise. It means that the millions of conformal field theories that have been discovered may be unstable and may decay into the true vacuum, where supersymmetry is broken and the low-energy behavior looks quite different from what perturbation theory tells us.

9.6. Nonperturbative Approaches

There are compelling, although not rigorous, grounds for believing that entirely new features will emerge in the nonperturbative regime, which makes it crucial to develop a consistent approach to a nonperturbative formulation of string theory. Nonperturbative methods, however, are notoriously difficult to formulate, much less solve. In the second half of this book, we will concentrate on several approaches that have been advocated. However, each has its own drawbacks, and none of them have been able to come close to selecting the true vacuum. However, there are grounds to believe that one or more of these methods will eventually give us insight into how the superstring selects its true vacuum. We will discuss the merits and faults of the following nonperturbative approaches that have appeared in the literature.

Universal Moduli Space

This approach [9] tries to fully exploit the beautiful modular properties of the Riemann surfaces appearing in the perturbation series. Usually, extraordinary symmetries must be present in a perturbation series in order for "miracles" to appear. Point particle Feynman graphs possess no such symmetries (except trivial Ward identities). However, string perturbation graphs do possess a powerful symmetry – invariance.

The modular properties that Riemann surfaces must obey are quite restrictive. Although it is prohibitive to sum the series, one can study the modular properties of Riemann surfaces with arbitrary, even infinite, genus, in hopes of being able to make quantitative statements about the nonperturbative behavior of perturbation theory. However, at present, this approach has proven to be prohibitively difficult. One approach to universal moduli space is to use Grassmannians, where a Riemann surface of genus g corresponds to a point in this space. One problem facing the Grassmannian approach, however, is the Schottky problem, which develops when one tries to use the period matrix as moduli parameters. (The Schottky problem was recently solved by mathematicians. However, the solution is so nonlinear that we do not, as yet, have any handle on the problem.) Thus, universal moduli space, although potentially a powerful formalism, is still too ill defined to make any definite statements about the nonperturbative behavior of strings.

Renormalization Group

The renormalization group method [10–19] in gauge theory has proven to be one of the most powerful approaches to analyzing the nonperturbative behavior of the theory. The new renormalization group approach to string theory is to postulate the existence of a new space, called "theory space," or the space of all quantum theories. In this theory space, conformal invariance may not be exact. However, the renormalization group flows in this theory space converge on fixed points, which may indicate the true vacuum of the theory. Thus, the equation of motion of the theory, instead of being parameterized in space–time, is now parameterized in theory space. In other words, the counterpart of the equations of motion of this approach is:

$$\beta = 0 \qquad\qquad (9.6.1)$$

that is, the renormalization group flows in theory space prefer conformal field theories.

Although this approach is quite novel, at present only the tree diagrams of the bosonic string have been reproduced. There is still much work to be done in the renormalization group approach, which will be explained in this chapter.

String Field Theory

String field theory [20, 21] is the most "conservative" of all the approaches made for nonperturbative string theory. The advantage of this approach is that it can effortlessly reproduce the entire perturbation theory, including loops. In a simple Lagrangian, one can express the entire information content of string theory. However, because it is so conservative, it may not be "crazy enough" for us to calculate nonperturbative effects. At present, the string field theory approach has failed to give us any reliable information about the nonperturbative regime of string theory, which is a big disappointment.

Matrix Models

The last approach we will explore is the matrix model approach [21], which has the great advantage of being the first theory that has successfully yielded nonperturbative information. In fact, matrix models can be solved exactly for a large array of actions.

The problem with matrix models, however, is that they only describe string theory in less than one dimension, that is, two-dimensional gravity coupled to $c \leq 1$ conformal matter. In dimensions less than one, in fact, one finds that the theory has finite degrees of freedom, which is the origin of why these models are exactly solvable. Efforts to push this limit beyond

$c \geq 1$ show that there are probably an infinite number of degrees of freedom and that a phase transition may occur.

Thus, the very feature that makes matrix models so attractive (exact solvability) is due to their finite degrees of freedom, which is also its main detraction. For higher dimensions, we expect that there are an infinite number of degrees of freedom, and hence, the model is once again unsolvable (except perhaps on computer).

Because matrix models have finite degrees of freedom, we can compare them with other models that also have finite degrees of freedom, such as topological field theories. In fact, one can formally show the equivalence between the matrix elements of matrix models and certain topological field theories.

9.7. Renormalization Group Approach

We will discuss the renormalization group approach in two parts. First, we will show how the renormalization group methods are useful in determining the background fields of string theory, treated as a sigma model on the sphere or plane. Second, we will try to generalize the approach into a program includes higher order interactions.

To actually perform the calculation of the β function to determine where it vanishes, we must make a power expansion. However, a power expansion in general relativity is not generally covariant. A Taylor expansion in fields, for example, gives an infinite series of derivatives, not covariant derivatives or curvatures. Thus, in order to preserve general covariance, we will use a particular choice of coordinates, called *Riemann normal coordinates* [10–12].

We wish to make the power expansion of the string variable around the classical co-ordinate X^μ, which satisfies the equations of motion. This is called the *background field method*, which is used extensively in relativity and gauge theories. We will use the expansion around the classical solution X^μ:

$$X^\mu \to x^\mu + \xi^\mu t + \cdots \qquad (9.7.1)$$

where t parameterizes the power expansion. We will then treat the resulting action as a quantum particle ξ^μ moving in a classical background, given by X^μ. Thus,

$$X^\mu(0) = x^\mu, \qquad X^\mu(1) = x^\mu + \xi^\mu \qquad (9.7.2)$$

Let $X^\mu(t)$ lie along a geodesic, so that it satisfies the equation:

$$\frac{d^2 X^\mu}{dt^2} + \Gamma^\mu_{\alpha\beta} \frac{dX^\alpha}{dt} \frac{dX^\beta}{dt} = 0 \qquad (9.7.3)$$

where $\Gamma^\mu_{\alpha\beta}$ are the Christoffel symbols:

$$\Gamma^{\mu}_{\alpha\beta} = \frac{1}{2}g^{\mu\nu}\left(g_{\alpha\nu,\beta} + g_{\nu\beta,\alpha} - g_{\alpha\beta,\nu}\right) \tag{9.7.4}$$

where the comma denotes an ordinary derivative. Physically, this means that ξ^{μ} is a vector pointing along the tangent vector at X to the geodesic passing through the points X and $X + \xi$.

If we plug the power expansion in t into the geodesic equation, we find, order by order in t:

$$X^{\mu}(t) = x^{\mu} + \xi^{\mu}t - \frac{1}{2}\Gamma^{\mu}_{\alpha\beta}\xi^{\alpha}\xi^{\beta}t^2 - \frac{1}{3!}\Gamma^{\mu}_{\alpha\beta\gamma}\xi^{\alpha}\xi^{\beta}\xi^{\gamma}t^3 + \cdots \tag{9.7.5}$$

where:

$$\begin{aligned}
\Gamma^{\mu}_{\alpha_1\alpha_2\alpha_3} &\equiv \partial_{\alpha_1}\Gamma^{\mu}_{\alpha_2\alpha_3} - \Gamma^{\nu}_{\alpha_1\alpha_2}\Gamma^{\mu}_{\nu\alpha_3} - \Gamma^{\nu}_{\alpha_1\alpha_2}\Gamma^{\mu}_{\alpha_3\nu} \equiv \nabla_{\alpha_1}\Gamma^{\mu}_{\alpha_2\alpha_3} \\
\Gamma^{\mu}_{\alpha_1\alpha_2\alpha_3\alpha_4} &\equiv \nabla_{\alpha_1}\Gamma_{\alpha_2\alpha_3\alpha_4}, \text{ etc.}
\end{aligned} \tag{9.7.6}$$

Notice that our equations are not necessarily covariant. To make the calculation easier, let us go into a specific frame in which the expansion coefficients vanish when symmetrized with respect to their lower indices:

$$\bar{\Gamma}^{\mu}_{(\alpha_1,\alpha_2,\dots,\alpha_n)} = 0 \tag{9.7.7}$$

The point of going into this frame is that the Riemann tensor becomes simpler, which in turn allows us to write all our expressions in fully covariant form. For example, the curvature tensor:

$$R^{\mu}_{\nu\alpha\beta} = \partial_{\alpha}\Gamma^{\mu}_{\nu\beta} - \partial_{\beta}\Gamma^{\mu}_{\nu\alpha} + \Gamma^{\lambda}_{\nu\beta}\Gamma^{\mu}_{\lambda\alpha} - \Gamma^{\lambda}_{\nu\alpha}\Gamma^{\mu}_{\lambda\beta} \tag{9.7.8}$$

can now be written as:

$$\begin{aligned}
\bar{R}^{\mu}_{\nu\alpha\beta} &= \partial_{\alpha}\bar{\Gamma}^{\mu}_{\nu\beta} - \partial_{\beta}\bar{\Gamma}^{\mu}_{\nu\alpha} \\
\partial_{\nu}\bar{\Gamma}^{\mu}_{\alpha\beta} &= \frac{1}{3}\left(\bar{R}^{\mu}_{\alpha\nu\beta} + \bar{R}^{\mu}_{\beta\nu\alpha}\right)
\end{aligned} \tag{9.7.9}$$

Similarly, we can always write the higher derivatives in terms of covariant derivatives of the curvature:

$$\begin{aligned}
\partial_{(\alpha_1}\partial_{\alpha_2}\bar{\Gamma}^{\mu}_{\alpha_3\nu)} &= -\frac{1}{2}D_{\alpha_1}\bar{R}^{\mu}_{\alpha_2\nu\alpha_3} \\
\partial_{(\alpha_1}\partial_{\alpha_2}\partial_{\alpha_3}\bar{\Gamma}^{\mu}_{\alpha_4)\nu} &= -\frac{3}{5}\left[D_{(\alpha_1}D_{\alpha_2}\bar{R}^{\mu}_{\alpha_3\nu\alpha_4)} + \frac{2}{9}\bar{R}^{\mu}_{(\alpha_1\alpha_2}{}^{\lambda}\bar{R}_{\alpha_3\lambda\alpha_4)\nu}\right]
\end{aligned} \tag{9.7.10}$$

The advantage of this expansion is that we can now covariantly expand the metric tensor $g_{\mu\nu}$ in terms of the small deviation ξ^{μ}:

$$\begin{aligned}
g_{\mu\nu}(x + \xi) = {}& g_{\mu\nu}(x) - \frac{1}{3}R_{\mu\alpha_1\nu\alpha_2}\xi^{\alpha_1}\xi^{\alpha_2} - \frac{1}{3!}D_{\alpha_1}R_{\mu\alpha_2\nu\alpha_3}\xi^{\alpha_1}\xi^{\alpha_2}\xi^{\alpha_3} \\
&+ \frac{1}{5!}\left[-6D_{\alpha_1}D_{\alpha_2}R_{\mu\alpha_3\nu\alpha_4}(x) + \frac{16}{3}R_{\alpha_1\nu\alpha_2}{}^{\lambda}R_{\alpha_3\mu\alpha_4\lambda}\right] \\
&\times \xi^{\alpha_1}\xi^{\alpha_2}\xi^{\alpha_3}\xi^{\alpha_4} + \cdots
\end{aligned} \tag{9.7.11}$$

Given this power expansion, we can now perform the integration over ξ^μ, which yields a series of loop diagrams. A few of the terms that appear in the action after the power expansion are:

$$
\begin{aligned}
S = S_0(x) + &\int d^2 z\, g_{\mu\nu}\, \partial_a x^\mu\, D^a \xi^\nu \\
+ \frac{1}{2} \int d^2 x &\Big(g_{\mu\nu}\, D_a \xi^\mu\, D^a \xi^\nu + R_{\mu\alpha_1\alpha_2\nu} \xi^{\alpha_1}\xi^{\alpha_2}\, \partial_a x^\mu\, \partial^a x^\nu \\
+ \frac{1}{3} &D_{\alpha_1} R_{\mu\alpha_2\alpha_3\nu}\, \xi^{\alpha_1}\xi^{\alpha_2}\xi^{\alpha_3}\, \partial_a x^\mu\, \partial^a x^\nu \\
+ \frac{4}{3} &R_{\mu\alpha_1\alpha_2\alpha_3} \xi^{\alpha_1}\xi^{\alpha_2}\, D_a \xi^{\alpha_3}\, \partial^a x^\mu + \cdots \Big)
\end{aligned}
\tag{9.7.12}
$$

where:

$$
D_a x^\mu = \partial_a x^\mu + \Gamma^\mu_{\alpha\beta} \xi^\alpha\, \partial_a \xi^\beta
\tag{9.7.13}
$$

In this fashion, any first quantized string action can be power expanded around classical fields X^μ, treating ξ^μ as the quantum field for which we must calculate the Feynman diagrams. Notice that the resulting action can be treated as a standard vector field ξ^μ propagating in two dimensions (where the μ index can be compared to "isospin").

So far, our discussion has been rather general, which can be applied to any sigma model. Now, let us specifically apply these methods to string theory. Let us begin with the first quantized action, where the string coordinate X^μ moves in a space that is curved [13–15]:

$$
\begin{aligned}
S = \frac{1}{4\pi\alpha'} \int d^2 z \Big[&\sqrt{g}\, g^{ab} G(X)_{\mu\nu} \partial_a X^\mu \partial_b X^\nu \\
&+ \epsilon^{ab} B_{\mu\nu} \partial_a X^\mu \partial_b X^\nu - \alpha' \sqrt{g}\, R^{(2)} \Phi(X) \Big]
\end{aligned}
\tag{9.7.14}
$$

The background metric is no longer flat, but is now curved via the symmetric space–time metric $G(X)^{\mu\nu}$, as well as the antisymmetric field $B(X)^{\mu\nu}$ and the dilaton field $\Phi(X)$. We will only treat the massless fields of the string theory in the background (in principle, we could have massive fields as well as part of the background).

The point is to treat this action as a standard σ model and then perform fluctuations around a curved background to calculate the β functions (one for each background field) associated with breaking scale invariance. To reestablish scale invariance, we must then set each of these β functions for each background field to zero. This establishes constraints on the background fields.

The trace of the energy-momentum tensor will become:

$$
\begin{aligned}
2\pi T^\mu_\mu = &\beta^G_{\mu\nu} \sqrt{g}\, g^{ab}\, \partial_a X^\mu\, \partial_b X^\nu \\
&+ \beta^B_{\mu\nu} \epsilon^{ab}\, \partial_a X^\mu\, \partial_b X^\nu + \beta^\phi \sqrt{g}\, R^{(2)}
\end{aligned}
\tag{9.7.15}
$$

where β^ϕ, $\beta^B_{\mu\nu}$, $\beta^G_{\mu\nu}$ represent the deviation from scale invariance. Using background field methods with Riemann normal coordinates, we find [14, 15]:

$$\beta^G_{\mu\nu} = R_{\mu\nu} - \frac{1}{4}H_\mu^{\lambda\sigma}H_{\nu\lambda\sigma} + 2\nabla_\mu\nabla_\nu\Phi + O(\alpha')$$

$$\beta^B_{\mu\nu} = \nabla_\lambda H^\lambda_{\mu\nu} - 2(\nabla_\lambda\Phi)H^\lambda_{\mu\nu} + O(\alpha') \tag{9.7.16}$$

$$\frac{\beta^\phi}{\alpha'} = \frac{D-26}{\alpha'48\pi^2} + \frac{1}{16\pi^2}\left[4(\nabla\Phi)^2 - 4\nabla^2\Phi - R + \frac{1}{12}H^2\right] + O(\alpha')$$

(If we compare this result with the standard commutation relations between components of the energy-momentum tensor, we find that the central term c is equal to β^ϕ. This may appear a bit surprising, because β can be operator valued, while the central term c is a constant. However, these equations are consistent because, using the Bianchi identities on the other β functions, we find:

$$0 = \nabla^\mu\left(R_{\mu\nu} - \frac{1}{4}H^2_{\mu\nu} + 2\nabla_\mu\nabla_\nu\Phi\right)$$

$$= \nabla_\nu\left[-2(\nabla\Phi)^2 + 2\nabla^2\Phi + \frac{1}{2}R - \frac{1}{24}H^2\right] \tag{9.7.17}$$

Thus, it is consistent to treat β^ϕ as a c number constant. This probably persists to higher loop levels.)

To make the meaning of these equations more transparent, let us rescale by:

$$G_{\mu\nu} \rightarrow e^{4\Phi/(D-2)}G_{\mu\nu} \tag{9.7.18}$$

then the equations take on the familiar form:

$$R_{\mu\nu} - \frac{1}{2}G_{\mu\nu}R - T^{\text{matter}}_{\mu\nu} = 0$$

$$\nabla^2\Phi + \frac{1}{6}e^{-8\Phi/(D-2)}H^2 = 0 \tag{9.7.19}$$

$$\nabla_\lambda\left[e^{-8\Phi/(D-2)}H^\lambda_{\mu\nu}\right] = 0$$

where:

$$T^{\text{matter}}_{\mu\nu} = \frac{1}{4}\left(H^2_{\mu\nu} - \frac{1}{6}G_{\mu\nu}H^2\right)e^{-8\Phi/(D-2)}$$

$$+ \frac{4}{D-2}\left[(\nabla_\mu\Phi)(\nabla_\nu\Phi) - \frac{1}{2}G_{\mu\nu}(\nabla\Phi)^2\right] \tag{9.7.20}$$

We recognize the meaning of these equations immediately. The first equation is the usual Einstein equation, in the presence of matter fields. The second and third equations guarantee the conservation of the energy-momentum tensor. As a consequence, we can compactly generate all three equations from a single action [14, 15]:

$$\int d^D x \sqrt{G} \left[R - \frac{4}{D-2} (\nabla \Phi)^2 - \frac{1}{12} e^{-8\Phi/(D-2)} H^2 \right] \qquad (9.7.21)$$

In summary, we have found the remarkable fact that the conditions for conformal invariance are precisely the classical equation of motion for these background fields. This testifies to the enormous self-consistency of string theory, that it can only self-consistently propagate in backgrounds that satisfy the correct equations of motion.

Although the renormalization group approach shows great self-consistency, so far we have only examined the free part of the string, that is, the free string as it moves on a simple Riemann surface. Now, we wish to formulate a more powerful approach in which one may derive interactions and, perhaps, the complete theory. The goal is to formulate a version of the renormalization group equations in "theory space," that is, the space of all theories, and then use the constraint $\beta = 0$ as an equation of motion. Although the formalism is far from achieving this ambitious goal, it can, in principle, reproduce many of the simpler results of string field theory.

Our approach follows point particle theory. In ordinary ϕ^4 theory, for example, we can use the Kadanoff–Wilson "block spin" formalism to derive the proper renormalization group equations. The idea behind this is simple. Consider the ϕ^4 theory with some cutoff, such as:

$$D_\Lambda(p^2) = \frac{1}{p^2 + m^2} K\left(\frac{p^2}{\Lambda^2}\right); \qquad K = \begin{cases} 1, & p^2 < \Lambda^2 \\ 0, & p^2 > \Lambda^2 \end{cases} \qquad (9.7.22)$$

The fundamental idea is that the theory should be independent of the cutoff. Specifically, if we lower the cutoff and then integrate out all modes in ϕ between the two cutoffs, the resulting equations should have the same form as the original equations, that is, the generating functional $Z(J)$ should obey:

$$\Lambda \frac{dZ(J)}{d\Lambda} = 0 \qquad (9.7.23)$$

where J is the source term. Explicitly, the generating functional is:

$$Z(J) = \int d\phi \exp \left\{ \int \frac{d^4 p}{(2\pi)^4} \left[-\frac{1}{2} \phi(p) \phi(-p)(p^2 + m^2) \right. \right.$$
$$\left. \left. \times \ K^{-1}\left(\frac{p^2}{\Lambda}\right) + J(p)\phi(-p) \right] + L_{\text{int}}(\phi) \right\} \qquad (9.7.24)$$

where L_{int} is an interaction Lagrangian given by:

$$L_{\text{int}} = \int d^4 x \left\{ -\frac{1}{2}(\delta m^2)\phi^2(x) - \frac{1}{2}(Z-1)\left[\partial_\mu \phi(x)\right]^2 - \frac{1}{4!}\lambda^0 \phi^4(x) \right\} \qquad (9.7.25)$$

Now, let us investigate how the generating functional varies with respect to the cutoff. (We will replace L_{int} with a more general L. Although we might have started initially with $L = L_{int}$ at cutoff Λ, at lower energy scales, the L may become quite complicated.) By differentiation with respect to Λ, we find:

$$\Lambda \frac{dZ}{d\Lambda} = \int d\phi \left\{ \int \frac{d^4k}{(2\pi)^4} \left[-\frac{1}{2}\phi(p)\phi(-p)(p^2 + m^2) \right. \right.$$
$$\left. \left. \times \; \Lambda \frac{\partial K^{-1}}{\partial \Lambda} \right] + \Lambda \frac{\partial L}{\partial \Lambda} \right\} \exp -S \tag{9.7.26}$$

Now, let us make the key assumption. We assume that the general L appearing in the above equation obeys the following:

$$\Lambda \frac{\partial L}{\partial \Lambda} = -\frac{1}{2} \int d^4p (2\pi)^4 (p^2 + m^2)^{-1} \Lambda \frac{\partial K}{\partial \Lambda}$$
$$\times \left[\frac{\partial L}{\partial \phi(-p)} \frac{\partial L}{\partial \phi(p)} + \frac{\partial^2 L}{\partial \phi(p) \partial \phi(-p)} \right] \tag{9.7.27}$$

Inserting this into the general expression, we find [16]:

$$\Lambda \frac{dZ}{d\Lambda} = \int d^4p \Lambda \frac{\partial K}{\partial \Lambda} \int d\phi \frac{\partial}{\partial \phi(p)}$$
$$\times \left\{ \left[\phi(p) K^{-1} + \frac{1}{2}(2\pi)^4 (p^2 + m^2)^{-1} \frac{\partial}{\partial \phi(-p)} \right] e^{-S} \right\} \tag{9.7.28}$$

Notice that the expression is equal to zero if the interaction term L obeys the previous assumption in Eq. (9.7.27).

The meaning of all this is that, if L obeys the assumption Eq. (9.7.27) that we postulated, the generating functional is indeed independent of the cutoff, so the N-point functions are left unchanged. In this sense, the assumption Eq. (9.7.27) we made for L can be taken as the "equation of motion" for L.

Similarly, we can take this formalism and carry it over to the σ model approach to string theory. As in the point particle theory case, we might begin with the action [17]:

$$S = \int \epsilon \partial X_\mu K(\epsilon^2 \partial^2) \epsilon \bar{\partial} X^\mu + S_{int} \tag{9.7.29}$$

where K^{-1} is a smooth function that vanishes at infinite momentum and equals one below the cutoff for $p < 1/\epsilon$.

As before, we demand that the generating functional for strings be independent of the cutoff $1/\epsilon$. We can generalize Eq. (9.7.27). The generating functional is cutoff independent if we make the following assumption:

$$\partial_t S_{int} = \int d^2z \, d^2w \, \partial_t G(z - w) \left[\frac{\delta S_{int}}{\delta X_\mu(z)} \frac{\delta S_{int}}{\delta X^\mu(w)} + \frac{\delta^2 S_{int}}{\delta X_\mu(z) \delta X^\mu(w)} \right] \tag{9.7.30}$$

where $t = \ln \epsilon$ and G is Green's function for the kinetic energy operator $\epsilon^2 \partial^2 K(\epsilon^2 \partial^2)$.

However, since the fields transform under a scale transformation as:

$$X(z)\mu \rightarrow (1/\epsilon) X_\mu(z/\epsilon) \tag{9.7.31}$$

we must also add the following to the right-hand side of the previous equation:

$$\int d^2 z \left(-\frac{1}{2} d + z \partial \right) X_\mu(z) \frac{\delta S_{\text{int}}}{\delta X_\mu} \tag{9.7.32}$$

Let us now try to solve some of these equations, at least in lowest order, to see if they agree with string field theory.

We choose a specific form for the propagator:

$$G(z_1, z_2) = -\ln|z_1 - z_2|^2 \hat{\theta}(|z_1 - z_2|^2 - 1) \tag{9.7.33}$$

where $\hat{\theta}$ is a smeared step function. It is 0 for arguments less than $-\frac{1}{2}$ and rises smoothly to 1 for arguments greater than $\frac{1}{2}$.

We will parameterize the Weyl invariance of the theory with a parameter $v(z)$, so that:

$$\delta z_a = v_a(z); \qquad \partial_a v_b + \partial_b v_a = \delta_{ab} \partial \cdot v \tag{9.7.34}$$

where the last condition is required for conformal invariance.

The point is to take the lowest order approximation to the theory. If we only take linear terms in S_{int}, then we can solve the system and compare it with string field theory. If we make a change of variables to $z + \epsilon v(z)$, then the propagator changes as:

$$\delta G = \left[\frac{v(z_1) - v(z_2)}{z_1 - z_2} + \text{h.c.} \right] \hat{\theta}(1 - |z_1 - z_2|^2) \tag{9.7.35}$$

and the linearized equations for Eq. (9.7.30) become:

$$\int d^2 z_1 \, d^2 z_2 \, \hat{\theta} \frac{v(z_1) - v(z_2)}{z_1 - z_2} \frac{\partial^2 S_{\text{int}}}{\partial X(z_1) \, \partial X(z_2)} + \int d^2 z \, v(z) \partial X \frac{\partial S_{\text{int}}}{\partial X(z)} \tag{9.7.36}$$

Now expand these terms in terms of a Taylor series:

$$v(z) = \sum \frac{v^{(n)} z^n}{n!}$$

$$X_\mu(z) = \sum \frac{X_\mu^{(n)} z^n}{n!} \tag{9.7.37}$$

Performing the integration, we have rewritten Eq. (9.7.36) as:

$$\sum_{n,m} \left[v^{(m)} X^{n+1} \frac{\partial S_{\text{int}}}{\partial X^{(m+n)}} + v^{(m+n+1)} \frac{\partial^2 S_{\text{int}}}{\partial X^{(n)} \, \partial X^{(m)}} \right] = 0 \tag{9.7.38}$$

But, if we write:

$$L_n \equiv \sum_m X^{(m)} \frac{\partial}{\partial X^{(n+m)}} - 2 \sum_{l+m=n} \frac{\partial^2}{\partial X^{(l)} \, \partial X^{(m)}} \qquad (9.7.39)$$

then our equations reduce to [18]:

$$L_n S_{\text{int}} = 0 \qquad (9.7.40)$$

which are the free field equations of string field theory.

The previous results may also be generalized to include ghosts as well as tree interactions, yielding roughly the same results as string field theory. The renormalization group method thus reproduces known results, especially those of string field theory, but has the defect that it cannot, as yet, go beyond known perturbative results. Furthermore, a considerable apparatus must be erected before we can reproduce results that are almost trivial in string field theory. Although the applications of this method have been primitive, the true power of this method has not yet been fully exploited.

9.8. Summary

In the first half of this book, we stressed that the perturbative vacuum of string theory was conformally invariant, and we developed an extensive list of exactly solvable conformal field theories. However, despite enormous progress, there are severe problems with the perturbative approach.

1. The perturbative expansion cannot break supersymmetry, which must be broken at low energies.
2. The dimension of space–time is perfectly stable. Perturbation theory cannot generate compactification.
3. The low-energy perturbative spectrum has problems (that is, dilaton mass).
4. Phenomenologically, although it is easy to get the **27** fermion representation of E_6, no one has yet found a conformal field theory that has *all* the desired properties of our low-energy world.

Thus, we now turn to a discussion of physics beyond the Planck scale, which, in some sense, is the "true home" of string theory. Several observations can be made about this remote region:

1. We can stretch the perturbative approach to the limit and examine compactified vacuums beyond the Planck length. Then, we discover a phenomenon called "duality," meaning that the classical solution is invariant under:

$$R/\sqrt{\alpha'} \to \sqrt{\alpha'}/R \qquad (9.8.1)$$

Philosophically, this means that a rapidly contracting universe, which has collapsed within the Planck length, is indistinguishable from an expanding universe. Thus, the Big Bang or Big Crunch can be viewed as a smooth event. Moreover, it means that there is a new Heisenberg uncertainty principle at work, stating that we cannot penetrate beyond the Planck scale. However, it is almost certain that nonperturbative quantum corrections will destroy this elegant picture at such high energies.

2. We can examine string theory in the high-temperature limit, where symmetry restoration occurs. At high temperatures, it is well known that the number of string states increases exponentially, giving us a limiting value of the temperature (Hagedorn temperature). This means that if we try to add energy to increase the temperature, the extra energies goes into making more particles, and hence, the temperature remains fixed. We can determine the proliferation of states at high energy by examining the partition function, which obeys:

$$P_D(n) \sim A' n^{-n(D+3)/4} \exp\left(2\pi\sqrt{nD/6}\right) \tag{9.8.2}$$

so that the limiting temperature becomes, in mass units,

$$m_0 = \frac{1}{\pi}\sqrt{\frac{3}{2\alpha' D}} \tag{9.8.3}$$

We will examine the free energy beyond the Planck scale:

$$F = -kT \ln Z \tag{9.8.4}$$

where:

$$Z = \int DX \exp\left(-\frac{1}{\hbar g^2}\int_0^{\hbar\beta} d\tau \int d^{D-1}x\, L\right) \tag{9.8.5}$$

We can calculate this to arbitrary orders in perturbation theory. If we expand the integration variable in terms of Abelian differentials ω_i:

$$dX^0 = \partial X^0 + \bar\partial X^0 \tag{9.8.6}$$

where:

$$\partial X^0 = \sum_{i=1}^g \lambda_i \omega_i; \qquad \bar\partial X^0 = \sum_{i=1}^g \bar\lambda_i \bar\omega_i \tag{9.8.7}$$

then the free energy becomes:

$$\frac{F_h}{VT} = \frac{4\pi^2 \alpha' T}{\hbar}\left(\frac{g^2 T^2 4\pi^2 \alpha'}{\hbar}\right)^{h-1} \Lambda_h \tag{9.8.8}$$

Simplifying the expression, we find:

$$F/VT \sim T \qquad (9.8.9)$$

This is a remarkable result, because it shows that the free energy grows slower than ordinary point particle field theory, where we expect the free energy to grow as T^{D-1}. This indicates that, at high temperatures, there is a phase transition to a new region that has fewer degrees of freedom than even a scalar particle theory, that is, finite degrees of freedom, which are typical of topological theories.

3. We can also take the approach of examining the high-energy limit of the theory $\alpha' \to \infty$, where we also expect to see symmetries restored. If we evaluate the first quantized path integral for higher genus, then the functional is dominated by:

$$X^\mu(\xi) = i \sum_i P_i^\mu G_{\hat{\mathbf{m}}}(\xi, \hat{\xi}_i) + O\left(\frac{1}{\alpha'}\right) \qquad (9.8.10)$$

where G is the multiloop Green's function defined on the Riemann surface.

Inserting the previous expression into the path integral, we find:

$$A_h(P_i) = g^{2h+2} \Gamma_h(P_i, \hat{\mathbf{m}}_{\mathbf{i}}, \hat{\xi}_i) \prod_i \tilde{V}_i(X^\mu, \hat{\xi}_j, P_j) + O\left(\frac{1}{\alpha'}\right) \qquad (9.8.11)$$

Taking the saddle point limit, our final result is:

$$A_{a_i}(P_i) = \prod_i \tilde{V}_{a_i}(X_{cl}^\mu D, P_i) A_{\text{tachyon}}(P_i) \left[1 + O(\alpha')\right] \qquad (9.8.12)$$

This is also remarkable, because it points to an infinite symmetry among all N-point scattering amplitudes that holds to any order in perturbation theory. It means that there must be a new symmetry at work that can link all amplitudes in terms of just one amplitude (for example, dilaton scattering). Thus, an enormous symmetry seems to opening at energies beyond the Planck scale.

4. We can also take the approach that perturbation theory is not only the wrong path and unreliable, but that it is also inherently unstable, that is, it is not Borel summable. If we look at the large genus h limit of the path integral:

$$\int d\mu_{\text{WP}} \, g^h Z(2) Z'(1)^{-13} \qquad (9.8.13)$$

we can evaluate the integral as a function of h by showing that the partition functions have a lower bound:

$$Z(2) > Z(1) > Z'(1)c^h \qquad (9.8.14)$$

and that the Weil–Petersson measure $d\mu_{\text{WP}}$ grows as $h!$ Multiplying them together, we find that the amplitudes grow as:

$$c^{3h-3}h!$$ (9.8.15)

so the theory is not Borel summable. (This calculation was done independent of the tachyon, so it is likely, but not proven, that superstring theory will also not be Borel summable.) This negative result could be a blessing in disguise, because it would eliminate the embarrassing possibility of millions upon billions of fully self-consistent perturbatively stable vacuums for strings.

Given the urgent necessity of a nonperturbative approach to string theory, let us list the various nonperturbative approaches that have been developed so far.

1. Universal moduli space is one of the more ambitious programs. It tries to fully exploit the modular properties of the perturbation series. The idea is to develop the theory of Riemann surfaces of an arbitrary and even infinite number of loops. One concrete approach is the Grassmannian formulation, in which a surface of genus g is represented as a single point in Grassmannian space. However, the mathematics of this approach is so difficult that real progress in this area has been slow.

2. The renormalization group approach exploits the fact that the string can move self-consistently only in backgrounds that satisfy the correct equations of motion. The idea is to start in "theory space" (that is, the space of all possible theories) and treat the statement of conformal invariance

$$\beta = 0$$ (9.8.16)

as an equation of motion. Thus, the theory will, via renormalization group flows, tend toward the correct vacuum. To see how this works to lowest order, let us treat string theory as a σ model and calculate the multiloop contributions to β. We make the expansion around a classical background:

$$\frac{d^2 X^\mu}{dt^2} + \Gamma^\mu_{\alpha\beta} \frac{dX^\alpha}{dt} \frac{dX^\beta}{dt} = 0$$ (9.8.17)

where $\Gamma^\mu_{\alpha\beta}$ are the usual Christoffel symbols:

$$\Gamma^\mu_{\alpha\beta} = \frac{1}{2} g^{\mu\nu} \left(g_{\alpha\nu,\beta} + g_{\nu\beta,\alpha} - g_{\alpha\beta,\nu} \right)$$ (9.8.18)

Expanded, we have:

$$X^\mu(t) = x^\mu + \xi^\mu t - \frac{1}{2} \Gamma^\mu_{\alpha\beta} \xi^\alpha \xi^\beta t^2 - \frac{1}{3!} \Gamma^\mu_{\alpha\beta\gamma} \xi^\alpha \xi^\beta \xi^\gamma t^3 + \cdots$$ (9.8.19)

The action then becomes:

$$S = S_0(x) + \int d^2z\, g_{\mu\nu}\, \partial_a x^\mu D^a \xi^\nu$$

$$+ \frac{1}{2} \int d^2x \left(g_{\mu\nu} D_a \xi^\mu D^a \xi^\nu + R_{\mu\alpha_1\alpha_2\nu} \xi^{\alpha_1} \xi^{\alpha_2}\, \partial_a x^\mu\, \partial^a x^\nu \right.$$

$$+ \frac{1}{3} D_{\alpha_1} R_{\mu\alpha_2\alpha_3\nu}\, \xi^{\alpha_1} \xi^{\alpha_2} \xi^{\alpha_3}\, \partial_a x^\mu\, \partial^a x^\nu$$

$$\left. + \frac{4}{3} R_{\mu\alpha_1\alpha_2\alpha_3} \xi^{\alpha_1} \xi^{\alpha_2} D_a \xi^{\alpha_3}\, \partial^a x^\mu + \cdots \right)$$

(9.8.20)

where:

$$D_a x^\mu = \partial_a x^\mu + \Gamma^\mu_{\alpha\beta} \xi^\alpha\, \partial_a \xi^\beta \qquad (9.8.21)$$

The Feynman rules can be read directly off the Lagrangian.

Now, let us apply this expansion to the following first quantized system, where the string moves in a background field specified by:

$$S = \frac{1}{4\pi\alpha'} \int d^2z \left[\sqrt{g}\, g^{ab} G(X)_{\mu\nu}\, \partial_a X^\mu\, \partial_b X^\nu \right.$$

$$\left. + \epsilon^{ab} B_{\mu\nu}\, \partial_a X^\mu\, \partial_b X^\nu - \alpha' \sqrt{g}\, R^{(2)} \Phi(X) \right]$$

(9.8.22)

Treating this as an ordinary two-dimensional theory, we can calculate the multiloops, and the expression for β becomes:

$$\beta^G_{\mu\nu} = R_{\mu\nu} - \frac{1}{4} H_\mu^{\lambda\sigma} H_{\nu\lambda\sigma} + 2\nabla_\mu \nabla_\nu \Phi + O(\alpha')$$

$$\beta^B_{\mu\nu} = \nabla_\lambda H^\lambda_{\mu\nu} - 2(\nabla_\lambda \Phi) H^\lambda_{\mu\nu} + O(\alpha')$$

$$\frac{\beta^\phi}{\alpha'} = \frac{D-26}{\alpha' 48\pi^2} + \frac{1}{16\pi^2} \left[4(\nabla\Phi)^2 - 4\nabla^2\Phi - R + \frac{1}{12} H^2 \right] + O(\alpha')$$

(9.8.23)

Notice that the equation $\beta = 0$ is precisely the same as the usual equations of motion for the background fields. Thus, it is possible to construct the effective action for these fields:

$$\int d^D x \sqrt{G} \left[R - \frac{4}{D-2} (\nabla\Phi)^2 - \frac{1}{12} e^{-8\Phi/(D-2)} H^2 \right] \qquad (9.8.24)$$

This approach, however, is defined only on a surface of genus zero. To find a more general approach, let us use the method of "block spin." Start with a renormalizable theory defined at some cutoff Λ and then lower the cutoff. By integrating the fields from the lower cutoff to the higher cutoff, we get an effective Lagrangian, such that the N-point functions should eventually be all the same.

In point particle physics, we start with a propagator with an explicit cutoff:

$$D_\Lambda(p^2) = \frac{1}{p^2 + m^2} K \left(\frac{p^2}{\Lambda^2} \right); \qquad K = \begin{cases} 1, & p^2 < \Lambda^2 \\ 0, & p^2 > \Lambda^2 \end{cases} \qquad (9.8.25)$$

We then demand that the generating functional remain invariant under changes in the cutoff:

$$\Lambda \frac{dZ(J)}{d\Lambda} = 0 \qquad (9.8.26)$$

where J is the source term. In order to make this happen, it means that the effective Lagrangian must obey:

$$\Lambda \frac{\partial L}{\partial \Lambda} = -\frac{1}{2} \int d^4 p (2\pi)^4 (p^2 + m^2)^{-1} \Lambda \frac{\partial K}{\partial \Lambda}$$
$$\times \left[\frac{\partial L}{\partial \phi(-p)} \frac{\partial L}{\partial \phi(p)} + \frac{\partial^2 L}{\partial \phi(p) \, \partial \phi(-p)} \right] \qquad (9.8.27)$$

This, then, is our starting point, an equation of motion for the effective Lagrangian.

Now apply this to string theory. The effective action for string theory must also obey the same equation. Taking into account the transformation of the string variable under a scale transformation, we have:

$$\partial_t S_{\text{int}} = \int d^2 z \, d^2 w \, \partial_t G(z - w) \left[\frac{\delta S_{\text{int}}}{\delta X_\mu(z)} \frac{\delta S_{\text{int}}}{\delta X^\mu(w)} + \frac{\delta^2 S_{\text{int}}}{\delta X_\mu(z) \, \delta X^\mu(w)} \right]$$
$$(9.8.30)$$

where G is Green's function which corresponds to the kinetic energy operator term $\epsilon^2 \partial^2 K(\epsilon^2 \partial^2)$.

This equation is nonlinear and difficult to solve. However, in the lowest approximation, this system can be solved, and we find:

$$L_n S_{\text{int}} = 0 \qquad (9.8.31)$$

Thus, in this approach, the conditions on the Fock space are now reinterpreted as equations of motion for the effective action. Similarly, the tree amplitudes can also be generated in this fashion. It is thus conjectured that the equations of the renormalization group approach reproduce those of string field theory.

3. String field theory is the most well-developed nonperturbative formalism. It is also the most conservative, as its origin lies in extracting the second quantized action by examining the perturbation series. Although a formidable apparatus has been created in this approach, we cannot, as yet, extract nonperturbative information about the theory. String field theory will be discussed in the next three chapters.

4. The matrix models approach is the only one to provide nonperturbative information concerning strings and two-dimensional gravity. For $D \leq 1$, in fact, because of the KdV hierarchy, the theory is exactly solvable. We can use it in the laboratory to test our ideas about strings, that is, it shows that the perturbation theory is not Borel summable at $D = 1$. However, there seems to be a fundamental block extending this approach beyond $D = 1$. The reason is simple: below $D = 1$, we have

a finite number of degrees of freedom, so the theory is, in some sense, topological and hence solvable. However, beyond $D = 1$, there are an infinite number of degrees of freedom, so the model no longer becomes exactly solvable. In some sense, the model is exactly solvable because it is trivial, that is, it has finite degrees of freedom. Matrix models will be described in a later chapter.

References

1. K. Kikkawa and M. Yamasaki, *Phys. Lett.* **149B**, 357 (1984).
2. N. Sakai and I. Senda, *Prog. Theor. Phys.* **75**, 692 (1984).
3. M. B. Green, J. H. Schwarz, and L. Brink, *Nucl. Phys.* **B198**, 474 (1982).
4. R. Hagedorn, *Nuovo Cim.* **56A**, 1027 (1968).
5. J. J. Atick and E. Witten, *Nucl. Phys.* **B310**, 291 (1988).
6. D. Gross, *Phys. Rev. Lett.* **60**, 1229 (1988).
7. D. Gross and P. F. Mende, *Nucl. Phys.* **B303**, 407 (1988); *Phys. Lett.* **B197**, 129 (1987).
8. D. Gross and V. Periwal, *Phys. Rev. Lett.* **60**, 2105 (1988).
9. D. Friedan and S. Shenker, *Phys. Lett.* **175B**, 287 (1986); *Nucl. Phys.* **B281**, 509 (1987).
10. L. Alvarez-Gaumé, D. Z. Freedman, and S. Mukhi, *Ann. of Phys.* **134** 85 (1981).
11. A. Sen, *Phys. Rev.* **D32**, 2102 (1985); *Phys. Rev. Lett.* **55**, 1846 (1985).
12. T. L. Curtright and C. K. Zachos, *Phys. Rev. Lett.* **53**, 1799 (1984).
13. C. Lovelace, *Nucl. Phys.* **B273**, 413 (1986).
14. C. G. Callan, D. Friedan, E. J. Martinec, and M. J. Perry, *Nucl. Phys.* **B262**, 593 (1985).
15. C. G. Callan, I. R. Klebanov, and M. J. Perry, *Nucl. Phys.* **B278**, 78 (1986).
16. J. Polchinski, *Nucl. Phys.* **B231**, 269 (1983).
17. T. Banks and E. Martinec, *Nucl. Phys.* **B294**, 733 (1987).
18. J. Hughes, J. Liu, and J. Polchinski, UTTG-13-88 (1988).
19. S. R. Das, S. Naik, and S. R. Wadia, *Mod. Phys. Lett.* **A4**, 1033 (1989); S. Jain, *Int. J. of Mod. Phys.* **A3**, 1759 (1988).
20. M. Kaku and K. Kikkawa, *Phys. Rev.* **D10**, 1110, 1823 (1974).
21. E. Witten, *Nucl. Phys.* **B268**, 253 (1986).
22. V. Kazakov, I. Kostov, and A. Migdal, *Phys. Lett.* **157B**, 295 (1985); F. David, *Nucl. Phys.* **B257** [**FS14**], 45, 543 (1985).

Chapter 10

String Field Theory

10.1. First versus Second Quantization

Although the methods of conformal field theory have given us a wealth of possible string vacuums and a framework in which to begin phenomenology, there are still severe deficiencies in this formulation. First, conformal field theory is necessarily a perturbative formulation. It is based on the *first quantized* string model propagating on various compactified manifolds. The problem is that the first quantized functional formulation [Eq. (1.3.1)] is based on the sum over conformally inequivalent Riemann manifolds of genus g, which yields a perturbative series of Feynman diagrams. The success of this formulation is that it yields a finite formulation of gravity interacting with quarks and leptons. However, its drawback is that millions of conformal field theories can be constructed using the methods presented in the previous chapters, and there is absolutely no concrete way in which to choose which, if any, of these millions of vacuums corresponds to our real world.

What is needed, however, is a *second quantized* string field theory [1, 2] that is not necessarily wedded to the sum over Riemann surfaces. We saw that perturbation theory by itself was not sufficient to compactify 10- or 26-dimensional space–time to a realistic four-dimensional manifold. Thus, an entirely new approach is required which allows us to calculate nonperturbative results, which we hope will be able to tell us which of the millions upon millions of conformal field theories are stable and which one, if any, our universe prefers.

Second, it is not clear whether the perturbation series makes any sense. Investigations of the high-energy behavior of the higher order graphs indicate that the perturbation series is not Borel summable. Ordinary gauge theories, such as QED, QCD, or the electro-weak theory, are also not Borel summable, which means that we must treat their perturbation series as an asymptotic one. Although QED, for example, rapidly converges to the correct value of the S matrix for electron photon processes at low orders in the coupling constant, eventually the perturbation series must diverge. This is not a problem for gauge theories, because we can always say that they must be embedded into a more realistic theory of the universe, which is Borel summable.

However, since string theory makes the pretense that it is the unifying theory of the universe, we cannot take refuge by embedding it into a higher theory. The meaning of this is that perturbation theory around conformal field theory is a potentially dangerous path and that the final formulation of string theory must necessarily be nonperturbative.

Third, the first quantized string at higher genus g is actually ill defined because of a century-old problem, the triangulation of moduli space. When we write the "sum over conformally inequivalent surfaces" in the path integral, the sum is actually ambiguous because of the problem of finding specific moduli for higher genus surfaces. The dimension of moduli space is well known, $6g - 6 + 2N$, but finding specific coordinates that implement this triangulation is exceedingly difficult to solve. For the past century, this problem in classical mathematics was unsolved. In the last few years, three triangulations of moduli space have been given.

1. Light cone coordinates: we will discover that the simplest string field theory, the light cone theory, solves this century-old problem in a simple way, via twists, string lengths, and propagation times [3].

2. Harer coordinates: we will find that covariant open string field theory, given by Witten, implements this set of coordinates [4].

3. Penner coordinates: so far, no string field theory can reproduce this set of coordinates [5].

In summary, we find that the first quantized functional in Eq. (1.3.1) is actually not well defined, but that the second quantized theory gives us explicit triangulations of moduli space. In fact, of the three known triangulations of moduli space in the mathematical literature, two of them come from string field theory. (The nonpolynomial theory, to be discussed in the next chapter, gives a fourth triangulation of moduli space, but the full details have yet to be worked out.)

At present, there have been various proposals for a nonperturbative formulation of string theory, which we will discuss. However, the most developed of these formulations is string field theory [1, 2, 6–10], which is a second quantized theory of strings. More speculative nonperturbative formulations will also be presented later. However, one must always stress that the nonperturbative theories of strings, in whatever formulation, is still in its infancy, and none of them have been able to give us the slightest hint of nonperturbative results in four dimensions. None of them can calculate the true vacuum of string theory. The problem, apparently, is that we still do not understand the underlying geometry of string theory. At present, it still resembles a seemingly random collection of folklore and rules of thumb.

For the general theory of relativity, the equivalence principle enabled Einstein to see that general covariance lay at the foundation of any theory of gravity. Then, it was straightforward to find the mathematical language in which to formulate the equivalence principle and general covariance.

At the present time, the string counterparts of the equivalence principle and general covariance are still not known, and hence, this is the heart of the problem in finding the right framework in which to formulate the theory. In this sense, string theory has been evolving backward, ever since its accidental discovery in 1968. With these remarks, we now begin with a discussion of string field theory, and the difference between first and second quantization.

A first quantized theory is formulated in terms of the *coordinates* describing a particle's motion. For a point particle, for example, its relativistic action is given by the invariant length swept out by its path. Let $x_\mu(\tau)$ represent a vector that points from the origin to the location of a particle. As the particle moves, it sweeps out a line, parameterized by τ. The action is:

$$S = -m \int d\tau \sqrt{\left(-\frac{dx_\mu}{d\tau}\right)^2} \sim \text{length} \qquad (10.1.1)$$

which is invariant under reparameterizations of the path:

$$\tau \to \bar{\tau}(\tau) \qquad (10.1.2)$$

This reparametrization invariance allows us to select a particular gauge choice, which we may choose to be:

$$x_0 = \tau \qquad (10.1.3)$$

in which case the action assumes the familiar nonrelativistic form in the limit of small velocities:

$$S \sim \int d\tau \, \frac{1}{2} m v_i^2 \qquad (10.1.4)$$

The advantage of this nonrelativistic formulation is that the theory is manifestly ghost free because all references to $x_0(\tau)$ have been eliminated.

The scattering amplitudes are defined by imposing, from the outside, the set of topologies over which the particle interacts, and then taking the Fourier transform:

$$A_N = \sum_{\text{Topologies}} \int d\mu \, Dx_\mu(\tau) e^{i \int d\tau L} \, e^{i \sum_j p_j \cdot x_j} \qquad (10.1.5)$$

where the sum over topologies represents a sum over predetermined Feynman graphs.

Several conclusions can be immediately drawn from this relatively simple example.

1. The first quantized formulation is necessarily perturbative. We must impose from the outside the set of Feynman paths over which to integrate, and each set of graphs represents a certain order in the perturbation theory. Nonperturbative phenomena cannot be seen to any finite order in perturbation theory.

2. The counting and coefficient of each graph is not clear. It is ambiguous which weights we assign to the various graphs and which graphs are included and which are excluded.

3. The first quantized formulation is not manifestly unitary. Although the free theory, via gauge fixing, can be seen to be totally free of ghosts, it is not clear that the final perturbative S matrix is unitary. (Hopefully, the constraint of unitarity will eventually fix the counting of all graphs.)

4. The first quantized formulation is basically on the mass shell. Thus, some of the most interesting questions are out of reach of the first quantized theory.

For string theory, there is also an additional complication at the level of perturbation theory. In principle, the "sum over all conformally inequivalent Riemann surfaces" appears to be an elegant statement of how string perturbation theory is constructed. However, it tells us virtually nothing about how to set up coordinates for these genus g surfaces. In fact, as we mentioned earlier, choosing moduli that can triangulate the higher Riemann surfaces is a notoriously difficult mathematical problem, dating back to the time of Riemann.

To remedy all these difficulties, we now pass to the second quantized theory. The first quantized formulation was based on the coordinates x_μ, which describe the motion of a point particle. The transition to the second quantized formulation begins when we introduce a *field* $\phi(x)$, which is a function of the coordinates.

In contrast to the first quantized string theory, the second quantized string field theory has an explicit dependence on Φ^3 or higher terms, meaning that the interactions are all fixed ahead of time. The weights and measures are hence uniquely fixed, and unitarity to all orders in perturbation theory is almost trivial to show. Because of the presence of explicit interaction terms, we now have an explicit triangulation of moduli space in terms of string field theory. Thus, a century-old mathematics problem, finding the correct moduli for genus g Riemann surfaces, is almost trivially solved.

The second quantized theory is also inherently an off-shell theory, so symmetry breaking, in principle, can be investigated by the theory. In addition, one does not have to resort to perturbation theory. In fact, quantum field theory is the only formalism in which a variety of techniques have been developed to handle nonperturbative phenomena.

In Chapter 1, we stressed that there are at least three ways in which a point particle or a string can be quantized, the Gupta–Bleuler approach (where Lorentz covariance is maintained and ghosts are allowed to propagate, but the physical states must be ghost free), the light cone approach (where the theory is formulated entirely in terms of ghost-free physical states, but Lorentz invariance must be carefully checked), and the BRST approach (where covariance and unitarity are maintained by ensuring that the physical states are BRST invariant.)

We start by trying to compute the canonical momenta corresponding to Eq. (10.1.1). We find that its momenta are not independent, but constrained:

$$p_\mu = \delta L/\delta \dot{x}_\mu, \qquad p_\mu^2 + m^2 \equiv 0 \qquad (10.1.6)$$

In the Gupta–Bleuler approach, we apply the constraint directly on the fields:

$$(p_\mu^2 + m^2)\phi(x) = 0 \qquad (10.1.7)$$

which is just the usual Klein–Gordon equation. This equation, in turn, can be derived from the standard covariant second quantized action:

$$S = \frac{1}{2} \int d^4x \, \phi(x) [\partial_\mu \partial^\mu - m^2] \phi(x) \qquad (10.1.8)$$

The next method is the light cone approach, where ghosts are explicitly eliminated. We start with the gauge invariant action:

$$S = \int d\tau \left[p_\mu \dot{x}^\mu - \frac{1}{2} e(p_\mu^2 + m^2) \right] \qquad (10.1.9)$$

which is invariant under:

$$\delta x_\mu = \epsilon \dot{x}_\mu, \qquad \delta p_\mu = \epsilon \dot{p}_\mu, \qquad \delta e = d(\epsilon e)/d\tau \qquad (10.1.10)$$

By calculating the equations of motion for e and p_μ and then eliminating them, we retrieve the usual first quantized action in terms of x_μ alone.

Let us now, however, select the light cone gauge:

$$x^+ = \tau \qquad (10.1.11)$$

and solve explicitly for p^- via the constraint

$$p_\mu^2 + m^2 = p_i^2 - 2p^-p^+ + m^2 = 0 \qquad (10.1.12)$$

If we apply the gauge on the action, we find that the term $p^- \dot{x}^+$ becomes p^-, that is, p^- is the new Hamiltonian in the light cone gauge. Thus, solving the constraint, we find that the Hamiltonian is given by:

$$H \equiv p^- = \frac{1}{2p^+}(p_i^2 + m^2) \qquad (10.1.13)$$

Let us now take the Fourier transform of the field with respect to x_-, so that the field becomes $\phi_{p^+}(x_i)$. Then, the equation of motion for the field, which is a function of only transverse fields, becomes:

$$\left\{ i \frac{\partial}{\partial \tau} - H \right\} \psi_{p^+}(x_i) = 0 \qquad (10.1.4)$$

and the new second quantized action is given by the Schrödinger-like equation:

$$\int Dx_i \, dp^+ \, \psi_{p^+}^\dagger(x_i) \left\{ i \frac{\partial}{\partial \tau} - H \right\} \psi_{p^+}(x_i) \qquad (10.1.15)$$

10.2. Light Cone String Field Theory

Next, we make the transition to the free string and find that there is a remarkable correspondence between the point particle and the string approach. In fact, at the free level, the equations can be practically transported from one to the other. (The major complication, we shall see, comes at the level of the interactions, which are highly nontrivial.)

The field theory of strings is based on $\Phi(X)$, which is a functional, that is, it is a function of every point $X_\mu(\sigma)$ along the string for all possible values of σ. Thus, the expression $\Phi[X(\sigma)]$ is actually incorrect. The correct functional dependence is given by:

$$\Phi(X_\mu) = \Phi\left[X_\mu(\sigma_1), X_\mu(\sigma_2), \ldots, X_\mu(\sigma_N)\right] \qquad (10.2.1)$$

where we let $N \to \infty$.

We can also decompose this string functional in any basis we wish. The most convenient basis contains Hermite polynomials. We can write:

$$\Phi(X) = \langle X | \Phi(x_0) \rangle \qquad (10.2.2)$$

where:

$$|\Phi(x_0)\rangle = \phi(x_0)|0\rangle + A_\mu(x_0)a_1^{\mu\dagger}|0\rangle + g_{\mu\nu}a_1^{\mu\dagger}a_1^{\nu\dagger}|0\rangle + \cdots \qquad (10.2.3)$$

where x_0 is the usual four vector representing ordinary space–time. Here, we see the explicit decomposition of the field functional in terms of the tachyon field $\phi(x_0)$, the Maxwell field $A_\mu(x_0)$, a massive graviton field $g_{\mu\nu}(x_0)$, etc.

Let us now construct free actions for the string field in each of the three gauge formalisms that we studied in Chapter 1. In the Gupta–Bleuler approach, we wish to impose the following conditions:

$$P_\mu^2 + \frac{X_\mu'^2}{\pi^2} = 0, \qquad P_\mu X'^\mu = 0 \qquad (10.2.4)$$

By taking the Fourier moments of these constraints, we arrive at:

$$L_n|\phi\rangle = 0, \qquad (L_0 - 1)|\phi\rangle = 0 \qquad (10.2.5)$$

We shall interpret the second of these constraints as the propagator of string states, so our action becomes:

$$S = \int DX_\mu \, \Phi(X)\{L_0 - 1\}\Phi(X) \qquad (10.2.6)$$

subject to the constraint that $L_n\Phi(X) = 0$, and where:

$$DX_\mu \equiv \prod_\mu \prod_n dX_{n\mu} = \prod_\mu \prod_\sigma dX_\mu(\sigma) \qquad (10.2.7)$$

Although the Gupta–Bleuler formalism is quite elegant, in actual practice, the elimination of the Virasoro constraints requires very difficult projection operators, whose complexity precludes their widespread application.

The light cone gauge, because it has eliminated all ghost modes, does not suffer from this difficulty. In the light cone gauge, we start with the first quantized action:

$$S = \int d\tau \, d\sigma \left[P_\mu \dot{X}^\mu - \lambda \left(P_\mu^2 + \frac{X_\mu'^2}{\pi^2} \right) + \rho \left(P_\mu X'^\mu \right) \right] \qquad (10.2.8)$$

As before, by eliminating λ, ρ, and P_μ via their equations of motion, we can show that the action is equal to the area swept out by the string in Eq. (1.2.7).

We wish to impose:

$$X^+ = p^+ \tau \qquad (10.2.9)$$

While solving explicitly for the constraints [Eq. (10.2.9)], we find that, as before, P^- becomes the new Hamiltonian:

$$H = \int_0^\pi d\sigma \, P^-(\sigma) = \frac{1}{2p^+} \int d\sigma \left(P_i^2 + \frac{X_i'^2}{\pi^2} \right) \qquad (10.2.10)$$

and the equation of motion therefore becomes:

$$\left(i \frac{\partial}{\partial \tau} - H \right) \Phi_{p^+}(X_i) = 0 \qquad (10.2.11)$$

and the free action becomes [1]:

$$\int DX_i \, dp^+ \, \Phi_{p^+}^\dagger(X_i) \left(i \frac{\partial}{\partial \tau} - H \right) \Phi_{p^+}(X_i) \qquad (10.2.12)$$

To generalize the light cone theory to interactions, however, requires a nontrivial extension of our results for the free theory. Some of the pioneers in quantum physics, such as Heisenberg and Yukawa, spent years trying to devise a nonlocal quantum theory. However, the problem with nonlocal theories is that they inevitably violate causality or relativity. When one vibrates one point in space, the interactions in these nonlocal theories travel faster than the speed of light.

The light cone theory solves this perplexing question of maintaining both causality and relativity. The theory is not a nonlocal theory, in the usual sense, but it is a *multilocal* theory. Interactions do not violate causality or Lorentz covariance because strings break instantaneously, and the interactions travel down the strings at speeds less than or equal to the speed of light.

The light cone interacting theory is based on the observation that open strings interact by breaking instantaneously at one point, or by reforming

at their endpoints with other open strings. In Fig. 10.1a, for example, we see
the topology of the scattering of several strings in the light cone gauge, such
that strings can only break in their interiors or reform at their endpoints.
To actually see that this yields the usual N-point amplitudes, consider
the conformal map that takes the upper half plane and maps it into the
configuration in Fig. 10.1b.

We take the Mandelstam map [11]:

$$\rho(z) = \sum_{i=1}^{N} \alpha_i \ln(z - z_i) \tag{10.2.13}$$

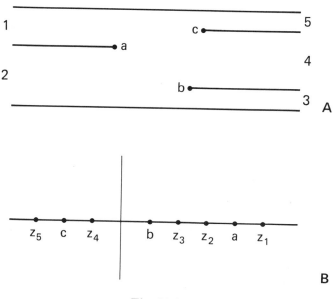

Fig. 10.1.

Let us now derive this conformal map, which takes us from the upper
half z plane to the complex $\rho = \tau + i\sigma$ plane. The simplest way is via the
Schwarz–Christoffel transformation.

We recall that the Schwarz–Christoffel equation transformations the z
plane into a polygon. Specifically, we wish to map the real axis onto the
perimeter of this polygon.

For an n-sided polygon, let z_i be n points along the real z axis. Each
z_i will be mapped to a point ρ_i in the ρ plane, which corresponds to the
corners of the polygon. Let α_i equal the interior angle of a corner of a

polygon at point ρ_i. For a square, for example, this angle is equal to $\pi/2$. Then, the map that takes us from the upper half z plane to the complex ρ plane is given by:

$$\frac{d\rho(z)}{dz} = k \prod_{i=1}^{N} (z - z_i)^{\alpha_i/\pi - 1} \tag{10.2.14}$$

or:

$$\rho(z) = k \int_{z_o}^{z} d\tilde{z} \prod_{i=1}^{N} (\tilde{z} - z_i)^{\alpha_i/\pi - 1} \tag{10.2.15}$$

where we have the condition:

$$\sum_{i=1}^{N} \alpha_i = (N - 2)\pi \tag{10.2.16}$$

To understand the last condition, let us take the limit as $z \to \infty$. In this limit, we wish the function ρ to be finite, therefore the exponent of z in this limit must be zero, which explains the previous condition.

We notice that as $z \to a_i$, the mapping becomes

$$d\rho(z)/dz \sim k(z - z_i)^{\alpha_i/\pi - 1} \tag{10.2.17}$$

Near z_i, let us assume that

$$z - z_i \sim \epsilon e^{i\theta} \tag{10.2.18}$$

for small ϵ.

We see that a line on the real axis that approaches z_i, hops over it via a small hemisphere, and then continues along the real axis is mapped into a bent line in the ρ plane and rotated by angle $\pi(\alpha_i/\pi - 1)$, which forms an interior corner of the polygon with angle α_i, as desired.

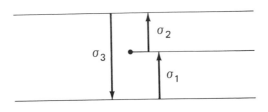

Fig. 10.2.

Next, we will write the interaction Lagrangian for the light cone string field theory. For open strings, an examination of Fig. 10.2 shows that strings can join at their endpoints (or break at an interior point). The interaction

Lagrangian is thus a Φ^3 term, with a Dirac delta function sandwiched in between [1]:

$$S_3 = \int \prod_{i=1}^{3} dp_r^i \, DX_i \, \delta \left(\sum_{r=1}^{3} p^{+r} \right) \delta_{123} \, \Phi_1(X_1)\Phi_2(X_2)\Phi_3^\dagger(X_3) + \text{h.c.}$$

(10.2.19)

where:

$$\delta_{123} = \prod_{i=1}^{2} \prod_{0<\sigma_i<\pi\alpha_i} \delta\left[X_3(\sigma_3) - \theta(\pi\alpha_1 - \sigma)X_1(\sigma_1) - \theta(\sigma - \pi\alpha_1)X_2(\sigma_2)\right]$$

(10.2.20)

where the string variables are defined as:

$$\begin{aligned}
\sigma_1 &= \sigma, & 0 &\leq \sigma \leq \pi\alpha_1 \\
\sigma_2 &= \sigma - \pi\alpha_1, & \pi\alpha_1 &\leq \sigma \leq \pi(\alpha_1 + \alpha_2) \\
\sigma_3 &= \pi(\alpha_1 + \alpha_2) - \sigma, & 0 &\leq \sigma \leq \pi(\alpha_1 + \alpha_2)
\end{aligned}$$

(10.2.21)

with the condition $\sum \alpha_i = 0$. Using the formalism developed by Mandelstam for light cone diagrams, we can then show that the above interaction is sufficient to derive most of the interacting amplitudes of string theory [11, 12].

The full open string theory, however, is more complicated than the closed string case in the light cone gauge. We necessarily must add four-string interactions and higher point interactions to the open string action. If we let Φ (Ψ) represent open (closed) strings, then the interactions for the open and closed strings symbolically have the structures [1]:

$$\begin{aligned}
L_{\text{open}} &= \Phi^3 + \Phi^4 + \Phi^2\Psi + \Psi^3 + \Phi\Psi \\
L_{\text{closed}} &= \Psi^3
\end{aligned}$$

(10.2.22)

In other words, the open string vertex function by itself cannot generate all string amplitudes, so we must necessarily include closed strings as well. Thus, even if we started out with an open string theory without any gravitons, we find that gravitons necessarily creep back into the theory. There is no choice: string theory is by its very nature a theory of quantum gravity.

There are several ways to see why the open string theory has five interactions and the closed string theory is cubic. The most direct way is to examine the string amplitudes to see if the postulated interactions reproduce the string theory. Let us take the real part of the Mandelstam map:

$$\operatorname{Re}\rho(z) = \tau = \sum_{i=1} \alpha_i \ln|z - z_i|$$

(10.2.23)

Notice that lines of equal τ correspond to equipotential lines created by charged sources placed at z_i with charges proportional to α_i. However, lines of equal τ in the ρ plane trace the topology of the interacting string.

Thus, by graphically examining the equipotential lines formed by charges placed on the perimeter of a circle or on the real line, we can trace the topology of interacting strings. The real part of the Mandelstam map is a map that takes the equipotential lines defined on a disk or the upper half plane and maps them into the vertical lines in the ρ plane.

By examining Fig. 10.3, we see that all five interactions must be present in the interacting string action. It is not hard to write the explicit form for all five interactions, since each is given by a Dirac delta function that describes the topological change described by the interaction. Thus, there is a finite region of moduli space that generates equipotential lines that cannot be described by three-string open vertices. This means that the integration region of the Koba–Nielsen variables z_i cannot be filled completely if we use only three-string vertices, that is, there are missing regions of the integration region that can only be filled by postulating four-string and higher interactions.

However, the closed string situation is dramatically different. We find that with cubic interactions alone, we can completely fill the integration region of the Shapiro–Virasoro amplitude. This is highly nontrivial because it gives us the first triangulation of moduli space in over a century, thereby solving a long-standing mathematical problem dating back to Riemann. The amplitudes generated by the light cone string theory have automatically subtracted the redundancy due to the mapping class group.

It may seem strange that the open and closed string light cone interactions have such different characteristics in string field theory. We will see that this situation becomes much more complicated with midpoint interactions, and that the closed string theory becomes nonpolynomial. One suspects that there must be a deeper, group theoretical reason why open and closed interactions have such startling different structures. We will see the reason for this in Chapter 12, when we discuss geometric string field theory.

10.3. Free BRST Action

The third method of passing from a first to a second quantized action is via the BRST method. Instead of quantizing the action expressed as the length of a point world line, we will quantize the following action instead:

$$S = \int d\tau \left(e^{-1}\dot{x}^2 - em^2 \right) \qquad (10.3.1)$$

which is invariant under reparametrization invariance, given by:

$$\delta x_\mu = \epsilon \dot{x}_\mu, \qquad \delta e = d(\epsilon e)/d\tau \qquad (10.3.2)$$

where e is a one-dimensional metric tensor on the world line of a moving point particle. We will choose the gauge:

Φ^4:

$\Phi\Psi$:

$\Phi^2\Psi$:

Ψ^3:

Φ^3:

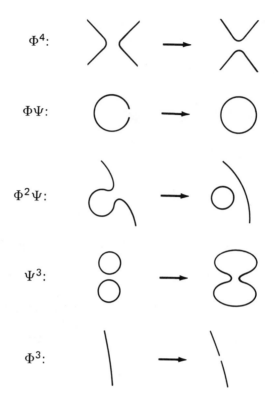

Fig. 10.3.

$$e = 1 \qquad (10.3.3)$$

and calculate the Faddeev–Popov ghost action that emerges from this choice. The Faddeev–Popov determinant is given by:

$$\Delta_{FP} = \det |\partial_\tau| = \int D\theta \, D\bar{\theta} \, \exp i \int d\tau \, \bar{\theta} \, \partial_\tau \theta \qquad (10.3.4)$$

Then, we calculate the BRST operator:

$$Q = \theta(\partial_\mu \partial^\mu - m^2) \qquad (10.3.5)$$

which must be applied onto physical states:

$$Q|\phi(x,\theta)\rangle = 0 \qquad (10.3.6)$$

Solving this constraint, one is left once again with only physical states. So, a natural choice for the BRST invariant action is given by:

$$\int Dx_\mu \, d\theta \, d\bar{\theta} \; \phi Q \phi \qquad (10.3.7)$$

which is invariant under:

$$\delta|\phi\rangle = Q|\lambda\rangle \qquad (10.3.8)$$

because Q is nilpotent.

Now, let us generalize our discussion to strings. Let us first take the conformal gauge:

$$g^{ab} = \delta^{ab} \qquad (10.3.9)$$

which eliminates reparameterization invariance and also local scale invariance. The Faddeev–Popov ghost factor can be exponentiated by introducing two sets of anticommuting ghosts, b and c. The resulting first quantized action becomes:

$$L = \frac{1}{\pi}\left(\partial_z X_\mu \, \partial_{\bar{z}} X^\mu + b\,\partial_{\bar{z}} c + \bar{b}\,\partial_z \bar{c}\right) \qquad (10.3.10)$$

which possesses a global BRST symmetry. Its generator is given by Q, so that the physical states must satisfy:

$$Q|\Phi(X,b,c)\rangle = 0 \qquad (10.3.11)$$

At this point, there are two possible BRST actions. The first is the straightforward generalization of the Gupta–Bleuler formalism, where we have [13–16]:

$$L = \langle\Phi(X,b,c)|(L_0^X + L_0^{gh} - 1)|\Phi(X,b,c)\rangle \qquad (10.3.12)$$

where we have explicitly split the L_0 operator in terms of its string and ghost oscillators, and where $\Phi(X,b,c)$ can have any ghost number.

However, there is also a second BRST formalism in which the gauge invariant is manifest. Let us choose the action [2]:

$$S = \int DX \, Db \, Dc \, D\bar{b} \, D\bar{c} \; \Phi Q \Phi \qquad (10.3.13)$$

where Φ has fixed ghost number $-\frac{1}{2}$. This field is therefore a truncation of the field in the previous BRST formalism.

The advantage of this second approach is that one can see explicitly the gauge invariance of the theory. The theory is invariant under:

$$\delta\Phi = Q\Lambda \qquad (10.3.14)$$

because Q is nilpotent.

To analyze the states within $|\Phi\rangle$, let us write the double vacuums $|\pm\rangle$ of the zero modes of the b and c oscillators:

$$c_0|+\rangle = 0; \qquad b_0|-\rangle = 0 \qquad (10.3.15)$$

where $|+\rangle = c_0|-\rangle$. Then, the field $|\Phi\rangle$ can be decomposed into two pieces:

$$|\Phi\rangle = \psi|-\rangle + \phi|+\rangle \qquad (10.3.16)$$

where the ψ and ϕ have no zero modes.

Let us now gauge fix the BRST gauge invariant theory in order to obtain the other two gauge fixed formalisms. We will first show that the BRST gauge invariant theory is, in fact, equivalent on shell to the light cone theory, which is defined totally in terms of physical transverse states [17, 18]. Also, we will discuss covariant gauges as well.

We begin by noting that any operator E, that can be written as a BRST commutator:

$$E = [Q, S]_\pm \qquad (10.3.17)$$

automatically vanishes on the BRST invariant states of the covariant field theory. To see this, we note that a BRST invariant state is one in which:

$$\begin{aligned} Q|\Phi\rangle &= 0 \\ |\Phi\rangle &\neq Q|\Lambda\rangle, \qquad \text{for all} \quad |\Lambda\rangle \end{aligned} \qquad (10.3.18)$$

Let us now apply E onto a BRST invariant state:

$$E|\Phi\rangle = QS|\Phi\rangle \mp SQ|\Phi\rangle = QS|\Phi\rangle = Q|\Lambda\rangle = 0 \qquad (10.3.19)$$

Thus, E annihilates BRST invariant states, modulo a gauge transformation.

Now, we use the fact that there exists an operator of this type that can be written as:

$$E \equiv N_T - N \qquad (10.3.20)$$

where N_T is the level number of a transverse state and N is the level number. The statement that

$$E|\Phi\rangle = 0 \qquad (10.3.21)$$

means that $N_T = N$ on such states or that $|\Phi\rangle$ is purely transverse. Our task, therefore, is to explicitly construct operators E and S that satisfy both Eqs. (10.3.17) and (10.3.20).

Let us define the following operators:

$$D_n = \oint \frac{dz}{2\pi i z} \frac{z^n}{k \cdot P(z)} \qquad (10.3.22)$$

where $P_\mu(z) = \sqrt{\pi} i z (d/dz) X_\mu(-i \ln z, 0)$. Then, it is possible to show the following explicit expressions for E and S:

$$E = (D_0 - 1)L_0 + \sum_{n=1}^{\infty} (D_{-n} L_n + L_{-n} D_n) - \sum_{n=1}^{\infty} n(c_{-n}\bar{c}_n + \bar{c}_{-n}c_n)$$

$$S = \sum_{-\infty}^{\infty} \bar{c}_{-n}(D_n - \delta_{n,0})$$

$$\qquad (10.3.23)$$

One can prove that E can be written in terms of transverse operators, as in Eq. (10.3.20), as well as in terms of D_n. Notice that both forms of E obey the following commutation relations:

$$[L_m, E] = -mL_m$$
$$[D_m, E] = -mD_m \qquad (10.3.24)$$
$$[V_m^i, E] = 0$$

where V_m^i is an operator that creates or destroys transverse states. However, it can be shown that the $\{D_m, L_m, V_m^i\}$ is a complete set of operators for the Fock space, and hence the two expressions for E in Eqs. (10.3.20) and (10.3.23) must be the same.

To apply this argument to string field theory, we notice that we can split the field $|\Phi\rangle$ into two pieces:

$$|\Phi\rangle = |\Phi\rangle_{\rm T} + |\Phi\rangle_{\rm L} \qquad (10.3.25)$$

where we have split the field into transverse (T) and longitudinal and ghost (L) sectors. By the gauge invariance $\delta|\Phi\rangle = Q|\Lambda\rangle$, we can (up to states that vanish on shell) choose a gauge that removes the longitudinal and ghost states $|\Phi\rangle_{\rm L}$.

Thus, the only part that is left after gauge fixing is:

$$\int DX_i \, Dc^0 \langle \Phi_{\rm T} | c^0 (L_0 - 1)_{\rm T} | \Phi \rangle_{\rm T} \qquad (10.3.26)$$

up to terms that vanish on shell. The integration over the ghost c^0 is trivial, so we are left with the usual light cone action in Eq. (10.2.12). (In the proof, we had to throw away pieces that vanished on shell. This means that off shell the BRST and light cone theories are actually *different*. However, this makes no difference to our discussion because we only need the equivalence on shell. Thus, when we calculate the expectation value between sets of BRST invariant asymptotic states at infinity, the actions for the BRST and light cone theory are different off shell, but they produce the same on shell matrix elements.)

Last, we wish to show that we can choose a covariant gauge so that the gauge invariant BRST action [Eq. (10.3.13)] becomes the gauge-fixed BRST action [Eq. (10.3.12)]. Let us choose the Siegel gauge [13]:

$$b_0|\Phi\rangle = 0 \qquad (10.3.27)$$

This eliminates about half the states within the field in Eq. (10.3.16).

However, whenever we fix any local gauge invariance, we must also add the Faddeev–Popov determinant factor. The Faddeev–Popov ghost determinant factor can be written as:

$$\langle \bar{\Lambda} | b_0 Q | \Lambda \rangle \qquad (10.3.28)$$

which arises from the gauge invariance $\delta|\Phi\rangle = Q|\Lambda\rangle$. Since $|\Phi\rangle$ has ghost number $-\frac{1}{2}$ and Q has ghost number one, this means that $|\Lambda\rangle$ has ghost number $-\frac{3}{2}$ and $\langle\bar{\Lambda}|$ has ghost number $\frac{3}{2}$.

Notice, however, that this ghost action, in turn, has its own gauge invariance:

$$\delta|\Lambda\rangle = Q|\Lambda_1\rangle \qquad (10.3.29)$$

which means that we must add yet another ghost term to the action, with ghost number equal to $-\frac{5}{2}$. In fact, every gauge fixing, in turn, yields yet another ghost action with a gauge invariance. This is the "ghosts-for-ghosts" effect, which introduces an infinite number of fields with differing values of the ghost number [19] [see Appendix I for a discussion of how to solve this problem].

The net effect of all of this is simple: we can introduce a single field $|\Phi\rangle$, which has *arbitrary* ghost number, such that the action is [13]:

$$\langle\Phi|(L_0^X + L_0^{bc} - 1)|\Phi\rangle \qquad (10.3.30)$$

which is just the gauge fixed action found earlier from Gupta–Bleuler quantization in Eq. (10.3.12).

10.4. Interacting BRST String Field Theory

We have seen the remarkable economy of string field theory at the free level. The entire theory of free open bosonic strings can be encapsulated into one simple action $\langle\Phi|Q|\Phi\rangle$. However, the situation with interactions is considerably more complicated. As we mentioned, the first quantized string theory summed over the set of all conformally inequivalent topologies. This conveniently concealed many difficult questions concerning how to place coordinates on Riemann surfaces. The principle problem is that, until recently, mathematicians have been unable to successfully triangulate moduli space for genus g Riemann surfaces, even after a century of experience with these surfaces. Remarkably, string field theory gives an exact triangulation of moduli space, thus solving a long-standing mathematical problem.

Let us begin our discussion by first requiring that open string field theory be a gauge theory that satisfies the axioms of gauge theory. Specifically, we need to postulate the existence of a derivative Q and a product operation $*$. We postulate the following five axioms [2]:

1. the existence of nilpotent derivative Q such that $Q^2 = 0$;
2. The associativity of the $*$ product:

$$[A * B] * C = A * [B * C] \qquad (10.4.1)$$

3. the Leibnitz rule:

$$Q[A * B] = QA * B + (-1)^{|A|} A * QB \qquad (10.4.2)$$

4.   the product rule:

$$\int A * B = (-1)^{|A||B|} \int B * A \qquad (10.4.3)$$

5.   the integration rule:

$$\int QA = 0 \qquad (10.4.4)$$

where $(-1)^{|A|}$ is -1 if A is Grassman odd and $+1$ if A is Grassmann even.

We postulate that the field A has the following transformation rule:

$$\delta A = Q\Lambda + A * \Lambda - \Lambda * A \qquad (10.4.5)$$

Then we can construct a curvature form given by:

$$F = QA + A * A \qquad (10.4.6)$$

such that:

$$\delta F = F * \Lambda - \Lambda * F \qquad (10.4.7)$$

It is easy therefore to show that the following is a total derivative:

$$\int F * F = \int Q \left[A * QA + \frac{2}{3} A * A * A \right] \qquad (10.4.8)$$

Therefore, the Chern–Simons form is gauge invariant [2]:

$$L = A * QA + \frac{2}{3} A * A * A \qquad (10.4.9)$$

The Chern–Simons form is preferable to the usual F^2 form found in ordinary gauge theory, because Q already has two derivatives contained within it.

     This formalism works for any gauge theory, not just strings. Our task is to find a multiplication operation that satisfies the postulates of the $*$ product. Then, gauge invariance is automatic, without any more work.

     We notice, first of all, that the $*$ operation is symmetric in all three strings. There is only one unique configuration that is symmetrical in all three fields, and that is given in Fig. 10.4, where the midpoint of the strings has been singled out.

     The multiplication operation:

$$|X_3\rangle = |X_1\rangle * |X_2\rangle \qquad (10.4.10)$$

simply means that we have exchanged the Fock spaces of strings 1 and 2 for string 3, such that the points along 1 and 2 have been identified with points along string 3. In analogy with Eq. (10.2.19) in the light cone theory, we will define the triple product (without ghosts) as a delta function:

$$\Phi * \Phi * \Phi = \int DX_1\,DX_2\,DX_3\,\Phi(X_1)\Phi(X_2)\Phi(X_3)$$

$$\times \prod_{r=1}^{3} \prod_{0 \le \sigma_r \le \pi/2} \delta\big[X_{r,\mu}(\sigma_r) - X_{r-1,\mu}(\pi - \sigma_{r-1})\big]\cdots$$

$$(10.4.11)$$

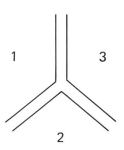

Fig. 10.4.

Let us now write the ghost number for all the operators in the theory. The c ghost has ghost number 1, the b ghost has ghost number -1, so that Q has ghost number 1. This, in turn, fixes the ghost number of the A field to be $-\frac{1}{2}$, since the action contains a term $\langle A|Q|A\rangle$, which must have total ghost number 0.

The ghost number of the gauge parameter Λ and the $*$ operation can be fixed by observing the variation of the A field in Eq. (10.4.5). In order for the left-hand side (with ghost number $-\frac{1}{2}$) to equal the ghost number of the right-hand side, the ghost number of Λ must be $-\frac{3}{2}$ and the ghost number of the $*$ operation must be $+\frac{3}{2}$.

Similarly, we can fix the ghost number of the $\int$ operation by demanding that the action have total ghost number zero. Putting everything together, we have the following set of ghost numbers [2]:

$$
\begin{array}{llll}
c: & 1 & *: & \dfrac{3}{2}\\[2mm]
b: & -1 & & \\[1mm]
Q: & 1 & \displaystyle\int: & -\dfrac{3}{2}\\[2mm]
A: & -\dfrac{1}{2} & \Lambda: & -\dfrac{3}{2}
\end{array}
\qquad (10.4.12)
$$

To enforce these ghost numbers poses no problem. On the A field, this means taking all possible sums of monomials constructed out of the c and b oscillators acting on the vacuum and then projecting out the $-\frac{1}{2}$ ghost number part.

More difficult is the $*$ operation. In addition to the Dirac delta functional for the X_μ in Eq. (10.4.11), we must also include the ghost part as well. The major complication (which becomes more severe as we progress to the superstring) is that there must be ghost insertion operators placed at the midpoint of the vertex function. This is because there is an anomaly in the ghost current.

There are two ways to correct for this, depending on whether we bosonize the ghost fields or not. If we bosonize the ghosts, we must note that the energy-momentum tensor for the bosonized field has a screening charge. To see this, let us bosonize the b, c ghost system with a weight 0 field σ, with

$$c = : e^{\sigma} :, \qquad b = : e^{-\sigma} : \qquad (10.4.13)$$

Let us calculate the energy-momentum tensor for this field, repeating our earlier discussion in Eqs. (3.1.29)–(3.1.33). We have:

$$T(z) = -\frac{1}{2}\partial\sigma\,\partial\sigma + k\,\partial^2\sigma \qquad (10.4.14)$$

and we take its operator expansion product to calculate its central charge:

$$T(w)T(z) \sim 2\left(\frac{1}{2}\right)^2 \langle\partial_w\sigma\,\partial_z\sigma\rangle + k^2\langle\partial_w^2\sigma\,\partial_z^2\sigma\rangle \sim \frac{1+12k^2}{2(w-z)^4}+\cdots \quad (10.4.15)$$

So, the central charge is:

$$c = 1 + 12k^2 \qquad (10.4.16)$$

The central charge for the b, c system is -26, so we have $k^2 = -\frac{9}{4}$. This gives us an imaginary value of k, which we can rectify by reversing the overall sign of the kinetic energy term. Thus, the final energy momentum tensor is:

$$T(z) = +\frac{1}{2}\partial\sigma\,\partial\sigma + \frac{3}{2}\partial^2\sigma \qquad (10.4.17)$$

We have a linear term in the energy-momentum tensor for the ghost field. This, in turn, means that the action itself, in terms of the bosonized field, must contain a linear term as well. Because we will be concerned with world sheets that have curvature singularities in them at the points where strings break, we will need the fully generally covariant generalization of this formalism. The full energy-momentum tensor of the bosonized ghost system, with general covariance put back in, is:

$$T_{ab} = \partial_a\sigma\,\partial_b\sigma + \frac{1}{2}g_{ab}(\partial^2\sigma) - \frac{i}{\pi}(\partial_a\,\partial_b - g_{ab}\,\partial^2)\sigma \qquad (10.4.18)$$

Our next task is to calculate the action that yields this energy-momentum tensor T_{ab}, via the definition:

$$\delta S = \frac{1}{2}\int d^2\xi\,\delta g^{ab}\,T_{ab} \qquad (10.4.19)$$

The final action is not too difficult to find. We recall from general relativity how to take the variation of the curvature tensor:

$$\delta \int d^2\xi\, gR = -\int d^2\xi\, \delta g^{ab}\left[(\partial_a\,\partial_b - g_{ab}\,\partial^2)\sigma\right] + \cdots \qquad (10.4.20)$$

Therefore, our final result for the action is given by [20]:

$$S = \frac{1}{2\pi}\int d^2\xi\, g^{ab}\,\partial_a\sigma\,\partial_b\sigma + \frac{3i}{2\pi}\left[\frac{1}{2}\int d^2\xi\, gR + \int_\partial dl\, k\sigma\right] \qquad (10.4.21)$$

where we have explicitly added the contribution from the boundary surface ∂ and where k is the extrinsic geodesic curvature of the boundary and dl is the line element along this boundary.

If σ is a constant, then this reduces to the Euler number of the surface:

$$\chi = \frac{1}{2\pi}\left[\frac{1}{2}\int d^2\xi\, gR + \int_\partial dl\, kg\right] \qquad (10.4.22)$$

The point of going through this exercise is that the curvature R is zero for most of the surface. However, for the breaking point of three strings, R has a delta function singularity, and the Euler number is equal to $-\frac{1}{2}$. Thus, for a self-consistent ghost system, we must insert an extra factor [2]:

$$e^{3i\sigma/2} \qquad (10.4.23)$$

into the vertex function precisely at the midpoint. Then, the rest of the ghost part of the vertex function is a Dirac delta functional, just as for the string variable, representing the continuity across the vertex. The modification of Eq. (10.4.11) therefore involves inserting Eq. (10.4.23) at the midpoint and multiplying by the Dirac delta functional for the σ field (in analogy with the three strings). A careful analysis of the resulting vertex function shows that it satisfies the correct properties of multiplication and that we can successfully reproduce the Veneziano model [21–14].

10.5. Four-Point Amplitude

We now begin a discussion of constructing the four-point Veneziano amplitude in Witten's string field theory. The four-point scattering amplitude will have the geometry as shown in Fig. 10.5. We wish to show that the scattering amplitude given by the field theory:

$$A_4(s,t) = \int_0^\infty d\tau \langle V_{125}|b_0 e^{-\tau(L_0-1)}|V_{534}\rangle + (s \leftrightarrow t) \qquad (10.5.1)$$

gives us back the familiar Veneziano formula. (Note the insertion of the factor b_0 in the propagator, which arises when we choose the $b_0\Phi = 0$ gauge.)

In order to perform this calculation, we need several ingredients, including:

1. the conformal map taking us from the upper half plane to the world sheet of the string scattering,
2. the Jacobian of the transformation from τ to the string world sheet to x of the Koba–Nielsen variables, and
3. the ghost contribution.

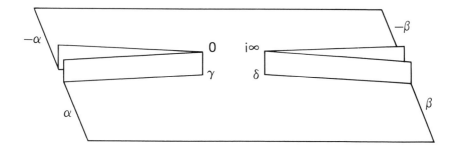

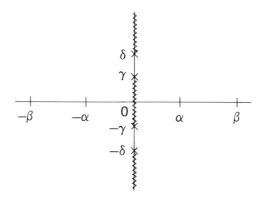

Fig. 10.5.

First, let us calculate the conformal map, which will take us from the upper half plane to the configuration shown in Fig. 10.5. In contrast to the light cone gauge, we need a conformal map that has a Riemann cut, as in the case of the three-vertex function. Using the Schwarz-Christoffel transformation, we find that the following map has the desired properties [21]:

$$\frac{dw}{dz} = \frac{N}{2} \frac{\sqrt{z^2 + \gamma^2} \sqrt{z^2 + \delta^2}}{(z^2 - \alpha^2)(z^2 - \beta^2)} \tag{10.5.2}$$

where the Riemann cut goes from $i|\gamma|$ to $-i|\gamma|$ and from $\pm i|\delta|$ to $\pm\infty$, as shown in the figure.

Now, let us place boundary conditions on the map so that all external strings have equal string lengths.

1. In order that the strip width at A equal that at B, we demand:

$$\alpha\beta = \gamma\delta \tag{10.5.3}$$

2. The strip width at A must be π. This gives:

$$\frac{dw}{dz} \rightarrow -\frac{1}{z - \alpha} \tag{10.5.4}$$

which gives us the normalization of N:

$$N = 2\alpha \frac{\beta^2 - \alpha^2}{\sqrt{\alpha^2 + \gamma^2} \sqrt{\alpha^2 + \delta^2}} \tag{10.5.5}$$

3. The segment FE has half the length of the strip at A. This condition is a bit more difficult, because it requires us to actually perform the integration from $(0, i\gamma)$ to $(i\gamma, i\delta)$.

    The integral in question requires the theory of first and third elliptic integrals. For example, we have the standard definition of a first elliptic integral:

$$\int_0^y \frac{dt}{\sqrt{(1 - t^2)(1 - k^2 t^2)}} = \int_0^\phi \frac{d\theta}{\sqrt{1 - k^2 \sin^2 \theta}}$$
$$= \int_0^{u_1} du = u_1 \equiv sn^{-1}(y, k) = F(\phi, k) \tag{10.5.6}$$

where $y = \sin\phi$ and $\phi = am\, u_1$.

The first complete elliptic integral is given by the definite integral:

$$K(k) = \int_0^{\pi/2} \frac{d\theta}{\sqrt{1 - k^2 \sin^2 \theta}}; \qquad k^2 = \frac{\gamma^2}{\delta^2}; \qquad k'^2 = 1 - k^2 \tag{10.5.7}$$

The integral of Eq. (10.5.2) for boundary values in (3.) is given by:

$$\frac{\pi}{2} = -\frac{N}{\delta} \left(\frac{\delta^2 - \gamma^2}{\beta^2 - \alpha^2} \right) \int_0^{K(k)} du$$
$$\times \left[\frac{1}{1 - \eta_1^2 sn^2(k, u)} - \frac{1}{1 - \eta_2^2 sn^2(k, u)} \right] \tag{10.5.8}$$

where:

$$\eta_1^2 = \frac{\gamma^2}{\delta^2}\frac{\beta^2 + \delta^2}{\beta^2 + \gamma^2}; \qquad \eta_2^2 = \frac{k^2}{\eta_1^2} \tag{10.5.9}$$

Performing the integral, the condition reduces to:

$$\frac{1}{2} = \Lambda_0(\theta_1, k) - \Lambda_0(\theta_2, k) \tag{10.5.10}$$

where:

$$\sin^2 \theta_1 = \frac{\beta^2}{\beta^2 + \gamma^2}; \qquad \sin^2 \theta_2 = \frac{\alpha^2}{\alpha^2 + \gamma^2} \tag{10.5.11}$$

and where Λ_0 is Heuman's lambda function given by:

$$\Lambda_0(\beta, k) = \frac{2}{\pi}[E(k)F(\beta, k') + K(k)E(\beta, k') - K(k)F(\beta, k')] \tag{10.5.12}$$

4. The segment DE has length τ. Once again, we must perform the integral of Eq. (10.5.2), which yields:

$$\tau/2 = K(k')[Z(\theta_2, k') - Z(\theta_1, k')] \tag{10.5.13}$$

and the Jacobi Zeta function is given by:

$$Z(\beta, k) = E(\beta, k) - (E/K)F(\beta, k) \tag{10.5.14}$$

We would also like the explicit relationship between the parameters in the map and x, the Koba–Nielsen variable that will appear in the Veneziano formula. We find:

$$x = \left(\frac{1 - \alpha^2}{1 + \alpha^2}\right)^2 \tag{10.5.15}$$

Also, we want the Jacobian that takes us from $d\tau$, defined on the string world sheet describing four-string scattering, and the parameters of the upper half plane. An explicit calculation gives us:

$$d\tau = -\frac{2\pi}{K(\gamma^2)} \frac{d\alpha}{\sqrt{1 + \alpha^2\gamma^2} \sqrt{\alpha^2 + \gamma}} \tag{10.5.16}$$

Last, we must add the contribution of the ghosts. Let $e^{-i\phi_+}$ represent the bosonized ghost contribution coming from the b ghosts. We find that the ghost part of the amplitude A_G is given by:

$$A_G = \int \frac{dz}{2\pi i} \frac{dz}{dw} \exp\left[-\sum_{j<k} \langle \phi(j)\phi(k) \rangle + \sum_{j} \langle \phi(j)\phi_+(z) \rangle\right] \tag{10.5.17}$$

It is easy to evaluate the contraction over these ghost fields. They are given by:

$$\langle \phi(j)\phi(k)\rangle = -\frac{1}{2}(\ln|z_j - z_k| + \ln|z_j - \bar{z}_k|)$$

$$\langle \phi(j)\phi_+(z)\rangle = -\frac{1}{2}\left[\ln(z_j - z) + \ln(z_j - \bar{z})\right]$$

(10.5.18)

Thus, the ghost contribution is given by:

$$A_G = (\alpha^{-2} - \alpha^2)^2 \int_C \frac{dz}{2\pi i}\frac{dz}{dw}\frac{1}{(\alpha^2 - z^2)(\alpha^{-2} - z^2)}$$

$$= \int \frac{1}{2\pi}\sqrt{\alpha^2 + \gamma^2}\sqrt{1 + \alpha^2\gamma^2}(1 - \alpha^4)\alpha^{-3}K(\gamma^2)$$

(10.5.19)

Now, let us put all factors together. Notice that the $K(\gamma^2)$ factor coming from the ghost cancels precisely the same factor coming from the Jacobian. In fact, once the contribution from the string itself is included, we have the final form of the s-channel graph [21]:

$$A_s = -2\int d\alpha \, \frac{1 - \alpha^4}{\alpha^3}\exp\left(-\sum_{j<k}P_j \cdot P_k\langle X(j)X(k)\rangle\right)$$

$$= -\frac{1}{4}\int_{1/2}^{1} dx \, x^{2P_1 P_2}(1 - x)^{2P_2 P_3}$$

(10.5.20)

When we add the t-channel contribution, the line integral goes from 0 to 1, and we retrieve the four-string Veneziano amplitude [Eq. (1.3.8)], as promised.

10.6. Superstring Field Theory

Buoyed by the relatively easy successes of the bosonic open string field theory, we now wish to generalize our discussion to the case of NS–R superstring field theory. Unfortunately, the ease with which the bosonic open string field theory was constructed rapidly disappears when we begin to generalize our discussion to the case of superstrings. Let us review the NS–R superstring theory and isolate where the problem lies.

We saw earlier that the NS–R theory has, in addition to the usual fermionic b, c ghost system, a bosonic ghost system β, γ in Eq. (1.4.19). The complication arises when we analyze the zero modes of these ghost operators. For the sake of convenience, let us combine both sets of ghosts into one set $\mathbf{b}$ and $\mathbf{c}$, such that ghost Lagrangian becomes [26]:

$$S = \frac{1}{\pi}\int d^2z \, \mathbf{b}\,\bar{\partial}\mathbf{c} + \text{c.c.}$$

(10.6.1)

where the $\mathbf{b}$ field has conformal weight λ and the $\mathbf{c}$ field has conformal weight $1 - \lambda$.

Then, the energy-momentum tensor is:

$$T(z) = -\lambda \mathbf{b}\,\partial \mathbf{c} + (1-\lambda)\partial \mathbf{b}\,\mathbf{c} \qquad (10.6.2)$$

where we have the normal mode decomposition:

$$\mathbf{b}(z) = \sum_{n\in\delta-\lambda+\mathbf{Z}} z^{-n-\lambda}\mathbf{b}_n$$

$$\mathbf{c}(z) = \sum_{n\in\delta+\lambda+\mathbf{Z}} z^{-n-1+\lambda}\mathbf{c}_n \qquad (10.6.3)$$

where $\delta = 0$ for NS and $\delta = 1/2$ for R boundary conditions.

For the b,c system, we have $\lambda = 2$, and for the β,γ system, we have $\lambda = \frac{3}{2}$. The ghost number current can be written as:

$$\mathbf{j}(z) = -\mathbf{bc} = \sum_n z^{-n-1}\mathbf{j}_n$$

$$\mathbf{j}_n = \sum_k \epsilon \mathbf{c}_{n-k}\mathbf{b}_k \qquad (10.6.4)$$

where $\epsilon = +1$ for Fermi statistics and $\epsilon = -1$ for Bose statistics.

Alternatively, we can form the ghost superfields:

$$B(z) = \beta(z) + \theta b(z)$$
$$C(z) = c(z) + \theta\gamma(z) \qquad (10.6.5)$$

and the ghost action can be written down as:

$$S = \frac{1}{\pi}\int d^2z\,d\theta\,d\bar\theta\; B\,\bar D C + \text{c.c.} \qquad (10.6.6)$$

and the ghost energy momentum is:

$$T(z) = -C\,\partial B + \frac{1}{2}DC\,DB - \frac{3}{2}\partial C\,B \qquad (10.6.7)$$

which can be decomposed as:

$$T_F(z) = -c\,\partial\beta - \frac{3}{2}\partial c\,\beta + \frac{1}{2}\gamma b$$

$$T_B(z) = c\,\partial b + 2\,\partial cb - \frac{1}{2}\gamma\,\partial\beta - \frac{3}{2}\partial\gamma\,\beta \qquad (10.6.8)$$

The BRST current can be written as

$$J_{\text{BRST}} = DC(C\,DB - \frac{3}{4}DC\,B) \qquad (10.6.9)$$

The problem arises when we try to define the ground state of the theory. For example, we have the freedom of defining several possible ground states:

$$\mathbf{b}_n|q\rangle = 0; \qquad n > \epsilon q - \lambda$$
$$\mathbf{c}_n|q\rangle = 0; \qquad n \geq -\epsilon q + \lambda \tag{10.6.10}$$

where:

$$\mathbf{j}_0|q\rangle = q|q\rangle$$
$$L_0|q\rangle = \frac{1}{2}\epsilon q(Q + q)|q\rangle \tag{10.6.11}$$

where $Q = \epsilon(1 - 2\lambda)$ and $\lambda = \frac{1}{2}(1 - \epsilon Q)$. For the b, c system, we have $\epsilon = 1$, $\lambda = 2$, $Q = -3$, and the central term $c = -26$. For the β, γ system, we have $\epsilon = -1$, $\lambda = \frac{3}{2}$, $Q = 2$, and $c = 11$. This is indeed puzzling because, in contrast to the usual Fock space associated with the bosonic string oscillators α_n, we apparently have the disaster of having an infinite set of different vacuums, each labeled by q!

For the bosonic string with its Fermi ghosts b, c, the situation is actually simple to analyze. By multiplying with various monomials in the oscillators, it is possible to show that the various $|q\rangle$ vacuums are actually redundant. We recall for the bosonic ghost system that we had two zero mode oscillators c_0 and b_0, which meant that the vacuum state was degenerate:

$$c_0|+\rangle = 0; \qquad b_0|-\rangle = 0$$
$$|+\rangle = c_0|-\rangle \tag{10.6.12}$$

We notice that there are two vacuums $|\pm\rangle$ that are related to each other by the multiplication of a monomial.

Now, we can compare the old vacuums with the new ones. We find:

$$|q = 1\rangle = |-\rangle, \qquad |q = 2\rangle = |+\rangle \tag{10.6.13}$$

We can calculate the scalar product between these vacuums by noticing that we can place j_0 between the scalar product of any two states. The values of the eigenvalues of j_0 change on $\langle q|$ states. This is because we have the strange identity:

$$\mathbf{j}_n^\dagger = -\mathbf{j}_{-m} - Q\delta_{m,0} \tag{10.6.14}$$

It is now easy to show that the only surviving nonzero scalar product is:

$$\langle -q - Q|q\rangle = 1 \tag{10.6.15}$$

This means:

$$\langle 2|1\rangle = 1 \tag{10.6.16}$$

This also means that the other vacuums $|q\rangle$ are actually not independent vacuums at all. For example, the vacuums $|0\rangle$ can be written as the product of a monomial and the "true" vacuum:

$$|0\rangle = b_{-1}|1\rangle, \qquad |0\rangle = b_{-1}b_0|2\rangle \tag{10.6.17}$$

Similarly, other "vacuums" can be written in terms of the vacuums $|\pm\rangle$:

$$|4\rangle = c_{-1}c_{-2}|2\rangle = c_0 c_{-1} c_{-2}|1\rangle$$
$$|3\rangle = c_{-1}|2\rangle = c_0 c_{-1}|1\rangle \tag{10.6.18}$$
$$|-1\rangle = b_{-1}b_{-2}|1\rangle = b_0 b_{-1} b_{-2}|2\rangle$$

Thus, we can always take the "true vacuum" to be $|-\rangle$, although this choice is not unique.

For the Bose sector, generated by the β, γ ghosts, the situation is much more difficult. We find, in fact, that the $|q\rangle$ are actually linearly independent and that no combination of monomials constructed from the oscillators can convert one vacuum into another. This means that the β, γ ghost sector of the NS–R superstring has an *infinite number of vacuums*!

However, the "vacuums" that come closest to the usual definition of vacuums can be defined in the NS sector as $|q = -1\rangle$ and in the R sector as either $|q = -\frac{1}{2}\rangle$ or $|q = -\frac{3}{2}\rangle$. This is because, for the NS sector, we have (for half-integer m, n):

$$\text{NS}: \quad \begin{cases} \beta_m|-1\rangle = 0; & m \geq \frac{1}{2} \\ \gamma_n|-1\rangle = 0; & n \geq \frac{1}{2} \end{cases} \tag{10.6.19}$$

Because these states act on the $|-1\rangle$ vacuum, we will call this the "-1 picture." For the R sector, we have (for integral m, n):

$$\text{R}: \quad \begin{cases} \beta_m|-\frac{1}{2}\rangle = 0; & m \geq 0 \\ \gamma_n|-\frac{1}{2}\rangle = 0; & n \geq 1 \end{cases} \tag{10.6.20}$$

and:

$$\text{R}: \quad \begin{cases} \beta_m|-\frac{3}{2}\rangle = 0; & m \geq 1 \\ \gamma_n|-\frac{3}{2}\rangle = 0; & n \geq 1 \end{cases} \tag{10.6.21}$$

Their scalar product are as follows:

$$\langle -\frac{3}{2}|-\frac{1}{2}\rangle = 1, \qquad \langle -1|-1\rangle = 1 \tag{10.6.22}$$

However, we also note that we could equally have taken the "zero picture" for the NS states based on the vacuum $|q = 0\rangle$. For the zero picture, we have:

$$\text{NS}: \quad \begin{cases} \beta_m|0\rangle; & n > -\frac{3}{2} \\ \gamma_n|0\rangle; & n \geq \frac{3}{2} \end{cases} \tag{10.6.23}$$

The existence of an infinite number of linearly independent vacuums for the Bose β, γ ghost sector at first seems like a disaster. However, there is a simple resolution for all of this. It is possible to show, *on-shell* that all these vacuums are actually equivalent. We will call these different sectors of the theory different *pictures*, and we can show that, on-shell, all pictures are equivalent.

10.7. Picture Changing

The simplest way of showing this is to bosonize the β, γ system. At first, this may seem a bit strange, since β, γ are already bosons, but we can write them as the product of two fermions and then bosonize these fermions. This can be implemented by the following definitions [25]:

$$\beta = e^{-\phi} \partial \xi, \qquad \gamma = e^{\phi} \eta \qquad (10.7.1)$$

or:

$$\eta = \partial \gamma \, e^{-\phi}, \qquad \partial \xi = \partial \beta \, e^{\phi} \qquad (10.7.2)$$

where we have written the bosons as the product of two fermions. We can then, in turn, bosonize once more:

$$\xi = e^{\chi}, \qquad \eta = e^{-\chi} \qquad (10.7.3)$$

At first, this strange bosonization may seem formal and a bit useless. But, there are several advantages to this bosonization. First, we can use this bosonization to interpolate between the NS and R sectors. Second, we can use this to create a fermion vertex function. Third, it will enable us to write a "picture changing" operator, which can be used to show that the S matrix is independent of the picture we use [26].

Notice that, in this formalism:

$$\mathbf{j}(z) = \epsilon \, \partial \phi(z) \qquad (10.7.4)$$

with the operator product expansion:

$$\mathbf{j}(z) e^{q\phi(w)} \sim \frac{q}{z - w} e^{q\phi(w)} \qquad (10.7.5)$$

and

$$T(z) e^{q\phi(w)} \sim \left[\frac{1}{2} \epsilon q(q + Q)(z - w)^{-2} + (z - w)^{-1} \partial_w \right] e^{q\phi(w)} \qquad (10.7.6)$$

This means that $e^{q\phi}$ has conformal weight given by:

$$\frac{1}{2} \epsilon q(q + Q) \qquad (10.7.7)$$

Because the states $|q\rangle$ can be defined via their L_0 and $\mathbf{j}$ eigenvalues and because multiplication by $e^{q\phi}$ can change these eigenvalues, we conclude that multiplication by $e^{q\phi}$ can actually change the eigenvalue q:

$$e^{q\phi(0)}|0\rangle = |q\rangle \qquad (10.7.8)$$

This is a rather remarkable formula. We saw in previous chapters that the NS and R sectors were based on distinct Fock spaces. Now, we see that multiplication by $e^{q\phi}$ can actually change NS vacuums into R vacuums and vice versa.

Now, we would like to rigorize some of these comments concerning picture changing. This analysis will be greatly facilitated by the introduction of several important operators. First, there is the "picture changing operator," which allows us to go from one picture to another:

$$
\begin{aligned}
X(z) &\equiv \{Q, \xi(z)\} \\
&= \frac{1}{2} e^{\phi} \psi^{\mu} \, \partial_z X_{\mu} + c \, \partial_z \xi - \frac{1}{4} b \, \partial_z \eta - \frac{1}{4} b \, \partial_z \eta \, e^{2\phi} - \frac{1}{4} \partial_z (b\eta e^{2\phi})
\end{aligned}
\tag{10.7.9}
$$

which raises the total ghost number by one and has zero conformal weight (and hence may be multiplied with any vertex function without changing its conformal weight).

Notice that X is automatically BRST invariant because it is written explicitly as a BRST commutator. (However, we should be careful to note that X is not, therefore, BRST trivial. Notice that ξ is not part of the usual Fock space of operators that we have defined, since Eq. (10.7.1) is defined only with $\partial \xi$. Thus, although X is written as a BRST commutator, it is not BRST trivial because ξ is part of a "big algebra" and not part of the "small algebra," which consists of the usual Fock space of operators.)

This operator can also be written as the BRST commutator of a step function $\theta(\beta)$:

$$
X(z) = \{Q, \theta[\beta(z)]\}
\tag{10.7.10}
$$

Once again, we see that X is BRST invariant because it is a BRST commutator, but it is not BRST trivial because we have introduced a new operator, the step function θ, which is not part of the usual small algebra of ordinary operators.

There is also the "inverse picture changing operator"

$$
Y(z) = 4c \, \partial_z \xi \, e^{-2\phi}
\tag{10.7.11}
$$

which performs the opposite function of $X(z)$, that is, it reduces the ghost number by one and also has zero conformal weight. The Y inverse picture changing operator can also be written as a Dirac delta function:

$$
Y(z) = -c(z)\delta'[\gamma(z)]
\tag{10.7.12}
$$

Formally, X and Y are inverses because:

$$
\lim_{z \to w} X(z)Y(w) \sim 1
\tag{10.7.13}
$$

Extreme care, however, must be exercised whenever using products of these operators, since they are potentially ill defined, especially when taken at the same point. For example, products of these operators are actually infinite when taken at the same point:

$$\lim_{z \to w} X(z)X(w) \sim \infty \qquad (10.7.14)$$

Products of these picture changing operators are actually only well defined on-shell on the physical Fock space, that is, on the cohomology class of Q. Products of these operators at the same point still diverge, but the divergence is outside the cohomology class of Q. Thus, if we restrict all of our manipulations within the cohomology class of Q (the physical subspace), then these operators are well defined.

To see this, we shall introduce the operators:

$$\begin{aligned}
: X_0 : &\equiv : \oint \frac{dz}{2\pi i} \frac{X(z)}{z} : \\
: Y_0 : &\equiv : \oint \frac{dz}{2\pi i} \frac{Y(z)}{z} :
\end{aligned} \qquad (10.7.15)$$

The key identity we want is:

$$: X_0 :^n : Y_0 :^n \sim 1 + [Q, \epsilon] \qquad (10.7.16)$$

where ϵ is some operator. Because we restrict all of our comments to the cohomology class of Q, we can drop the second term (which may be infinite), meaning that we have an infinite number of picture changing operators and their inverses. On-shell, however, all these pictures are equivalent. To go between any of the various pictures (on-shell), one simply multiplies by $: X_0 :^n$ or Y_0^m.

For completeness, we list some more identities involving the ghost operators:

$$\begin{aligned}
\eta(z) &= z \, \partial\gamma \, \delta[\gamma(z)] \\
e^{-\phi} &= \delta[\gamma(z)] \\
e^{\phi} &= \delta[\beta(z)]
\end{aligned} \qquad (10.7.17)$$

which can be proven by showing that the left and right side of the equations have the same operator product expansions.

10.8. Superstring Action

Now, let us apply this technology to the superstring action. Let us first work in the "-1 picture," where the A field has ghost number $-\frac{1}{2}$ as usual, and the ψ field has ghost number zero.

The only complication we have in counting ghost numbers is the vertex function. If we bosonize all the ghosts, then there are three ghost contributions, given by σ, ϕ, and χ. We can proceed as before, calculating their contributions to the energy-momentum tensor and then extrapolating back to obtain the final action. Repeating the same steps as for the bosonic ghost, we find:

$$S = \frac{1}{2\pi} \int d^2\xi \, g^{ab} \left(\partial_a \chi \, \partial_b \chi - \partial_a \phi \, \partial_b \phi \right)$$
$$+ \frac{1}{2\pi} \left[\frac{1}{2} \int d^2\xi \, R(\chi + 2\phi) + \int dl \, k(\chi + 2\phi) \right] \qquad (10.8.1)$$

From this, knowing that the curvature tensor R is zero except at the midpoint of three strings, we can read off the ghost contribution at the midpoint:

$$e^{i\phi + i\chi/2} \qquad (10.8.2)$$

Including the ghost contribution from σ, this gives us a total ghost number of $\frac{1}{2}$ for a new vertex $*$. This ghost number, however, causes some technical problems with regard to the counting. We can, for example, postulate the following variation of fields:

$$\delta A = Q\Lambda + X \left(A * \Lambda - \Lambda * A \right) + \Psi * \chi - \chi * \Psi$$
$$\delta\Psi = Q\chi + A * \chi - \chi * A + X(\Psi * \Lambda - \Lambda * \Psi) \qquad (10.8.3)$$

Notice that, in order to get the ghost numbers correct, we had to insert a new operator X at the midpoint. This operator has ghost number $+1$ and conformal weight 0. Fortunately, such an operator does exist, and it is the picture changing operator we met before.

Similarly, there is now a problem with getting the ghost number of the free fermions to come out correctly. Naively, the net ghost number of $\int \Psi * Q\Psi$ equals $+1$, so we have to insert yet another operator with ghost number -1 at the midpoint. The most likely choice is the inverse picture changing operator Y_{-1}. The ghost numbers are:

$$
\begin{array}{llll}
\gamma: & 1 & \int: & -\frac{1}{2} \\
\beta: & -1 & & \\
A: & -\frac{1}{2} & \Lambda: & -3/2 \\
 & & \chi: & -1 \\
\Psi: & 0 & X: & 1 \\
*: & \frac{1}{2} & Y: & -1
\end{array}
\qquad (10.8.4)
$$

With this choice of ghost numbers, the action in the -1 picture becomes [27]:

$$S_{-1} = \frac{1}{2} \int A * QA + \frac{1}{3} \int X(A * A * A)$$
$$+ \frac{1}{2} \int Y_{-1}\Psi * Q\Psi + \int A * \Psi * \Psi \qquad (10.8.5)$$

It is easy to check that the action is invariant under the postulated transformations.

At first, we seem to have a self-consistent field theory with a new supergauge invariance. However, this is not true. There is a fundamental flaw in the above action, the Wendt anomaly [28].

The problem arises because the picture changing operator X, defined at the same point, diverges:

$$\lim_{z \to w} X(z)X(w) \sim \infty \qquad (10.8.6)$$

This divergence does not appear in the action, or in the gauge variation of the action, but it appears everywhere else, that is, in the amplitudes, the closure of the algebra, and the variation of the equations of motion [28]. For example, when calculating the bosonic four-string interaction, we are led to construct propagators such as:

$$X \frac{b_0}{L_0 - 1} X \qquad (10.8.7)$$

which are inserted between two vertex functions. By commuting the X past the propagator, we find that we can cancel the $1/(L_0 - 1)$ term, in which case the two picture changing operators appear on top of each other. Similarly, by taking the multiple gauge variation of the fields, we find that the X's appear on top of each other. To remedy the situation, we will go to the "0 picture" [29, 30], in which case the only changes in the ghost number are:

$$\text{gh}(A) \to +\frac{1}{2}; \qquad \text{gh}(\Lambda) \to -\frac{1}{2} \qquad (10.8.8)$$

All other ghost numbers remain the same. In the 0 picture, the new variations are:

$$\begin{aligned}
\delta A &= Q\Lambda + A * \Lambda - \Lambda * A + X(\Psi * \chi - \chi * \Psi) \\
\delta \Psi &= Q\chi + A * \chi - \chi * A + \Psi * \Lambda - \Lambda * \Psi
\end{aligned} \qquad (10.8.9)$$

Notice that if we take multiple variations of the fields or if we vary the equations of motion we never have the problem with X^2. The new action in the 0 picture is [29, 30]:

$$\begin{aligned}
S_0 = {}& \frac{1}{2} \int Y_{-2} A * QA + \frac{1}{3} \int Y_{-2} A * A * A \\
&+ \frac{1}{2} \int Y_{-1} \Psi * Q\Psi + \int Y_{-1} A * \Psi * \Psi
\end{aligned} \qquad (10.8.10)$$

The important point is that the collision of two X's does not occur in the 0 picture.

To see this, let us make a gauge variation of the equations of motion to see if any divergent operators occur. The equations of motion are:

$$\begin{aligned}
QA + A * A + X(\Psi * \Psi) &= 0 \\
Q\Psi + A * \Psi + \Psi * A &= 0
\end{aligned} \qquad (10.8.11)$$

A gauge variation of these equations yields:

$$\delta\big[QA + A * A + X(\Psi * \Psi)\big] = [QA + A * A + X(\Psi * \Psi), \Lambda]$$
$$+ X\big[Q\Psi + A * \Psi + \Psi * A, \chi\big]$$
$$\delta\big(Q\Psi + A * \Psi + \Psi * A\big) = [Q\Psi + A * \Psi + \Psi * A, \Lambda]$$
$$+ [QA + A * A + X(\Psi * \Psi), \chi]$$

(10.8.12)

In fact, the only potentially singular operator identity we have when we iterate the gauge transformation is

$$Y_{-2}X = Y_{-1} \tag{10.8.13}$$

To complete the zero picture, we need to construct an explicit operator expression for Y_{-2}, which satisfies the above equation. There are actually several possibilities [29, 30]. The simplest uses a trick introduced in Ref. 31, that is, we introduce operators defined in the lower half complex plane:

$$Y_{-2} = Y(i)Y(-i) \tag{10.8.14}$$

where i represents the intersection midpoint of the string, and $-i$ symbolically represents the point with opposite chirality (that is, it exists in the lower half plane). Because the point $-i$ never meets the point i (because they are on opposite sides of the real axis), there is never any problem with divergent operator expansions.

However, this is not the only possibility. We can, for example, write *all* possible operators that are BRST invariant (but not BRST trivial), have the correct ghost number, and obey the previous expression. There are 15 possible operators that have the correct ghost and picture constraints. When we demand BRST invariance, this list reduces to 5 possible operators (and linear combinations of them), which then satisfy all possible constraints.

At this point, it appears that we have too many possible candidates for the operator Y_{-2} which have all the desired characteristics. Fortunately, one can show that all 6 possibilities of $Y_{-2}^{(i)}$ (and their combinations) are BRST equivalent to the original Y_{-2}, that is,

$$Y_{-2}^{(i)} = Y_{-2} + \{Q, y^{(i)}\} \tag{10.8.15}$$

for some operator $y^{(i)}$.

In summary, we have now constructed a large class of open superstring field theories, each represented by a particular choice of $Y_{-2}^{(i)}$, but all of them on-shell equivalent. Thus, the on-shell theory, which is the only physically relevant theory, is unique.

10.9. Summary

The first quantized approach suffers from many problems, including the following.

1.   Millions of possible vacuums are known, but the first quantized theory cannot choose which, if any, is the correct vacuum.
2.   The first quantized theory cannot break supersymmetry, spontaneously compactify from 10 to 4 dimensions, or give us a completely successful low-energy phenomenology.
3.   The first quantized approach may not be Borel summable.
4.   The first quantized approach is not manifestly unitary.
5.   Last, the first quantized path integral is not well defined, because the sum over conformally inequivalent surfaces is a classical, unsolved problem in Riemann surface theory.

Within the last few years, three explicit triangulations of moduli space have been discovered, two of them arising from string field theory:

1.   light cone coordinates, found in the light cone string field theory;
2.   Harer coordinates, used in Witten's string field theory; and
3.   Penner coordinates, which cannot, as yet, be written as a string field theory.

String field theory is the most promising comprehensive method of formulating string theory nonperturbatively (although it still cannot compute the correct nonperturbative vacuum). String field theory is based on defining a multilocal field functional simultaneously defined at all points along the string:

$$\Phi(X) \equiv \lim_{N \to \infty} \Phi\left[X_\mu(\sigma_1) \cdots X_\mu(\sigma_N)\right] \qquad (10.9.1)$$

The simplest string field theory is the light cone theory, which begins by solving the Virasoro constraints:

$$P_\mu^2 + \frac{X_\mu'^2}{\pi^2} = 0, \qquad P_\mu X'^\mu = 0 \qquad (10.9.2)$$

In the light cone gauge, P^- becomes the new Hamiltonian:

$$H = \int_0^\pi d\sigma \, P^-(\sigma) = \frac{1}{2p^+} \int d\sigma \left(P_i^2 + \frac{X_i'^2}{\pi^2} \right) \qquad (10.9.3)$$

so that the free action reads:

$$\int DX_i \, dp^+ \, \Phi_{p^+}^\dagger(X_i) \left(i \frac{\partial}{\partial \tau} - H \right) \Phi_{p^+}(X_i) \qquad (10.9.4)$$

To compute interactions, we first use the conformal map between the upper half z plane and the string surface:

$$\rho(z) = \sum_{i=1}^{N} \alpha_i \ln(z - z_i) \qquad (10.9.5)$$

Given all possible maps that one can define in this fashion, we find that the structure of light cone string field theory can be symbolically written as follows, for open (closed) string fields represented by $\Phi(\Psi)$:

$$L_{\text{open}} = \Phi^3 + \Phi^4 + \Phi^2\Psi + \Psi^3 + \Psi\Phi$$
$$L_{\text{closed}} = \Psi^3 \qquad (10.9.6)$$

By contrast, the covariant BRST open string field theory is based on the simple observation that Q is nilpotent. Thus, the action:

$$S = \int DX\, Db\, Dc\, D\bar{b}\, D\bar{c}\, \Phi Q\Phi \qquad (10.9.7)$$

where Φ has the fixed ghost number $-\frac{1}{2}$, is invariant under:

$$\delta\Phi = Q\Lambda \qquad (10.9.8)$$

The equation of motion $Q\Phi = 0$ yields the correct Fock space for the string. If we take the gauge:

$$b_0|\Phi\rangle = 0 \qquad (10.9.9)$$

then the Faddeev–Popov ghosts have ghosts. In fact, we have an infinite sequence of "ghosts-for-ghosts" whose quantization can be best described by the Batalin–Vilkovisky (BV) quantization in Appendix 1. The net effect of all these Faddeev–Popov ghosts is simple: we simply relax the ghost number on Φ to include all possible ghost numbers, with the action:

$$L = \langle \Phi(X, b, c) | (L_0^X + L_0^{gh} - 1) | \Phi(X, b, c) \rangle \qquad (10.9.10)$$

The fully interacting open string field theory can be written in exact parallel with the postulates of ordinary gauge theory:

1. the existence of the nilpotent derivative Q such that $Q^2 = 0$;
2. the associativity of the $*$ product:

$$[A * B] * C = A * [B * C] \qquad (10.9.11)$$

3. the Leibnitz rule:

$$Q[A * B] = QA * B + (-1)^{|A|} A * QB \qquad (10.9.12)$$

4. the product rule:

$$\int A * B = (-1)^{|A||B|} \int B * A \qquad (10.9.13)$$

5. the integration rule:

$$\int QA = 0 \qquad (10.9.14)$$

where $(-1)^{|A|}$ is -1 if A is Grassman odd and $+1$ if it is Grassmann even.

We postulate that the field A has the following transformation rule:

$$\delta A = Q\Lambda + A * \Lambda - \Lambda * A \qquad (10.9.15)$$

So, the Chern–Simons form is gauge invariant:

$$L = A * QA + \frac{2}{3} A * A * A \qquad (10.9.16)$$

The $*$ multiplication rule is carried out via the Dirac delta function. The bosonic vertex (without ghosts) is given by:

$$\int DX_1\, DX_2\, DX_3\, \Phi(X_1) * \Phi(X_2) * \Phi(X_3)$$
$$= \int DX_1\, DX_2\, DX_3\, \Phi(X_1)\Phi(X_2)\Phi(X_3) \qquad (10.9.17)$$
$$\times \prod_{r=1}^{3} \prod_{0 \le \sigma_r \le \pi/2} \delta\big[X_{r,\mu}(\sigma_r) - X_{r-1,\mu}(\pi - \sigma_{r-1})\big]$$

The ghost part of the vertex function is a bit more tricky. The b, c system can be bosonized in terms of a scalar field σ, which has the action:

$$S = \frac{1}{2\pi} \int d^2\xi\, g^{ab}\, \partial_a\sigma\, \partial_b\sigma + \frac{3i}{2\pi} \left[\frac{1}{2} \int d^2\xi\, gR + \int_\partial dl\, k\sigma \right] \qquad (10.9.18)$$

Notice that if the world sheet of the interacting string is flat, there are delta function curvature singularities at isolated points (the midpoints). This contributes a factor:

$$e^{3i\sigma/2} \qquad (10.9.19)$$

at the midpoint of three strings. Thus, the $*$ multiplication rule must have a ghost insertion of $\frac{3}{2}$ ghost number at the midpoint. This, then, gives us the ghost numbers for all fields and operators:

$$
\begin{array}{ll}
c: \quad 1 & \\
b: \quad -1 & *: \quad 3/2 \\
Q: \quad 1 & \int: \quad -3/2 \qquad (10.9.20) \\
A: \quad -\dfrac{1}{2} & \Lambda: \quad -3/2
\end{array}
$$

We can also show that the BRST theory reproduces the four-string scattering amplitude:

$$A_4(s,t) = \int_0^\infty d\tau \langle V_{125} | b_0 e^{-\tau(L_0-1)} | V_{534} \rangle + (s \leftrightarrow t) \qquad (10.9.21)$$

The conformal map that takes us from the upper half z plane to the w plane of the interacting strings is:

$$\frac{dw}{dz} = \frac{N}{2} \frac{\sqrt{z^2 + \gamma^2} \sqrt{z^2 + \delta^2}}{(z^2 - \alpha^2)(z^2 - \beta^2)} \qquad (10.9.22)$$

Then, it is straightforward to calculate the Jacobian taking us from τ defined on the world sheet to x, the Koba–Nielsen variable. We find:

$$d\tau = -\frac{2\pi}{K(\gamma^2)} \frac{d\alpha}{\sqrt{1 + \alpha^2 \gamma^2} \sqrt{\alpha^2 + \gamma}} \qquad (10.9.23)$$

Then, the ghost contribution can be written as:

$$A_G = \int \frac{dz}{2\pi i} \frac{dz}{dw} \exp\left[-\sum_{j<k} \langle \phi(j)\phi(k) \rangle + \sum_j \langle \phi(j)\phi_+(z) \rangle \right] \qquad (10.9.24)$$

so that the final result is:

$$A_s = -2 \int d\alpha \frac{1-\alpha^4}{\alpha^3} \exp\left[-\sum_{j<k} P_j \cdot P_k \langle X(j)X(k) \rangle \right]$$
$$= -\frac{1}{4} \int_{1/2}^1 dx\, x^{2P_1 P_2} (1-x)^{2P_2 P_3} \qquad (10.9.25)$$

which gives us the Veneziano model when we add the contribution from the other channel.

The superstring field theory, however, is more involved. We start with the β, γ ghost action, which can be written as:

$$S = \frac{1}{\pi} \int d^2 z\, \mathbf{b}\, \bar{\partial}\mathbf{c} + \text{c.c.} \qquad (10.9.26)$$

where the $\mathbf{b}$ field has conformal weight λ and the $\mathbf{c}$ field has conformal weight $1 - \lambda$. Then, the energy-momentum tensor is:

$$T(z) = -\lambda \mathbf{b}\, \partial\mathbf{c} + (1 - \lambda)\partial\mathbf{bc} \qquad (10.9.27)$$

where we have the normal mode decomposition:

$$\mathbf{b}(z) = \sum_{n \in \delta - \lambda + \mathbf{Z}} z^{-n-\lambda} \mathbf{b}_n$$
$$\mathbf{c}(z) = \sum_{n \in \delta + \lambda + \mathbf{Z}} z^{-n-1+\lambda} \mathbf{c}_n \qquad (10.9.28)$$

For the b, c system, we have $\lambda = 2$, and for the β, γ system, we have $\lambda = \frac{3}{2}$.

The problem is that there are an infinite number of degenerate vacuums for the β, γ system:

$$\begin{aligned} \mathbf{b}_n|q\rangle = 0; & \qquad n > \epsilon q - \lambda \\ \mathbf{c}_n|q\rangle = 0; & \qquad n \geq -\epsilon q + \lambda \end{aligned} \qquad (10.9.29)$$

where:

$$\begin{aligned} \mathbf{j}_0|q\rangle &= q|q\rangle \\ L_0|q\rangle &= \frac{1}{2}\epsilon q(Q + q)|q\rangle \end{aligned} \qquad (10.9.30)$$

where $Q = \epsilon(1 - 2\lambda)$ and $\lambda = \frac{1}{2}(1 - \epsilon Q)$. For example, for the -1 picture, we have:

$$\text{NS}: \quad \begin{cases} \beta_m| - 1\rangle = 0; & \quad m \geq \frac{1}{2} \\ \gamma_n| - 1\rangle = 0; & \quad n \geq \frac{1}{2} \end{cases} \qquad (10.9.31)$$

Because these states act on the $| - 1\rangle$ vacuum, we will call this the -1 picture.

We can bosonize these ghosts by a trick. First, we break up each boson into the product of two fermions, and we bosonize each fermion:

$$\beta = e^{-\phi} \partial \xi, \qquad \gamma = e^{\phi} \eta \qquad (10.9.32)$$

or:

$$\eta = \partial \gamma \, e^{-\phi}, \qquad \partial \xi = \partial \beta \, e^{\phi} \qquad (10.9.33)$$

This gives us the possibility of writing two new operators, the picture changing operator X and the inverse picture changing operator Y:

$$\begin{aligned} X(z) &\equiv \{Q, \xi(z)\} \\ &= \frac{1}{2} e^{\phi} \psi^{\mu} \, \partial_z X_{\mu} + c \, \partial_z \xi - \frac{1}{4} b \, \partial_z \eta - \frac{1}{4} b \, \partial_z \eta \, e^{2\phi} - \frac{1}{4} \partial_z (b\eta \, e^{2\phi}) \end{aligned} \qquad (10.9.34)$$

and:

$$Y(z) = 4c \, \partial_z \xi \, e^{-2\phi} \qquad (10.9.35)$$

In the original -1 picture, there was the problem of $X^2 = \infty$ appearing in the amplitudes, as well as in the iteration of the gauge invariance. To solve this problem, we go to the 0 picture where the action is:

$$\begin{aligned} S &= \frac{1}{2} \int Y_{-2}A * QA + \frac{1}{3} \int Y_{-2}A * A * A \\ &+ \frac{1}{2} \int Y_{-1}\Psi * Q\Psi + \int Y_{-1}A * \Psi * \Psi \end{aligned} \qquad (10.9.36)$$

This hinges upon developing a new operator, Y_{-2}, with the property:

$$Y_{-2}X = Y_{-1} \qquad (10.9.37)$$

One candidate for this is:

$$Y_{-2} = Y(i)Y(-i) \qquad (10.9.38)$$

There are actually many possible solutions to the above equation, but all of them are on-shell equivalent. Thus, in the 0 picture, the theory seems to be free of any anomalies.

References

1. M. Kaku and K. Kikkawa, *Phys. Rev.* **D10**, 1110, 1823 (1974).
2. E. Witten, *Nucl. Phys.* **B268**, 253 (1986).
3. S. Giddings and S. Wolpert, *Comm. Math. Phys.* **109**, 177 (1987).
4. J. Harer, *Invent. Math.* **72**, 221 (1982); *Ann. Math.* **121**, 215 (1985); J. Harer and D. Zagier, *Invent. Math.* **85**, 457 (1986).
5. R. C. Penner in *Mathematical Aspects of String Theory*, S.T. Yau, ed., World Scientific, Singapore (1986); *Comm. Math. Phys.* **113**, 299 (1987).

For reviews, see Refs. 6–10.

6. M. Kaku, "String Field Theory," *Int. J. of Mod. Phys.* **A2**, 1 (1987).
7. W. Siegel, *Introduction to String Field Theory*, World Scientific, Singapore (1988).
8. T. Banks, SLAC-PUB 3996 (1986).
9. P. West, CERN/TH-4660 (1986).
10. C. Thorn, *Phys. Rep.* **175**, 1 (1989).
11. S. Mandelstam, *Nucl. Phys.* **B64**, 205 (1973); **B69**, 77 (1974).
12. E. Cremmer and J. L. Gervais, *Nucl. Phys.* **B76**, 209 (1974); *Nucl. Phys.* **B90**, 410 (1975).
13. W. Siegel, *Phys. Lett.* **142B**, 276 (1984); **151B**, 391, 396 (1985).
14. T. Banks and M. Peskin, *Nucl. Phys.* **B264**, 513 (1986).
15. W. Siegel and B. Zwiebach, *Nucl. Phys.* **B263**, 105 (1985).
16. H. Hata, K. Itoh, T. Kugo, H. Kunitomo, and K. Ogawa, *Phys. Lett.* **175B**, 138 (1986).
17. M. D. Freeman and D. I. Olive, *Phys. Lett.* **175B**, 151 (1986).
18. M. Peskin and C. B. Thorn, *Nucl. Phys.* **B269**, 509 (1986).
19. M. Bochicchio *Phys. Lett.* **193B**, 31 (1987).
20. O. Alvarez, *Nucl. Phys.* **B216**, 125 (1983).
21. S. Giddings, *Nucl. Phys.* **B278**, 242 (1986).
22. S. Giddings and E. Martinec, *Nucl. Phys.* **B278**, 91 (1986).
23. S. Samuel, *Phys. Lett.* **181B**, 249 (1986).
24. D. Gross and A. Jevicki, *Nucl. Phys.* **B282**, 1 (1987).
25. E. Cremmer, C. B. Thorn, and A. Schwimmer, *Phys. Lett.* **179B**, 57 (1986).
26. D. Friedan, E. Martinec, and S. Shenker, *Nucl. Phys.* **B271**, 93 (1986).
27. E. Witten, *Nucl. Phys.* **B276**, 291 (1986).
28. C. Wendt, *Nucl. Phys.* **B314**, 209 (1989).
29. C. R. Preitschoff, C. B. Thorn, and S. A. Yost, UFIFT-HEP-89-19; I. Ya. Aref'eva, P. B. Medvedev, and A. P. Zubarev, SMI-10-1989.
30. O. Lechtenfeld and S. Samuel, *Nucl. Phys.* **B310**, 254 (1988).

Chapter 11

Nonpolynomial String Field Theory

11.1. Four-String Interaction

String field theory, so far, has been relatively clean and simple. For example, the light cone string field theory for closed strings [1] was purely cubic, yet it successfully reproduced the highly nonlinear theory of Einstein. The covariant version of the open string field theory [2] was even simpler, being just a Chern–Simons term.

The covariant closed string theory, however, is where the real physics lies. The heterotic string, the most physical of all the string theories, is necessarily a closed string theory. However, the generalization of Ref. 2 to the covariant closed string case has proven to be unexpectedly difficult. Several groups have attempted to generalize the midpoint-type interactions to the closed string case [3, 4]. In Fig. 11.1, we see the symmetric three-string vertex (which has the geometry of a cookie cutter).

This symmetric configuration of three closed strings has caused considerable confusion in the literature, with several claims that this three-string cubic interaction is sufficient to yield a successful closed string theory field. However, this cubic interaction fails to reproduce the amplitudes of string theory. Higher interactions, in fact, are required to yield a successful theory. The correct approach is to formulate a nonpolynomial action for the closed string field theory. The nonpolynomial action was developed by Kaku [5, 6] and also by the Kyoto/MIT group [7]. It is easy to see, using pictures alone, that this interaction is not gauge invariant and that it does not reproduce the usual Shapiro–Virasoro amplitude [8, 9].

The symmetric configuration cannot be gauge invariant because the midpoint of four or more closed strings is no longer a common point. To see this, let us return to the case of open strings. If we make successive gauge variations of the string field, we find that all strings share a common point, the midpoint. The common boundary between adjacent strings is always $\pi/2$. Products of these strings are therefore associative. Thus, the transformation $\delta\Phi = \Lambda * \Phi - \Phi * \Lambda$ creates a series of terms that cancel each other, because the minus sign coming from one graph cancels the plus sign coming from another graph.

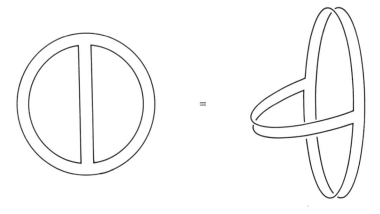

Fig. 11.1.

For N closed strings, however, this is no longer the case. The reason for this is that the string field must be multiplied by the factor:

$$P = \int_0^{2\pi} d\theta \, e^{i(L_0 - \tilde{L}_0)\theta} \qquad (11.1.1)$$

which rotates the string, guaranteeing that the origin of the closed string is not a special point. If we now apply this gauge transformation on the closed string, we arrive at terms like:

$$\delta L \sim \Phi \star \Phi \star (\Phi \star \Lambda) + \cdots \qquad (11.1.2)$$

where the string in parentheses is rotated by an angle θ from the other two strings. Thus, the midpoint is no longer a common point among the four strings, and cancellations do not occur.

In particular, *associativity is violated* if we insist on keeping the parameterization length of all closed strings equal to 2π. For example, in the product $(A \star B) \star C$, we see that the common boundary between string A and B must be π, but that string C may rotate at any angle with respect to the product $A \star B$ and, hence, may have a variable boundary with strings A and B. However, in the product $A \star (B \star C)$, we find that A and B no longer share a common boundary of π, but that strings B and C do. Therefore, the product is not associative, and it is impossible to get a cancellation between these two terms when we make a gauge variation of the fields. The beautiful axioms of cohomology in Eqs. (10.4.1)–(10.4.4) are clearly violated for closed strings.

The cubic interaction is also incorrect because it does not properly reproduce the Shapiro–Virasoro amplitude. To see this, let us return to the case of open string scattering. Let us imagine that two strings come in from the left, as in Fig. 11.2a, and that two other strings come in from the right. This graph is part of the s-channel interaction. These two sets of cookie cutters collide and then rescatter back to the left and right. In Fig. 11.2b, we see the configuration for t-channel scattering, where two pairs of strings come in from the top and bottom. The two cookie cutters also collide and rescatter to the top and bottom.

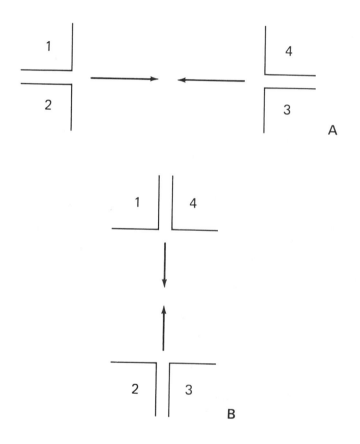

Fig. 11.2.

Notice that the s-channel configuration can instantly be deformed into the t-channel configuration because, at the instant of collision, they have the same geometry. By adding the s- and t-channel graphs, we find that we can fill up the entire region of integration for the Veneziano amplitude. Thus, the s-channel graph occupies the region from 0 to $\frac{1}{2}$ in Eq. (10.5.20), the t-channel graph occupies the region from $\frac{1}{2}$ to 1, and the instantaneous deformation of four strings takes place at the very center, at $x = \frac{1}{2}$. (This is in contrast to the light cone case, shown in Fig. 11.2c and 11.2d, where the t- and u-channel scattering graphs have a totally different geometry and hence cannot be smoothly connected without adding a new interaction, the four-string interaction. In fact, we find a finite region of moduli space along the x axis that is missing if we only have three-string light cone interactions in the action. This is, in fact, how the four-string interaction was discovered in Ref. 1.)

Notice that for four closed strings, this argument no longer holds [10, 11]. Imagine two closed strings coming in from the left, as in Fig. 11.3a, meeting two other closed strings coming in from the right, such that they are at a relative twist of angle $\theta = 0$. This is the s-channel contribution. Notice that, for this angle, the four closed strings can instantly rearrange themselves, such that two closed strings leave in the up direction and two closed strings leave in the down direction. This is the t-channel contribution, shown in Fig. 11.3b. Therefore, at first it appears that the s-channel graph can instantly be deformed into the t-channel graph, leaving no missing region.

However, this is not true. In Fig. 11.4, we see that the s-channel interaction can take place where the two cookie cutters are displaced by a nonzero angle θ between them. Now, when they collide, it is impossible to instantly deform the rotated s-channel graph into a t-channel graph. It is impossible to make this transformation, keeping all string lengths equal.

One problem is that, for closed strings, moduli space for the four-string interaction is two dimensional. Thus, the angle θ, which never appears in the open string case, separates the two sets of closed strings and hence prevents the instantaneous deformation of s-channel into t-channel graphs.

The solution to the problem is that there is a missing region of moduli space for the four-string scattering amplitude. Fortunately, it is possible to precisely fill this missing region if we add a new elemental interaction to the theory, a four-string interaction [10–11].

To understand how a four-string interaction can fill up the missing region, let us first identify what the missing region looks like [10, 11]. The moduli space of the Shapiro–Virasoro amplitude, we recall, is simply the entire complex plane. This two-dimensional plane, in turn, can be stereographically mapped to the sphere.

Let us analyze Fig. 11.5. In regions I, II, and III, we have the usual s-, t-, and u-channel scattering amplitudes. Notice that these scattering amplitudes fill up a two-parameter region of the sphere. This is because, for four-closed string scattering, there are two parameters that can describe the interaction: τ, which is the distance separating the two pairs of closed strings, and θ, the relative twist angle between these two sets of closed strings. Thus, regions I, II, and III can each be parameterized by the two moduli $0 \leq \theta \leq 2\pi$ and $0 \leq \tau \leq \infty$.

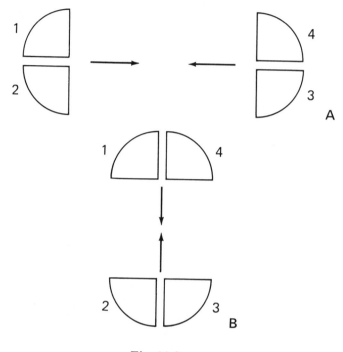

Fig. 11.3.

Notice, however, that regions I, II, and III do not fill up the sphere, but leave a large, triangular portion of the northern and southern hemisphere absent. This is the missing region. To understand this, imagine once again the collision of two sets of closed strings as in Fig. 11.3, such that the angle θ separating them is again set to zero. As we mentioned earlier, only at this point do we have the ability to instantly deform the s-channel and t-channel graphs into each other.

On the sphere, these symmetrical points correspond to three points A, B, C along the equator, which separate the regions I, II, and III. This, in turn, has the topology of a Rubik's cube. Imagine a simplified Rubik's cube (consisting of eight smaller cubes) where we have the ability of rotating the bottom half with respect to the top half or of rotating the left half with respect to the right half. We have two independent rotations we can make along a horizontal or vertical axis, in analogy with Fig. 11.4. Let us say that each rotation must be 180°. It is easy to see that there are three distinct orientations of the Rubik's cube that we can form by making successive 180° rotations.

Therefore, we will call these three points A, B, C along the equator

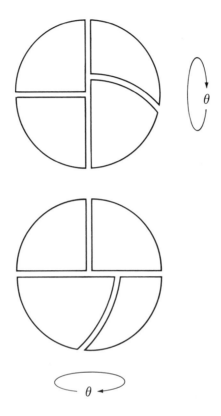

Fig. 11.4.

the Rubik's cube points. As expected, at each Rubik's cube point, there are four ways in which we can rotate the Rubik's cube (either clockwise or counterclockwise along a horizontal or vertical axis). These four ways of rotating each of the three Rubik's cube points correspond to the 12 lines that form the boundary of regions I, II, and III in Fig. 11.5.

Notice that, as we perform various rotations away from the Rubik's cube points, we are only modifying the graph by a single parameter, given by the twist angle θ. However, we know that moduli space is actually two dimensional, therefore we are moving from the s-channel to the t-channel graph by a one-parameter family of rotations, which is not sufficient to fill the missing region.

To smoothly deform an s-channel amplitude to a t-channel scattering amplitude requires successive rotations, which are forbidden on the Rubik's cube. For example, imagine rotating the Rubik's cube 45° along the vertical axis and then rotating it again 45° along the horizontal axis. Clearly, the Rubik's cube will break, which means that the rotations that we can perform on the Rubik's cube only correspond to the boundary of regions I, II, and III, not the interior.

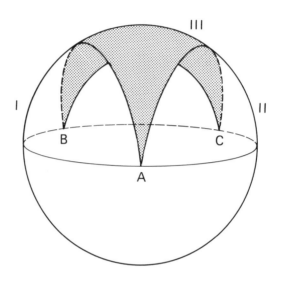

Fig. 11.5.

To see in detail how these manipulations of the missing region can occur, let us begin with a tetrahedron, as in Fig. 11.6, such that each of the four triangular sides has equal perimeter, given by 2π [10, 11]. Each of these four sides will represent the four closed strings, which have just collided.

Let the index i represent each of the four faces. Let a_{ij} represent the common distance between the ith and jth triangle. We wish to set the perimeter of each triangle to be 2π:

$$a_{12} + a_{13} + a_{14} = 2\pi$$
$$a_{21} + a_{23} + a_{24} = 2\pi$$
$$a_{31} + a_{32} + a_{34} = 2\pi \qquad (11.1.3)$$
$$a_{41} + a_{42} + a_{43} = 2\pi$$

We have 6 unknown a_{ij} and 4 constraints on them, so we have 2 degrees of freedom in which to describe the tetrahedron. Notice that this is also precisely the number of degrees of freedom in Koba–Nielsen space.

Then, the three Rubik's cube points A, B, C in Fig. 11.5 can be represented as:

$$A: \quad a_{12} = \pi, \quad a_{14} = \pi$$
$$B: \quad a_{12} = \pi, \quad a_{14} = 0 \qquad (11.1.4)$$
$$C: \quad a_{12} = 0, \quad a_{14} = \pi$$

The three rotations (along one hemisphere) that can take us from one Rubik's cube point to another are then represented by:

$$A \rightarrow B: \quad a_{12} = \pi, \quad a_{14} = \pi \rightarrow 0$$
$$A \rightarrow C: \quad a_{12} = \pi \rightarrow 0, \quad a_{14} = \pi \qquad (11.1.5)$$
$$B \rightarrow C: \quad a_{12} + a_{14} = \pi$$

To find the missing region (that is, the region allowed by the constraints on the tetrahedron), let us take a_{12} and a_{13} as the two independent variables. Then, solving the constraints, the missing region is given by [10, 11]:

$$\text{missing region}: \quad \begin{cases} a_{12}, a_{13} \leq \pi \\ \\ a_{12} + a_{13} \geq \pi \end{cases} \qquad (11.1.6)$$

(We have also placed the constraint $a_{ij} \leq \pi$, whose meaning we will discuss later.)

Let us now plot the missing region on a graph. In the a_{12}–a_{13} plane, we see that the missing region, as expected, is nothing but a right triangle. When this triangle is mapped stereographically onto the sphere, we find that it spreads over the northern hemisphere, completely filling the missing region. (By permuting the legs, we also obtain the southern hemisphere.) By a simple analysis of the tetrahedron graph, we find that we can completely fill the missing region M_R of the Koba–Nielsen plane.

To understand how the boundary of the missing region ∂M_R connects various Mandelstam channels, let us first analyze the pole structure of the various amplitudes. For the four-string scattering amplitude, for example, we have the following pole structure:

$$A(s, t, u) = \sum_I A_{12}^I \frac{1}{s_{12} - m_I^2} A_{34}^I + (2 \leftrightarrow 3) + (2 \leftrightarrow 4) + M_R(s, t, u) \quad (11.1.7)$$

where I represents an infinite tower of Reggeons, s_{ij} represents the energy squared of the Reggeon in the channel formed by the ith and jth strings, and the last term represents the tetrahedron graph, which has no poles.

Let us now capsulize our results. If we let $\tau \rightarrow 0$ for the scattering amplitude for two cookie cutters, we find that the topology of the four colliding cookie cutters forms a graph, called G_4, as in Fig. 11.4 This graph has the topology of a degenerate tetrahedron, that is, one of its legs, for example, a_{12}, has a length equal to zero:

$$G_4 = \lim_{a_{12} \rightarrow \pi} \mathbf{P_4} \qquad (11.1.8)$$

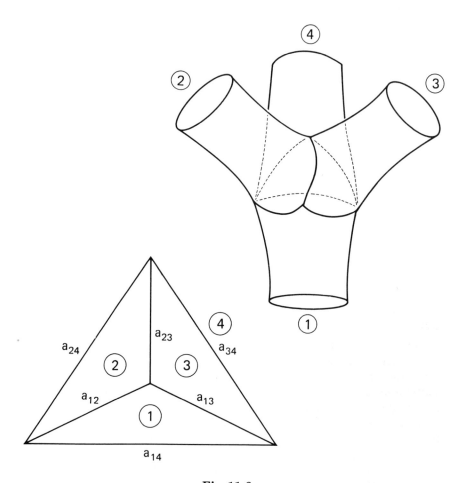

Fig. 11.6.

Notice that each graph G_4 is parameterized by an angle θ, that is, the angle at which the two cookie cutters collide. Thus, there is a one-to-one correspondence between graphs G_4 and points in the missing region:

$$G_4 \leftrightarrow \partial M_{R_4} \qquad (11.1.9)$$

Now that we have practiced with the four-string interaction, let us consider the higher interactions. For five-point scattering, the poles are represented via the energy variable s_{ij}. We have:

$$
\begin{aligned}
A(s_{ij}, s_{kl}) = &\sum_{\text{permutations}} \sum_{IJ} A^I_{12} \frac{1}{s^2_{12} - m^2_I} A^{IJ}_3 \frac{1}{s_{45} - m^2_J} A^J_{45} \\
&+ \sum_{\text{permutations}} \sum_{I} A^I_{12} \frac{1}{s_{12} - m^2_I} M^I_{345} + M(s_{ij}, s_{kl})
\end{aligned}
\qquad (11.1.10)
$$

where M_{345} and $M(s_{ij}, s_{kl})$ represent the tetrahedron and the prism graphs, respectively, which have no poles.

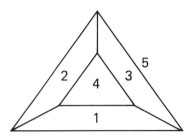

Fig. 11.7.

Let us work out the detailed structure for $N = 5$ in Fig. 11.7. The missing region is given by [6]:

$$M_{R_5} = \begin{cases} a_{12} \leq a_{35} \leq a_{12} + a_{24} \leq a_{35} + \pi \\ a_{13} \leq a_{24} \leq a_{13} + a_{35} \leq a_{24} + \pi \\ a_{12} + a_{13} \leq \pi \end{cases} \tag{11.1.11}$$

Now, let us analyze the missing region. Unfortunately, because the missing region is four-dimensional, it is impossible to visualize its boundaries. However, by setting one of the edges to π, we can reconstruct the three-dimensional figure corresponding to ∂M_{R_5}.

Let us set $a_{24} = \pi$. (Notice that there are 10 permutations we can make on this choice). For this particular boundary, we now have:

$$\partial M_{R_5} = \lim_{a_{24} \to \pi} M_{R_5} = \begin{cases} a_{12} \leq a_{35} \\ a_{13} + a_{35} \geq \pi \\ a_{12} + a_{13} \leq \pi \end{cases} \tag{11.1.12}$$

The Rubik's cube points for the 5-faced prism graph are more complicated than for the tetrahedron. First, we have the Rubik's cube points for each of the tetrahedrons contained within the prism, corresponding to setting $a_{ij} = \pi$. They have previously been studied in detail. More interesting is the Rubik's cube points for the prism graph. For a particular permutation of the external lines, there are 12 ways in which to obtain the Rubik's cube points:

$$\text{Rubik's cube points} : \begin{cases} a_{14} = a_{13} = a_{25} = a_{23} = 0 \\ 12 \text{ permutations} \end{cases} \tag{11.1.13}$$

[The total number of Rubik's cube points for all possible permutations of external lines, but excluding the lower order Rubik's cube points, is given by $\frac{1}{2}(N-1)!$]

The $N = 5$ case, however, is too simple-minded to see the next major complication, the fact that there is more than 1 polyhedron at each level. For example, we have been able to identify at least 2 distinct polyhedra at the 6th level, 5 distinct polyhedra at the 7th level, and 14 distinct polyhedra at the 8th level.

In Figs. 11.8a and 11.8b, we see how the labeling for the polyhedra is given for up to $N = 6$. There are two polyhedra at this level, which we call $(6)_1$ and $(6)_2$. To solve for the missing region, we need only to set $a_{ij} < \pi$ and the internal perimeter for 3 contiguous polygons to be greater than 2π. The complication in finding the missing region is that there are many hidden identities among the various legs that one must factor out in order to find the following for the missing region. If we choose $a_{12}, a_{25}, a_{34}, a_{36}, a_{46}, a_{16}$ as our set of independent variables, then we find for missing region $M_{R_{6(1)}}$ of the cube [6]:

$$a_{36} \leq a_{12} + a_{25} \leq a_{16} + a_{36} + a_{46} \leq \pi + a_{12} + a_{25}$$
$$a_{12} + a_{25} \leq 2a_{36} + a_{34} + a_{16} + a_{46} - \pi \leq \pi + a_{12} + a_{25}$$
$$a_{12} + a_{25} \leq a_{46} + a_{34} + a_{36} \leq \pi + a_{12} + a_{25}$$
$$a_{25} \leq a_{34} + a_{46} + a_{36} + a_{16} - \pi \leq \pi + a_{25}$$
$$a_{25} \leq a_{36} + a_{16} \leq \pi + a_{25} \tag{11.1.14}$$
$$a_{25} \leq a_{34} + a_{36}$$
$$a_{36} + a_{34} + a_{46} \leq 2\pi$$
$$a_{12} \leq a_{46} + a_{36}$$

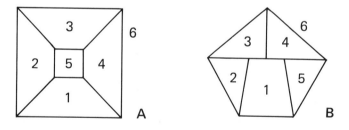

Fig. 11.8.

Let us now analyze the structure of this missing region. The missing region, of course, is 6 dimensional and cannot be visualized. However, it is possible to analyze the boundary of the missing region rather simply. If we let a_{15} go to zero and $a_{14} = \pi$, then this is equivalent to making the polygon formed by the 1st and 4th strings have circumference 2π, that is, the 1st and 4th strings form a cookie cutter and hence can be removed from the polygon, creating a 5-faced prism. In this way, we can reduce the missing region of the cubic polyhedron into the prism graph or lower. We find:

$$\lim_{a_{15}\to 0; a_{14}\to \pi} M_{R_{6(1)}} = M_{R_5} \qquad (11.1.15)$$

Since we have already analyzed the boundary of the missing region of the prism graph, we have now decomposed the missing region of the cubic graph.

There is one more limit one can take on the missing region, and that is to take the perimeter surrounding 3 faces to be equal to 2π. Then, the cubic splits into 2 tetrahedrons. For example, we have:

$$\lim_{P_{125}\to 2\pi} M_{R_{6(1)}} = M_{R_4} \oplus M_{R_4} \qquad (11.1.16)$$

where P_{125} is the perimeter that encloses the 1st, 2nd, and 5th strings in the cube. Thus, once again, we found have that the boundary of the missing region can always be decomposed into lower order polyhedra, so there are no new surprises.

Let us analyze the other, asymmetric 6-faced polyhedra shown in Fig. 11.8b. Again, there are many hidden identities that prevent a simple analysis of this figure. However, we find that a basic set of independent variables is given by $a_{16}, a_{13}, a_{56}, a_{26}, a_{34}, a_{46}$ and that the dependent variables are given in terms of this independent set [6]:

$$
\begin{aligned}
a_{36} &= 2\pi - a_{16} - a_{26} - a_{46} - a_{56} \\
a_{23} &= -a_{13} - a_{34} + a_{16} + a_{26} + a_{46} + a_{56} \\
a_{12} &= 2\pi - 2a_{26} + a_{13} + a_{34} - a_{16} - a_{46} - a_{56} \\
a_{15} &= a_{26} - a_{13} + a_{46} \\
a_{14} &= -a_{34} + a_{56} + a_{26} - a_{13} \\
a_{54} &= 2\pi - a_{56} - a_{26} + a_{13} - a_{46}
\end{aligned}
\qquad (11.1.17)
$$

The restriction that all perimeters that bisect the polyhedron have lengths greater than or equal to 2π then serves to determine the entire missing region $M_{R_{6(2)}}$:

$$
\begin{aligned}
a_{13} &\le a_{26} \le a_{13} + a_{34} \\
a_{13} &\le a_{26} + a_{46} \le \pi + a_{13} \\
a_{13} + a_{34} &\le a_{46} + a_{56} + a_{26} \le 2\pi + a_{13} \\
a_{13} &\le a_{16} + a_{56} + a_{26} + a_{46} \le 2\pi + a_{13} \\
a_{34} &\le a_{16} + a_{56} + a_{26} \le a_{34} + 2\pi \\
\pi &\le a_{16} + a_{26} + a_{46} + a_{56} \le 2\pi
\end{aligned}
\qquad (11.1.18)
$$

$$
\begin{aligned}
a_{16} + a_{56} + a_{26} + a_{46} &\le \pi + a_{13} + a_{34} \\
a_{26} + a_{16} + a_{46} + a_{56} &\le 2\pi + a_{13} + a_{34} - a_{26} \\
&\le \pi + a_{26} + a_{16} + a_{46} + a_{56}
\end{aligned}
$$

It is now straightforward to analyze the boundary of the missing region for the $(6)_2$ polyhedra and show that we can also reduce it to lower polyhedra by taking specific values of the legs. For example, by taking a_{12} to be π and a_{13} to be 0, we can substitute these values into the missing region $M_{R_{6(2)}}$ and show that it reproduces the prism graph, that is,

$$\lim_{a_{12}\to\pi; a_{13}\to 0} M_{R_{6(2)}} = M_{R_5} \tag{11.1.19}$$

For this configuration, we see that the 1st string has disappeared, reducing the 6-faced polyhedra down to a 5-faced polyhedra.

Similarly we can show that, by taking the perimeter that surrounds the 1st, 2nd, and 3rd polygons to be equal to 2π, the polyhedron $(6)_2$ splits into two smaller tetrahedrons:

$$\lim_{P_{123}\to 2\pi} M_{R_{6(2)}} = M_{R_4} \oplus M_{R_4} \tag{11.1.20}$$

These two reductions of the polygon $(6)_2$ into lower polyhedra simply represent the fact that we are taking clusters of polygons to represent external strings with length 2π, thus confirming again that the boundary of missing regions always connects different Mandelstam channels corresponding to midpoint scattering, that is, scattering of cookie cutters and clusters of cookie cutters.

11.2. N-Sided Polyhedra

Let us generalize some of this to the N-sided polygon case. Let N be the number of faces of this polygon. Then, the number of vertices or corners C and the number of edges E can be written as follows:

$$\begin{aligned} N &= \#\text{ faces} \\ E &= 3(N-2) = \#\text{ edges} \\ C &= 2(N-2) = \#\text{ corners} \end{aligned} \tag{11.2.1}$$

Let us label each of these faces by i, which numbers from 1 to N. The total number of variables appearing in our formulas given by a_{ij} is equal to the number of edges E.

Let us now calculate the number of independent variables within the N-sided polyhedra. We set the total perimeter of each side (corresponding to an external closed loop) equal to 2π. This constraint can be easily enforced by setting:

$$\sum_{i=1}^{N} a_{ij} = 2\pi \tag{11.2.2}$$

Then, the total number of independent variables is equal to the number of a_{ij}, or edges E, minus the number of constraints, or N. Thus, the total number of variables (or Koba–Nielsen variables) is equal to

$$E - N = 2N - 6 = \# \text{ Koba} - \text{Nielsen variables} \qquad (11.2.3)$$

which is the correct counting. (For the N-point scattering amplitude, there are $2N$ Koba–Nielsen variables z_i, but 6 of them can be eliminated by choosing 3 of them to be the points $0, 1, \infty$, leaving the number $2N - 6$ for the independent Koba–Nielsen variables.)

Now, let us compare this with the number of variables appearing in the conformal map. The map is [5, 6]:

$$\frac{d\rho}{dz} = N \frac{\prod_{i=1}^{N-2} \left[(z - w_i)(z - \tilde{w}_i)\right]^{1/2}}{\prod_{j=1}^{N}(z - \gamma_j)} = \frac{f(z)}{g(z)} \qquad (11.2.4)$$

The unknowns coming from γ_i are directly related to the Koba–Nielsen variables. Notice that we have $2N$ variables contained within the complex γ_i. However, we know that three of them can be fixed to be $0, 1, \infty$, so we really only have $2N - 6$ variables within the γ_i, which is precisely the number of Koba–Nielsen variables.

The total number of remaining unknowns is given by the $2 \times 2 \times (N-2)$ variables contained within the complexes w_i and $\tilde{w}_i$, as well as two coming from N. Thus, the total number of remaining unknowns is equal to $4N - 6$.

These remaining unknowns can be fixed by placing external constraints on the theory. Notice that we must set N external strings to have length 2π, which can be enforced, as before, as:

$$2\pi = \lim_{z \to \gamma_i} \frac{f(z)(z - \gamma_i)}{g(z)} \qquad (11.2.5)$$

This gives us a total of $2N$ constraints.

Last, we have the constraints coming from the fact that the collision of N strings takes place simultaneously at a constant value of τ in the z plane. Thus, we wish to set the real parts of all interacting points to be the same:

$$\text{Re } \rho(w_i) = \text{Re } \rho(\tilde{w}_j) \qquad (11.2.6)$$

for all i and j. Since there are $N - 2$ pairs of interacting points, this gives us $2 \times (N - 2)$ constraints, minus two. Altogether, we have $2N + 2(N - 2) - 2$ constraints, for a total of $4N - 6$ constraints. Notice that this is precisely equal to the number of unknowns in the mapping.

Let us now summarize how the counting proceeds for the N-sided polyhedra and also the conformal map. For the Koba–Nielsen variables, we have:

$$\text{Koba} - \text{Nielsen variables} = 2N - 6 = \begin{cases} E - N \\ \#(\gamma_i) - 6 \end{cases} \qquad (11.2.7)$$

while for the unknowns or the constraints, we have:

$$4N - 6 \text{ unknowns}: \quad (N, z_i)$$
$$4N - 6 \text{ constraints}: \quad \left[\rho(\gamma_i), \text{Re}\,\rho(z_i) \text{ the same}\right] \tag{11.2.8}$$

Now that we have determined precisely the relationship between constraints and unknowns, we must tackle the more difficult constraint of setting limits on the range of the a_{ij}. The key to this is to realize that, for the cases of $N = 4$ and $N = 5$, we had the curious constraint $a_{ij} \leq \pi$.

To understand this curious constraint, let us introduce the idea of "slicing" the polygon in half, that is, dividing up the N faces into two sets, such that the faces in each set are contiguous. Let us say that the number of ways we can partition the polygon into two sets is labeled by I. Call the set on the left L with elements labeled by i and the set on the right as R with elements labeled by j. The number of contiguous faces within L or R must be greater or equal to 2. Then, define P_I as the perimeter of the slice:

$$P_I = \sum_{i \in L; j \in R} a_{ij} \tag{11.2.9}$$

We will now generalize the curious constraint as follows for the arbitrary N-sided polyhedra [5–7]:

$$P_I \geq 2\pi, \qquad \text{for all } I \tag{11.2.10}$$

If we define M_N to be the region in a_{ij} space defined by the N-sided polyhedron, then the boundary of M_4 corresponds to $P_I = 2\pi$:

$$\partial M_N = \lim_{P_I = 2\pi} M_N \tag{11.2.11}$$

For example, for the $N = 4$ case, there are 3 ways in which we can slice the tetrahedron, with 2 faces in L or R.

Let us take the partition so that $i = 1$ and $i = 2$ faces are within L. But, demanding that the perimeter P_I of the four-sided figure in L be greater than 2π is equivalent to fixing a constraint on the common boundary between these faces $a_{12} \leq \pi$. Thus, the origin of the constraint $a_{ij} \leq \pi$, first found for $N = 4$, is simply the constraint that $P_I \geq 2\pi$ for the tetrahedron, that is,

$$a_{ij} \leq \pi \leftrightarrow P_I \geq 2\pi, \qquad \text{for } N = 4 \tag{11.2.12}$$

11.3. Nonpolynomial Action

Let us now write the nonpolynomial action that obeys all these constraints [5–7]. There are, however, several major complications in addition to those found for the open string field theory.

First, we must deal with the ghost counting problem. Since the Fock space of the closed string field $\Psi(X)$ is composed of products of the left-moving and right-moving states of the open string field $\Phi(X)$, the ghost number of the closed string field must be integral. By carefully examining the cohomology of Ψ, it can be shown that it contains two complete transverse closed string Fock spaces, with ghost numbers 0 or -1. For either case, however, the naive closed string action must vanish because the ghost numbers do not sum to zero:

$$\langle \Psi | Q \Psi \rangle \equiv 0 \qquad (11.3.1)$$

If the Ψ field has ghost number 0 (-1), then the naive action has ghost number $+1$ (-1), so the action vanishes in either case.

We have two ways in which to get the ghost numbers to match. We can insert the operators $b_0 - \bar{b}_0$ or $c_0 - \bar{c}_0$ into the action if Ψ has ghost number 0 or -1. The action is then nonzero. But, there is the second complication, which is that we must somehow obtain the constraint:

$$(L_0 - \tilde{L}_0)|\Psi\rangle = 0 \qquad (11.3.2)$$

which states that the string field should have no dependence on the origin of the coordinate axis. Since this constraint does not emerge from the action, it must be imposed from the outside, as an additional constraint (although attempts have been made to derive this from an additional ghost constraint). [To deal with these questions, the philosophy we take in this chapter is that the nonpolynomial action is inherently a gauge fixed action. Because reparameterization invariance lies at the heart of string theory and since the nonpolynomial action breaks reparameterization invariance (for example, because the string has fixed parameterization length 2π), we will treat the action as the by-product of gauge fixing a higher action, so that the imposition of Eq. (11.3.2) from the outside poses no problem. If certain rules for the ghost modes seem a bit artificial, it is because the nonpolynomial action is the gauge fixed by-product of a reparameterization invariant action.]

There are, therefore, several ways in which to construct the closed string action at the free level that lead eventually to the same cohomology and hence identical transverse states. We thus make the following choices. We define the string field as follows:

$$\Psi = c_0^- |\phi\rangle + c_0^- c_0^+ |\psi\rangle + |\chi\rangle + c_0^+ |\eta\rangle \qquad (11.3.3)$$

where the physical transverse states reside in the lowest excitation of $|\phi\rangle$ and where $c_0^+ = \frac{1}{2}(c_0 + \bar{c}_0)$, $c_0^- = c_0 - \bar{c}_0$, $b_0 = b_0 + \bar{b}_0$, and $b_0^- = \frac{1}{2}(b_0 - \bar{b}_0)$.

So, the action reads:

$$\langle \Psi | Q b_0^- | \Psi \rangle \qquad (11.3.4)$$

where we define the string field Ψ to have the factor P of Eq. (11.1.1) inserted at all times. We will simply abbreviate this action by the usual notation $\langle \Psi | Q\Psi \rangle$, where it is understood that we apply all these conventions. (In general, we will omit the ghost insertions for convenience. They can be easily re-inserted.)

When we generalize this action to the interacting case, we find yet another complication, which is that there are many polyhedron at the Nth level that satisfy all the constraints. Let us label by the index i the various polyhedra at the Nth level that satisfy all constraints.

For example, for $N = 4$ and $N = 5$, we have only 1 polyhedron that satisfies these constraints. However, for the $N = 6$ case, we have 2 polyhedra. At the $N = 7$ case, we have 5 distinct polyhedra, and at the $N = 8$ level, we have 17 different polyhedra.

From now on, we will label the ith polygon with n faces as $(n)_i$. As we did in the light cone and covariant open string cases, we can define the following vertex function (without ghosts), which satisfies the constraints of the $(n)_i$ polygon:

$$\prod_{i=1}^{n} \left[X_i(\sigma_i) - \sum_{j=1}^{n} \theta_{ij} X_j(\sigma_j) \right] |V_{(n)_i}\rangle = 0 \qquad (11.3.5)$$

where:

$$\theta_{ij} = \eta_{ij} \theta(\sigma_i - b_{ij}) \theta(c_{ij} - \sigma_i) \qquad (11.3.6)$$

and:

$$\eta_{ij} = \begin{cases} +1 & \text{if} \quad a_{ij} \neq 0 \\ 0 & \text{otherwise} \end{cases} \qquad (11.3.7)$$

If the ith and jth polygons share no boundary, then $a_{ij} = 0$, and Eq. (11.3.6) is equal to 0. Also, b_{ij} and c_{ij} for the $(n)_i$ polyhedra are defined to be the common boundary between the adjacent polyhedra and are a rather obvious generalization of the three-string vertex in Eq. (10.2.20) in the light cone case.

Then, let us define, for the $(n)_i$ polyhedra, the following string functional:

$$\langle \Psi^n \rangle_i \equiv (n!)^{-1} \langle \Psi_{[1}| \langle \Psi_2| \langle \Psi_{n]}| \dots V_{(n)_i}\rangle \qquad (11.3.8)$$

Now, let us define the field functional for the n-sided polyhedron by summing over the index i:

$$\langle \Psi^n \rangle \equiv \sum_i c(n)_i \langle \Psi^n \rangle_i \qquad (11.3.9)$$

where the coefficients $c(n)_i$ tell us how to weight the ith polygon with n sides.

Finally, we write the action for the nonpolynomial theory:

$$L = \langle \Psi | Q\Psi \rangle + \sum_{n=3}^{\infty} \alpha_n \langle \Psi^n \rangle \qquad (11.3.10)$$

which we demand is invariant under:

$$\delta|\Psi\rangle = Q|\Lambda\rangle + \sum_{n=1}^{\infty} \beta_n|\Psi^n\Lambda\rangle \qquad (11.3.11)$$

Our goal is to find explicit values for α_n, β_n, and especially $c(n)_i$ by demanding that the action be invariant under the variation. One complication is that external strings are always defined such that their length is 2π, while lines appearing within the vertex function may have lengths different from 2π. Thus, to differentiate between these two cases, we will use the double bars $||$ whenever the contraction is over states with length 2π.

The variation of the vertex function is thus:

$$\delta\langle\Psi^n\rangle = n\langle\Psi^{n-1}||\delta\Psi\rangle \qquad (11.3.12)$$

We have used double bars here, because the length of the string $\delta|\Psi\rangle$ is 2π. Now, we can take variation of the entire action:

$$\begin{aligned}
\delta L &= 2\langle Q\Psi||\delta\Psi\rangle + \sum_{n=3}^{\infty} n\alpha_n\langle\Psi^{n-1}||\delta\Psi\rangle \\
&= 2\left\langle Q\Psi||\sum_{n=1}^{\infty}\beta_n\Psi^n\Lambda\right\rangle + \sum_{n=3}^{\infty} n\alpha_n\langle\Psi^{n-1}||Q\Lambda\rangle \qquad (11.3.13) \\
&\quad + \sum_{n=3,m=1} n\alpha_n\beta_m\langle\Psi^{n-1}||\Psi^m\Lambda\rangle
\end{aligned}$$

This variation is equal to zero if we have:

$$\begin{aligned}
&2\langle Q\Psi||\beta_{n-1}\Psi^{n-1}\Lambda\rangle + (n+1)\alpha_{n+1}\langle\Psi^n||Q\Lambda\rangle \\
&+ \sum_{p=1}^{n-2}(n-p+1)\alpha_{n-p+1}\beta_p\langle\Psi^{n-p}||\Psi^p\Lambda\rangle = 0
\end{aligned} \qquad (11.3.14)$$

Expressed in this way, the formula may not be that transparent, so let us write it in the form:

$$(-1)^n\langle\Psi^n||Q\Lambda\rangle + n\langle Q\Psi||\Psi^{n-1}\Lambda\rangle + \sum_{p=1}^{n-2} C_p^n\langle\Psi^{n-p}||\Psi^p\Lambda\rangle = 0 \qquad (11.3.15)$$

where the unknowns are now encoded within C_p^n and $c(n)_i$.

We now notice another complication not found in the open string case. We realize that the vertex function in Eq. (11.3.14) *cannot* be BRST invariant:

$$\sum_{i=1}^{n} Q_r|V_{(n)_i}\rangle \neq 0 \qquad (11.3.16)$$

This means, of course, that the theory cannot be written with the standard cohomological axioms [Eqs. (10.4.1)–(10.4.4)] that made the open string theory so elegant and simple. Any naive attempt to fit the closed string field theory into the cohomological framework is doomed to fail because of the failure of BRST invariance of the vertex function.

From Eq. (11.3.14), we can immediately write the first equalities:

$$\beta_n = \frac{1}{2}(-1)^{n-1}(n+1)(n+2)\alpha_{n+2}$$
$$C_p^n = \frac{(n-p+1)\alpha_{n-p+1}\beta_p}{(n+1)\alpha_{n+1}} \tag{11.3.17}$$

Let us, for the moment, keep only the C_1^n as unknowns. Then, we can define:

$$\bar{n}! \equiv \prod_{l=2}^{n} C_1^l \tag{11.3.18}$$

so that:

$$\alpha_n = \frac{2^{n-1}g^{n-2}}{n\,\overline{n+1}!}$$
$$\beta_n = \frac{(-1)^{n-1}2^n(n+1)g^n}{\overline{n+1}!} \tag{11.3.19}$$
$$C_p^n = \frac{(-1)^{p-1}\bar{n}!(p+1)}{\overline{n-p}!\,\overline{p+1}!}$$

where g is a one-parameter degree of freedom within the constraints that corresponds to the coupling constant.

Now comes the more difficult part, actually calculating the various coefficients $c(n)_i$ that appear within the expansion. The calculation is long and arduous, so we will only mention the important aspects of the calculation. Let us define $s(n)_i$ as the number of ways that the n_i polyhedron can be rotated into itself. For example, for the lower polyhedra, it is easy to see (Fig. 11.9):

$$s(4) = 12$$
$$s(5) = 6$$
$$s(6)_1 = 24, \qquad s(6)_2 = 2 \tag{11.3.20}$$
$$s(7)_1 = 10, \qquad s(7)_2 = 3, \quad s(7)_3 = 3$$
$$s(7)_4 = s(7)_5 = 2$$

It turns out that the key to the entire calculation lies in the identity:

$$s(n)_i c(n)_i = s(n)_j c(n)_j \tag{11.3.21}$$

for any i and j. Then, we can write everything in terms of $s(n)_1$ and $c(n)_1$:

$$\alpha_n = 2^{n-2}g^{n-2}/s(n)_1 c(n)_1$$
$$s(n)_1 = 2(n-2) \tag{11.3.22}$$

If we factor out $s(n)_1$ in the vertex, then we can write a new vertex, called the symmetrized vertex:

$$\langle \Psi^n \rangle_{\text{SYM}} = \frac{1}{s(n)_1} \langle \Psi^n \rangle \qquad (11.3.23)$$

Then, the final result is:

$$L = \frac{1}{2} \langle \Psi | Q \Psi \rangle + \sum_{n=3}^{\infty} 2^{n-2} g^{n-2} \langle \Psi^n \rangle_{\text{SYM}} \qquad (11.3.24)$$

This action can be written in a slightly more transparent fashion. Set $g = \frac{1}{2}$, and write the interaction explicitly in terms of the $(n)_j$ polyhedra. Then, the action becomes:

$$L = \frac{1}{2} \langle \Psi | Q \Psi \rangle + \sum_{n=3}^{\infty} \frac{\langle \Psi^n \rangle_j}{s(n)_j} \qquad (11.3.25)$$

This is our final form for the action. It seems rather surprising that the overall coefficient of the interaction corresponding to the polyhedra $(n)_j$, modulo rotations, is exactly equal to 1! (The origin of this rather strange fact is revealed in Chapter 12, when we discuss reparameterizations.)

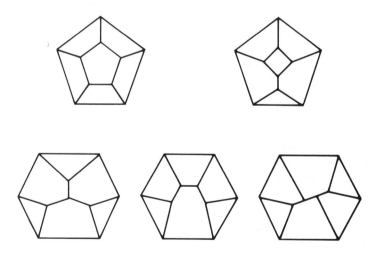

Fig. 11.9.

11.4. Conformal Maps

For the scattering of four closed strings (with equal circumference) at the tree and one-loop level, the conformal map is easy to write. For the tree amplitude, the Schwarz-Christoffel transformation gives us the following map:

$$\frac{d\rho}{dz} = N \frac{\prod_{i=1}^{4}(z - \nu_i)^{1/2}}{\prod_{j=1}^{4}(z - \gamma_j)} \tag{11.4.1}$$

where the points γ_i are mapped to infinity and represent external strings, while Riemann cuts connecting the various ν_i represent the interaction, that is, the line along which two closed strings merge into a third.

To obtain the one-loop, four-puncture scattering amplitude, one simply replaces the various factors $(z - z_i)$ appearing in the above map with $\theta_1(z - z_i)$. However, for higher loops, replacing θ_1 with the generalized Θ function fails to yield the conformal map for the genus g conformal map. The generalization to the arbitrary case is nontrivial.

As a result, we will derive the conformal map $d\rho = \omega_z dz$ for the g-handle, p-puncture graph by carefully analyzing uniqueness arguments.

We wish to have

1. a conformal map whose square transforms as a quadratic differential;
2. a map whose only singularities are double poles at p points γ_i, and whose only zeros are $2p + 4g - 4$ points ν_i; and
3. a periodic function defined on the surface so that the point z going around any a cycle or b cycle will return back to the same point.

Our task is to use these restrictions to find the unique conformal map taking us from a Riemann surface with g handles and p punctures to the flat two-dimensional world sheet describing a Feynman-like diagram with g internal loops and p external legs.

To find the unique conformal map, it will be useful to first review some essentials concerning holomorphic and meromorphic functions defined on a Riemann surface. On a genus g Riemann surface, we can define g first Abelian differentials ω_i on the torus that have no singularities.

We can now define Ω as the period matrix, which obeys:

$$\Omega_{ij} = \oint_{b_j} \omega_i; \qquad \delta_{ij} = \oint_{a_i} \omega_j \tag{11.4.2}$$

The period matrix is conformally invariant. To each distinct period matrix there is a distinct Riemann surface.

Now, we wish to define periodic functions on this Riemann surface. Let us review how this was done for the single-loop torus. For that case, we know that we can deform the a cycle and b cycle so that they intersect at a common point on the surface. If we cut the torus along the a cycle and b cycle and then unravel the surface, we find a parallelogram, as in Chapter 4, whose opposite sides are identified. If we copy this parallelogram an infinite number of times on the complex plane, then we obtain a lattice. Then, it is straightforward to define a ϑ function on this surface that has the correct periodicity properties. This periodicity is achieved by summing a function over the infinite lattice, so that displacements along the lattice leave the function invariant.

Let us repeat these same steps for the genus g Riemann surface. There are now $2g$ cycles. Now, take a point on this surface and deform the cycles so that they all intersect this point once. Cut along these $2g$ cycles that intersect this point. Unravel the surface and find a polygon whose sides are identified. By traveling across any side of this polygon, we return to the polygon, but from another side. Now, extend this polygon periodically in all directions, so that we have a lattice of polygons.

Let us label the sites of the lattice. Let bold-faced indices $\mathbf{n}$ be g-component vectors with integer entries. Then, the generalized Θ function can be defined on the lattice. Its periodicity property arises because it is summed over all lattice sites. It can be defined as [12–15]:

$$\Theta(\mathbf{z}|\Omega) = \sum_{n \in Z^g} \exp\left(i\pi \mathbf{n}^T \Omega \mathbf{n} + 2\pi i \mathbf{n}^T \mathbf{z}\right) \tag{11.4.3}$$

The Θ function is no longer defined as a simple function of the complex parameter z. Instead, it is defined in terms of a vector $\mathbf{z}$ on the lattice:

$$\mathbf{z} = \int_{p_0}^{z} \omega \tag{11.4.4}$$

where p_0 is an arbitrary point (which will disappear when we form the conformal maps).

As we know from our discussion of the single-loop torus, we can also define a spin structure on the surface, depending on whether we have periodic or antiperiodic boundary conditions when we move completely around a cycle. There are four possible spin structures for the torus. Likewise, we can define a generalized Θ function on the lattice with a spin structure:

$$\Theta\begin{bmatrix}\alpha\\\beta\end{bmatrix}(\mathbf{z}|\Omega) = \sum_{\mathbf{r}-\alpha \in Z^g} \exp\left[i\pi \mathbf{r}^T \Omega \mathbf{r} + 2\pi i \mathbf{r}^T(\mathbf{z}+\beta)\right] \tag{11.4.5}$$

where the spin structures $[\alpha]$ and $[\beta]$ are two component spinors with g entries.

There are 2^{2g} possible spin structures, corresponding to the different ways in which we can transport a two-dimensional spinor across the various boundaries or cycles of the polygon, picking up factors of $+1$ or -1 in the process. Under a shift in the lattice,

$$\mathbf{z} \rightarrow \mathbf{z} + \Omega \cdot \mathbf{n} + \mathbf{m} \tag{11.4.6}$$

we find that the generalized Θ function transforms as:

$$\Theta \begin{bmatrix} \alpha \\ \beta \end{bmatrix} (\mathbf{z} + \Omega \cdot \mathbf{n} + \mathbf{m} | \Omega) = e^{2\pi i \alpha \cdot \mathbf{m} - i\pi \mathbf{n} \cdot \Omega \cdot \mathbf{n} - 2\pi i \mathbf{n} \cdot (\mathbf{z} + \beta)} \Theta \begin{bmatrix} \alpha \\ \beta \end{bmatrix} (\mathbf{z} | \Omega) \quad (11.4.7)$$

We can also define the function that generalizes the function $z_i - z_j$ on the complex plane. It is called the "prime form," and is represented by Θ functions as follows:

$$E(z', z) = \frac{\Theta \begin{bmatrix} \alpha \\ \beta \end{bmatrix} \left(\int_{z'}^{z} \omega | \Omega \right)}{\sqrt{h \begin{bmatrix} \alpha \\ \beta \end{bmatrix} (z) h \begin{bmatrix} \alpha \\ \beta \end{bmatrix} (z')}} \quad (11.4.8)$$

where:

$$h \begin{bmatrix} \alpha \\ \beta \end{bmatrix} (z) = \sum_{i=1}^{g} \frac{\partial \Theta \begin{bmatrix} \alpha \\ \beta \end{bmatrix} (0 | \Omega)}{\partial z_i} \omega_i(z) \quad (11.4.9)$$

where α and β label the spin structure on the Riemann surface. (The prime form's dependence on the spin structure will drop out.)

We will also make use of the celebrated Riemann vanishing theorem [12, 13], which allows us to compute the zeros of the Θ function. It states that the function:

$$\Theta \left(\mathbf{e}_0 + \int_{z_0}^{z} \omega | \Omega \right) \quad (11.4.10)$$

either vanishes identically or has g zeros, which are located at the points z_i, which satisfy:

$$\sum_{i=1}^{g} \int_{z_0}^{z_i} \omega = -\mathbf{e}_0 + \Delta_{z_0} + \mathbf{n} + \Omega \mathbf{m} \quad (11.4.11)$$

where:

$$(\Delta_{z_0})_j = \frac{1}{2} - \frac{\Omega_{jj}}{2} + \sum_{i=1; i \neq j}^{g} \oint_{a_i} \omega_i(z) \int_{z_0}^{z} \omega_j(w) \quad (11.4.12)$$

Unfortunately, with Θ and $E(z, z')$ alone, we cannot satisfy constraints (1), (2), and (3). We need yet one more function defined on the lattice, given by the $\sigma(z)$ function of Ref. 13 and15:

$$\sigma(z) = \exp \left[-\sum_{i=1}^{g} \oint_{a_i} \omega_i(z') \ln E(z', z) \right] \quad (11.4.13)$$

In Ref. 16, these three functions were used to construct the conformal map for the multiloop open string case. We will generalize this map for the genus g closed string case.

Let us now analyze the singularity and conformal properties of these three functions. We note that the prime form $E(z', z)$ transforms as a $-1/2$ differential in z, that Θ is locally a zero differential, and that σ transforms as a $g/2$ differential.

$E(z', z)$ has a zero at $z' = z$ but no poles, σ has no zeros or poles, and Θ, by the Riemann vanishing theorem, can have g zeros [12–13]. Then, there is one unique combination of Θ, prime form, and the σ function that transforms as a quadratic differential and has the correct zero and pole structure [16, 17]:

$$\left(\frac{d\rho}{dz}\right)^2 = N^2 \Theta \left(\Delta_{z_o} - \sum_{i=1}^{g} \int_{z_o}^{\nu_i} \omega + \int_{z_o}^{z} \omega \Big| \Omega\right)$$
$$\times \frac{\prod_{i=g+1}^{2p+4g-4} E(z, \nu_i | \Omega)}{\prod_{j=1}^{p} E(z, \gamma_j | \Omega)^2} |\sigma(z)|^3$$

(11.4.14)

By counting the differential order of each factor, we see that the conformal map has the correct order of 2:

$$-\frac{1}{2}(2p + 3g - 4) + p + \frac{3g}{2} = 2$$

(11.4.15)

The conformal map is then found by taking the integral of the square root of the map.

Let us now relate the singularity of the map with the world sheet of the closed string scattering amplitudes. Because of the prime forms $E(z', z)$ in the denominator, the double poles, corresponding to the p punctures, are located at γ_i. Furthermore, the $2p + 4g - 4$ zeros at ν_i come from the Θ function and the prime forms in the numerator. The double poles, after taking the square root, then correspond to the p external lines of the string scattering world sheet. Each pair of zeros of the map, in turn, correspond to the merger of three closed strings. (At first, the map seems to be unsymmetrical with respect to the various ν_i, since some of them are to be found within the Θ function and others within the prime form. However, by simply redefining the constant N, we can interchange the various ν_i and show that the function is really symmetrical in all ν_i.)

Now that we have explicitly constructed the conformal map for the nonpolynomial theory for arbitrary genus g and puncture p, let us count modular parameters and verify that we have the correct counting. In general, we want the total number of unknowns, contained within the complex parameters in the map, to be equal to the sum of the dimension of moduli space $(6g - 6 + 2p)$ plus the number of constraints we place on the complex parameters to fix the overall shape of the conformal surface, that is,

$$\# \text{ unknowns} = \# \text{ constraints} + \# \text{ moduli}$$

(11.4.16)

This equation is easily checked. The total number of unknowns in the conformal map is given by the complex variables N, γ_i, and ν_i, which respectively total $2 + 2p + 2(2p + 4g - 4) = 6p + 8g - 6$ unknowns. Furthermore, the dimension of moduli space is equal to $6g - 6 + 2p$. The number of constraints can be broken down as follows: (1) 6 come from fixing the overall projective transformations on the complex plane; (2) $2p$ come from fixing the residue of the pole at each γ_i to be a real number 2π, which fixes the circumference of each cylinder; (3) $2(p+g-2)$ come from fixing the real and imaginary parts of the various Riemann cuts to conform to the geometry of a closed string scattering amplitude. If we pair off the points ν_i, then

$$\operatorname{Re}\rho(\nu_i) = \operatorname{Re}\rho(\nu_j) \qquad (11.4.17)$$

for all i and j in a pair. This places the Riemann vertically in the complex ρ plane. Then, we also have:

$$\operatorname{Im}\rho(\nu_i) - \operatorname{Im}\rho(\nu_j) = \pm\pi \qquad (11.4.18)$$

which fixes the overlap between the ith and jth string to be π.

Last, we must subtract two from the number of constraints. This is because the system is actually overconstrained. The sum of the residues at γ_j plus the line integral around the Riemann cuts equal zero. [If we take a line integral around an infinitesimally small circle in the complex plane, the residue is zero. Now, expand this small circle until it engulfs all pairs of Riemann cuts and extends out to infinity. Because the line integral is still zero, this means that the sum of the residues at γ_i do not sum to zero (as in the light cone case), but cancels against the sum of the line integrals around the Riemann cuts.]

Putting everything together, we now have $4p + 2g$ constraints, so the total number of unknowns $6p + 8g - 6$ equals the sum of the number of moduli plus the number of constraints, as expected.

The conformal map also accommodates the possibility of arbitrary polyhedra occurring within a loop amplitude. Polyhedra occur when the real parts of several pairs of Riemann cuts coincide and the difference of their imaginary parts no longer equals π. Since each ν_i is mapped onto a vertex of a polyhedron, then the various edges within the polyhedron can have varying lengths, corresponding to varying differences between the imaginary parts of $\rho(\nu_i)$.

For example, the conformal map can create a polyhedron with M vertices when the real parts of $M/2$ pairs of Riemann cuts all have the same real part $\operatorname{Re}\rho(\nu_i)$. By varying the differences within their imaginary parts, one can vary the lengths of the edges of the polyhedron. In this way, by clustering pairs of Riemann cuts located at ν_i into different groups, we can create polyhedra of arbitrary complexity within a loop diagram.

(As an aside, notice that the map is so general that we can also accommodate all possible light cone configurations as well by changing boundary conditions. By letting the external strings have arbitrary circumferences and by collapsing all pairs of Riemann cuts into single points, we find that the sum of the residues of the poles now sums to zero, thus reproducing the boundary conditions of the light cone theory. Furthermore, it is commonly thought that one cannot smoothly distort a theory where strings interact at their midpoints into a theory where strings interact at their endpoints, as in the light cone theory, that is, the light cone limit is singular. However, this is incorrect. By explicit computation, one can show that one can smoothly take the limit in the conformal map and reach the light cone theory, as predicted in geometric string field theory [6].)

11.5. Tadpoles

The conformal map [Eq. (11.4.14)] is general enough to include all possible multiloop graphs, including the one-loop tadpole, where the residues of the poles do not have to sum to zero. Unfortunately, we will show that these tadpoles violate modular invariance (giving us an infinite overcounting of moduli space) and hence pose problems for our original action [Eq. (11.3.24)]. Originally, it was thought that the violation of modular invariance for the tadpole was sufficient to kill all possible closed string field theories. We will see that there is a simple resolution to this puzzle, which is that Eq. (11.3.24) must be treated as a classical action and that quantum corrections must be added to it to restore modular invariance. (These tadpoles do not occur in the light cone field theory because the string length is proportional to the momentum p^+. Since momentum is conserved, so is string length across a three-string vertex, so tadpole graphs are forbidden. However, in the nonpolynomial theory, because string lengths are fixed and therefore independent of the momentum, tadpole graphs must be considered.)

Let us specialize to the case of $g = 1$. Then,

$$\sigma(z) = 1; \qquad \Delta_{z_o} = \frac{1}{2} - \frac{\tau}{2} \tag{11.5.1}$$

and the prime form reduces to:

$$E(z, w) = \frac{2\pi i \sqrt{zw} \Theta\left[\frac{1}{2}\right]\left(\frac{\ln(z/w)}{2\pi i}\middle| \Omega\right)}{\Theta\left[\frac{1}{2}\right]\left(0\middle| \Omega\right)} \tag{11.5.2}$$

Then, the map for the $g = 1$, p-puncture diagram is

$$\frac{d\rho}{dz} = N \frac{\sqrt{\prod_{i=1}^{p} \theta_1(z - \nu_i)\theta_1(z - \tilde{\nu}_i)}}{\prod_{i=1}^{p} \theta_1(z - \gamma_i)} \tag{11.5.3}$$

where ν_i and $\tilde{\nu}_i$ are the splitting points, γ_i the punctures, and

$$N = \frac{\theta_1'(0) \prod_{i=2}^{p} \theta_1(\gamma_1 - \gamma_i)}{\sqrt{\prod_{i=1}^{p} \theta_1(\gamma_1 - \nu_i)\theta_1(\gamma_1 - \tilde{\nu}_i)}} \tag{11.5.4}$$

In this form, the poles and zeros of the map are manifest. However, from the theory of conformal maps, we also know that Riemann surfaces with genus $g \leq 2$ can be written as hyperelliptic functions (without using Θ functions) that are formed by gluing conformal planes together across several Riemann cuts. Let us check, therefore, that our map can be written as a hyperelliptic function for the one-loop, one-puncture tadpole, which can be written as [19]:

$$\frac{d\rho}{dz} = N \frac{\sqrt{z-y}}{\sqrt{z(z-1)(z-x)}}, \qquad N = \frac{1}{2} \qquad (11.5.5)$$

where we join the two surfaces along cuts that go from 0 to 1 and x to ∞ (see Fig. 11.10).

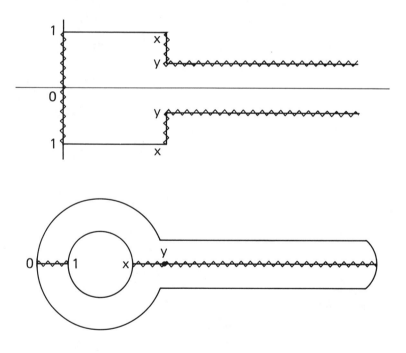

Fig. 11.10.

To show the equivalence between Eqs. (11.5.3) and (11.5.5), let us first make a series of changes of variables. We define $z = t^2$ and:

$$k^2 = \frac{1}{x} = \frac{\theta_2^4}{\theta_3^4}$$

$$t = \frac{\theta_3}{\theta_2}\frac{\theta_1(\nu)}{\theta_4(\nu)}$$

$$\nu = \frac{\mu}{\theta_3^2} \tag{11.5.6}$$

$$\mu = \int_0^t \frac{dt}{\sqrt{(1-t^2)(1-k^2t^2)}}$$

$$\theta_i = \theta_i(0)$$

Using standard theta-function relations, one finds:

$$1 - t^2 = \frac{\theta_2^2(\nu)\theta_4^2}{\theta_2^2\theta_4^2(\nu)}; \qquad 1 - k^2t^2 = \frac{\theta_3^2(\nu)\theta_4^2}{\theta_3^2\theta_4^2(\nu)}; \qquad dt = \frac{\theta_3}{\theta_2}\frac{\theta_4^2\theta_2(\nu)\theta_3(\nu)}{\theta_2^2(\nu)}\,d\nu \tag{11.5.7}$$

Also, define

$$y = \left[\frac{\theta_3}{\theta_2}\frac{\theta_1(\nu')}{\theta_4(\nu')}\right]^2 = \left[\frac{\theta_3}{\theta_2}\frac{\theta_4(\nu'+\tau/2)}{\theta_1(\nu'+\tau/2)}\right]^2 \tag{11.5.8}$$

Put everything together, we find:

$$\frac{d\rho}{d\nu} = \frac{\theta_1'(0)}{\theta_1(\nu'+\tau/2)}\frac{\sqrt{-\theta_1(\nu+\nu'+\tau/2)\theta_1(\nu-\nu'-\tau/2)}}{\theta_1(\nu)} \tag{11.5.9}$$

which is identical to Eq. (11.5.3), as desired. Because Eq. (11.5.5) is a hyperelliptic map, it is formed by sewing two Riemann sheets together across two cuts.

To conform with the usual string world sheet, we will parameterize the surface by two variables, t, which represents the circumference of the loop, and θ, which represents the twist angle within the loop. We now place the constraints: $(T = t + i\theta)$:

$$2\pi = \int_0^1 dz \sqrt{\frac{y-z}{z(1-z)(x-z)}}$$

$$T = \int_1^x dz \sqrt{\frac{y-z}{z(z-1)(x-z)}} \tag{11.5.10}$$

where $t \in [0, \infty]$ is the length of the loop, while $\theta \in [-\pi/2, \pi/2]$ is the twist angle.

To study the modular properties of this map [Eq. (11.5.5)], we must determine the region in τ space that corresponds to the tadpole moduli space given by:

$$0 \leq \theta \leq 2\pi$$

$$0 \leq T \leq \infty \tag{11.5.11}$$

The problem arises in the dangerous region $T \to 0$, which corresponds to $\tau \to 0$. We will be interested in the region $x \to 1$ and $y \to 1$. By a direct power expansion of Eq. (11.5.10) for small x, we can show:

$$T(x) \sim -\frac{\pi^2}{2} \frac{1}{\ln(x-1)} + O\left[(x-1)\ln(x-1)\right] \tag{11.5.12}$$

Now, let us expand τ as a power of x. We know that the period matrix τ can be written in terms of the integrals over the first Abelian differentials, as in Eq. (11.4.2). The hyperelliptic surface is created by gluing two sheets together along the cuts 0 to 1 and also between x to ∞. The holomorphic differential is then dz/y. By taking the integrals over the cycles, we find:

$$\tau = \frac{\omega_1}{\omega_2}; \qquad \omega_1 = \int_1^x \frac{dz}{y}; \qquad \omega_2 = \int_0^1 \frac{dz}{y} \tag{11.5.13}$$

where $y^2 = z(z-1)(z-x)$ as in Eq. (11.5.5).

We can write the relation between x and τ explicitly:

$$x(\tau) = \frac{\theta_3^4(0|\tau)}{\theta_1^4(0|\tau)} \tag{11.5.14}$$

Power expanding in small x, we find:

$$\tau(x) \sim -\frac{i\pi}{\ln[(x-1/16]} + O\left[(x-1)/\ln(x-1)\right] \tag{11.5.15}$$

Comparing Eqs. 11.5.12) and (11.5.15), we find that the final result linking T and τ is [19]:

$$\tau \sim \frac{2i}{\pi} T(\tau) \tag{11.5.16}$$

for small T. This is bad for modular invariance. Because moduli space [Eq. (11.5.11)] includes a finite region around $T \sim 0$, this also means that it includes a finite region around $\tau \sim 0$. But, the origin of the τ plane includes an infinite number of copies of the modular region, and hence Eq. (11.5.11) (which comes from string field theory) maps into an infinite number of copies of the single-loop amplitude.

This unfortunate situation is not just particular to the one-puncture graph. For example, for the map of Eq. (11.4.14), we can also take the case of $g = 1$ and $p = 2$. By changing variables to hyperelliptic coordinates, we find that the map is given by [18]:

$$\frac{d\rho}{dz} = N \frac{\sqrt{(z-y)(z-\tilde{y})}}{(z+y_2)\sqrt{z(z-1)(z-x)}} \tag{11.5.17}$$

For different parameters, this map will correspond to the three different diagrams that one can write for a one-loop tadpole with two punctures. The circumference of the loop, the length of the stem of the tadpole, the twist angle, etc., can be formed by fixing the various line integrals along the cut. A careful analysis of this graph shows that when the length of the neck of the tadpole graph becomes infinite, we reproduce the original tadpole map with one puncture, which is known to violate modular invariance. Thus, this violation of modular invariance by tadpoles is a persistent feature of the action.

In sum, we find that our original nonpolynomial action successfully reproduces all tree graphs for the closed string theory, but fails at the loop level. This means that the action is really a *classical* action and that a fully *quantized* action must be modified by a second series of nonpolynomial graphs, which contain loop contributions that kill the tadpole divergences.

Furthermore, we can also show that the functional measure $D\Phi$ is not invariant under the gauge variation [Eq. (11.3.11)]. In other words, the original action [Eq. (11.3.24)] suffers from two problems, the lack of modular invariant multiloop graphs and the presence of anomalies in the variation of the functional measure. Actually, this problem is really a blessing in disguise. By adding a second nonpolynomial series of terms into the action, we can kill both problems [21].

At first, this may seem strange. The presence of anomalies in the functional measure is a local property of the field theory, while the lack of modular invariance is a global property of the modular group. It seems strange that by adding a second series of nonpolynomial terms that we can kill both contributions.

The origin of this is because the breakdown of modular invariance creates an infinite number of copies of the loop graphs, which in turn is responsible for the ultraviolet divergences of the theory. However, as shown by Shapiro, one can simply take one cover of moduli space and thus eliminate the ultraviolet divergences. Thus, the lack of ultraviolet divergences in string theory lies in modular invariance. Similarly, the presence of anomalies in the functional measure also arises from ultraviolet divergences. We find, therefore, that both the presence of anomalies and the lack of modular invariance have the common origin, the ultraviolet behavior of string theory. Thus, it is not surprising that a second nonpolynomial set of terms is required to make Eq. (11.5.24) complete and that it can solve both problems simultaneously.

11.6. Summary

The ease and elegance with which the covariant open string field theory could be written has to be balanced against the difficulty and frustration faced in attempts to construct the covariant closed string field theory. However, since the heterotic string is a closed string theory, it is important to solve this pressing problem. The origin of this problem lies deep within a classical mathematical problem, the triangulation of moduli space. After a century of unsuccessful attempts to find specific coordinates that could cover moduli space once and only once, only three triangulations have been discovered, and two of them arise naturally from string field theory.

The old light cone field theory, for example, successfully triangulated moduli space with only cubic actions. However, attempts to repeat the success of the open string field theory (with fixed parameterization length) failed. Specifically, the four-point scattering amplitudes created by gluing three symmetric closed string vertices together failed to cover the Koba–Nielsen space.

The missing region can be seen to have the topology of a tetrahedron. If a_{ij} is the length that adjoins the ith and jth leg of this tetrahedron, then we have 6 unknown a_{ij} and 4 constraints (because the circumference of each closed string equals 2π). The perimeter of this missing region is represented by the motions one can execute on a Rubik's cube, that is, by making rotations around the horizontal x or vertical z axis

$$
\begin{aligned}
A \to B : & \quad a_{12} = \pi, & a_{14} = \pi \to 0 \\
A \to C : & \quad a_{12} = \pi \to 0, & a_{14} = \pi \\
B \to C : & \quad a_{12} + a_{14} = \pi
\end{aligned}
\tag{11.6.1}
$$

The missing region is the triangular region contained within these boundaries:

$$
\text{missing region} : \quad \begin{cases} a_{12}, \ a_{13} \le \pi \\ a_{12} + a_{13} \ge \pi \end{cases}
\tag{11.6.2}
$$

It is easy to see that the missing region smoothly connects the s, t, and u Mandelstam channels together, in the same way that the open four-string graph smoothly connected the t and u channel graphs of the Veneziano amplitude in the light cone gauge.

By explicit calculation, we can then divide the Veneziano amplitude into several pieces, corresponding to scattering in the various Mandelstam channels and the missing region (which has no poles):

$$
A(s,t,u) = \sum_I A_{12}^I \frac{1}{s_{12} - m_I^2} A_{34}^I + (2 \leftrightarrow 3) + (2 \leftrightarrow 4) + M_R(s,t,u)
\tag{11.6.3}
$$

Not surprisingly, this missing region persists for the higher point functions. The five-point function has the missing region:

$$
M_{R_5} = \begin{cases} a_{12} \le a_{35} \le a_{12} + a_{24} \le a_{35} + \pi \\ a_{13} \le a_{24} \le a_{13} + a_{35} \le a_{24} + \pi \\ a_{12} + a_{13} \le \pi \end{cases}
\tag{11.6.4}
$$

For higher point graphs, we must add one more constraint. If we slice the polyhedron in half into two smaller polygons, then P_I, the perimeter of each polygon, must satisfy:

$$
P_I = \sum_{i \in L; j \in R} a_{ij}
\tag{11.6.5}
$$

(We need this constraint in order for the missing region to adjoin various Mandelstam channels, since each channel contains a P_I equal to 2π.) This missing region also shows up explicitly in the conformal map we use to map the complex plane to the world sheet of the closed string scattering amplitudes:

$$\frac{d\rho}{dz} = N\frac{\prod_{i=1}^{N-2}\left[(z-w_i)(z-\tilde{w}_i)\right]^{1/2}}{\prod_{j=1}^{N}(z-\gamma_j)} = \frac{f(z)}{g(z)} \qquad (11.6.6)$$

This map contains square roots, so the Riemann cuts correspond to the overlap between three closed strings. The counting of Koba–Nielsen variables for the tree diagrams comes out correct:

$$\text{Koba} - \text{Nielsen variables} = 2N - 6 = \begin{cases} E - N \\ \#(\gamma_i) - 6 \end{cases} \qquad (11.6.7)$$

while for the unknowns/constraints, we have:

$$\begin{aligned} 4N - 6 \text{ unknowns}: & \quad (N, z_i) \\ 4N - 6 \text{ constraints}: & \quad \left[\rho(\gamma_i),\ \text{Re}\,\rho(z_i)\ \text{the same}\right] \end{aligned} \qquad (11.6.8)$$

The action for the closed string field theory has several differences between the open string case.

1. Ghost counting comes out incorrectly, which necessitates inserting a ghost factor into the free action.
2. The constraint $(L_0 - \tilde{L}_0)|\Psi\rangle$ must be added in by hand (or the projection operator must be inserted everywhere in the action).
3. There is more than one polyhedron at each level.
4. The vertex function is not BRST invariant, so we cannot use the cohomology axioms found for the open string field theory.

As a first guess, let us postulate the action:

$$L = \langle\Psi|Q\Psi\rangle + \sum_{n=3}^{\infty}\alpha_n\langle\Psi^n\rangle \qquad (11.6.9)$$

which we demand is invariant under:

$$\delta|\Psi\rangle = Q|\Lambda\rangle + \sum_{n=1}^{\infty}\beta_n|\Psi^n\Lambda\rangle \qquad (11.6.10)$$

Inserting the variation into the action, we find that the resulting terms must all vanish, which reduces to the following:

$$(-1)^n\langle\Psi^n||Q\Lambda\rangle + n\langle Q\Psi||\Psi^{n-1}\Lambda\rangle + \sum_{p=1}^{n-2}C_p^n\langle\Psi^{n-p}||\Psi^p\Lambda\rangle = 0 \qquad (11.6.11)$$

where the unknowns are now encoded within C_p^n and $c(n)_i$.

After a lot of hard work, we find that the final action is quite simple:

$$L = \frac{1}{2}\langle\Psi|Q\Psi\rangle + \sum_{n=3}^{\infty} \frac{\langle\Psi^n\rangle_j}{s(n)_j} \qquad (11.6.12)$$

that is, the overall coefficient of each graph (modulo rotations) is equal to 1!

To generalize these results to loops, we want a conformal map that satisfies the following properties. We want

1. a conformal map whose square transforms as a quadratic differential;
2. a map whose only singularities are double poles at p points γ_i, and whose only zeros are $2p + 4g - 4$ points ν_i; and
3. a function defined on the Picard torus, so that the point z going around any a cycle or b cycle will return back to the same point.

Fortunately, we can construct periodic and quasi-periodic functions on the genus g Riemann surface using Θ functions, which are defined as:

$$\Theta(\mathbf{z}|\Omega) = \sum_{n\in Z^g} \exp\left(i\pi\mathbf{n}^T\Omega\mathbf{n} + 2\pi i\mathbf{n}^T\mathbf{z}\right) \qquad (11.6.13)$$

where $\mathbf{n}$ labels an integer-valued g-dimensional vector defined on the lattice formed by unraveling the Riemann surface by cutting along the various cycles. In terms of the prime form $E(z, z')$, the σ function, and the Θ function, we can write the complete conformal map:

$$\left(\frac{d\rho}{dz}\right)^2 = N^2\Theta\left(\Delta_{z_o} - \sum_{i=1}^{g}\int_{z_o}^{\nu_i}\omega + \int_{z_o}^{z}\omega\Big|\Omega\right)$$

$$\times \frac{\prod_{i=g+1}^{2p+4g-4} E(z, \nu_i|\Omega)}{\prod_{j=1}^{p} E(z, \gamma_j|\Omega)^2}|\sigma(z)|^3 \qquad (11.6.14)$$

There are problems, however, at the one-loop level. Tadpole graphs (which do not appear in the light cone theory because of momentum conservation) appear in the nonpolynomial theory. By taking $g = 1$ and $p = 1$ in the conformal map, we have:

$$\frac{d\rho}{dz} = N\frac{\sqrt{z-y}}{\sqrt{z(z-1)(z-x)}}, \qquad N = \frac{1}{2} \qquad (11.6.15)$$

Unfortunately, this map shows that modular invariance is violated. If we define T to be the circumference of the loop in the tadpole graph, then we find

$$\tau \sim \frac{2i}{\pi}T(\tau) \qquad (11.6.16)$$

Since the region around complex T is part of the parameter space of the field theory, this means that the field theory includes a finite region around $\tau \sim 0$, which is known to have an infinite number of copies of moduli space.

This means that our original action is actually only a classical action and that a second set of nonpolynomial terms must be added that can serve two functions. This second set

1. kills the anomalies that arise when we take the gauge variation of the functional measure $D\Psi$ and
2. eliminates the overcounting due to tadpoles.

In summary, we now have a successful closed string field theory that is the generalization of Einstein's equations when power expanded around the graviton. The nonpolynomial action reproduces all tree graphs by filling up all missing regions of moduli space, but it has to be supplemented by a second set of nonpolynomial terms that can kill all unwanted divergences and overcountings.

The difficulty of writing the nonpolynomial theory, however, leads us to suspect that it is the gauge fixed version of a higher theory. Since reparameterizations lie at the heart of string theory and since the nonpolynomial theory breaks reparameterization invariance (since the parameterization length of the string is 2π), we are led in the direction of trying to gauge reparameterizations, which we shall do in Chapter 12.

References

1. M. Kaku and K. Kikkawa, *Phys. Rev.* **D10**, 1110, 1823 (1974).
2. E. Witten, *Nucl. Phys.* **B268**, 253 (1986).
3. J. Lykken and S. Raby, *Nucl. Phys.* **B278**, 256 (1986).
4. A. Strominger, *Phys. Rev. Lett.* **58**, 629 (1987); *Nucl. Phys.* **B294**, 93 (1987).
5. M. Kaku in *Functional Integration, Geometry, and Strings*, 25th Karpacz Winter School, Feb. 20–Mar. 5, 1989, Z. Haba and J. Sobcyk, eds., Birkhaeuser, Basel (1989).
6. M. Kaku, *Phys. Rev.* **D41**, 3734 (1990); Osaka preprint OU-HET 121 (1989).
7. T. Kugo, H. Kunitomo, and K. Suehiro, *Phys. Lett.* **226B**, 48 (1989); T. Kugo and K. Suehiro, KUNS 988 HE(TH) 89/08 (1989); M. Saadi and B. Zwiebach, *Ann. Phys.* **192**, 213 (1989).
8. M. A. Virasoro, *Phys. Rev.* **177**, 2309 (1969).
9. J. Shapiro, *Phys. Lett.* **33B**, 361 (1970).
10. M. Kaku and J. Lykken, *Phys. Rev.* **D38**, 3067 (1988); [The missing region was first conjectured in: S. Giddings and E. Martinec, *Nucl. Phys.* **B278**, 256 (1986).]
11. M. Kaku, *Phys. Rev.* **D38**, 3052 (1988).
12. D. Mumford, *Tata Lectures on Theta*, Birkhaeuser, Basel (1983).
13. J. Fay, *Theta Functions on Riemann Surfaces*, Lecture Notes in Mathematics, Vol. 352, Springer-Verlag, Berlin and New York (1973).
14. L. Alvarez-Gaumé, G. Moore, and C. Vafa, *Comm. Math. Phys.* **106**, 1 (1986).
15. E. Verlinde and H. Verlinde, *Nucl. Phys.* **B288**, 357 (1987).
16. S. Samuel, CCNY preprint (1989).
17. L. Hua and M. Kaku, *Phys. Rev.* **D41**, 3748 (1987).
18. L. Hua and M. Kaku, *Phys. Lett.* **250B**, 56 (1990).

19. G. Zemba and B. Zwiebach, *J. Math. Phys.* **30**, 2388 (1989); H. Sonoda and B. Zwiebach, *Nucl. Phys.* **B331**, 592 (1990); B. Zwiebach, *Phys. Lett.* **241B** 343, (1990); B. Zwiebach, MIT-CTP 1909, 1910, 1911, 1912.
20. M. Saadi, *Mod. Phys. Lett.* **A5**, 551 (1990).
21. M. Kaku, *Phys. Lett.* **250B**, 64 (1990).

Chapter 12

Geometric String Field Theory

12.1. Why So Many String Field Theories?

One of the embarrassments of string field theory [1, 2] is that it has *fewer* symmetries than the first quantized approach. Reparameterization invariance is nowhere to be seen, which accounts for the fact that the midpoints and endpoints of strings become special points. Worse, the light cone theory seems to be totally incompatible with the midpoint theory. They seem to triangulate moduli space in entirely different ways. Each is totally different off-shell, each has a different set of interactions, yet they miraculously yield the same S matrix. For example, for closed strings, the nonpolynomial closed string field theory triangulates moduli space in a highly nontrivial fashion, while the light cone theory triangulates moduli space in an almost trivial fashion via strings that fission. Our goal is to postulate a geometric string field theory [3–7] in which reparameterization symmetry is built in from the start, that is, we wish to gauge the reparameterization group. In this way, one should be able to select as gauge choices either the endpoint or midpoint as special points along the string. Then, we should be able to view the various string field theories as gauge choices, that is, the "midpoint gauge" and "endpoint gauge. This, in turn, would give us a unifying principle by which all self-consistent string field theories could be seen as gauge choices of a single theory.

This will make sense out of the various string field theory actions, which can be symbolically represented as follows (where Φ represents an open string field and Ψ a closed string field). We will find that the geometric theory has a cubic interaction, but that, in various gauges, it can become quite complicated, with higher point interactions. From this cubic interaction, we will be able to derive the full gauge fixed theory, either in the midpoint or endpoint gauge.

In Fig. 10.3, for example, we saw the five fundamental open string interactions in the light cone string field theory for the endpoint gauge. In Fig. 10.4, however, we saw the simple cubic interaction for the midpoint gauge. The fact that both actions can be derived by gauge fixing the geometric theory can be summarized symbolically as:

$$(\Phi^3)_{\text{geometric}} = \begin{cases} (\Phi^3 + \Psi^3 + \Phi^2\Psi + \Phi^4 + \Psi\Phi)_{\text{endpoint}} \\ (\Phi^3)_{\text{midpoint}} \end{cases} \tag{12.1.1}$$

Similarly, in the last chapter, we saw that the covariant string field theory (in the midpoint gauge) was nonpolynomial, while the light cone string field theory was purely cubic. The fact that both of these theories can be derived from geometric string field theory can be summarized symbolically as:

$$(\Psi^3)_{\text{geometric}} = \begin{cases} (\Psi^3)_{\text{endpoint}} \\ (\Psi^3 + \Psi^4 + \cdots)_{\text{midpoint}} \end{cases} \tag{12.1.2}$$

The goal of this approach is to start with a group, develop its representations, and then construct curvatures and the action itself, in much the same way that ordinary gauge theories are constructed.

In ordinary gauge theory, the starting point is a group, such as $SU(N)$. Next, we demand that the connections transform as representations of the group. Then, we construct the curvatures and then the action itself. Gauge theories also allow us the freedom of choosing a continuum of gauges. For example, the R_ξ gauge allows us to continuously go from the renormalizable gauge to the unitary gauge, which superficially have nothing to do with each other. In the renormalizable gauge, unitarity is very obscure, and in the unitary gauge renormalizability is not obvious. However, the existence of the R_ξ gauge is enough to guarantee the existence of a theory that is both renormalizable and unitary.

In ordinary gauge theory, we also see the origin of why higher interactions occur in various gauges. Consider ordinary QED, for example, in the Coulomb gauge. The action contains the terms:

$$A_0 \nabla^2 A_0 + g A_0 \bar{\psi} \gamma_0 \psi \tag{12.1.3}$$

Notice that the propagator is not gauge invariant, and its time component can be completely eliminated. We can, therefore, eliminate the A_0 by functionally integrating it. After integration, the action contains the term [6]:

$$(\bar{\psi}\gamma_0\psi)\frac{1}{\nabla^2}(\bar{\psi}\gamma_0\psi) \tag{12.1.4}$$

This is called the instantaneous four-fermion *Coulomb term*, which arises because of a gauge choice. (It appears to violate relativity, because the $1/\nabla^2$ acts instantaneously between the pairs of fermions. However, this nonrelativistic effect is canceled by other terms, making the entire theory relativistic.)

The lesson here is that the propagator of a gauge theory is not gauge invariant; by changing the gauge, it is possible to eliminate the time dependence of the propagator completely for certain redundant gauge fields. By integrating over these redundant gauge fields, we obtain instantaneous four-fermion interactions.

Precisely the same thing happens in geometric string field theory. Because reparameterizations are now part of the gauge group, the propagator $D = 1/(L_0 - 1)$ is not gauge invariant. In fact, we can gauge it away. Thus, contact terms emerge in the action. By functionally integrating over some of the redundant fields, we can generate higher order interactions. Symbolically, we can write this as:

$$U_1 D U_2^{-1} = D + \int d\mu \qquad (12.1.5)$$

where U is the generator of reparameterizations and $\int d\mu$ represents the instantaneous "zipper" found in the previous chapter.

In contrast to QED, string theory can become fully nonpolynomial. This is because in QED, the A_0 field coupled linearly with only fermion fields. However, in string theory, the string field Φ is coupled to itself non-linearly. Thus, the cubic interaction of string theory can generate nonpolynomial interactions as we functionally integrate over the various fields.

To see how a polynomial theory can be come nonpolynomial, consider the simple case of N Yukawa mesons, interacting via an action:

$$\sum_a \partial_\mu \phi^a \, \partial^\mu \phi^a + \sum_{abc} v_{abc} \phi^a \phi^b \phi^c \qquad (12.1.6)$$

where the constant coefficient v_{abc} is symmetric and $v_{aab} = 0$. Now, let us functionally integrate over one of the fields, say ϕ^e. The integration is trivial to perform, because the action is quartic in ϕ^e:

$$\int d^4x \, d^4y \sum_{abcd} \phi^a(x) \phi^b(x) v_{abcd} \phi^c(y) \phi^d(y) \qquad (12.1.7)$$

where:

$$v_{abcd} = \sum_e v_{abe} \, D(x - y) v_{ecd} \qquad (12.1.8)$$

As expected, the new term in the action is quartic in the fields. The coefficient of this quartic term is actually the four-scalar meson scattering amplitude. However, the quartic form of the action is really an illusion because of the presence of the propagator, which makes the term nonlocal. Thus, we really do not have a quartic contact term in the action.

Notice that we can also perform the functional integration over two fields ϕ^e and ϕ^f $(e \neq f)$ and arrive at:

$$\int d^4x \, d^4y \, [\phi \times \phi \quad \phi \times \phi]_{x,i} M_{xy,ij}^{-1} \begin{bmatrix} \phi \times \phi \\ \phi \times \phi \end{bmatrix}_{y,j} \qquad (12.1.9)$$

where:

$$[\phi \times \phi]_e = v_{abe} \phi^a \phi^b \qquad (12.1.10)$$

and:

$$M^{-1}_{xy,ij} = \begin{bmatrix} D(x-y) & -\phi\times \\ -\phi\times & D(x-y) \end{bmatrix}^{-1}_{xy,ij} \qquad (12.1.11)$$

Thus, a local polynomial action has become a nonpolynomial action by functional integration. Although the action is formally nonpolynomial, this procedure is actually a trick because the action is nonlocal. We see, therefore, that ordinary polynomial point particle theories can become nonpolynomial, but only at the price of making them become nonlocal in the process.

In generalizing this simple example to string field theory, there is a crucial difference. Because the propagator is not a gauge invariant quantity, it is possible to gauge the propagator $D(x-y)$ to one, in which case the theory becomes (locally) nonpolynomial. This is an important point. It is precisely in this way that the geometric string field theory, which has a cubic interaction, can become nonpolynomial when we choose the endpoint gauge as a gauge choice.

12.2. The String Group

The entire purpose of geometric string field theory is to construct a group theoretic formalism by which to derive all of string field theory from first principles. For example, both relativity and the Yang-Mills theory can be derived from two simple principles.

Global symmetry: The fields of the theory A^i_μ and ω^{ab}_μ transform as irreducible, ghost-free representations of $SU(N)$ and the Lorentz group.

Local symmetry: The theory is locally $SU(N)$ and general covariant.

The physics is contained in the first principle. It selects the specific representations for the connection fields and, by demanding that the fields be ghost free, forces the theory to have only two derivatives. The second principle of local gauge invariances then fixes the action uniquely. (The second principle by itself is too weak to fix the theory; there are an infinite number of actions compatible with local gauge invariance.) The unique solution to both principles is given by:

$$L = -\frac{1}{4}\sqrt{-g}\,F^i_{\mu\nu}F^{i\mu\nu} - \frac{1}{2\kappa^2}\sqrt{-g}\,R_{\mu\nu}g^{\mu\nu} \qquad (12.2.1)$$

By analogy, we now wish to postulate the principles for string field theory that will uniquely fix the action. We postulate [3–5] the following.

Global symmetry: The fields of the theory $A^{\{\alpha\}}_\sigma$ transform as irreducible ghost-free representations of $\mathrm{Diff}(S_1)$ and the Lorentz group.

Local symmetry: The theory must be locally invariant under the universal string group.

The goal of this chapter is to find the unique action compatible with these two principles and then show that we can derive the light cone and nonpolynomial actions from them. Not surprisingly, almost all of the work in defining geometric string field theory is reduced to finding the representations of the universal string group. Let us begin by postulating an algebra defined on the space of *unparameterized strings*. This means that the usual starting point of string theory, which is the string variable $X_\mu(\sigma)$, must now be treated as a gauge choice.

In the space of unparameterized strings, we place our strings in an arbitrary background. We want a theory without reference to a metric or space–time, that is, a theory that is topological. Let C be an unparameterized, oriented string, existing in some unspecified background. We wish to construct an operator L_C, which is indexed by C. What is remarkable is that in this theory, where there is no light cone, no metric, no concept of space–time, we can define a well-behaved algebra.

Let us start with the following definitions. In a space of unparameterized strings in an unspecified background, the only topological statement we can make about two strings C_1 and C_2 is whether they are separated or whether they overlap or touch. Thus, the only topological structure constants we can define on this space is related to strings that touch.

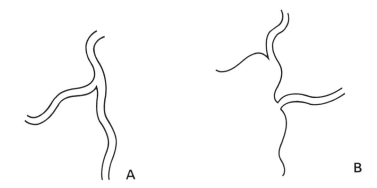

Fig. 12.1.

We define three oriented open strings $\{C_1, C_2, C_3\}$ to form a *triplet* if they can be arranged cyclically as in Fig. 12.1a. They form an *antitriplet* of strings if we reverse the orientation of the strings. Now, let us define a structure constant on this space, which has no structure or metric:

$$f^{C_3}_{C_1, C_2} = \begin{cases} +1 & \text{for triplets} \\ -1 & \text{for antitriplets} \\ 0 & \text{otherwise} \end{cases} \qquad (12.2.2)$$

We now define an algebra among the generators as follows:

$$[L_{C_1}, L_{C_2}] = f_{C_1, C_2}^{C_3} L_{C_3} \qquad (12.2.3)$$

The topological nature of this algebra is easy to see. Let $\bar{C}$ be *conjugate* to C if both $\bar{C}$ and C belong in the same triplet. (A string obviously has an infinite number of conjugates.) Then, the effect of the string group is to transform a string C into its conjugates $\bar{C}$.

What is remarkable about this algebra is that the Jacobi identities close properly, even though the space on which they are defined has no structure. If we take the multiple commutator of three strings, we find:

$$\left[L_{C_1}, [L_{C_2}, L_{C_3}]\right] + \text{perm} = 0 \qquad (12.2.4)$$

This can be written as:

$$f_{C_1, C_2}^{C_4} f_{C_4, C_3}^{C_5} + \text{perm} = 0 \qquad (12.2.5)$$

The fact that the sum equals zero can be seen in Fig. 12.1.b, where we take three arbitrary strings and calculate the Jacobi identity for them. It is easy to see that the Jacobi identity reduces to:

$$+1 - 1 + 0 = 0 \qquad (12.2.6)$$

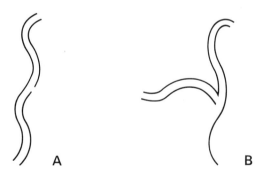

A   B

Fig. 12.2.

At first, because the space has no structure or metric, one might suspect that this is a trivial result. However, it is possible to see that this result is nontrivial by taking Fig. 12.2a as the definition of a triplet. If we then calculate the Jacobi identity for this triplet, we have:

$$+1 + 0 + 0 \neq 0 \qquad (12.2.7)$$

that is, the Jacobi identity fails, and we cannot form a consistent algebra. (To close the algebra, one necessarily needs four-string and higher interactions to cancel the diagram in Fig. 12.2b. In fact, this is precisely the origin of the four-string interaction found in the light cone theory. In order to create a gauge theory for the light cone theory, one has to add the four-string interaction to cancel precisely this diagram in Fig. 12.2.b).

At first, one might suspect that the generalization of this result to closed strings is straightforward. However, this is not so. Closed strings, topologically speaking, are totally different in structure to open strings. Specifically, an attempt to use Fig. 12.3a as the definition of a triplet leads to problems.

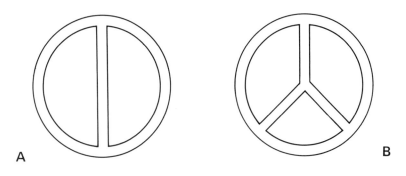

A   B

Fig. 12.3.

1. A fully symmetric structure constant does not satisfy the Jacobi identity. The multiple commutator leads to the following:

$$+1 + 1 + 1 \neq 0 \qquad (12.2.8)$$

which does not cancel. The remaining diagram is shown in Fig. 12.3b.

2. The structure constant should be antisymmetric in C_1 and C_2, but closed strings are necessarily symmetric among the strings (since we can simply rotate the entire configuration in a circle). (This is, in fact, the reason why the closed string theory of Chapter 11 is nonpolynomial. Four-string interactions are required to cancel the previous graph, which in turn requires five-string graphs to cancel them, etc., until the gauge group becomes nonpolynomial.)

As a result, we must make a radical departure from the open string case. First, we must make L_C a Grassmann odd operator in order to get the statistics to come out correct. Second, we cannot use Fig. 12.3a as the definition of the triplet, but instead must use Fig. 12.4. We then define this to be the basic triplet with the following symmetric structure constant:

$$f^{C_3}_{C_1 C_2} = \begin{cases} +1(-1) & \text{for triplets if } C_3 \text{ equals outer (inner) string} \\ 0 & \text{otherwise} \end{cases}$$

(12.2.9)

Then, the fundamental identity is given by:

$$\{L_{C_1}, L_{C_2}\} = \hat{f}^{C_3}_{C_1, C_2} L_{C_3}$$

$$\hat{f}^{C_3}_{C_1, C_2} = \theta_{C_1 \cap C_2 \cap C_3} f^{C_3}_{C_1, C_2}$$

(12.2.10)

where θ is a Grassmann odd constant defined at the intersection point of the three strings.

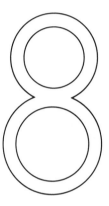

Fig. 12.4.

The important point is that this new definition of the algebra satisfies the Jacobi identity. A careful analysis of the various terms generated by the Jacobi identity show that it is satisfied:

$$[L_{C_1}, \{L_{C_2}, L_{C_3}\}] + \text{perm} = 0$$

(12.2.11)

We now see the rough origin of why string field theory has the basic structure shown earlier in the various gauges. The interaction predicted by the string algebra is cubic if we define the triplets as in Fig. 12.1a and Fig. 12.4, but becomes much more complicated when we change parameterizations and use Figs. 12.2a and 12.3a as the basic triplet.

Next, we wish to calculate invariants for this algebra. Let ϕ_C transform as a covariant vector and ϕ^C as a contravariant vector. (The raising and lowering operator is just the operator that reverses the orientation of the string.) Under the string group, they transform as:

$$\delta\phi^{C_1} = f^{C_1}_{C_2, C_3} \phi^{C_2} \Lambda^{C_3}$$

$$\delta\phi_{C_1} = f^{C_3}_{C_2, C_1} \Lambda^{C_2} \phi_{C_3}$$

(12.2.12)

where Λ is a small parameter.

By taking a multiple variation of the these fields, we can show that they form a representation of the algebra:

$$\delta_{1,2}\phi^{C_1} = f^{C_1}_{C_2,C_3} \Sigma^{C_3} \phi^{C_2} \tag{12.2.13}$$

where:

$$\Sigma^{C_3} = f^{C_3}_{C_2,C_1} \Lambda_2^{C_2} \Lambda_1^{C_1} \tag{12.2.14}$$

and the raising and lowering operator for the index C is given by reversing the orientation of the open string. Now, it is an easy matter to construct invariants under the string group:

$$\phi \times \phi \equiv \sum_C \phi_C \phi^C$$

$$\phi \times \phi \times \phi \equiv \sum_{C_1,C_2,C_3} \phi_{C_1} \phi_{C_2} \phi_{C_3} f^{C_1,C_2,C_3} \tag{12.2.15}$$

We will shortly see that these invariants will become the basic building blocks of the action.

12.3. Universal String Group

Up to now, we have defined our string group in purely topological terms, without any reference to any background metric, space–time, or parameterizations. We would now like to generalize the string group, which is defined topologically, to include the symmetries of the bosonic string, which include reparameterizations in space–time.

We remind ourselves that the symmetries of the first quantized string include two types of reparameterizations included within $\text{Diff}(S_1)$:

1. $\text{Diff}(S_1)_0$, generated by L_n, which preserves the parameterization length of the string; and
2. scale transformations, generated by $\delta/\delta\alpha$, which rescale the overall parameterization length α of the string.

(Strictly speaking, the L_n by itself can rescale the overall parameterization length of the string. However, the Jacobian of the map that implements this rescaling has an infinite determinant. Thus, it is more convenient to split off the generator of scale transformations from the L_n, although it is not absolutely necessary to do so.)

In this section, we will discuss only the first symmetry, $\text{Diff}(S_1)_0$, generated by the L_n, which preserves the parameterization length of the string. In the next section, we will include the second symmetry, scale transformations, which are generated by $\delta/\delta\alpha$. Thus, the generalization of the string group to include the string symmetries will be the universal string group, defined via:

$$SG = USG/\text{Diff}(S_1)$$
$$\text{Diff}(S_1) = \text{Diff}(S_1)_0 \otimes \text{scale} \tag{12.3.1}$$

We define the universal string group (USG) as the group that maps strings into their conjugates and strings into themselves. Thus, we postulate that the underlying gauge group of string field theory is given by:

$$\text{USG}: \quad \begin{cases} C \to C \\ C \to \bar{C} \end{cases} \tag{12.3.2}$$

As a first step in finding the generators of this algebra, let us replace the string C with $X_\mu(\sigma)$. This also means replacing L_C with L_X, where L_X is a multilocal functional simultaneously dependent on all points σ_i along the string.

$$L_C \to L_X \equiv L_{\{X_\mu(\sigma_1), X_\mu(\sigma_2), \ldots, X_\mu(\sigma_N)\}} \tag{12.3.3}$$

We now define the open string structure constant as:

$$f^{X_3}_{X_1, X_2} = \prod_{i=1}^{3} \prod_{0 \le \sigma_i \le \sigma_i^0} \delta\left[X_\mu(\sigma_i) - X_\mu(\pi\alpha_{i-1} - \sigma_{i-1})\right] \tag{12.3.4}$$

where the ith string has parameterization length $\pi\alpha_i$ and σ_i^0 is the point at which the strings meet.

We also wish to define the effect of reparameterizations on the string. Let L_σ represent the Virasoro generator whose moments correspond to $L_n - L_{-n}$. Then, the effect of reparameterizing a string:

$$\sigma \to \sigma + \epsilon(\sigma) \tag{12.3.5}$$

can be represented the Virasoro operator L_σ, where we smear L_n across a string. Under this transformation, the fields $\Phi(X)$ transform as:

$$\delta\Phi(X) = \epsilon(\sigma) X'^\mu(\sigma) \, \partial_{\mu\sigma} \Phi(X) \tag{12.3.6}$$

where $\partial_{\mu\sigma} = \delta/\delta X^\mu(\sigma)$.

Then, a subalgebra of the USG can be represented as:

$$[L_X, L_Y] = f^Z_{XY} L_Z$$
$$[L_\sigma, L_X] = X'^\mu(\sigma) \, \partial_{\mu\sigma} L_X \tag{12.3.7}$$
$$[L_\sigma, L_\rho] = f^\omega_{\sigma,\rho} L_\omega$$

where the first equation is a parameterized version of the string algebra, the second equation shows that the USG is the semidirect product between the string group and the reparameterization group, and the last equation is just the usual Virasoro algebra (smeared over strings parameterized by σ, ω, ρ).

To gauge the reparameterization group, we will follow the example of the general theory of relativity, where vierbeins and connections are used to implement general covariance. In relativity, we introduce the vierbein e^a_μ, which takes us from curved space indices μ to flat space indices a. For our purposes, the reparameterization is over the string variable $X^\mu(\sigma)$, so the vierbein has many more indices: $e^{\nu\omega}_{\mu\sigma}$, where μ and ν are space–time indices and σ and ω are continuous indices defined over the length of the string.

The vierbein is needed to construct invariants by soaking up the terms that arise form the transformation of differentials and derivatives:

$$
\begin{aligned}
dX^{\mu\sigma} &= \frac{\partial X^{\mu\sigma}}{\partial \bar{X}^{\nu\rho}} d\bar{X}^{\nu\rho} \\
\frac{\partial X^{\mu\sigma}}{\partial \bar{X}^{\nu\rho}} &= \frac{\partial X^{\mu\sigma}}{\partial \tilde{X}^{\lambda\omega}} \frac{\partial \tilde{X}^{\lambda\omega}}{\partial \bar{X}^{\nu\rho}}
\end{aligned}
\tag{12.3.8}
$$

where μ, ν, and λ are Lorentz indices and ω, ρ, and σ are continuous indices. For example, we need the vierbein to make a generally covariant measure:

$$
DX = \prod_{\mu\sigma} dX^{\mu\sigma} \to \det\left(e^{\nu\omega}_{\mu\sigma}\right) \prod_{\nu\rho} d\bar{X}^{\nu\rho}
\tag{12.3.9}
$$

To make the derivative $\partial_{\mu\sigma}$ covariant, we must also add in a connection field:

$$
\partial_{\mu\sigma} \to \nabla_{\mu\sigma} \equiv \partial_{\mu\sigma} + \omega^{\{\beta\}}_{\mu\sigma\{\alpha\}}
\tag{12.3.10}
$$

where the indices of the connection field depend on which tensor it is operating on.

(Normally, the introduction of these two fields will create many more degrees of freedom than are found in string theory. To eliminate all but the essential ones, we must place an additional constraint on the theory consistent with gauge invariance. We will demand that the curvature associated with this connection field and the covariant derivative of the vierbein both vanish. This means that the vierbein is just a function of a vector field ζ^σ. This is fortunate, because it means that we have to integrate over all reparameterizations of the vertex given by $\sigma_i \to \sigma_i + \zeta^{\sigma_i}$. Functionally integrating over the vierbein is thus equivalent to integrating over all parameterizations of the string, with each new parameterization labeled by ζ^σ. For more details, see Refs. 3–5.)

We are, however, not yet complete. We must still add the mysterious "ghost sector" of the gauge algebra. In a geometric theory, where the entire theory is derived from a single gauge group, there is no room for ad hoc devices, such as Faddeev–Popov ghosts, which are by-products of gauge fixing. Thus, the geometric theory must somehow explain the presence of these ghosts.

To see the origin of this strange ghost sector, we remind ourselves that in ordinary gauge theory the Yang–Mills field $A^i_\mu(x)$ or the connection field in relativity $\omega_\mu^{ab}(x)$ has a tangent space defined at every point x. The indices in the tangent space do not transform under general coordinate transformations.

The same situation applies in string field theory. As our first postulate, we demanded that the connection fields of the theory transform as irreducible representations of $\text{Diff}(S_1)$, that is, they must be Verma modules. Thus, the basic field $\Phi^{\{\alpha\}}$ must be indexed by the letters contained in $\{\alpha\}$, which is in one-to-one correspondence with the elements of a Verma module. The Verma module transforms covariantly under the reparameterization group.

Let L_σ be the Virasoro generator that shifts the string parameterization at σ. We can calculate the following Clebsch–Gordon coefficient:

$$|\mathbf{e}^{\{\alpha\}}\rangle \equiv L_{-n_1}^{\alpha_1} L_{-n_2}^{\alpha_2} \cdots L_{-n_M}^{\alpha_M} |0\rangle$$
$$L_\sigma |\mathbf{e}^{\{\alpha\}}\rangle = f_{\sigma,\{\beta\}}^{\{\alpha\}} |\mathbf{e}^{\{\beta\}}\rangle \tag{12.3.11}$$

where L_{-n} represents an abstract set of Virasoro generators, independent of X^μ. Because the L_n acting on an element of a Verma module simply maps it into another element of the module, the Clebsch–Gordon coefficients f are numerically well defined.

Then, a field indexed by a Verma module transforms as follows:

$$\delta \Phi^{\{\alpha\}} = \epsilon(\sigma) f_{\sigma,\{\beta\}}^{\{\alpha\}} \Phi^{\{\beta\}} + \epsilon(\sigma) X'^\mu(\sigma) \, \partial_{\mu\sigma} \Phi^{\{\alpha\}} \tag{12.3.12}$$

We can also construct the connection field $A_\sigma^{\{\alpha\}}$, which is a mixed tensor such that the top index labels a Verma module and σ is a continuous index that parameterizes the string. It transforms as

$$\delta A_\sigma^{\{\alpha\}} = \epsilon(\sigma) f_{\sigma,\rho}^\omega A_\omega^{\{\alpha\}} + \epsilon(\sigma) f_{\sigma,\{\beta\}}^{\{\alpha\}} A_\sigma^{\{\beta\}} + \epsilon(\sigma) X'^\mu(\sigma) \, \partial_{\mu\sigma} A_\sigma^{\{\alpha\}} \tag{12.3.13}$$

Let us now generalize our algebra to include these Verma modules. We wish to construct a three-vertex function defined entirely on Verma modules:

$$\langle \mathbf{e}^{\{\alpha\}}| \langle \mathbf{e}^{\{\beta\}}| \langle \mathbf{e}^{\{\gamma\}}|V\rangle = f^{\{\alpha\}\{\beta\}\{\gamma\}} \tag{12.3.14}$$

where each Verma module $\langle \{\alpha\}|$ is defined by a series of L_{-n} operators acting on the vacuum $\langle 0|$. We do not need the exact representation for $|V\rangle$. It is sufficient to establish how it transforms under L_n. We define $|V\rangle$ to obey the Goto–Naka continuity equations, that is, the L_σ defined over the three strings obey the same continuity conditions as the string variable in Eq. (12.3.4). Notice that this is sufficient to numerically calculate all $f^{\{\alpha\}\{\beta\}\{\gamma\}}$. Then we push each L_n contained within the Verma module to the right. It transforms into L_m, which then moves to the left, eventually annihilating on the vacuum, but creating many constant terms in the process. We successively move all L_n to the right until they are all removed, leaving only the constant term $f^{\{\alpha\}\{\beta\}\{\gamma\}}$.

We wish to change our generator L_X into $L_X^{\{\alpha\}}$, which will alter the generators as follows:

$$[L_X^{\{\alpha\}}, L_Y^{\{\beta\}}] = f_{\{\alpha\}\{\beta\}\{\gamma\}} f_{XY}^Z L_Z^{\{\gamma\}}$$
$$[L_\sigma, L_X^{\{\alpha\}}] = X'^\mu(\sigma)\partial_{\mu\sigma} L_X^{\{\alpha\}} + f_{\sigma\{\beta\}}^{\{\alpha\}} L_X^{\{\beta\}} \qquad (12.3.15)$$
$$[L_\sigma, L_\rho] = f_{\sigma\rho}^\omega L_\omega$$

This the complete USG (without the effect of the rescaling, which we will mention in the next section).

Before we end our discussion of $\mathrm{Diff}(S_1)_0$, let us mention the importance of having our basic field $A_\sigma^{\{\alpha\}}$ transform as an element of a Verma module. It remains to be seen whether our connection fields, which are all Verma modules, can reproduce the field usually encountered in string field theory, which is the $-\frac{1}{2}$ ghost sector of the field generated by all products of the c and b ghosts, that is,

$$\Phi_{\mathrm{BRST}} = P_{-1/2}\Phi[X, c(\sigma), b(\sigma)] \qquad (12.3.16)$$

where $P_{-1/2}$ projects out the $-\frac{1}{2}$ ghost sector. We can show, however, that the two formulations are exactly the same.

The simplest way to see this is to calculate the character of the BRST field. We know from previous chapters that the character of an irreducible Verma module is given by:

$$\mathrm{ch}\, V = \prod_n (1 - x^n)^{-1} = \sum_n p(n)x^n \qquad (12.3.17)$$

Let us now show that the character of the BRST field is also the same. We begin by showing that the level number R and the ghost number G of the BRST field are given by

$$R = \sum_{n=1}^\infty (n\theta_{-n}\theta^n + n\theta^{-n}\theta_n)$$
$$G = \sum_{n=1}^\infty (\theta^{-n}\theta_n - \theta_{-n}\theta^n) + \frac{1}{2}(\theta^0\theta_0 - \theta_0\theta^0) \qquad (12.3.18)$$

The ghost number $-\frac{1}{2}$ projection operator can be represented as:

$$\delta_{G,-1/2} = \frac{1}{2\pi i}\oint dz\, z^{2G} \qquad (12.3.19)$$

Putting everything together, we find that the character of the BRST field can be expressed as follows:

$$\mathrm{ch}\, \Phi_{\mathrm{BRST}} = \sum_n \dim \Phi_n x^n = \mathrm{Tr}(\delta_{G,-1/2} x^R) \qquad (12.3.20)$$

To perform the actual trace, we need a few more identities:

$$\mathrm{Tr}\, x^{\theta^{-n}\theta_n + \theta_{-n}\theta^n} = 1 - x - x - x^2$$

$$\mathrm{Tr}_n(z^{2T} x^R) = 1 - x^n(z^2 + z^{-2}) + x^{2n}$$

$$\langle\, +\, |\delta_{G,-1/2}|\, -\, \rangle = 1$$

$$\langle\, -\, |\delta_{G,-1/2}|\, +\, \rangle = 0$$

$$(12.3.21)$$

Putting everything together, we can now calculate the trace of the BRST field:

$$
\begin{aligned}
\mathrm{ch}\, \Phi_{\mathrm{BRST}} &= \sum_{n=1}^{\infty} \frac{1}{2\pi i} \oint dz (z^{-1} - z)\big[1 - x^n(z^2 + z^{-2}) + x^{2n}\big] \\
&= \int \frac{dz}{2\pi i}(z^{-1} - z) \sum_{n=1}^{\infty}(1 - x^n)^{-1}\, \theta_1(\nu, q)(\sin^{-1}\pi\nu)q^{-1/4} \\
&= -\sum_{n=1}^{\infty}(1 - x^n)^{-1} \frac{1}{2\pi i}\oint dz\, q^{-1/4} \sum_{n=-\infty}^{n=\infty}(-1)^n q^{(n-1/2)^2} z^{-2n+1} \\
&= \prod_{n=1}^{\infty}(1 - x^n)^{-1}
\end{aligned}
$$

$$(12.3.22)$$

where we have used:

$$x = q^2; \qquad z = e^{-\pi i \nu} \tag{12.3.23}$$

(Notice that the last step is possible because only the $n = 1$ term survives the integration.) Thus, the BRST field, with its strange dependence on Faddeev–Popov ghosts, is now viewed as an irreducible representation of $\mathrm{Diff}(S_1)$ in the geometric theory. This yields a simple interpretation of the ghost sector of string theory.

12.4. How Long Is a String?

In the previous section, we have discussed how $\mathrm{Diff}(S_1)$ can be broken down into two parts, the reparameterizations, which leave the parameterization length invariant, and the scale transformations, which change the parameterization length. Before, we treated the first type of symmetry. In this section, we will include a discussion of reparameterizations that rescale the string length.

The question of what is meant by the "string length" has caused considerable confusion in the literature. The origin of this problem lies in the fact that there are actually three different versions of string length, and we must carefully distinguish between them.

First, there is the *invariant* string length given by the integral:

$$L = \int_0^{\pi\alpha} d\sigma \sqrt{X_\mu'^2(\sigma)} \qquad (12.4.1)$$

which remains invariant under a scale change. The value of L is independent of α. Rescaling the value of α in the integral does not change the invariant length. Second, we have the *parameterization* string length given by:

$$\int_0^{\pi\alpha} d\sigma \qquad (12.4.2)$$

which should be a gauge artifact. The scattering amplitudes should be independent of the choice of α. Third, we have the *world sheet* length of the string, which is a function of the metric tensor g_{ab}. If we, for example, make a scale transformation $\sigma \to s\sigma$, for some s, then the two-dimensional derivatives $\partial_a \to (1/s)\partial_a$ if a points in the σ direction. The rescaling of the derivatives, in turn, can be absorbed into a rescaling of the metric tensor since the action contains the term $\partial_a X g^{ab} \partial_b X$.

The problem with scale invariance, however, is that the usual Q_{BRST} operator is defined with a fixed parameterization length. The original derivation of Q_{BRST} fixed the overall parameterization length to be π.

We recall that, given any Lie algebra, one can define a nilpotent operator Q. We wish, therefore, to derive this operator by reinserting the original scale invariance of the string. We recall that the Lagrangian of the string can be written as:

$$L = P_\mu \dot{X}^\mu - H$$
$$H_0 = \lambda_1 \left(P_\mu^2 + \frac{1}{\pi^2} X_\mu'^2 \right) + \lambda_2 (P_\mu X'^\mu) \qquad (12.4.3)$$

where the λ_i are functions of the metric g_{ab}, that is, $\lambda_0 \sim 1/\sqrt{g}g^{00}$ and $\lambda_1 \sim g^{01}/g^{00}$. The Hamiltonian consists entirely of the constraints, which in turn form the Virasoro algebra. However, this form of the Lagrangian has tacitly removed all mention of the original scale invariance of theory.

The Lagrange multipliers λ_i in the previous expression have no canonical momentum associated with them. To rectify this, let us define the canonical moment as π_i. Then, the complete Lagrangian is:

$$L = P_\mu \dot{X}^\mu + \pi_i \dot{\lambda}^i - H$$
$$H = \omega^i \pi_i + H_0 \qquad (12.4.4)$$

The two Lagrangians have the same field content. By integrating out the ω_i Lagrange multipliers, we find that $\pi_i = 0$, so that the $\pi_i \lambda_i$ term vanishes, leaving us with just the original H_0. Although this seems like a trivial change, the important fact is that the gauge group of the theory is now enlarged to include an Abelian sector generated by the π_i. The gauge algebra now includes the commutators:

$$[\pi_i, \pi_j] = 0, \qquad [\pi_i, L_n] = 0 \qquad (12.4.5)$$

As expected, the operators π_i commute among each other and with the Virasoro generators.

Since the operators of the theory now depend on two new variables, we will find it convenient to define them to be equal to $\delta/\delta\alpha(\sigma)$ and $\delta/\delta\tau(\sigma)$, where $\alpha(\sigma)$ and $\tau(\sigma)$ are two new fields that are functions of the g_{ab}. We will define the zero modes of $\alpha(\sigma)$ and $\tau(\sigma)$ to be α, the parameterization length, and τ, a new parameter that will eventually become proper time.

Given the algebra of the USG, we must now construct the connection fields and the action. The structure constants of the group give us a multiplication operation on strings:

$$\times \equiv f^{\{\alpha\}\{\beta\}\{\gamma\}} f^Z_{XY} \qquad (12.4.6)$$

and they also define the connection field for us, given by $A_\sigma^{\{\alpha\}}$. The $\times$ symbol is a Clebsch–Gordon coefficient representing the structure constants of the USG. The covariant derivative is now given by:

$$\nabla_\sigma = L_\sigma(X) + aL_\sigma + A_\sigma^{\{\alpha\}} \qquad (12.4.7)$$

where a is a constant, $L_\sigma(X)$ contains the complete set of conformal Virasoro generators defined over X space, and L_σ is an abstract Virasoro generator acting only on the tangent space indices $\{\alpha\}$.

Because the fields are defined on Verma modules while the covariant derivative transforms with conformal weight 2 and is indexed by the continuous index σ, we need a Clebsch–Gordon coefficient that represents the tensor product between these two different representations. We now define the Clebsch–Gordon coefficient:

$$\gamma^\sigma_{\{\alpha\}\{\beta\}} = \langle \mathbf{e}_{\{\alpha\}} | \gamma^\sigma | \mathbf{e}_{\{\beta\}} \rangle \qquad (12.4.8)$$

This Clebsch–Gordon coefficient is uniquely defined. Since γ^σ transforms with a fixed conformal weight, we can move the L_n to the right and L_{-m} to the left, until they all annihilate, leaving a numerical value for the Clebsch–Gordon coefficient.

We can now write the action. Since $\delta^{\sigma\rho}$ is not a true constant tensor of $\mathrm{Diff}(S_1)$, we must use another to contract on the tensors of the theory. The unique action is:

$$L = \epsilon^{\sigma\rho\omega} \left(A_\sigma \times \nabla_\rho A_\omega + \frac{2}{3} A_\sigma \times A_\rho \times A_\omega \right) \qquad (12.4.9)$$

where we have suppressed tangent space indices and $\epsilon^{\sigma\rho\omega} = \gamma^\sigma \gamma^\rho \gamma^\omega$.

Notice that the field $A_\sigma^{\{\alpha\}}$ only comes in the combination $\gamma^\sigma A_\sigma^{\{\alpha\}}$; we can treat it as one field $\Phi^{\{\alpha\}}$. Written entirely in terms of $\Phi^{\{\alpha\}}$, the final action is:

$$\Phi \gamma^\sigma \nabla_\sigma \Phi + \frac{2}{3} \Phi \times \Phi \times \Phi \qquad (12.4.10)$$

Although this looks like the action found in the BRST formalism, there are several major differences. All fields are functions of X as well as $\alpha(\sigma)$, $\tau(\sigma)$, γ^α, and γ^τ. We also integrate over the vierbein. Last, $\gamma^\sigma \nabla_\sigma$ is almost the nilpotent operator Q, with key differences that we now discuss.

12.5. Cohomology

The addition of these new operators π_i to the gauge algebra means that the usual nilpotent Q has to be modified. By the standard prescription, we find that the modified $\bar{Q}$ equals:

$$\gamma^\sigma \nabla_\sigma = \bar{Q} = Q_0 + Q_1$$
$$Q_1 = i \int d\sigma \left[\gamma^\alpha(\sigma) \frac{\delta}{\delta\alpha(\sigma)} + \gamma^\tau(\sigma) \frac{\delta}{\delta\tau(\sigma)} \right] \tag{12.5.1}$$

where Q_0 is the usual BRST operator at fixed parameterization length. We have added a new set of ghosts γ^α and γ^τ to match the new generators π_i, which we have written as $\delta/\delta\alpha$ and $\delta/\delta\tau$.

Now, let us determine the cohomology of this new operator $\bar{Q}$. We suspect that since scale invariance commutes with the Virasoro generators, the scale sector decouples completely from the states, leaving us with the usual physical states of the string field.

Our fields are dependent on all the modes α_n and τ_n smeared over the string. However, by simple arguments, one can show that all these modes decouple from the theory if one can show that the zero modes of these oscillators decouple. If we only take the zero modes of these oscillators, then the theory is only dependent on α and τ, and the theory (with its ghosts) has an additional (nonphysical) symmetry $Osp(26, 2/2)$ [9–10].

Our fields Φ and the gauge parameters Λ must now be functions of γ^α and γ^τ, which can be decomposed as follows:

$$\Phi = \phi_0 + \gamma^\tau \phi_\tau + \gamma^\alpha \phi_\alpha + \gamma^\tau \gamma^\alpha \phi_{\alpha\tau}$$
$$\Lambda = \Lambda_0 + \gamma^\tau \Lambda_\tau + \gamma^\alpha \Lambda_\alpha + \gamma^\tau \gamma^\alpha \Lambda_{\alpha\tau} \tag{12.5.2}$$

The physical states must therefore be solutions to $Q\Phi = 0$, which can be decomposed into the following set of four equations:

$$Q_0 \phi_0 = 0$$
$$i\partial_\alpha \phi_0 + Q_0 \phi_\alpha = 0$$
$$i\partial_\tau \phi_0 + Q_0 \phi_\tau = 0 \tag{12.5.3}$$
$$-i\partial_\alpha \phi_\tau + i\partial_\tau \phi_\alpha + Q_0 \phi_{\alpha\tau} = 0$$

(We will find that the physical string states are hidden within ϕ_α.) This condition is subject to the variation $\delta\Phi = \bar{Q}\Lambda$, which can be decomposed as:

$$\delta\phi_0 = Q_0\Lambda_0$$
$$\delta\phi_\alpha = i\partial_\alpha\Lambda_0 + Q_0\Lambda_\alpha$$
$$\delta\phi_\tau = i\partial_\tau\Lambda_0 + Q_0\Lambda_\tau \qquad (12.5.4)$$
$$\delta\phi_{\alpha\tau} = -i\partial_\alpha\Lambda_\tau + i\partial_\tau\Lambda_\alpha + Q_0\Lambda_{\alpha\tau}$$

Now, let us begin the process of removing redundant gauge degrees of freedom to see what the physical states look like. First, we can invert the τ derivative and use Λ_α to set $\phi_{\alpha\tau}$ to zero and use Λ_0 to set ϕ_τ to zero, that is, symbolically:

$$\Lambda_\alpha \rightarrow \phi_{\alpha\tau} = 0, \qquad \Lambda_\tau \rightarrow \phi_\tau = 0 \qquad (12.5.5)$$

Reexamining our cohomology, we find that $\partial_\tau\phi_0 = 0$ and $\partial_\tau\phi_\alpha = 0$; they are now τ independent.

Although we have removed the τ dependence of the system, there is still a residual symmetry, consisting of τ independent gauge transformations. The τ independent fields are still invariant under the variation:

$$\delta\phi_0 = Q_0\Lambda_0^{(0)}$$
$$\delta\phi_\alpha = i\partial_\alpha\Lambda_0^{(0)} + Q_0\Lambda_\alpha^{(0)} \qquad (12.5.6)$$

We now use $\Lambda_0^{(0)}$ to eliminate ϕ_0, since we can show that ϕ_0 consists totally of BRST redundant states. (Usually, the operator Q_0 is not invertible. However, in this case, we can effectively invert Q_0 because ϕ_0 has a nonzero ghost number.) However, there is still one last gauge invariance, given by:

$$\delta\phi_\alpha = i\partial_\alpha\bar{\Lambda}_0^{(0)} + Q_0\Lambda_\alpha^{(0)} \qquad (12.5.7)$$

We can use this last symmetry to eliminate the nonphysical modes within ϕ_α and also to eliminate its α degree of freedom.

In summary, we find that the physical states satisfy:

$$\frac{\delta}{\delta\alpha}\phi_\alpha = 0$$
$$\frac{\delta}{\delta\tau}\phi_\alpha = 0 \qquad (12.5.8)$$
$$Q_0\phi_\alpha = 0$$

We have thus shown that the α and τ sectors of the string field have completely decoupled from the theory, leaving only the usual physical states.

So far, we have been only working with the on-shell system, that is, the equations of motion. There are problems, however, when we try to apply this formalism to the Lagrangian itself. In fact, if we quantize incorrectly, we find that the theory is riddled with ghosts, or the theory vanishes entirely. We must thus be very careful in setting up the Lagrangian for the system.

We begin with the action $\Phi Q\Phi$ integrated over γ^τ and γ^α, which can be expanded as follows:

$$L = i\phi_0^\dagger \dot{\phi}_\alpha + \phi_{\alpha\tau}^\dagger Q_0\phi_0 - i\phi_0^\dagger \partial_\alpha \phi_\tau - \phi_\alpha^\dagger Q_0\phi_\tau + h.c. \qquad (12.5.9)$$

where † simply refers to reversing the sign of α.

There are several things that appear incorrect in this action. First, we notice that the action is nondiagonal in the fields. Thus, it appears that the theory has ghosts. For example, if we have an action with two fields ϕ_1 and ϕ_2, then we can decompose the following action so that there is a ghost:

$$\phi_1 Q \phi_2 = (\phi_a Q \phi_a) - (\phi_b Q \phi_b) \qquad (12.5.10)$$

where we have made the substitution $\phi_1 = \phi_a + \phi_b$ and $\phi_2 = \phi_a - \phi_b$. The presence of the extra -1 sign in front of the second term indicates the presence of unwanted ghosts. Second, if we set $\phi_{\alpha\tau}$ and ϕ_τ equal to zero, we find that the Lagrangian collapses to a trivial factor of $\phi_0\phi_\alpha$. The theory, it appears, is empty.

So we are faced with two unpleasant choices. Either the theory is vacuous, or else the theory contains ghosts as well as the usual physical states. Actually, there are simple answers to both problems. The argument that ghosts are always present in a nondiagonal theory is faulty because we are dealing with a gauge theory, in which one of the fields is a redundant gauge field. The ghost field ϕ_b, which causes us problems, may be eliminated because it is redundant.

The second problem, that the action is zero, is actually more sophisticated and requires some analysis. To see where the answer lies, let us review some basics of gauge fixing in ordinary Maxwell theory. The Maxwell action contains the term:

$$L = A_0 \nabla \cdot \mathbf{E} + \cdots \qquad (12.5.11)$$

Normally, the Coulomb gauge consists of setting $\nabla \cdot \mathbf{A} = 0$. Treating the A_0 field as a Lagrange multiplier, we can then eliminate it and find Gauss's law: $\nabla \cdot \mathbf{E} = 0$. In the Coulomb gauge, we can therefore impose the following three conditions:

$$\nabla \cdot \mathbf{A} = 0$$
$$A_0 = 0 \qquad (12.5.12)$$
$$\nabla \cdot \mathbf{E} = 0$$

The problem occurs when we try to choose another Coulomb-type gauge, the temporal gauge $A_0 = 0$. By setting the Lagrange multiplier to zero, we can no longer impose Gauss's law. We have used up all our gauge constraints, yet we are not able to retrieve the previous set of constraints. Thus, it appears as a mystery how we can rederive the original three conditions found in the standard Coulomb gauge.

The answer to this puzzle is that the longitudinal part of $\mathbf{E}$ is still redundant; $\mathbf{E}$ commutes with the Hamiltonian, so once $\mathbf{E}$ is set to zero initially, it remains zero throughout the history of the system. Another way to see this is to decompose the electric field and the A_i field into transverse and longitudinal parts. Reinserting them into the action, we find that the longitudinal parts decouple from the action, so they can be removed, leaving only the transverse components in the action. The lesson is that whenever we set the Lagrange multiplier to zero, we can no longer impose the counterpart of Gauss's law. However, the system still has residual symmetries that allow us to impose Gauss's law.

A similar situation applies to the string gauge system. By demanding that ϕ_τ and $\phi_{\alpha\tau}$ be set to zero, the system collapses. This is only an illusion, however. We are still allowed to set:

$$Q_0\phi_0 = 0, \qquad Q_0\phi_\alpha + i\partial_\alpha\phi_0 = 0 \qquad (12.5.13)$$

To see that the system is not vacuous, let us impose another gauge constraint on the system. We impose the condition:

$$\phi_{\alpha\tau} = 0, \qquad \phi_\tau = \phi_\alpha\delta(\tau)\delta(\alpha - \pi) \qquad (12.5.14)$$

The system then reduces to two terms: a diagonal, τ-independent term

$$\phi_\alpha^\dagger Q_0\phi_\alpha \qquad (12.5.15)$$

at fixed string length π and a term dependent on α and τ. Notice that these second terms, like the longitudinal part of $\mathbf{E}$, decouple from the system, and hence can be thrown away. The moral of this discussion is that the final action resulting from eliminating the α and τ sectors of the theory is the usual one found in Chapter 10.

12.6. Interactions

Next, let us give an explicit representation of the structure constants of the theory, which are the vertex functions. To prove that this geometric formulation actually exists, we wish to find a conformal map that smoothly interpolates between the light cone map and the midpoint map. This is nontrivial, because at first one may suspect that the light cone gauge limit is a singular one (because the region of overlap between two of the legs goes to zero). However, by explicit construction, we will show that the interpolating map exists and is well behaved in the limit that it approaches the light cone gauge.

The map we desire is given by [3–5]:

$$\rho(z) = -\alpha_1 \ln(z - 1) - \alpha_2 \ln z - \sum_{r=1}^{3} \alpha_r \ln \left[(az^2 + bz + c)^{1/2} + a_r z + b_r\right]$$

$$(12.6.1)$$

where:

$$a_1 = \frac{\alpha_1^2 - \alpha_2^2 + \alpha_3^2}{2\alpha_1}; \qquad a_2 = \frac{-\alpha_1^2 + \alpha_2^2 + \alpha_3^2}{2\alpha_2}; \qquad a_3 = \alpha_3$$

$$b_1 = \frac{\alpha_1^2 + \alpha_2^2 - \alpha_3^2}{2\alpha_1}; \qquad b_2 = -\alpha_2; \qquad b_3 = \frac{\alpha_1^2 - \alpha_2^2 - \alpha_3^2}{-2\alpha_3} \qquad (12.6.2)$$

$$a = \alpha_3^2; \qquad b = \alpha_1^2 - \alpha_2^2 - \alpha_3^2; \qquad c = \alpha_1^2 - \alpha_2^2 - \alpha_3^2$$

One important feature of this map is that the sum of the parameterization lengths does not sum to zero, as in the light cone gauge. Instead, we find:

$$\sum_{r=1}^{3} \alpha_r = a_{12} \neq 0 \qquad (12.6.3)$$

The proof that we can smoothly go to the light cone limit is a bit tricky, because we have to take the limit as $a_{12} \to 0$, and potential infinities do arise in this limit. However, all cancel precisely. In fact, in the light cone limit, we find:

$$\lim_{a_{12} \to 0} \rho(z) = \alpha_1 \ln(z - 1) + \alpha_2 \ln z + a_{12} \ln a_{12} - a_{12} \ln(\alpha_3 z + \alpha_2)$$
$$- \ln \left(|\alpha_2 \alpha_3 \alpha_1^{-1}|^{-\alpha_1} |\alpha_1 \alpha_3 \alpha_2^{-1}|^{-\alpha_2} |\alpha_1 \alpha_2 \alpha_3^{-1}|^{-\alpha_3} \right)$$
$$(12.6.4)$$

As we carefully take the limit as $a_{12} \to 0$, we find that the map converges to the usual light cone map.

The proof that the map goes continuously to the midpoint gauge is easier. If we take the limit $\pi|\alpha_r| \to \pi$, then it is tedious, but straightforward, to show that the map approaches the limit:

$$\rho(z) = \frac{(z^2 - z + 1)^{3/2} - (z - 1/2)(z - 2)(z + 1)}{z(z - 1)} \qquad (12.6.5)$$

which is precisely the midpoint gauge conformal map.

Because the limit is continuous, we also know that the limit of the vertex functions, which are constructed out of these maps, is also smooth:

$$\lim_{|\alpha_r| = \pi} |V_{\text{interpolating}}\rangle = |V_{\text{midpoint}}\rangle$$
$$\lim_{\sum \alpha_r = 0} |V_{\text{interpolating}}\rangle = |V_{\text{endpoint}}\rangle \qquad (12.6.6)$$

Furthermore, it is possible to transform one vertex function into another via Virasoro operators. If we let $|V_{\alpha_1 \alpha_2 \alpha_3}\rangle$ represent a vertex with arbitrary parameterization lengths, then we have:

$$|V_{\alpha_1 \alpha_2 \alpha_4}\rangle = U_{\alpha_1 \alpha_2 \alpha_3}^{\bar{\alpha}_1 \bar{\alpha}_2 \bar{\alpha}_3} |V_{\bar{\alpha}_1, \bar{\alpha}_2, \bar{\alpha}_3}\rangle \qquad (12.6.7)$$

where:

$$U = \exp \sum_{n=1}^{\infty} (L_n - L_{-n}) \zeta_n \qquad (12.6.8)$$
$$U_{\alpha_1 \alpha_2 \alpha_3}^{\bar{\alpha}_1 \bar{\alpha}_2 \bar{\alpha}_3} = U_{\alpha_1}^{\bar{\alpha}_1} U_{\alpha_2}^{\bar{\alpha}_2} U_{\alpha_3}^{\bar{\alpha}_3}$$

where ζ_n are the Fourier modes of the field ζ^σ, which is the remnant of the vierbein after we set the curvature to zero. Notice that by integrating over all possible ζ^σ, we are in essence integrating over all possible parameterizations of the triplets, that is,

$$f_{X_1,X_2}^{X_3} = \prod_{i=1}^{3} \prod_{0 \le \sigma_i \le \sigma_i^0} \delta \left[X_\mu(\sigma_i + \zeta^{\sigma_i}) - X_\mu(\pi\alpha_{i-1} - \sigma_{i-1} - \zeta^{\sigma_{i-1}}) \right]$$

(12.6.9)

The interaction vertex $|V\rangle$ must also be modified to reflect the presence of the metric tensor g_{ab} in the theory. We can insert the following factor, which reflects the boundary conditions of the endpoint vertex, into the usual vertex function:

$$|V\rangle = \delta(\tau_1 - \tau_3)\delta(\tau_2 - \tau_3)\delta(\gamma_1^\tau - \gamma_3^\tau)\delta(\gamma_2^\tau - \gamma_2^\tau)\,\delta(\sum_i \alpha_i)\delta(\sum_i \gamma_i^\alpha)\cdots$$

(12.6.10)

It is easy to show that the sum of the Q_1 for the three strings vanishes when acting on this vertex function:

$$\sum_i^3 Q_1^i |V\rangle = 0 \qquad\qquad (12.6.11)$$

12.7. From Polynomial to Nonpolynomial String Theory

Last, we would like to show that we can smoothly go from the geometric action, Eq. (12.4.10), which is purely cubic, to either the polynomial endpoint gauge theory or the nonpolynomial midpoint gauge theory. Our goal is to show that we can eliminate the dependence on α, τ, γ^α, and γ^τ so that we are left with either the light cone theory or the nonpolynomial theory.

Endpoint Gauge

First, we would like to derive a covariant theory with endpoint type vertices. This is relatively easy because the vertex is already defined with the sum of the parameterization lengths to be zero. We can simply set the vierbein ζ^σ to zero.

However, there is one crucial complication. *It is impossible to simply remove the α and τ degrees of freedom from the theory.* This is because our naive gauge choice $\phi_\tau = \delta(\alpha - \pi)\delta(\tau)\phi_\alpha$ is not allowed. With a vertex as in Fig. 12.2a, where the parameterization lengths sum to zero, we cannot fix the three string lengths to be equal. The interacting gauge variation of

ϕ_τ does not permit taking such a gauge (or else the vertex function equals zero).

Therefore, we are forced to keep the τ and α parameters intact. They cannot be removed from the theory as long as we have the vertex defined in the endpoint gauge. Instead, we will fix the gauge to be:

$$\phi_\tau = b_0 \phi_\alpha \qquad (12.7.1)$$

In this gauge, the free action reduces to:

$$\phi_\alpha \left[i\partial_\tau - (L_0 - 1) \right] \phi_\alpha \qquad (12.7.2)$$

because $\{Q, b_0\} = (L_0 - 1)$. Although this looks like the original light cone action, we emphasize that L_0 is fully covariant and the usual b, c ghost sector is intact.

At first, one might naively suspect that this reduced theory is the "covariantized light cone theory" [11]. However, this is not true. There are, in fact, severe problems with naively taking the action $\phi Q_0 \phi + \phi^3 + \phi^4$ and integrating over α. The problems with the covariantized light cone theory are as follows.

1. Because α is integrated over but is not a gauge parameter, we have an infinite overcounting of states at the loop level. All fields are now indexed by a fictitious parameter α, giving us an unwanted redundancy in the fields.

2. Because proper time is missing in the covariantized light cone theory, the loop diagrams do *not* have the geometry of the usual light cone world sheet. For example, in the usual light cone theory, the two internal lines forming a single loop diagram have the same propagation times because of the existence of the proper time parameter τ, which is single valued on the world sheet. However, in the covariantized light cone theory, the propagation times of the two internal legs are entirely different, and a single-valued parameter τ does not exist on this world sheet. This means that the number of moduli for genus g loop diagrams is equal to $7g$, not $6g$, as in the usual light cone theory. The extra modulus comes from the unequal propagation times of internal legs, which are now all independent.

3. The theory is not really gauge invariant. Superficially, the action seems to obey a nonlinear gauge invariance because of the presence of the ϕ^4 term. However, this is an illusion. The measure of integration $D\phi$ is *not* invariant under this nonlinear gauge transformation, and hence, the theory has gauge anomalies. (In the light cone theory, because of proper time and conservation of string length, the potentially anomalous terms arising from a Lorentz transformation of the measure all cancel.) For these three reasons, the covariantized light cone theory is unacceptable. Our endpoint gauge action differs from this incorrect theory because of the presence of a proper time parameter τ and also the ghosts γ. The

proper time coordinate τ produces world sheet configurations that are identical to the usual light cone theory, which is known to be modular invariant. Also, the theory in the endpoint gauge satisfies the boundary conditions of Eq. (12.5.8) at infinity, so all dependence on α and τ eventually disappears from the action.

Midpoint Gauge

In the endpoint gauge, we saw that it was forbidden to impose the gauge condition of Eq. (12.5.14), which would remove all dependency on τ and α from the very outset. This is because the vertex function has unequal parameterization lengths, and hence, a common parameterization length is forbidden. Our goal, however, is to derive the nonpolynomial theory of the previous chapter, which has no dependence on α or τ at all. The only way in which to enter the midpoint gauge, we see, is to change the string lengths of the field, so that all dependency on α disappears in the vertex function.

In contrast to the endpoint gauge, in the midpoint gauge, we have the possibility of eliminating the dependence on α and τ at the beginning. However, we shall have to use the vierbein to convert the endpoint vertex to the midpoint vertex.

As in the toy model action of Eq. (12.1.6), we can functionally integrate over one of the string fields at fixed α. As before in Eq. (12.1.7), the functional integration generates a nonlocal quartic theory, with the interaction term:

$$\langle \Psi \times \Psi | D(X - Y) | \Psi \times \Psi \rangle \tag{12.7.3}$$

when we integrate over the external string lengths of the four fields.

Successive integrations over two fields Ψ_i and Ψ_j yield:

$$\int DXDY \left(\Psi \times \Psi \quad \Psi \times \Psi \right)_{X,i} M_{XY,ij}^{-1} \begin{pmatrix} \Psi \times \Psi \\ \Psi \times \Psi \end{pmatrix}_{Y,j} \tag{12.7.4}$$

where:

$$M_{XY}^{-1} = \begin{pmatrix} D(X - Y) & -\Psi \times \\ -\Psi \times & D(X - Y) \end{pmatrix}_{XY}^{-1} \tag{12.7.5}$$

As in the point particle case, this process creates a nonlocal nonpolynomial action whose individual terms are equal to the scattering amplitude for N-strings in the endpoint gauge. The overall coefficient of each scattering amplitude is equal to one.

As it stands, this formula is useless, because the interaction is non-local. However, we now point out the crucial point: there is a vierbein in the vertex function $\times$. By changing the vierbein to convert the endpoint vertex into the midpoint vertex, the vertex function is defined for strings of equal parameterization length, and we can then eliminate the α dependence entirely.

However, by changing to the midpoint vertex, we are left with extra factors of the U matrix in Eq. (12.6.7), which contains the vierbein.

The elimination of these extra U factors is the key to the calculation. We saw that the string propagator D is not invariant under reparameterizations, as in Eq. (12.1.5). Under a reparameterization, the propagator D with one string length turns into another D (with a different string length) plus another term that is proportional to an integration over moduli space. By carefully treating all such terms, we find that *the leftover term in Eq. (12.1.5) will generate the nonpolynomial action.*

When we reinsert the vierbein back into the scattering amplitudes and make the transition to the midpoint gauge, we can extract the polyhedra out of the integrals. Since each scattering amplitude occurs in the action in the endpoint gauge with the coefficient one, we find that the coefficient of each polyhedra (modulo symmetries) is also equal to one, to all orders.

Since our cubic action describes both open, and closed strips, let us first focus on a simpler problem, the generation of the open four-string interaction from open string field theory. It is important to notice that the coefficient of the four-string (nonlocal) interaction appearing in Eq. (12.7.3) is just the four-string scattering amplitude. Our problem is thus reduced to studying this four-string scattering amplitude in the midpoint gauge and then making the transformation to the endpoint gauge.

The coefficient of the four-string term Φ^4 appearing in the functional integral is the four-string scattering matrix for t- and u-channel scattering. Let us now insert the trivial factor $UU^{-1} = 1$ into this scattering amplitude. This U operator is chosen to convert the midpoint vertex into the endpoint matrix. For the process $1 + 4 \to 2 + 3$ and $1 + 3 \to 2 + 4$, we have:

$$
\begin{aligned}
\langle V_{145}|D_5|V_{523} &>_{\text{midpoint}} +\langle V_{135}|D_5|V_{524}\rangle_{\text{midpoint}} \\
&= \langle V_{\alpha_1\alpha_4\alpha_5}|D_5|V_{\alpha_5\alpha_2\alpha_3}\rangle_{\text{endpoint}} \\
&\quad + \langle V_{\alpha_1\alpha_3\alpha_5}|D_5|V_{\alpha_5\alpha_2\alpha_4}\rangle_{\text{endpoint}} \\
&\quad + \int_{M_R} d\tilde{\alpha}_5 \mu \langle V_{\alpha_1\alpha_4\tilde{\alpha}_5}|V_{\tilde{\alpha}_5\alpha_2\alpha_3}\rangle_{\text{interpolating}} \\
&\quad + \int_{M_R} d\tilde{\alpha}_5 \mu \langle V_{\alpha_1\alpha_3\tilde{\alpha}_5}|V_{\tilde{\alpha}_5\alpha_2\alpha_4}\rangle_{\text{interpolating}}
\end{aligned}
\tag{12.7.6}
$$

(we will defer the proof of this until the next section), where the last two contact terms combine to give the integration region $\alpha_2 - \alpha_4 \leq \tilde{\alpha}_5 \leq \alpha_2 + \alpha_4$, where $\alpha_4 \leq \alpha_1, \alpha_2 \leq \alpha_3$. Notice that the contact term is the square of the interpolating vertex, that is, the parameterization lengths neither sum to zero nor are they equal.

The essential point is that, under a gauge transformation and the insertion of the U matrix, the midpoint scattering amplitude turns into the endpoint scattering amplitude and a second term. This new term arises because the propagator is not an invariant. Thus, the propagator D (in the midpoint gauge) has split into two pieces. One piece has been transformed back into D (in the endpoint gauge), while the other part has been transformed into the number one (generating an instantaneous four-string

Coulomb term). If we evaluate the topology of this Coulomb term, its measure of integration, its region of integration, etc., we find that it is precisely the four-string interaction first introduced in Ref. 2. It is, in fact, precisely the square of the interpolating vertex, just as the Coulomb term is the square of the two-fermion vertex.

Thus, the four-string interaction is a gauge artifact. It is the result of choosing a gauge in which the USG is nonlinearly realized. By changing the vierbein to a gauge in which the USG is linearly realized (the midpoint gauge), the four-string interaction vanishes. In between these two extremes, we have the interpolating gauge, where the four-string interactions can vary between zero (the midpoint gauge) and the full four-string interaction found in the endpoint gauge.

Now, let us reexpress in the language of string field theory, given the fact that the four-string amplitude is the (nonlocal) coefficient generated by functionally integrating one string field over a cubic action. Let us use the symbol $\times$ when we are multiplying strings in the endpoint gauge and the symbol $*$ when we are multiplying strings in the midpoint gauge.

The $\times$ vertex is converted into the $*$ vertex by the action of the reparameterization matrix U (which changes only the X and b, c ghost parameterizations. Dropping L_n gauge operators, we have:

$$\langle \Phi * \Phi | D(X - Y) | \Phi * \Phi \rangle_{\text{midpoint}}$$

$$= \langle \Phi \times \Phi | D(X - Y) | \Phi \times \Phi \rangle_{\text{endpoint}} + \int d\mu \langle \Phi \times \Phi | \Phi \times \Phi \rangle \qquad (12.7.7)$$

The second term is the counterpart of the instantaneous Coulomb term and has the precise topology of the four-string interaction found in Chapter 10. We see that the four-string interaction of the light cone theory is therefore a gauge artifact, a by-product of transforming from the midpoint gauge to the endpoint gauge.

Now, let us generalize the discussion to the closed string case. The s-channel contribution to the scattering amplitude can be written as:

$$\langle V_{\alpha_1 \alpha_2 \alpha_5} | D_5 | V_{\alpha_5 \alpha_3 \alpha_4} \rangle_{\text{endpoint}} = \langle V_{125} | D_5 | V_{534} \rangle_{\text{midpoint}}$$

$$+ \int da_{12} \, da_{13} \, \mu \langle V_{12\tilde{\alpha}_5} | e^{ia_{13}(L_o - \tilde{L}_o)} | V_{\tilde{\alpha}_5 34} \rangle_{\text{interpolating}} \qquad (12.7.8)$$

where $\tilde{\alpha}_5 = 4\pi - 2a_{12}$,

$$D_5 = b_o \tilde{b}_o \int_0^1 dr \int_0^{2\pi} d\theta \, r^{L_o + \tilde{L}_o - 2} e^{i\theta(L_o - \tilde{L}_o)} \qquad (12.7.9)$$

and μ is a measure term that will be calculated in the next section. It eventually becomes the Jacobian between the polyhedron's parameters and the Koba–Nielsen variables.

Since the coefficient of the nonlocal quartic term in the string action is the four-string scattering amplitude, we can rewrite this formula as:

$$\langle \Psi \times \Psi | D(X - Y) | \Psi \times \Psi \rangle_{\text{endpoint}}$$

$$= \langle \Psi * \Psi | D(X - Y) | \Psi * \Psi \rangle_{\text{midpoint}} + \int d\mu \frac{\langle \Psi^4 \rangle}{s(4)} \qquad (12.7.10)$$

This is the formula that we desire. We have no converted the theory from the endpoint gauge (where the theory is purely cubic) to the midpoint gauge, where the theory has generated a four-string action to lowest order. This four-string action has precisely the topology of the tetrahedron graph that we studied in the last chapter.

There is one last remaining step, however. Although the X-dependent terms in the vertex appearing in the last expression are defined purely with equal parameterization lengths, the terms that are not affected by L_n still remain. In other words, the α and τ dependences in the vertex function still remain, but as purely dummy variables.

In the last step, we can use the Parisi–Sourlas [16–17] mechanism, which states that a system with $Osp(D + 2/2)$ space–time symmetry is equivalent to ordinary (D) symmetry, that is, a fermion corresponds to a "negative" dimension that cancels against the positive dimension of a boson. Since the X-dependent terms in the vertex function are now totally free of the parameterization lengths, only the α and τ enter the vertex function multiplicatively, through Eq. (12.6.9).

To see how this works, let us analyze the simplest case of the single-loop diagram with two external legs. We will show that the integration over τ and α cancels the integration over θ and $\bar\theta$. To show this, we will use an old trick due to Feynman, which is to introduce auxiliary parameters. Consider the example of an ordinary point particle ϕ^3 theory.

Let Δ be the propagator of the point particle. The Feynman rules for the single-loop graph give us the product of two propagators, integrated over space–time. Let us now make a sequence of transformations by inserting a series of Feynman parameters into the amplitude:

$$\frac{1}{\Delta_1 \Delta_2} = \int_0^1 \frac{dx}{\left[x\Delta_1 + (1 - x)\Delta_2 \right]^2}$$

$$= \alpha_3 \int_0^{\alpha_3} \frac{d\alpha_1 \, d\alpha_2 \, \delta(\alpha_1 + \alpha_2 - \alpha_3)}{(\alpha_1 \Delta_1 + \alpha_2 \Delta_2)^2}$$

$$= \alpha_3 \int_0^{\alpha_3} \frac{d\alpha_1 \, d\alpha_2 \, \delta(\alpha_1 + \alpha_2 - \alpha_3) d^2\theta}{(\alpha_1 \Delta_1 + \alpha_2 \Delta_2 + \bar\theta\theta)} \qquad (12.7.11)$$

$$= \alpha_3 \int_0^{\alpha_3} \prod_i d\alpha_i \, d^2\theta_i \, d\tau \, \delta(\alpha_1 + \alpha_2 - \alpha_3) \delta(\theta_1 + \theta_2 - \theta_3)$$

$$\times \exp -\tau(\alpha_1 \bar\Delta_1 + \alpha_2 \bar\Delta_2)$$

where $\bar\Delta = \Delta + \bar\theta\theta$.

In the first line, we have the Feynman propagators of the loop rewritten as an integral over the Feynman parameter x. In the second line, we have rewritten x in terms of fictitious parameters α_i. Next, by introducing two Grassmann variables θ, we have reduced the power of the denominator by one. Last, because we have a single factor in the denominator, we can introduce one more dummy variable τ, which converts the denominator into an exponential.

Notice that τ, α, and θ are all dummy variables and were introduced purely via a series of identities. However, we will give a physical interpretation to them. These and other identities (for higher loops) show that we can start with a ϕ^3 theory with no mention of α or τ, such that the Feynman loop amplitude simply consists of a product of propagators Δ_i integrated over space–time. Using the super-Feynman trick, we can convert this product of Δ_i into an integration over proper times and α and θ, such that α and θ are conserved at each vertex, and τ is a common parameter for all legs of the loop diagram that start and finish at the same point. This means we can make the following replacement, from a theory defined totally in terms of $\phi(x)$ to a theory with the fictitious parameters α, τ, and θ. We can make the following replacement:

$$
\int dx \left(\frac{1}{2} \partial_\mu \phi \, \partial^\mu \phi + g \phi^3 \right)
$$

$$
\rightarrow \int d\tau \, d\alpha \, d^2\theta \, \phi(x, \tau, \alpha, \theta, \bar{\theta}) \big(\alpha \partial_\tau + \Delta \big) \phi(x, \tau, \alpha, \theta, \bar{\theta})
$$

$$
+ g \int dx \, d\tau \, \delta \left(\sum_i \alpha_i \right) \prod_i d\alpha_i \, d^2\theta_i \delta^2 \left(\sum_i \theta_i \right) \phi^3(\alpha_i, \tau, x, \theta_i, \bar{\theta}_i)
$$

$$
\tag{12.7.12}
$$

with the condition that the external states are independent of these fictitious parameters. Obviously, we can do the same for string fields with equal parameterization lengths.

Let us now summarize the various steps we used to gauge fix the geometric theory down to the nonpolynomial theory of the last chapter.

1. First, we eliminated the σ index of the $A_\sigma^{\{\alpha\}}$ field, leaving us with a cubic action in Ψ.

2. Second, we integrated over one of the fields at fixed α. This generated a four-string, nonlocal term.

3. Third, to make the nonlocal term into a local term, we used the vierbein in the U function to change the X and b, c dependent terms in the vertex function from the endpoint gauge to the midpoint gauge.

4. Fourth, since the propagator was not invariant under U transformations, we generated new "instantaneous terms," which were precisely the tetrahedron and higher graphs.

5. Fifth, we used the Parisi–Sourlas mechanism [16–17] to eliminate the dummy variables α, τ, γ^τ, and γ^σ. To retrieve the original nonpolynomial action, we simply "undo" the functional integrations over the string fields.

Last, we notice that we can generate polyhedra of arbitrary complexity by making successive integrations over the string field. For example, the five-point string function is generated by multiple functional integrations:

$$\langle \Psi \times \Psi | D(X - Y) | \Psi | D(Y - Z) | \Psi \times \Psi \rangle_{\text{endpoint}}$$

$$= \langle \Psi * \Psi | D(X - Y) | \Psi | D(Y - Z) | \Psi * \Psi \rangle_{\text{midpoint}} + \int d\mu \frac{\langle \Psi^5 \rangle}{s(5)} \tag{12.7.13}$$

12.8. Method of Reflections

In this section, we will prove the crucial identity Eq. (12.7.6) linking midpoint and endpoint amplitudes, using the "method of reflections." We begin by introducing the U matrix, which contains the vierbein that changes the parameterization length of a string:

$$|V_{\alpha_1 \alpha_2 \alpha_3}\rangle = \prod_{i=1}^{3} U_{\alpha_i}^{\tilde{\alpha}_i} |V_{\tilde{\alpha}_1 \tilde{\alpha}_2 \tilde{\alpha}_3}\rangle \sim \prod_{r=1}^{3} [1 + \delta \epsilon_r^n (L_n - \tilde{L}_{-n})] |V_{\tilde{\alpha}_1 \tilde{\alpha}_2 \tilde{\alpha}_3}\rangle \tag{12.8.1}$$

(In this discussion, we will only use the infinitesimal version of the U matrix. However, there are technical complications to making it fully finite.) Notice that we can convert all expressions to factors containing only L_{-m} by using the fact that L_n, acting on a vertex function, "reflects" and turns into L_{-m}:

$$\left(L_n^r / \alpha_r - \sum n \tilde{N}_{nm}^{rs} L_{-m}^s / \alpha_s - c_n^r \right) |V_{\alpha_1 \alpha_2 \alpha_3}\rangle = 0 \tag{12.8.2}$$

where:

$$c_n^r = \frac{D}{2} \sum_{p=1}^{n-1} p(n-p) N_{p,n-p}^{rr} + \sum_{p=1}^{n-1} (p^2 - n^2) \tilde{N}_{p,n-p}^{rr} - n^2 \tilde{N}_{no}^{rr} \tag{12.8.3}$$

Also, when L_n passes through a propagator, it picks up a factor of x^n:

$$L_n x^{L_o} = x^n x^{L_o} L_n \tag{12.8.4}$$

By analogy, the elimination of the vierbein can be compared to a light beam reflecting off two parallel mirrors. The light beam is like L_n, and the parallel mirrors are like two vertex functions. The light beam (L_n) bounces back and forth an infinite number of times between the two mirrors (vertex functions), each time getting weaker in intensity (because x is less than 1) until the light beam eventually disappears. Fortunately, these terms converge and can be summed exactly to all orders. After an infinite number of reflections, we find a simple result:

$$\langle V_{\alpha_1 \alpha_5 \alpha_5} | L_n x_5^{L_o} | V_{\alpha_5 \alpha_2 \alpha_3} \rangle = f(x_5) \langle V_{\alpha_1 \alpha_4 \alpha_3} | \tilde{x}_5^{L_o} | V_{\alpha_5 \alpha_2 \alpha_5} \rangle \tag{12.8.5}$$

for the t-channel process $1 + 4 \rightarrow 5 \rightarrow 2 + 3$.

We see that the effect of adding L_n between the vertices is to generate a factor of $f(x_5)$ and to change $x_5^{L_0}$ to $\tilde{x}_5^{L_0}$. In other words, we have the transformation:

$$x_5^{L_0-1}dx_5 \rightarrow \tilde{x}_5^{L_0} x_5^{-1} f(x_5)dx_5 = \tilde{x}_5^{L_0-1}d\tilde{x}_5 \qquad (12.8.6)$$

Now, let us split the propagator D into two pieces, $D_{\rm I}$ and $D_{\rm II}$, where the first piece contains an integral from 0 to $1 - \epsilon$, and the second piece contains an integral from $1 - \epsilon$ to 1:

$$D = \left(\int_0^{1-\epsilon} + \int_{1-\epsilon}^1 \right) dx_5 \, x_5^{L_0-1} = D_{\rm I} + D_{\rm II} \qquad (12.8.7)$$

Let $A_{\rm I}$ be the scattering amplitude for the t-channel process with the propagator $D_{\rm I}$. Then, we have:

$$
\begin{aligned}
A_{\rm I} &= \left\langle V_{\alpha_1\alpha_4\alpha_5} | D_{\rm I} | V_{\alpha_5\alpha_2\alpha_3} \right\rangle_{\rm midpoint} \\
&= \left\langle V_{\tilde{\alpha}_1\tilde{\alpha}_4\tilde{\alpha}_5} | U_{\alpha_1\alpha_4\alpha_5}^{\tilde{\alpha}_1\tilde{\alpha}_4\tilde{\alpha}_5} \, D_{\rm I} \, U_{\alpha_5\alpha_2\alpha_3}^{\alpha_5\alpha_2\alpha_3} \, V_{\alpha_5\alpha_2\alpha_3} \right\rangle \\
&= \left\langle V_{\tilde{\alpha}_1\tilde{\alpha}_4\tilde{\alpha}_5} | \int_0^1 \tilde{x}_5^{L_0-1} d\tilde{x}_5 | V_{\tilde{\alpha}_5\tilde{\alpha}_2\tilde{\alpha}_3} \right\rangle_{\rm endpoint}
\end{aligned}
\qquad (12.8.8)
$$

The important point is to notice that $D_{\rm I}$, which only has a piece of the integration region, has now changed into an integral over $\tilde{x}_5$ between 0 to 1. In fact, we see that the new propagator is simply D.

More important is the contribution from $D_{\rm II}$ to the matrix element, which we call $A_{\rm II}$. We find that x_5 again transforms into $\tilde{x}_5$ after the method of reflections. The difference now is that we set $\tilde{x}_5 \equiv 1$. This, in turn, changes the value of the intermediate string length α_5 into a new variable, called α_6:

$$
\begin{aligned}
A_{\rm II} &= \left\langle V_{\alpha_1\alpha_4\alpha_5} | D_{\rm II} | V_{\alpha_5\alpha_2\alpha_3} \right\rangle_{\rm midpoint} \\
&= \left\langle V_{\tilde{\alpha}_1\tilde{\alpha}_4\tilde{\alpha}_6} | U_{\alpha_1\alpha_4\alpha_6}^{\tilde{\alpha}_1\tilde{\alpha}_4\tilde{\alpha}_6} \, D_{\rm II} \, U_{\alpha_6\alpha_2\alpha_3}^{\alpha_6\alpha_2\alpha_3} \, V_{\alpha_6\alpha_2\alpha_3} \right\rangle \\
&= \int f(x_6) x_6^{-1} dx_6 \left\langle V_{\tilde{\alpha}_1\tilde{\alpha}_4\tilde{\alpha}_6} | V_{\tilde{\alpha}_6\tilde{\alpha}_2\tilde{\alpha}_3} \right\rangle
\end{aligned}
\qquad (12.8.9)
$$

It is important to notice that the last term can become a matrix element without any propagator sandwiched between them. The propagator has completely disappeared, leaving only an integration over x_6. This term will eventually become the four-string interaction.

Now, let us sum both the t-channel and s-channel contributions to the scattering amplitude. The two channels add up to give a contact term integrated over the whole region of the usual four-string interaction. We find:

$$\langle V_{\alpha_1\alpha_4\alpha_5}|D_{\mathrm{I}}|V_{\alpha_5\alpha_2\alpha_3}\rangle_{\mathrm{midpoint}} + \langle V_{\alpha_2\alpha_4\alpha_7}|D_{\mathrm{I}}|V_{\alpha_7\alpha_1\alpha_3}\rangle_{\mathrm{midpoint}}$$

$$= \langle V_{\alpha_1\alpha_4\alpha_5}|D_{\mathrm{I}}|V_{\alpha_5\alpha_2\alpha_3}\rangle_{\mathrm{endpoint}} + \langle V_{\alpha_2\alpha_4\alpha_7}|D_{\mathrm{I}}|V_{\alpha_7\alpha_1\alpha_3}\rangle_{\mathrm{endpoint}}$$

$$+ \int_{\alpha_1-|\alpha_4|}^{\alpha_1+|\alpha_4|} \mu\, d\alpha \langle V_{\alpha_1\alpha_4\alpha}|V_{\alpha_2\alpha_3\alpha}\rangle$$

$$(12.8.10)$$

Since the four-string scattering amplitude is the coefficient of the four-string interaction in the Lagrangian, we find that the midpoint gauge (where a four-string interaction is absent) has transformed into the endpoint gauge, where a four-string interaction emerges because D is not invariant under a reparameterization. Comparing this with what happened earlier with the scalar meson example, we see that Eq. (12.1.7) for the point particle case has generated a genuine four-string interaction when we generalize to the string case.

Likewise, it is straightforward to generalize this analysis to the closed string case, in which case we generate a new four-string (tetrahedron) graph. For the general case, where we may have a vertex function with an arbitrary number of legs, it seems prohibitively difficult to apply this simple method to extract out the polyhedra graphs. In general, the method of reflections changes the variable of integration as follows for the closed string propagator:

$$\tilde{z} = z\big[1 + \xi(z)\big] \qquad (12.8.11)$$

where:

$$\xi(z) = \sum_{r=1}\sum_{i,j,k,\ldots}\sum_{n_i}\sum_{n_j} \delta\alpha_r^{n_i}\,\tilde{N}_{n_in_j}^{r_ir_j}\,z^{n_j}\,n_j\,\tilde{N}_{n_jn_k}^{r_jr_k}\,z^{n_k}\,n_k\,\tilde{N}_{n_kn_l}^{r_kr_l} \qquad (12.8.12)$$

where the values of $i,j,k,\ldots$ can be determined by following the paths of light beams that can reflect in any direction from a series of mirrors placed at the location of each vertex function. Thus, the path traced out by $i,j,k,\ldots$ touches all vertex functions sequentially.

At first, it may seem to be a hopeless task to make sense out of this series. However, fortunately, it can be almost trivially summed to all orders. Two key observations make the sum almost totally transparent. First, let us analyze the following contraction for N-string scattering:

$$\prod_{i,j,k}\left\langle V_{\alpha_i\alpha_j\alpha_k}\left|\prod_l D_l \prod_{p,q,r}\right|V_{\alpha_p\alpha_q\alpha_r}\right\rangle \qquad (12.8.13)$$

The vertex functions are given by:

$$\prod_{nm,rs} \exp\left(\frac{1}{2}\alpha_{-n}^r N_{nm}^{rs}\alpha_{-m}^s + b_{-n}^r n \tilde{N}_{nm}^{rs} c_{-m}^s\right)|0\rangle \qquad (12.8.14)$$

multiplied by the appropriate insertion factors at the joints where strings meet,

$$N_{nm}^{rs} = \frac{1}{mn(2\pi)^2} \oint_{z_r} \oint_{z_s} (z - z')^{-2} e^{-n\zeta_r(z) - m\zeta_s(z')}$$
(12.8.15)

and the conformal map from the complex plane to the sheet describing scattering in the midpoint gauge is given by [15]:

$$\frac{d\rho}{dz} = N \frac{\prod_{i=1}^{2(N-2)}(z - z_i)^{1/2}}{\prod_{r=1}^{N}(z - \gamma_r)}$$
(12.8.16)

where N and the γ_r are determined, for the missing region, by setting the external lengths to be 2π and by setting all the real parts of $\rho(z_i)$ to be the same.

Using coherent states, we can evaluate all the matrix elements in Eq. (12.8.13) explicitly. All matrix elements are special cases of the following formula:

$$\langle 0| \exp\left[\frac{1}{2}(a_i|M_{ij}^{(2)}|a_j) + (a_i|L_i^{(2)})\right] \exp\left[\frac{1}{2}(a_i^\dagger|M_{ij}^{(1)}|a_j^\dagger) + (a_j^\dagger|L_j^{(1)})\right] |0\rangle$$

$$= \det^{-D/2}\left(1 - M^{(2)}M^{(1)}\right)_{ij} \exp\left(L_i^{(1)}\left|\left(\frac{1}{1 - M^{(2)}M^{(1)}}\right)_{ij}\right|L_j^{(2)}\right)$$

$$\times \exp\frac{1}{2}\left(L_i^{(2)}\left|M_{ij}^{(1)}\left(\frac{1}{1 - M^{(2)}M^{(1)}}\right)_{jk}\right|L_k^{(2)}\right)$$

$$\times \exp\frac{1}{2}\left(L_i^{(1)}\left|\left(\frac{1}{1 - M^{(2)}M^{(1)}}\right)_{ij}\right|M_{jk}^{(2)}L_k^{(1)}\right)$$
(12.8.17)

where this formula allows us to express the Neumann function defined over an entire N-string scattering Riemann in terms of elemental three-string Neumann functions.

Now, let us study the power expansion implied by the above equation. Notice that the power expansion in the $M^{(2)}M^{(1)}$ matrices is *precisely* the expansion found for the method of reflections in Eq. (12.8.12). Thus, it is rather trivial to sum the entire series and obtain the Neumann function of the N-string scattering matrix. The only reason why the expansion was rather complicated looking was because we were writing the N-string Neumann function in terms of three-string Neumann functions.

Now that we can reexpress the $\tilde{x}$ variables in terms of the Neumann function defined over the N-string scattering matrix, we use the second trick. We show that the shift in the propagator variable is simply given by a conformal transformation that changes the external dimensions of the N-string scattering amplitude. This second trick gives us a nice physical interpretation for the method of reflections.

Let us first define the Neumann function over a Riemann surface corresponding to the scattering of N closed strings:

$$\nabla^2 N(\sigma, \tau; \sigma', \tau') = 2\pi\delta(\sigma - \sigma')\delta(\tau - \tau')$$
(12.8.18)

Now, let us multiply both sides by τ' and integrate over both σ' and τ'. The right-hand side then becomes τ. Solving, we find:

$$2\pi\tau = \int d\tau'\,d\sigma'\,\nabla^2 N = \int d\tau'\,d\sigma'\,\tau'\,\partial^2_{\tau'}N$$

$$= \int d\sigma'\,d\tau'\,[\partial_{\tau'}\,(\tau'\,\partial_{\tau'}N) - \partial_{\tau'}N]$$

(12.8.19)

Normally, in the simpler case of the light cone formalism, this integral can be evaluated knowing only the value of the Neumann function at infinity. We find:

$$\tau = \frac{1}{2\pi}\left[\sum_{r=1}\int d\sigma'_r\,N(\sigma,\tau;-\infty,\sigma'_r)d\sigma'_r\right.$$

$$\left.-\sum_{s=1}\int d\sigma'_s\,N(\sigma,\tau;\infty,\sigma'_s)\right]_{\text{endpoint}}$$

(12.8.20)

where the sum over $r(s)$ is taken over incoming (outgoing) strings. We have ignored certain boundary terms in Eq. (12.8.14) because we know that:

$$\tau'\,\partial N \sim \tau'n N_{nm}e^{-n\tau'} \to 0$$

(12.8.21)

For our case, however, we can no longer use this formula because of the presence of the crucial Riemann cuts, which define the topology of string interactions. In particular, the integral over σ' now contributes a term that occurs in the middle of the Riemann sheet, rather than at infinity. In Fig. 14, for example, we see that there are vertical flaps that rise from the Riemann sheet for both open and closed strings scattering in the midpoint gauge, so that partial integration in the τ' variable must necessarily pick up these contributions.

The integration is a bit tricky, because several terms contribute to the integral. First, let us analyze the term $\tau'N$ in Eq. (12.8.14). We find that this term actually vanishes at the location of the Riemann cut τ_c because:

$$\tau'_1\,\partial_{\tau'_1}N\big|_{\tau'_1=\tau_c} + \tau'_2\,\partial_{\tau_{2'}}N\big|_{\tau'_2=\tau_c} = 0$$

(12.8.22)

Notice that this term vanishes because there are two contributions to the line integral from infinity to the Riemann cut $\tau = \tau_c$ from either side of the Riemann cut. Fortunately, they cancel precisely (because $d\tau$ and ∂N have opposite parity under reflection across τ_c).

We are left, therefore, with the final answer for τ:

$$\tau = \frac{1}{2\pi}\left[\sum_{p,q,r,s=1}\sum_{u=p,q,r,s}\int d\sigma'_u\,N(\sigma,\tau;\tau'_u,\sigma'_u)(-1)^{n_u}\right]$$

(12.8.23)

where:

$$\tau_r' = -\infty; \qquad n_r = -1$$
$$\tau_s' = \infty; \qquad n_s = 1$$
$$\tau_p' = \tau_{c_p}; \qquad n_p = -1 \qquad (12.8.24)$$
$$\tau_q' = \tau_{c_q}; \qquad n_q = 1$$

where the location of the Riemann cuts for incoming strings is given by τ_{c_p} and for outgoing strings by τ_{c_q}.

Similarly, we can set τ in the previous expression to be at the turning point τ_c. We can also calculate the modular parameter, the proper time between two vertices, by taking the difference between the τ_c at two different turning points.

Now, let us make small variations in the external lengths of the strings at $\tau = \pm\infty$ and at $\tau = \tau_c$. By this fashion, we can calculate precisely how the modular parameter, the proper time between interacting points, changes as we slowly make the transition between endpoint and midpoint gauges. Notice that the expression for $\delta\tau$ after making these variations in Eq. (12.8.23) is *precisely* equal to the absolute value of $\delta\bar{z}$ in Eq. (12.8.12). Thus, this acts as a powerful check on the correctness of our results.

This process also explains how to make two-dimensional conformal transformations in string field theory, which is necessarily defined at one τ slice. We know that if we make a small change in the Riemann sheet at one point, the entire Riemann sheet immediately changes in response. However, this seems to precisely violate the spirit of string field theory, which chops up the Riemann surface into elemental vertices. The key to understanding how one-dimensional reparameterizations in σ space become two-dimensional reparameterizations in σ and τ space is through the vierbein. By inserting the vierbein into our theory and using the method of reflections, the entire Riemann surface responds to a one-dimensional reparameterization at one vertex.

The same analysis can be carried out to all orders in the open string theory. The only difference is that, at higher orders when we make the transition from the midpoint gauge to the endpoint gauge, we find that the nonplanar one-loop graph exhibits a new feature: the pole structure of the nonplanar scattering amplitude begins to change. To represent this change in the pole structure of the amplitude, we insert a new set of Fock space operators into the path integral:

$$1 \equiv \int D^2\Psi \, e^{-\int \Psi^\dagger(X)\Psi(X)DX} \qquad (12.8.25)$$

where Ψ is a closed string field. In this way, we can explicitly calculate the contribution of the (open $\to$ closed string) transition graph to the action.

It is then a simple matter to convert all midpoint gauge string fields into the endpoint gauge in the action. Fortunately, by analyzing the pole structure of the nonlocal series, we find that the new action is actually polynomial (with the usual five interaction terms). Last, because the endpoint gauge generates purely planar Riemann surfaces, we can gauge fix the theory to the light cone, and *we retrieve the entire theory of Ref. 1, starting from the midpoint theory of Ref. 2.*

In summary, we have now succeeded in carrying out our original plan: postulating a set of first principles, constructing a unique cubic action based on these principles, and then showing that we can obtain either the midpoint or the endpoint theories by appropriately fixing the gauge. However, we should emphasize that the geometric theory is still in its infancy, that we are unable to carry out any nonperturbative analysis of the true vacuum of string theory. Much more effort will be needed to bring the geometric theory to the level now enjoyed by nonperturbative point particle quantum field theory.

We will now turn to a new approach, one that, for the first time, actually yields nonperturbative information about string theory: matrix models. We will, however, have to pay a price. In exchange for exact nonperturbative information about string theory, we will have to fix the dimension of space time to be $D \leq 1$.

Matrix models, however, are not divorced from string field theory. In fact, it is actually possible to summarize the constraint equations of matrix models as Ward identities on string field theory. Thus, string field theory gives us by far the simplest way of reformulating matrix models to reveal their group structure. String field theory gives us a compact and transparent way of seeing the origin of the mysterious constraint equations of matrix models.

12.9. Summary

Geometric string field theory is an attempt to derive string field theory, and hence all of string theory, from first principles. The first question facing such a theory is why are there so many string field theories. For open strings, the light cone theory contains five elemental interactions, while the midpoint theory contains only one:

$$(\Phi^3)_{\text{geometric}} = \begin{cases} (\Phi^3 + \Psi^3 + \Phi^2\Psi + \Phi^4 + \Psi\Phi)_{\text{endpoint}} \\ (\Phi^3)_{\text{midpoint}} \end{cases} \qquad (12.9.1)$$

The situation reverses itself for closed strings. There, we have one elemental interaction in the light cone gauge, but an infinite number of them in the midpoint theory:

$$(\Psi^3)_{\text{geometric}} = \begin{cases} (\Psi^3)_{\text{endpoint}} \\ (\Psi^3 + \Psi^4 + \cdots)_{\text{midpoint}} \end{cases} \qquad (12.9.2)$$

Our goal is to gauge the reparameterization group, so that the curious puzzle of why there are so many string field theories can be reduced to finding the "midpoint gauge" and "endpoint gauge" of a higher, geometric string field theory.

It seems curious that a theory can be polynomial in one gauge but nonpolynomial in another. However, there is ample precedent for such a phenomenon in gauge theory. In QED, for example, in the Coulomb gauge, the propagator of the A_0 field has no time component and hence can be eliminated entirely, generating the instantaneous four-fermion interaction:

$$(\bar{\psi}\gamma_0\psi)\frac{1}{\nabla^2}(\bar{\psi}\gamma_0\psi) \tag{12.9.3}$$

A cubic theory has thus become quartic because the propagator is not a gauge invariant object. By eliminating the time component of the propagator, we can eliminate redundant fields, generating higher order interactions.

Similarly, consider a scalar meson theory with the action:

$$\sum_a \partial_\mu \phi^a \, \partial^\mu \phi^a + \sum_{abc} v_{abc} \phi^a \phi^b \phi^c \tag{12.9.4}$$

By integrating over two fields, we find an effective quartic term emerging:

$$\int d^4x \, d^4y \, (\phi \times \phi \quad \phi \times \phi)_{x,i} M_{xy,ij}^{-1} \begin{pmatrix} \phi \times \phi \\ \phi \times \phi \end{pmatrix}_{y,j} \tag{12.9.5}$$

where:

$$(\phi \times \phi)_e = v_{abe} \phi^a \phi^b \tag{12.9.6}$$

and

$$M_{xy,ij}^{-1} = \begin{bmatrix} D(x-y) & -\phi\times \\ -\phi\times & D(x-y) \end{bmatrix}_{xy,ij}^{-1} \tag{12.9.7}$$

We have traded a local cubic theory for a nonlocal nonpolynomial theory. At the point particle level, this is all we can do. However, the real difference comes when we generalize our results to geometric string field theory, where the propagator is not gauge invariant and may be gauged to one. Then, a local cubic theory can be transformed into a local nonpolynomial theory.

The advantage of geometric string field theory is that it is based on two simple principles.

Global Symmetry: the fields of the theory $A_\sigma^{\{\alpha\}}$ transform as irreducible ghost-free representations of $\text{Diff}(S_1)$ and the Lorentz group.

Local Symmetry: The theory must be locally invariant under the universal string group.

The universal string group, in turn, consists of several parts, including reparameterizations.

Let us first define the string group, which is defined topologically on unparameterized strings $\{C\}$. Topologically, strings either do or do not overlap. Let us define a triplet of three oriented strings as in Fig. 12.1a. The structure constant of the string algebra is given by:

$$f_{C_1, C_2}^{C_3} = \begin{cases} +1 & \text{for triplets} \\ -1 & \text{for antitriplets} \\ 0 & \text{otherwise} \end{cases} \qquad (12.9.8)$$

Remarkably, this structure constant satisfies the Jacobi identities for this definition of a triplet. This is a nontrivial result, because the Jacobi identities are not satisfied for Fig. 12.2a.

Similarly, for a triplet of three closed strings defined as in Fig. 12.4, the structure constant is defined as:

$$f_{C_1 C_2}^{C_3} = \begin{cases} +1(-1) & \text{for triplets if } C_3 \text{ equals outer (inner) string} \\ 0 & \text{otherwise} \end{cases}$$
$$(12.9.9)$$

The point, however, is that the Jacobi identity is not satisfied if we take Fig. 12.3a as our definition of a triplet, and hence, the midpoint gauge is an inconvenient gauge with which to work, that is, the gauge group will not close on three-string vertices.

The relation between the string group, which is topologically defined without any reference to space–time or parameterizations, to the universal string group is given as follows:

$$SG = \frac{USG}{\text{Diff}(S_1)} \qquad (12.9.10)$$
$$\text{Diff}(S_1) = \text{Diff}(S_1)_0 \otimes \text{scale}$$

This can be expressed by saying that the universal string group consists of all maps of a string into itself and all its conjugates:

$$USG: \quad \begin{cases} C \to C \\ C \to \bar{C} \end{cases} \qquad (12.9.11)$$

If, for the moment, we ignore scale transformations and the "ghost sector" (which becomes the tangent space of the geometric theory), we find that the algebra of the universal string group consists of:

$$\{L_{C_1}, L_{C_2}\} = \hat{f}_{C_1, C_2}^{C_3} L_{C_3}$$
$$\hat{f}_{C_1, C_2}^{C_3} = \theta_{C_1 \cap C_2 \cap C_3} f_{C_1, C_2}^{C_3} \qquad (12.9.12)$$

where:

$$f_{X_1, X_2}^{X_3} = \prod_{i=1}^{3} \prod_{0 \le \sigma_i \le \sigma_i^0} \delta\left[X_\mu(\sigma_i) - X_\mu(\pi \alpha_{i-1} - \sigma_{i-1})\right] \qquad (12.9.13)$$

To implement reparameterizations, we must include a string vierbein into the theory. For example, the measure transforms with the determinant of the vierbein:

$$DX = \prod_{\mu\sigma} dX^{\mu\sigma} \to \det\left(e^{\nu\omega}_{\mu\sigma}\right) \prod_{\nu\rho} d\bar{X}^{\nu\rho} \tag{12.9.14}$$

One complication is that the fields of the theory transform as a Verma module, labeled by the index $\{\alpha\}$. This means that the algebra must be enlarged to include these Verma module indices:

$$[L_X^{\{\alpha\}}, L_Y^{\{\beta\}}] = f_{\{\alpha\}\{\beta\}\{\gamma\}} f_{XY}^Z L_Z^{\{\gamma\}}$$
$$[L_\sigma, L_X^{\{\alpha\}}] = X'^\mu(\sigma)\partial_{\mu\sigma} L_X^{\{\alpha\}} + f_{\sigma\{\beta\}}^{\{\alpha\}} L_X^{\{\beta\}} \tag{12.9.15}$$
$$[L_\sigma, L_\rho] = f_{\sigma\rho}^\omega L_\omega$$

Last, we must include the effect of scale transformations. If we requantize the string, we find that the components of g_{ab} enter into the definition of the gauge group. The familiar action:

$$L = P_\mu \dot{X}^\mu - H$$
$$H_0 = \lambda_1 \left(P_\mu^2 + \frac{1}{\pi^2} X_\mu'^2\right) + \lambda_2 (P_\mu X^{2\mu}) \tag{12.9.16}$$

must be generalized to include the conjugates π_i to the metric tensor. There is, therefore, an Abelian sector to the theory with an algebra:

$$[\pi_i, \pi_j] = 0$$
$$[\pi_i, L_n] = 0 \tag{12.9.17}$$

Once we have defined the universal string group, we find that the action is uniquely specified by:

$$L = \epsilon^{\sigma\rho\omega} \left(A_\sigma \times \nabla_\rho A_\omega + \frac{2}{3} A_\sigma \times A_\rho \times A_\omega\right) \tag{12.9.18}$$

Gauge fixing involves a long series of steps. If we solve for the cohomology of the system, we find that the physical states of the theory obey:

$$\frac{\delta}{\delta\alpha}\phi_\alpha = 0, \qquad \frac{\delta}{\delta\tau}\phi_\alpha = 0, \qquad Q_0\phi_\alpha = 0 \tag{12.9.19}$$

that is, the Abelian sector of the theory decouples, as expected.

To go back and forth between the endpoint gauge and the midpoint gauge, we need to define an intermediate gauge, called the interpolating gauge, which interpolates between the midpoint and endpoint vertices:

$$\lim_{|\alpha_r|=\pi} |V_{\text{interpolating}}\rangle = |V_{\text{midpoint}}\rangle$$
$$\lim_{\sum \alpha_r = 0} |V_{\text{interpolating}}\rangle = |V_{\text{endpoint}}\rangle \tag{12.9.20}$$

It is possible to show this exactly.

Last, we can show that, by integrating over one of the open string fields, a new four-string contact term is generated. Like the case of the interacting meson theory, we find that a quartic term is generated as follows:

$$\langle \Phi * \Phi | D(X - Y) | \Phi * \Phi \rangle = \langle \Phi \times \Phi | D(X - Y) | \Phi \times \Phi \rangle + \int d\mu \langle \Phi \times \Phi | \Phi \times \Phi \rangle$$
(12.9.21)

The important point here is that the propagator is not gauge invariant, so that it is possible to gauge the propagator to one, giving us a four-string contact term. In fact, one can show that this contact term is precisely the four-string interaction of the light cone theory.

Similarly, integrating over the closed string field and then reparameterizing yields the tetrahedron graph:

$$\langle \Psi \times \Psi | D(X - Y) | \Psi \times \Psi \rangle_{\text{endpoint}}$$
$$= \langle \Psi * \Psi | D(X - Y) | \Psi * \Psi \rangle_{\text{midpoint}} + \int d\mu \frac{\langle \Psi^4 \rangle}{s(4)}$$
(12.9.22)

By iterating this procedure, we can generate polyhedra of arbitrary complexity. Thus, we recover the original nonpolynomial theory of the last chapter.

References

1. M. Kaku and K. Kikkawa, *Phys. Rev.* **D10**, 1110, 1823 (1974).
2. E. Witten, *Nucl. Phys.* **B268**, 253 (1986).
3. See M. Kaku, *Introduction to Superstrings*, Springer-Verlag, Berlin and New York (1988), Chapter 8.
4. M. Kaku, *Int. J. Mod. Phys.* **A2**, 1 (1987); *Phys. Lett.* **200B**, 22 (1988); *Int. J. Mod. Phys.* **A5**, 659 (1990).
5. M. Kaku, *Phys. Rev.* **D38**, 3052 (1988); L. Hua and M. Kaku, *Phys. Rev.* **D41**, 3748 (1990).

For alternative geometric approaches, see Refs. 6 and 7.

6. I. Bars and S. Yankielowicz, USC preprint.
7. K. Bardakci, LBL 21736-UCB-PTH 86/18.
9. A. Neveu and P. C. West, *Nucl. Phys.* **B293**, 266 (1987).
10. S. Uehara, *Phys. Lett.* **B190**, 76 (1987).
11. H. Hata, K. Itoh, T. Kugo, H. Kunitomo, and K. Ogawa, *Phys. Rev.* **D34**, 2360 (1986); *Phys. Lett.* **176B**, 186 (1986); *Phys. Rev.* **D35**, 1318 (1987); A. Neveu and P. C. West, *Nucl. Phys.* **B268**, 125 (1986).
12. M. Kaku, *Phys. Rev.* **D41**, 3734 (1990).
13. M. Kaku and J. Lykken, *Phys. Rev.* **D38**, 3067 (1988).
14. T. Kugo, H. Kunitomo, and K. Suehiro, *Phys. Lett.* **226B**, 49 (1989).
15. M. Saadi and B. Zwiebach, *Ann. Phys.* **192**, 48 (1989).
16. G. Parisi and N. Sourlas, *Phys. Rev. Lett.* **43**, 744 (1979).
17. W. Siegel, *Phys. Lett.* **142B**, 276 (1984).

Chapter 13

2D Gravity and Matrix Models

13.1. Exactly Solvable Strings

To any finite order in perturbation theory, one does not see any of the interesting nonperturbative properties of gauge theories, such as confinement, tunneling, formation of strings, etc. As a consequence, two approximations have been developed, large N methods and lattice gauge theory, to analyze gauge theories in the nonperturbative regime. However, both approaches are still in their infancy, and neither has given us definitive results.

The same situation may eventually apply to string theory. Continuum methods, such as string field theory, are still much too difficult, with too many degrees of freedom, to analyze nonperturbative string phenomena. However, lattice models have emerged as a surprisingly simple way in which to extract nonperturbative information from strings, using a discrete approximation to the Riemann surfaces of string theory.

The idea behind this nonperturbative approach is simple, and we can proceed in at least two ways. The first approach is to place the original Polyakov action on a lattice, where we discretize the two-dimensional world sheet. Since the lattice is two-dimensional, it can be analyzed either analytically or by computer.

The second approach is to use matrix models [1–10]. The essential breakthrough in matrix models is that there is a well-defined limit in which a certain class of solvable point particle gauge theories can approximate the dual string theory for dimensions less than or equal to one. In fact, not only do matrix models allow us to use ordinary point particle Feynman diagrams to reproduce string theory amplitudes, they also allow us to calculate all Green's functions to all orders in perturbation theory.

To see how point particle gauge theory can miraculously reproduce string theory, we begin by studying a class of Feynman diagrams called fishnets [11]. Given a point particle scalar ϕ^n theory, the fishnet diagram has the topology of a two dimensional lattice, where the lines within the lattice are given by Feynman propagators Δ. The fishnet consists of a simple product of Feynman propagators connecting the points of a lattice, which can be exponentiated:

$$\prod_{ij} \Delta[(x_i^\mu - x_j^\mu)^2] = \exp \sum_{ij} \ln \Delta[(x_i^\mu - x_j^\mu)^2] \qquad (13.1.1)$$

For a fishnet, the points x_i and x_j appearing in the product are neighboring points, so we can power expand:

$$(x_i^\mu - x_j^\mu)^2 \rightarrow \epsilon^2 \left(\frac{dx^\mu}{d\sigma}^2 + \frac{dx^\mu}{d\tau}^2 \right) + \cdots \qquad (13.1.2)$$

where ϵ is a small parameter measuring the distance between neighboring points. Notice that we have made the transition from i and j, which are discontinuous co-ordinates which label the neighboring points of the fishnet, to continuous co-ordinates σ and τ, which parametrize the two-dimensional world sheet.

Now assume that we can factor out the light cone singularity $\Delta(0)$ which appears in the Feynman propagator when we power expand the logarithm:

$$\ln \Delta[(x_i - x_j)^2] \sim \ln \left[\Delta(0)(1 + \epsilon^2 \frac{\Delta'(0)}{\Delta(0)}(x_i - x_j)^2 + \cdots \right] \qquad (13.1.3)$$

With this important assumption, we can write the product over Feynman propagators as a surface integral over the two dimensional worldsheet:

$$\int \prod_{k,\nu} dx_k^\nu \prod_{ij} \Delta[(x_i^\mu - x_j^\mu)^2] \rightarrow \int Dx^\nu \exp \left[-k \int d\sigma d\tau \left(\frac{dx^\mu}{d\sigma}^2 + \frac{dx^\mu}{d\tau}^2 \right) \right]$$
$$(13.1.4)$$

where $\epsilon^2 \sim d\sigma d\tau$ and k is a constant.

Thus, assuming that we can remove the light cone singularity, the fishnet diagram smoothly approaches the familiar functional integral over a Riemann surface. The key step was replacing the Feynman propagators defined over a fishnet with a Gaussian, which is only possible if we can throw away the light cone singularity.

By itself, however, this observation, while interesting, is essentially useless, since fishnet diagrams do not necessarily dominate the S-matrix of field theory. Thus, although fishnet diagrams may have interesting properties, in general they have nothing to do with the final scattering amplitudes, which are the only physically relevant quantities.

However, the important exception to this are gauge theories, where it is possible to make a power expansion in some parameter in which these fishnet diagrams do, indeed, dominate the S-matrix. As noticed by 't Hooft, this important parameter is $1/N^2$ appearing in $SU(N)$ gauge theory [12].

For example, in $SU(N)$ gauge theory coupled to quarks, the parameter N in a Feynman diagram only appears when we contract a series of delta functions onto themselves in a loop, that is $\sum \delta_i^i = N$. Since the vertex functions of gauge theory consist of a series of delta functions, the parameter N appears whenever we contract these delta functions around a loop.

Thus, a large fishnet diagram is proportional to:

$$A \sim (g)^{V_3 + 2V_4} N^I \qquad (13.1.5)$$

where g is the coupling constant, V_i is the number of i-point vertices appearing in the Feynman diagram, and I is the number of gauge loops.

We now use a classical result due to Euler, who showed that a polyhedron with P edges, V corners, F faces, and H holes satisfies the following topological relation:

$$V - P + F = 2 - 2H \qquad (13.1.6)$$

We now treat a fishnet Feynman diagram as a large polyhedron. In the language of Feynman diagrams, P becomes the number of propagators, V the number of vertices, F the number of loops, H the number of holes in the fishnet, and $F = L + I$, where L is the number of quark loops in the Feynman diagram and I the number of gauge loops. Using the relation $2P = \sum nV_n$ and $V = \sum V_n$, then we have:

$$A \sim (g)^{2P - 2V} N^L \sim (g^2 N)^{\frac{1}{2} V_3 + V_4} (N)^{2 - 2H - L} \qquad (13.1.7)$$

Now take the limit $N \to \infty$, $g \to 0$, and $g^2 N \to$ const. We find that the leading diagram which survives has $H = 0$ and $L = 1$, that is, they are planar diagrams with the quark line surrounding the edge of the diagram.

Furthermore, if we take the limit of small but finite $1/N^2$, then we find that a power expansion in $1/N^2$ is equivalent to a power expansion in the number of loops in the fishnet diagram. The usefulness of the $1/N^2$ power expansion, therefore, is that it converts a point particle perturbation expansion into an expansion over the genus of a Riemann surface found in string theory.

In actual practice, however, Yang-Mills theory is quite complicated, so we will find it more convenient to solve a simpler problem, a theory where the fundamental field is a hermitian $N \times N$ matrix called M. These are called matrix models.

The advantage of matrix models is that they are so simple that they can be solved exactly in the large N limit and shown to be equivalent to string theory for dimensions less than or equal to one. (More precisely, by $D \leq 1$ string theory, we actually mean two-dimensional gravity in zero dimensions, without the presence of any string variable, coupled to $c \leq 1$ conformal matter.)

Although the large N power expansion is a perturbation series in the genus g of the surface, certain recursion relations exist that allow us to find exact solutions for the models, giving us, for the first time, nonperturbative information about the theory. Thus, these models are so simple that the transition from a perturbation series in two-dimensional surfaces to a fully nonperturbative theory is not such a great barrier.

We will see that matrix models in the "double scaling limit" are exactly solvable for $c \leq 1$. This limit means taking $N \to \infty$ and carefully adjusting the cosmological constant (related to the area of the surface) to be at a critical point of the theory. These critical points, in turn, are indexed by an integer k. We will find that, for $k = 2$, the matrix model approximates ordinary two-dimensional gravity. For $k = 3, 4, \ldots$, we find that matrix models approximate two-dimensional gravity coupled to nonunitary, conformal matter. For the general case of multimatrix models, we can reproduce two-dimensional gravity coupled to the (BPZ) minimal series of (p, q) conformal matter.

The attractive feature, then, of the lattice and matrix model approaches is that they provide nonperturbative information about string theory using old, well-established methods. For the first time, nonperturbative features of string theory are emerging (such as the instability, that is, non-Borel summability, of perturbation theory).

The disadvantage of these models, as we have pointed out, is that they cannot realistically describe string theories in 26, 10, or 4 dimensions. There are certain mathematical barriers preventing analysis beyond one dimension. In fact, we will see that these models can only be solved for low, unrealistic values of the dimension. Beyond the $c = 1$ barrier, many of our approximations break down (for example, real constants become complex and potentials become unphysical).

In addition, matrix models only have a finite number of degrees of freedom, and hence we can usually find exact results. Beyond $c > 1$, the system has an infinite number of degrees of freedom, and we do not expect the system to be exactly solvable. Thus, in some sense the price we pay for exact solubility of the string theory is low dimensionality and finite degrees of freedom, which do not describe the real world. However, there is hope that these theories, being the first to provide nonperturbative information about strings, will give us valuable insight into this previously forbidden yet crucially important region.

13.2. 2D Gravity and KPZ

Usually, string theory is defined at the critical dimension of 10 or 26, where the theory is scale invariant and the metric g_{ab} can be set to δ_{ab}. For values of the dimension other than 26 or 10, we have the noncritical string theory of Polyakov, where the string variable $X_\mu(\sigma, \tau)$ interacts with the two-dimensional metric tensor g_{ab}. For noncritical string theory, there is a conformal anomaly, so the three fields within the metric cannot be eliminated totally. The metric reduces to just one field, the Liouville field ϕ, where $g_{ab} = e^\phi \delta_{ab}$.

Usually, for critical strings in 26 or 10 dimensions, we have the freedom to eliminate the Liouville field completely. However, in these low-dimensional matrix models, which are noncritical, we must leave the Liouville mode intact. Thus, in this off-critical picture, the two-dimensional gravitational field does not vanish but becomes a key player, and the string field actually reduces to matter fields coupled to two-dimensional (Liouville) gravity.

We begin by isolating the contribution of the conformal anomaly, which we will calculate by taking the variation of the functional measure under a scale transformation. First, the X_μ integration gives us the determinant of the Laplacian:

$$\int D_g X \equiv \int DX \, \exp\left(-\int d^2 z \sqrt{g} \, g^{ab} \, \partial_a X^\mu \, \partial_b X_\mu\right)$$
$$= \left[\frac{2\pi}{\int d^2 z \sqrt{g}} \, \det(-\nabla^2)\right]^{-D/2} \tag{13.2.1}$$

We must calculate how this term in the functional measure transforms under a scale transformation of the metric in order to isolate the conformal anomaly. We can use heat kernel methods to calculate the change in the determinant of the Laplacian under a scale transformation. The result is:

$$D_{e^\sigma g} X = e^{(D/48\pi) S_L(\sigma;g)} D_g X$$
$$S_L(\sigma, g) = \int d^2 z \sqrt{g} \left(\frac{1}{2} g^{ab} \, \partial_a \sigma \, \partial_b \sigma + R\sigma + \mu e^\sigma\right) \tag{13.2.2}$$

The action S_L is called the "Liouville action." The classical equations of motion for the σ field are of form $\partial_a \partial^a \sigma \sim e^\sigma$ and are quite difficult to solve. Under this scale transformation, we also find that the b, c ghost system is not invariant, but transforms nontrivially under a scale transformation. We will define:

$$\int D_g b \, D_g c \equiv \int \Delta_{FP} Db \, Dc \tag{13.2.3}$$

where Δ_{FP} includes the metric-dependent and b, c-dependent factors, that is, the Faddeev–Popov terms. Using heat kernel methods, we find that this term transforms as the following under a scale transformation:

$$D_{e^\sigma g} b \, D_{e^\sigma g} c = e^{(-26/48\pi) S_L(\sigma, g)} D_g b \, D_g c \tag{13.2.4}$$

When we multiply the two contributions Eqs. (13.2.2) and (13.2.4), together, we notice that the action S_L occurs with the coefficient $D - 26$:

$$\left[(D-26)/48\pi\right] S_L(\sigma, g) \tag{13.2.5}$$

which vanishes when $D = 26$. This is the choice usually taken in string theory, which is called critical string theory, where the two-dimensional metric can be eliminated entirely. Thus, in 26 dimensions, we never have to worry about the Liouville degree of freedom contained within the two-dimensional metric tensor.

However, since $D = 26$ string theory is too difficult to solve, we will be interested in solving simpler theories, such as the case of low dimensions. In this chapter, we will explore ways to solve this problem exactly. From the perspective of conformal field theory, we are interested in studying the case of two-dimensional gravity coupled to D identical copies of conformal matter with weight 0 (that is, the string X_μ).

One complication, however, is that in lower dimensions, the conformal anomaly, Eq. (13.2.5), does not disappear. This means that the difficult Liouville mode σ must be carefully included in all our discussions. However, it is still possible to extract information about the theory exactly. For example, from the work of Knizhnik, Polyakov, and Zamolodchikov (KPZ) [13, 14], we know the asymptotic form of the partition function for two dimensional gravity. If the area A of the two-dimensional world sheet is defined as:

$$A = \int d^2\xi \sqrt{\det g_{ab}} \tag{13.2.6}$$

then the partition function is:

$$
\begin{aligned}
Z(A) = \int DX(\xi) \int Dg_{ab}(\xi)\delta\left(\int d^2\xi\sqrt{g} - A\right) \\
\times \exp\left(-\int d^2\xi\sqrt{g}\,g^{ab}\partial_a X_\mu\,\partial_b X_\mu\right)
\end{aligned}
\tag{13.2.7}
$$

Using light-cone quantization of the two-dimensional theory, KPZ found that the partition function behaves asymptotically as follows as the area A approaches infinity:

$$Z(A) \sim A^{-3+\gamma}\exp(kA) \tag{13.2.8}$$

where γ, the string susceptibility, can be shown to equal [13, 14]:

$$\gamma = \frac{1}{12}\left[D - 1 - \sqrt{(D-1)(D-25)}\right] \tag{13.2.9}$$

Although this form of the string susceptibility was originally derived in the light cone gauge, it can also be derived in the conformal gauge [15, 16].

To see this, we first assume that after regularization the final result for the Jacobian after a rescaling is equal to:

$$S(\phi, \hat{g}) = \frac{1}{8\pi}\int d^2z\left(\sqrt{\hat{g}}\,\hat{g}^{ab}\partial_a\phi\,\partial_b\phi - \frac{1}{4}Q\sqrt{\hat{g}}\,\hat{R}\phi + \mu_1\sqrt{\hat{g}}\,e^{\alpha\phi}\right) \tag{13.2.10}$$

where we have rescaled $g = e^{\alpha\phi}\hat{g}$. We have chosen this expression, involving the undetermined Q, α, μ_1, because it is the most general form of the regularized action consistent with the symmetries of the system.

Actions of this type (with the crucial factor containing Q) often occur when we bosonize a fermionic system, as in the Feigin–Fuchs free field formalism in Eq. (7.1.2). It is easy therefore to calculate the contribution to the anomaly of the ϕ field, which is:

$$c_\phi = 1 + 3Q^2 \tag{13.2.11}$$

Imposing the fact that the entire system has zero anomaly, we find that the anomaly is the sum of three terms, all of which add up to zero:

$$c = c_\phi + D - 26 = 0 \tag{13.2.12}$$

or:

$$Q = \sqrt{(25 - D)/3} \tag{13.2.13}$$

The next step is to calculate the value of α. This is easy because we demand that $g = e^{\alpha\phi}\hat{g}$ be invariant, which means that $e^{\alpha\phi}$ has conformal weight 1. Again, from ordinary conformal field theory, we also know how to compute the conformal weight of operators such as $e^{\alpha\phi}$. If the energy-momentum tensor is given by:

$$T_{zz} = -\frac{1}{2}(\partial\phi\,\partial\phi + Q\,\partial^2\phi) \tag{13.2.14}$$

then the conformal weight of $e^{q\phi}$ is:

$$\mathrm{wt}(e^{q\phi}) = -\frac{1}{2}q(Q + q) \tag{13.2.15}$$

Setting this equal to 1, we then find:

$$\alpha = (-1/2\sqrt{3})\left[\sqrt{25 - D} \pm \sqrt{(1 - D)}\right] \tag{13.2.16}$$

Using the explicit values for Q and α, let us now calculate the behavior of $Z(A)$ under a rescaling. If we shift by a constant value:

$$\phi \to \phi + \rho/\alpha \tag{13.2.17}$$

then the action of Eq. (13.2.10) shifts by the integral of the curvature tensor $\hat{R}$ over the Riemann surface:

$$S \to S - Q(1 - h)\rho/\alpha \tag{13.2.18}$$

where h is the genus of the surface. The shift in the δ function is given by:

$$\delta\left(\int e^{\alpha\phi}\sqrt{\hat{g}}\,d^2z - A\right) \to e^{-\rho}\delta\left(\int e^{\alpha\phi}\sqrt{\hat{g}}\,d^2z - e^{-\rho}A\right) \tag{13.2.19}$$

Putting everything together, we find that the partition function scales as:

$$Z(A) \sim A^{(1-h)Q/\alpha - 1} \tag{13.2.20}$$

so that the susceptibility given in Eq. (13.2.8) must therefore be [16]:

$$\gamma(h) = (1/12)(1 - h) \left(D - 25 \pm \sqrt{25 - D} \sqrt{1 - D} \right) + 2 \qquad (13.2.21)$$

Setting $h = 0$, we retrieve the original result of Eq. (13.2.9).

Notice that the string susceptibility becomes complex for the dimension D between 1 and 25, meaning that attempts to naively extend the matrix model approach beyond $D = 1$ will inevitably have severe problems. This is, in fact, perhaps the most important roadblock facing this formalism, preventing a realistic, nonperturbative formulation of string theory.

However, assuming that this low-dimensional barrier can be eventually surmounted, these matrix models may make the transition from being simplistic "toy" models to being relatively realistic nonperturbative descriptions of string theory. The most optimistic outcome would be that string theory might be *defined* to be the large N double scaling limit of matrix models, thereby replacing the Riemann surface description.

To begin our discussion, we note that there are two ways in which to proceed. First, we can work directly with the Polyakov functional, either analytically or by computer. For example, on computer, we can approximate the functional as:

$$Z = \sum_{S} \int \prod_{i=1}^{D} DX_i \, \exp \left(- \sum_{ij} (X_i - X_j)^2 - \beta E - h \sum_{i} \sigma_i \right) \qquad (13.2.22)$$

where we sum over all triangulations S of the surface, i and j label the vertices on the surface, β is the inverse Ising temperature, and h is the magnetic field. However, we will explore the second approach, matrix models, which yields the exact analytic solution to the problem for $c \leq 1$.

13.3. Matrix Models

The second way to proceed is to make the connection between the two-dimensional gravitational world sheet and the topology swept out by the Feynman diagrams of a matrix model. The correctness of this approach will be evident when we calculate the string susceptibility γ and independently check Eq. (13.2.21). In this spirit, let us analyze the simplest description of these matrix models, defined in terms of a field M, which is an $N \times N$ matrix.

We start with the action:

$$L = \sum_{\mu=1}^{D} \text{Tr}(\partial_\mu M \, \partial^\mu M^\dagger) + \text{Tr}(MM^\dagger) + \frac{g}{N} \text{Tr}(MM^\dagger MM^\dagger) \qquad (13.3.1)$$

We can set $M^\dagger = M$ and also generalize to an interaction with arbitrary powers of the matrix M:

$$U = \sum_{i=3}^{\infty} V_i = g_3 \operatorname{Tr} M^3 + g_4 \operatorname{Tr} M^4 + \cdots \tag{13.3.2}$$

so the generating functional is given by (for $D = 0$):

$$Z(\beta) = \int DM \, \exp\left[-\beta \operatorname{Tr} U(M) \right] \tag{13.3.3}$$

Let us now analyze the Feynman diagrams emerging from this Lagrangian. In general, a diagram will have P propagators, V vertices, and I closed loops. As before, we have the Euler relation: $V - P + I = 2 - 2H$ where H is the number of holes in the surface on which the polyhedron is drawn. Notice that this number is a topological invariant, dependent only on the topological nature of the surface upon which we draw the polyhedron.

For the Feynman diagrams arising from a hermitian matrix model, for fixed β/N, each vertex, propagator, and loop contributes the following factors:

$$\text{vertex} \to N$$
$$\text{propagator} \to 1/N \tag{13.3.4}$$
$$\text{loop} \to N$$

Using Eq. (13.3.3), by multiplying these factors in a Feynman expansion, we arrive at:

$$\ln Z \sim N^{2(1-H)} \left(N/\beta \right)^A \tag{13.3.5}$$

where A is the area of the random surface.

Then, we see that the vacuum energy graph, for example, divided by N^2 has a finite limit in the planar $(H = 0)$ limit. We see that the overall factor is damped by a factor $1/N^{2H}$, so that a perturbation in $1/N^2$ is actually a perturbation in the number of holes in the surface. This gives us the justification for comparing the matrix model in the large N limit with the string theory perturbation series.

The interesting features appear when we take the double scaling limit:

$$N \to \infty, \qquad \beta/N \to \text{const.} \tag{13.3.6}$$

In this limit, we find that the free energy is independent of many of the details of the nature of the potential U, that is, we find a universal scaling behavior:

$$\ln Z(\beta) \to -F(t) \tag{13.3.7}$$

where the scaling variable is defined to be:

$$t \equiv (\beta - N)\beta^{-1/(2k+1)} \tag{13.3.8}$$

We will find that there are k parameters in the potential $U(\phi)$ that we can adjust, giving us the multicritical behavior for the specific heat, defined by

$$f(t) = d^2 F(t)/dt^2 \qquad (13.3.9)$$

Remarkably, we will be able to derive an exact expression for $f(t)$ in terms of a differential equation, independent of the perturbation expansion. However, we can check the correctness of our results against the perturbation expansion, where the coupling constant is:

$$g^2_{\text{string}} = t^{-(2+1/k)} \qquad (13.3.10)$$

To make this discussion concrete, we now use some tricks pioneered by Brezin, Itzykson, Parisi, and Zuber [17–20] to solve the matrix model.

If we are, for example, interested in the vacuum energy, we wish to find a solution for the following path integral (in zero dimensions):

$$\exp[-N^2 E^{(0)}](g) = \lim_{N \to \infty} \int \prod_{ij} dM_{ij}$$
$$\times \exp - \left(\frac{1}{2} \operatorname{Tr} M^2 + \frac{g}{N} \operatorname{Tr} M^4 \right) \qquad (13.3.11)$$

Notice that we have dropped the kinetic term entirely, meaning that we are only analyzing the simplified case of $D = 0$.

One crucial observation is that the functional measure over the matrix M,

$$\prod_{ij} dM_{ij} \equiv \prod_i dM_{ii} \prod_{i<j} d(\operatorname{Re} M_{ij}) d(\operatorname{Im} M_{ij}) \qquad (13.3.12)$$

can be simplified by diagonalizing the matrix M. Then, we are left with an integration over the eigenvalues λ_i of M and a matrix Π. This means that we can write the measure in terms of the eigenvalues and the matrix Π:

$$\prod_{ij} dM_{ij} = \prod_i d\lambda_i \prod_{i<j} (\lambda_i - \lambda_j)^2 \, d\Pi_{ij} \qquad (13.3.13)$$

The advantage of this approach is that the action is written entirely in terms of traces over products of M, so that the dependence on the diagonalizing Π matrix completely disappears. This means that we can trivially integrate over Π, leaving only an integration over the eigenvalues λ_i:

$$\exp[-N^2 E^{(0)}(g)] = \lim_{N \to \infty} \int \prod_i d\lambda_i \prod_{i<j} (\lambda_i - \lambda_j)^2$$
$$\times \exp - \left(\frac{1}{2} \sum \lambda_i^2 + \frac{g}{N} \sum \lambda_i^4 \right) \qquad (13.3.14)$$

We evaluate this integral by the method of steepest descent. The energy $E^{(0)}$ can now be explicitly calculated in terms of the classical variable λ_i. In terms of this classical variable, we find:

$$E^{(0)}(g) = \lim_{N \to \infty} \frac{1}{N^2} \left[\sum_i \left(\frac{1}{2}\lambda_i^2 + \frac{g}{N}\lambda_i^4 \right) - {\sum}' \ln|\lambda_i - \lambda_j| \right] + \cdots$$

$$\lambda_i/2 + (2g/N)\lambda_i^3 = {\sum_j}' \frac{1}{\lambda_i - \lambda_j}$$

(13.3.15)

where the prime means we do not sum over $i = j$.

Let us now take the large N limit. We can replace the classical λ_i with a new, continuous variable $\lambda(x)$, defined by:

$$\lambda_i = \sqrt{N}\,\lambda(i/N) \tag{13.3.16}$$

In the large N limit, we can then write the vacuum energy as:

$$E^{(0)}(g) = \int_0^1 dx \left[\frac{1}{2}\lambda^2(x) + g\lambda^4(x) \right] - \int_0^1 \int_0^1 dx\,dy \, \ln|\lambda(x) - \lambda(y)| \tag{13.3.17}$$

and $\lambda(x)$ obeys the constraint:

$$\frac{1}{2}\lambda(x) + 2g\lambda^3(x) = P \int_0^1 \frac{dy}{\lambda(x) - \lambda(y)} \tag{13.3.18}$$

where P represents taking the principle part of the integral. This last constraint is actually sufficient to determine the function $\lambda(x)$, given some assumptions on its analytic behavior.

To solve for $\lambda(x)$, we introduce two new functions, $u(x)$ and $F(x)$, defined via:

$$\frac{dx}{d\lambda} \equiv u(\lambda); \qquad \int_{-2a}^{+2a} d\lambda\, u(\lambda) = 1$$

$$F(\lambda) \equiv \int_{-2a}^{+2a} d\mu\, \frac{u(\mu)}{\lambda - \mu} \tag{13.3.19}$$

(We will assume that $u(\lambda)$ vanishes outside some support $[-2a, 2a]$. We can determine $F(\lambda)$ by analytical arguments. We can show $F(\lambda)$ is analytic in the complex λ plane cut along this interval, behaves as $1/\lambda$ when $|\lambda|$ goes to ∞, is real for real λ outside this interval, and when λ approaches this interval, then $F(\lambda \pm i\epsilon) = \lambda/2 + 2g\lambda^3 \mp i\pi u(\lambda)$.)

Using the analytic properties of these functions along the interval $[-2a, +2a]$ and their behavior at infinity, we can obtain the solution [17]:

$$F(\lambda) = \frac{1}{2}\lambda + 2g\lambda^3 - \left(\frac{1}{2} + 4ga^2 + 2g\lambda^2 \right) \sqrt{\lambda^2 - 4a^2}$$

$$u(\lambda) = \frac{1}{\pi} \left(\frac{1}{2} + 4ga^2 + 2g\lambda^2 \right) \sqrt{4a^2 - \lambda^2} \tag{13.3.20}$$

Plugging these expressions back into the formula for the vacuum energy, we find:

$$E^{(0)}(g) - E^{(0)}(0) = \int_0^{2a} d\lambda\, u(\lambda) \left(\frac{1}{2}\lambda^2 + g\lambda^4 - 2\ln\lambda\right) - (g = 0)$$

$$= \frac{1}{24}(a^2 - 1)(9 - a^2) - \frac{1}{2}\ln a^2$$

$$(13.3.21)$$

where:

$$a^2 = \frac{1}{24g}[(1 + 48g)^{1/2} - 1] \qquad (13.3.22)$$

Inserting the value of a as a power expansion in g, we now have:

$$E^{(0)}(g) - E^{(0)}(0) = -\sum_{p=1}^{\infty}(-12)^p \frac{(2p-1)!}{p!(p+2)!} \qquad (13.3.23)$$

$$= 2g - 18g^2 + 288g^3 - 6048g^4 + \cdots$$

Remarkably, the simplicity of the $D = 0$ matrix model gives us a simple expression for the energy.

Now, let us find the critical point for this expansion and compare our results with the values given by KPZ. At the critical value, we can show that the vacuum energy goes as $(\beta - \beta_c)^{5/2}$ in the limit of infinite N (that is, genus h of the surface is zero. In fact, by repeating the previous steps, a careful expansion of the vacuum energy as a function of $1/N^2$ shows that the next few terms in the Taylor expansion behave as $\ln(\beta - \beta_c)$ and $(\beta - \beta_c)^{-5/2}$ for $h = 1$ and $h = 2$, respectively.)

Our goal is to calculate the string susceptibility. We define the susceptibility as the second derivative of the free energy $F(\beta)$:

$$\chi(\beta) = F''(\beta) \qquad (13.3.25)$$

and the exponent of this susceptibility at the critical point is given by γ:

$$\chi \sim (\beta - \beta_c)^{-\gamma} \qquad (13.3.26)$$

By inserting the behavior of $F(\beta)$ into this expression, we find that

$$\gamma = 2 - \frac{5}{2}(h - 1); \qquad h = 0, 1, 2 \qquad (13.3.27)$$

Now, compare the above expression for the exact result for the exponent of susceptibility [Eq. (13.2.21)] appearing in the KPZ formalism. We find a precise agreement for $D = 0$, which gives us great confidence in the power of the matrix models to approximate two-dimensional gravity.

13.4. Recursion Relations

At this point, these results have been interesting, but largely perturbative. If this were all that one could do with matrix models, then it would be very disappointing. The key, of course, is to extend our perturbative results to the nonperturbative regime. To accomplish this nontrivial feat, we will use the method of recursion relations, which will allow us to write differential equations that extend our analysis to a fully nonperturbative theory.

To do this, we will introduce the method of orthogonal polynomials. As before, we notice that we can reduce any of these matrix models to an integration over the eigenvalues λ_i, with a measure proportional to the Vandermonde determinant appearing in Eq. (13.3.13):

$$\Delta_N = \prod_{i<j}^{N} (\lambda_i - \lambda_j) \tag{13.4.1}$$

so that the partition function can be expressed entirely in terms of the eigenvalues:

$$Z(\beta) = \int \prod_{i=1}^{N} d\lambda_i \, \Delta_N^2 \exp\left[- \sum_{i=1}^{N} \beta U(\lambda_i) \right] \tag{13.4.2}$$

where β represents the "temperature" of the Coulomb gas with a potential given by $U(\lambda)$.

Now let us use the key step that will make the matrix models solvable. We will introduce polynomials $P_n(\lambda)$ [18, 19], which are orthogonal with respect to the weight $d\mu(\lambda) = \exp\left[- \beta U(\lambda) \right] d\lambda$, that is, these polynomials are functions of the potential $U(\lambda)$. By construction, these polynomials obey the relations:

$$\int d\lambda \exp\left[- \beta U(\lambda) \right] P_n(\lambda) P_m(\lambda) = \delta_{n,m} h_n \tag{13.4.3}$$

For example, it is possible to write an explicit representation of these polynomials in terms of the potential U as:

$$P_n(\lambda) \equiv Z_n^{-1} \int \prod_{i=1}^{n} d\lambda_i \exp\left[- \beta U(\lambda_i) \right] (\lambda - \lambda_i) \Delta_n^2$$
$$Z_n \equiv \int \prod_{i=1}^{n} d\lambda_i \exp[-\beta U(\lambda_i)] \Delta_n^2 \tag{13.4.4}$$

(Notice that we obtain the full generating functional if we set $n = N$: $Z_N(\beta) = Z(\beta)$.)

It is possible to construct these polynomials so that they satisfy a recursion relation for any potential $U(\lambda)$:

$$\lambda P_n(\lambda) = P_{n+1}(\lambda) + S_n P_n(\lambda) + R_n P_{n-1}(\lambda) \tag{13.4.5}$$

where the polynomials h_n, R_n, and S_n will be determined shortly. (S_n can be set equal to zero because we assume that $U(-\phi) = U(-\phi)$.)

Now we wish to show that Z_n can be written entirely in terms of h_i.

We first notice that $P_n(\lambda) = \lambda^n$ plus lower powers of λ. Next, we notice that the integral of P_n times λ^i over $d\mu(\lambda)$ is equal to zero if $i < n$. This means that we can always power expand P_n and integrate over each power of λ^i, where most integrals actually vanish.

Using (13.4.4), we now observe that:

$$
\begin{aligned}
h_n &= \int d\mu(\lambda) P_n^2(\lambda) = \int d\mu(\lambda) P_n(\lambda) \lambda^n \\
&= \int \frac{d\mu(\lambda)}{Z_n} \prod_{i=1}^{n} d\mu(\lambda_i)(\lambda - \lambda_i) \Delta_n^2 \lambda^n \\
&= [(n+1)Z_n]^{-1} \int \prod_{i=1}^{n+1} d\mu(\lambda_i) \Delta_{n+1}^2 \\
&= \frac{Z_{n+1}}{(n+1)Z_n}
\end{aligned}
\tag{13.4.6}
$$

where we have set $\lambda = \lambda_{n+1}$ and have written Δ_{n+1}^2 in terms of Δ_n^2.

Using the recursion relation Eq. (13.4.5), we can also show:

$$
\begin{aligned}
h_{n+1} &= \int d\lambda \, \exp\left[-\beta U(\lambda)\right] P_{n+1} \lambda P_n \\
&= \int d\lambda \, \exp\left[-\beta U(\lambda)\right] (P_{n+2} + S_{n+1} P_{n+1} + R_{n+1} P_n) P_n \\
&= R_{n+1} h_n
\end{aligned}
\tag{13.4.7}
$$

Using Eq. (13.4.6), we can now write the partition function entirely in terms of the functions:

$$
Z_N = N! \prod_{i=0}^{N-1} h_i
\tag{13.4.8}
$$

This is the desired expression.

It shows that all the information concerning the system is encoded within the h_n, or within the R_n and S_n. The precise form of the potential U determines these polynomials, which in turn determines the partition function exactly.

From these recursion relations, we can now actually solve for the free energy of the system. Because $\lambda P_n' \sim n\lambda^n$, we can rewrite the recursion relation as:

$$
\begin{aligned}
n h_n &= \int d\lambda \, e^{-\beta U} \lambda P_n' P_n = \int d\lambda \, e^{-\beta U} P_n' (P_{n+1} + R_n P_{n-1}) \\
&= R_n \int d\lambda \, e^{-\beta U} P_n' P_{n-1} = R_n \int d\lambda \, e^{-\beta U} \beta U' P_n P_{n-1}
\end{aligned}
\tag{13.4.9}
$$

where we have integrated by parts. Fortunately, the last integral can be performed exactly by using the explicit power expansion of the potential U:

$$
\int d\lambda\, e^{-\beta U} P_n U' P_{n-1}
$$

$$
= \int d\lambda\, e^{-\beta U} P_n \left[\sum_{p=0} 2(p+1)g_{p+1}\lambda^{2p+1} \right] P_{n-1} \tag{13.4.10}
$$

$$
= h_n \left[\sum_{p=0} 2(p+1)g_{p+1} \sum_{\text{paths}} R_{\alpha_1}\cdots R_{\alpha_p} \right]
$$

where we have expanded

$$
U = \sum_{p=1} g_p \lambda^{2p} \tag{13.4.11}
$$

and the sum over *paths* can be defined as follows.

Imagine a staircase which zigzags up and down, such that it begins at height $(n-1)$ and ends at height n. The staircase has $p+1$ steps that go up, and p steps that go down, so it has $2p+1$ steps altogether. There are $(2p+1)!/p!(p+1)!$ such possible staircases. For each of these possible staircases, associate a product of R's. A down step from k to $k-1$ generates a factor of R_k, and an up step generates the number one. Thus, the sum over *paths* is the sum over all possible staircases in Eq. (13.4.10), such that each staircase is associated with a product of R's that label the number of down steps from k to $k-1$.

For example, for $p = 1$, the sum can be represented as:

$$
\sum_{\text{paths}} = R_{n-1} + R_n + R_{n+1} \tag{13.4.12}
$$

For $p = 2$, the sum becomes:

$$
\sum_{\text{paths}} = R_{n-2}R_{n-1} + R_{n-1}^2 + 2R_{n-1}R_n
$$

$$
+ R_{n-1}R_{n+1} + R_n^2 + 2R_n R_{n+1} + R_{n+1}^2 + R_{n+1}R_{n+2} \tag{13.4.13}
$$

Inserting the sum over paths into our original recursion relation Eqs. (13.4.9) and (13.4.10), we now have:

$$
\frac{n}{\beta} = 2R_n \sum_{p=0}(p+1)g_{p+1} \sum_{\text{paths}} R_{\alpha_1}\cdots R_{\alpha_p} \tag{13.4.14}
$$

Let us analyze this remarkable set of equations with a few examples.

Example: k = 2

For our first case, let us begin with the simple potential:

$$U = g_1 \lambda^2 + g_2 \lambda^4 \tag{13.4.15}$$

We can now perform the sum over the paths explicitly. Using Eqs. (13.4.12) and (13.4.14), we obtain (for $1 \le n \le N$ and $x = n/\beta$) [21]:

$$x = 2R_n[g_1 + 2g_2(R_{n+1} + R_n + R_{n-1})] \tag{13.4.16}$$

Now, let us take the large N limit (for fixed x), in which case we have:

$$R_n(x) = R(x) + [x(n - N)/N] R'(x) + (1/2)[x(n - N)/N]^2 R''(x) + \cdots \tag{13.4.17}$$

and Eq. (13.4.14) becomes:

$$\lim_{N \to \infty} x = 2 \sum_{p=0}^{\infty} \frac{(2p + 1)!}{(p!)^2} g_{p+1} R^{p+1} \equiv w(R) \tag{13.4.18}$$

In the double scaling limit, Eq. (13.3.7), we define the multicritical points x_c and R_c as the points where the following conditions are satisfied:

$$w'(R_c) = w''(R_c) = \ldots = w^{(k-1)}(R_c) = 0 \tag{13.4.19}$$

for some integer k. Let us investigate the simplest case $k = 2$ (which will correspond to pure two-dimensional gravity). In this case, $w(R) - x_c$ is proportional to $(R - R_c)^2$.

Now we must calculate corrections to Eq. (13.4.18) to higher orders in $1/N^2$. By carefully calculating these higher order corrections in Eq. (13.4.16) via (13.4.17), we have:

$$x = w(R) + N^{-2}4g_2 x^2 RR''(x) + O(N^{-4}) \tag{13.4.20}$$

At the multi-critical point, we find that $x - x_c$, $w(R) - x_c$, and $N^{-2}R''$ all have the same order of magnitude.

We now introduce a function f, which is the scaling function. For R_n near R_c, we find that it approaches the following scaling limit:

$$R_n - R_c = N^{-2/(2k+1)} f(t) \tag{13.4.21}$$

where $t = (x_c - x)N^{2k/(2k+1)}$.

Plugging all these values into Eq. (13.4.20), we find the following equation for the scaling function in the multicritical limit:

$$t = f^2 + f'' \tag{13.4.22}$$

whose solution is a Painlevé transcendental of the first kind [21–23].

Example: k = 3

For our next example, we will extend our results for the next multi-critical point. For $k = 3$, the calculation is only a bit more difficult (but for arbitrary k, we will have to use a different method). In this case, we can repeat the same steps as before:

$$U = g_1\lambda^2 + g_2\lambda^4 + g_3\lambda^6 \qquad (13.4.23)$$

This gives us [21]:

$$w(R) = 2g_1 R + 12g_2 R^2 + 60g_3 R^3 \qquad (13.4.24)$$

The tricritical point $(w' = w'' = 0)$ gives us:

$$R_c = -g_2/15g_3 \qquad (13.4.25)$$

In the large N limit, a Taylor expansion gives us:

$$\begin{aligned} x = w(R) + N^{-2}x^2 w''(R)R''/6 + 30N^{-2}x^2 R'^2 g_3 \\ + N^{-4}RR''''x^4(g_2 + 33g_3 R)/3 + O(N^{-6}) \end{aligned} \qquad (13.4.26)$$

At the tricritical point, the above equation becomes the following equation for the scaling function [21–23]:

$$t = f^3 + ff'' + (f')^2/2 + f''''/10 \qquad (13.4.27)$$

By comparing this equation and its critical exponents with those found for two-dimensional gravity coupled to (nonunitary) conformal matter, we can determine that the $k = 3$ case corresponds to the Yang–Lee edge singularity.

For general values of k, we will shortly find that the multicritical matrix models represent two-dimensional gravity coupled to nonunitary conformal matter with:

$$c = 1 - \left[6(p-q)^2/pq\right] \qquad (13.4.28)$$

where $(p, q) = (2k - 1, 2)$, or:

$$c = 1 - \frac{3(2k-3)^2}{2k-1} \qquad (13.4.29)$$

Using the method of orthogonal polynomials, we see that we have been able to find exact solutions for the scaling functions in the large N limit.

However, we see that the equation for the scaling function for higher k becomes prohibitively difficult because of the sum over the staircases in Eq. (13.4.14) found in the recursion relations for the orthogonal polynomials. We will thus need a more powerful formalism.

13.5. KdV Hierarchy

Let us now switch to an operator language, which will simplify our calculation of the h_n and allow us to generalize the previous results for arbitrary values of k [22–24]. In particular, we will see the (KdV) hierarchy emerge.

Let us define the scalar product as:

$$\langle A|B\rangle \equiv \int d\lambda \exp\big[-\beta U(\lambda)\big]A(\lambda)B(\lambda) \tag{13.5.1}$$

For example, if we define the vector $|n\rangle$ via the polynomial

$$\bar{P}_n(\lambda) \equiv P_n(\lambda)/\sqrt{h_n} \rightarrow |n\rangle \tag{13.5.2}$$

then we have:

$$\langle n|m\rangle = \delta_{n,m} \tag{13.5.3}$$

From Eqs. (13.4.3) and (13.4.4), we have:

$$\langle n|\hat{\lambda}|m\rangle = \delta_{n,m+1}\sqrt{R_n} + \delta_{m,n+1}\sqrt{R_m} + \delta_{n,m}S_m \tag{13.5.4}$$

where the function λ is now written as an operator $\hat{\lambda}$.

Now consider the following relations:

$$\int d\mu(\lambda)\bar{P}_n(\lambda)\frac{d}{d\lambda}\bar{P}_n = 0$$
$$\int d\mu(\lambda)\bar{P}_{n-1}\frac{d}{d\lambda}\bar{P}_n = n\sqrt{\frac{h_{n-1}}{h_n}} = \frac{n}{\sqrt{R_n}} \tag{13.5.5}$$

where $d\mu(\lambda) = \exp\big[-\beta U(\lambda)\big]d\lambda$. These relations, which follow from the orthonormality of $\bar{P}_n$, can be written as:

$$\langle n|U'(\hat{\lambda})|n\rangle = 0$$
$$\langle n-1|U'(\hat{\lambda})|n\rangle = n/(\beta\sqrt{R_n}) \tag{13.5.6}$$

where we have integrated by parts. The operator $d/d\lambda$ pulls down a factor of $U'(\lambda)$, which then becomes an operator. These equations contain the same information as Eq. (13.4.14). We now take the large N limit; this means that we will replace the discontinuous variable n with the continuous variable $x = n/\beta$ and replace R_n by $R(x)$.

To solve the system of equations, Eq. (13.5.6), exactly in this limit, we must convert the bra vector $\langle n-1|$ into $\langle n|$. This is most easily done by introducing two conjugate operators $\hat{n}$ and θ, such that $\langle n|\hat{n} = \langle n|n/\beta$. We can express these conjugate operators as $\hat{n} = (-i/\beta)d/d\theta$ and $\hat{\theta} = (i/\beta)d/d\hat{n}$. With these operators, we can show that $\langle n-1| = \langle n|e^{-i\hat{\theta}}$. Now we can eliminate the presence of $\hat{\theta}$ by taking the large β limit, where we can power expand $e^{-i\hat{\theta}}$. The only remnant of this operator is a term $d^2/d\hat{n}^2$, which can be rewritten as d^2/dx^2.

Collecting all terms, we then find that we can rewrite Eq. (13.5.6) as:

$$x\beta = \langle x|U'(2-\hat{H})|x\rangle$$
$$\hat{H} = -\frac{1}{\beta^2}\frac{d^2}{dx^2} + \big[1 - R(x)\big] \tag{13.5.7}$$

where $R_c = 1$ and we have introduced the effective Hamiltonian of the system $\hat{H}$ coming $d^2/d\hat{n}^2$.

We will find it convenient to define U as the following:

$$U(\phi) = B\left(\frac{1}{2}, -\frac{1}{2} - \nu\right)(2 - \phi)^{\nu+(1/2)} + (\phi \leftrightarrow -\phi) \tag{13.5.8}$$

where B is the Euler beta function. (There is a certain amount of arbitrariness in this choice, since most of the details of the potential are washed out when we approach a critical point.) Written in this form, we can now insert the expression for U into Eq. (13.5.7), and we find:

$$\beta x = -2\nu B\left(\frac{1}{2}, \frac{1}{2} - \nu\right)\langle x|\hat{H}^{\nu-(1/2)}|x\rangle \tag{13.5.9}$$

Rescaling $x \to 1 - t\beta^{-2k/(2k+1)}$, we can write:

$$\begin{aligned} t &= -2\nu B\left(\frac{1}{2}, \frac{1}{2} - \nu\right)\langle t|\hat{H}^{\nu-(1/2)}|t\rangle \\ &= -2\nu B\left(\frac{1}{2}, \frac{1}{2} - \nu\right)\left\langle t\left|\left[-(d/dt)^2 + f(t)\right]^{\nu-(1/2)}\right|t\right\rangle \\ &= 2\nu B\left(\frac{1}{2}, \frac{1}{2} - \nu\right)\oint \frac{d\omega}{2\pi i}\, \omega^{\nu-(1/2)}\left\langle t\left|\frac{1}{-\omega + \hat{H}}\right|t\right\rangle \end{aligned} \tag{13.5.10}$$

Fortunately, at this point we have reduced the problem to solving for the matrix element of the inverse of the Schrödinger operator $\hat{H}$, which can be solved via the methods of Gelfand and Dikii [25]. We are interested in solving for the matrix element:

$$R(t, t'; \xi) \equiv \left\langle t\left|\frac{1}{\xi + \hat{H}}\right|t'\right\rangle \tag{13.5.11}$$

The solution of this equation can be written in terms of the KdV hierarchy. For example, let us define the following operator:

$$K[f(t), \nabla_t] \equiv -\frac{1}{2}\frac{d^2}{dt^2} + f(t) + \frac{1}{\nabla_t}f(t)\nabla_t \tag{13.5.12}$$

Then, we can show that the matrix element of the inverse of the Schrödinger Hamiltonian can be given as:

$$R(t, t; \xi) = \sum_{l=0}^{\infty} \frac{R_l(f)}{\xi^{l+1/2}}$$

$$R_l(f) = \left\{-\frac{1}{2}K[f(t), \nabla_t]\right\}^l \cdot \frac{1}{2} \tag{13.5.13}$$

$$R'_{l+1} = \frac{1}{4}R''_l - uR'_l - \frac{1}{2}u'R_l$$

[To prove this, we note that R satisfies $(\hat{H} + \xi)_t R = (\hat{H} + \xi)_{t'} R = \delta(t - t')$. Therefore, $c(\xi) \equiv R R_{tt'} - R_t R_{t'}$ is a constant independent of t and t'. Given the behavior of R at infinity, we can set $c(\xi)$ to zero. If we then set $t = t'$, we have the differential equation $-R''' + 4(u + \xi)R' + 2u'R = 0$. Then Eq. (13.5.13) follows.]

We now have all the tools necessary to solve for t. By inserting the expression for the resolvant $R(t, t; \xi)$ into Eq. (13.5.10) for t, we find [22]:

$$t = \frac{k!}{(2k-1)!!} \{K[f(t), \nabla_t]\}^k \cdot 1 \tag{13.5.14}$$

It is now straightforward to calculate the values of the various R_l from Eq. (13.5.13), which can be written as:

$$R_0 = \frac{1}{2}, \qquad R_1 = -\frac{1}{4}f$$

$$R_2 = \frac{1}{16}(3f^2 - f'')$$

$$R_3 = -\frac{1}{64}[10f^3 - 10ff'' - 5(f')^2 + f'''']$$

$$R_4 = \frac{1}{256}[35f^4 - 70f(f')^2 - 70f^2 f'' \tag{13.5.15}$$
$$+ 21(f'')^2 + 28f'f''' + 14ff'''' - f'''''']$$

Expanding Eq. (13.5.14), we find:

$$k = 1: \quad t = f$$

$$k = 2: \quad t = f^2 - \frac{1}{3}f''$$

$$k = 3: \quad t = f^3 - ff'' - \frac{1}{2}(f')^2 + \frac{1}{10}f^{(4)} \tag{13.5.16}$$

$$k = 4: \quad t = f^4 - 2f(f')^2 - 2f^2 f'' + \frac{3}{5}(f'')^2$$
$$+ \frac{4}{5}f'f''' + \frac{2}{5}ff^{(4)} - \frac{1}{35}f^{(6)}$$

Notice that, for $k = 2$ and $k = 3$, we retrieve the results for Eqs. (13.4.22) and (13.4.27) (modulo trivial minus signs and rescalings) which were found by explicitly summing over the lower order staircases in Eq. (13.4.14). (We can also check the correctness of these results perturbatively. We know that, in the planar limit, that is, genus 0, the derivative terms in the expansion of the Hamiltonian vanish. Setting all primes to zero in the previous expression, we find that:

$$\lim_{g_{\text{string}} \to 0} f(t) = t^{1/k} \tag{13.5.17}$$

which is the correct perturbative expression.)

Although we have found the differential equations that define the solution to our problem, we find that we need more data before we can uniquely calculate their behavior. This is because perturbation theory does not specify the initial conditions for these higher order differential equations in Eq. (13.5.16). These initial conditions cannot be determined perturbatively, that is, they arise only because we have entered a fully nonperturbative domain of two-dimensional gravity.

Now that we have found explicit relations that the scaling function obeys, it is also straightforward to calculate the correlation functions of the theory. We recall that the presence of the potential U in the matrix model Lagrangian created a new function $w(R)$ in Eq. (13.4.18) that determined the scaling properties of the free energy.

More precisely, if the potential U was given by:

$$U(\phi) = \sum_k U_{2k}\phi^{2k} \tag{13.5.18}$$

for some coefficients U_{2k}, then the corresponding $w(R)$ is given by:

$$w(R) = \sum_k \frac{(2k)!}{k!(k-1)!}U_{2k}R^k \tag{13.5.19}$$

This tight relationship between U and $w(R)$ can be also be written in terms of line integrals:

$$w(R) = \oint \frac{dz}{2\pi i}U'\left(z + \frac{R}{z}\right)$$
$$U(\phi) = \int_0^1 \frac{du}{u} w[u(1-u)\phi^2] \tag{13.5.20}$$

Near the k critical point, the $k-1$ derivatives of $w(R)$ vanished. Near this point, therefore,

$$w(R) = 1 - (R_c - R)^k \tag{13.5.21}$$

where we can set $R_c = 1$. [At this point, we note that the potential U_k corresponding to w at the critical point is a sum of factors $(-1)^{l-1}\phi^{2l}$, summed from $l = 1$ to k. This means that the pure gravitational case, corresponding to $k = 2$, has a potential that is not bounded. This case, strictly speaking, is not well defined. This means that we must modify the potential in this case, that is, add corrections to the potential that bend the potential upward at infinity, so that the good features of the model are preserved.]

We now wish to calculate correlation functions among a new set of operators, called O_l, which will have useful scaling properties. Specifically, if we are in the kth critical mode, we demand that, by adding this function O_l to the potential U, we create a corresponding perturbation within $w(R)$ corresponding to the lth critical mode, that is,

$$w(R) \to w_k(R) + \epsilon[w_l(R) - 1] \tag{13.5.22}$$

for small ϵ.

Because of the tight relationship between U and $w(R)$, we can find an explicit representation of this operator O_l that generates the above change in $w(R)$:

$$O_l(\phi) \sim \text{Tr} \int_0^1 \frac{du}{u}[1 - u(1-u)\phi^2]^l \qquad (13.5.23)$$

The above expression is actually divergent, but its divergent piece vanishes in the scaling limit.

Let us now find the operator associated with this small perturbation. Recalling that $\phi \sim 2 - \hat{H}$, we can write:

$$O_l \sim \hat{H}^{l+1/2} \qquad (13.5.24)$$

We can take the matrix element of this operator, recalling that the matrix element of $\hat{H}$ is related to the KdV operator $K[f(t)]$. We can normalize O_l as follows:

$$\langle O_l \rangle = \frac{l!!}{(2l+1)!!} \int_t^\infty dt' \, K[f(t')]^{l+1} \cdot 1 \qquad (13.5.25)$$

Notice that this correlation function is exact. No approximations, other than the double scaling limit, have been made. Now, let us go to the perturbative limit. As before, to lowest order in the string coupling constant g_{string}, the derivatives disappear in the Hamiltonian, that is, the primes disappear within the KdV hierarchy. Thus,

$$\lim_{g_{\text{string}} \to 0} K^{l+1} \cdot 1 = \frac{(2l+1)!!}{(l+1)!} f^{l+1} \qquad (13.5.26)$$

and $f(t) = t^{1/k}$. Solving for the value of this and higher correlation functions in the spherical (genus 0) limit, we find [22]:

$$\langle O_1 \rangle = -\frac{k}{(l_1+1)(l_1+k+1)} t^{1+(l_1+1)/k}$$

$$\langle O_1 O_2 \rangle = -\frac{1}{(l_1+l_2+1)} t^{(l_1+l_2+1)/k}$$

$$\langle O_1 O_2 O_3 \rangle = -\frac{1}{k} t^{-1+(l_1+l_2+l_3+1)/k}$$

$$\langle O_1 O_2 O_3 O_4 \rangle = -\frac{1}{k^2}(l_1+l_2+l_3+l_4+1-k) t^{-2+(l_1+l_2+l_3+l_4+1)/k}$$

$$(13.5.27)$$

Correlations such as the above will be useful in establishing the equivalence between the matrix models in the double scaling limit and topological field theory, which will be discussed in Chapter 14.

Last, before leaving the one-matrix model, let us make some comments about transitions between multicritical points. In principle, we should examine the possibility of transitions between different values of k in our formalism. Consider, for example, the string equation:

$$t = \sum_l \left(l + \frac{1}{2}\right) t_l R_l(u) \tag{13.5.28}$$

where we have now generalized Eq. (13.5.14) by summing over many multicritical points and introducing the variable t_l for each critical point. (Normalizations can be absorbed into t_l.) The variable u is now a function of $t = t_0$ and t_l.

Equation (13.5.27) gives us a way in which to analyze the relationship between many critical points. Surprisingly enough, we find this same equation appearing in topological field theory, which once again supports the claim that matrix models and topological field theories are the same.

One way of implementing many critical points is to add new terms to the original action proportional to $t_l O_l$, so that taking variations of the correlation functions with respect to t_l brings down factors of O_l. This is a convenient formalism, but we have to physically interpret the meaning of O_0. From our original definition of O_l as the correction to the potential that induces Eq. (13.5.22), we see that the effect of O_0 is to change w by a constant, so that $t = t_0$ is shifted. But, t_0 is conjugate to the area of the random surface, so we can think of O_0 as a "puncture operator" P that does nothing but pick out a marked point on the surface. Symbolically, we can now summarize this by:

$$\frac{\partial}{\partial t_0} \langle \prod_i O_i \rangle = \langle P \prod_i O_i \rangle \tag{13.5.29}$$

Since the susceptibililty can be written in terms of the second derivative of the free energy, we also have:

$$u = \langle PP \rangle = \frac{\partial^2}{\partial t_0^2} F \tag{13.5.30}$$

From Eqs. (13.5.13) and (13.5.25), we also see that the expectation value of O_l is proportional to the integral of R_{l+1} with respect to t_0. With a suitable normalization, we can write:

$$\frac{\partial}{\partial t_0} \langle O_l \rangle = R_{l+1} \tag{13.5.31}$$

Let us put this all together. Since a derivative with respect to t_0 brings down a P, and a derivative with respect to t_l brings down an O_l, we now have, symbolically,

$$\partial/\partial t_0 = P, \qquad \partial/\partial t_\ell = O_\ell \tag{13.5.32}$$

This, in turn, implies the identity:

$$\langle O_l PP \rangle = \frac{\partial}{\partial t_l} u = \frac{\partial}{\partial t_0} R_{l+1} \tag{13.5.33}$$

This last equation shows that the matrix models generate the $l+1$ flow of the KdV hierarchy. In summary, when analyzing transitions between critical points, the fundamental equation governing the behavior of the system is given by the string equation, Eq. (13.5.28), in terms of functions R_l, which obey the flows of the KdV hierarchy, Eq. (13.5.32) and the recursion relation, Eq. (13.5.13).

13.6. Multimatrix Models

So far, we have analyzed the one-matrix model, which successfully reproduced the behavior of two-dimensional gravity coupled to nonunitary conformal matter, labeled by an index k. However, by generalizing the theory to include multimatrix models, we should be able to reproduce two-dimensional gravity coupled to a much wider class of conformal models, such as the minimal (p, q) series.

The generalization to higher matrix models is straightforward. Let M_i represent $q - 1$ distinct $N \times N$ matrices. The $q - 1$ matrix model is defined by the functional integral:

$$Z = \ln \int \prod_{i=1}^{q-1} dM_i \, \exp \left\{ -\mathrm{Tr} \left[\sum_{i=1}^{q-1} V_i(M_i) - \sum_{i=1}^{q-2} c_i M_i M_{i+1} \right] \right\} \quad (13.6.1)$$

As before, we diagonalize the M_i matrix in terms of its eigenvalues and integrate over its angular part. Repeating the steps for the one-matrix model, we find:

$$Z = \ln \int \prod_{\substack{i=1,q-1 \\ p=1,N}} d\lambda_i^p \Delta(\lambda_1) \Delta(\lambda_{q-1}) \times \exp \left\{ - \left[\sum_{i,p} V(\lambda_i^p) - \sum_{i,p} c_i \lambda_i^p \lambda_{i+1}^p \right] \right\} \quad (13.6.2)$$

where:

$$\Delta(\lambda_i) = \prod_{p_1 < p_2} (\lambda_i^{p_1} - \lambda_i^{p_2}) \quad (13.6.3)$$

Now, define Q_i and P_i to be operators that represent insertions of λ_i and $d/d\lambda_i$, respectively, into the functional integral. Although these operators are actually quite complicated, they must obey the simple relation:

$$[P_i, Q_i] = 1 \quad (13.6.4)$$

because λ and $d/d\lambda$ have the same commutation relations. In the double scaling limit, experience shows that the insertion of an extra λ into the integral can be accomplished via q-th order differential operators of the form:

$$Q = d^q + \{v_{q-2}(x), d^{q-2}\} + \cdots + 2v_0(x) \quad (13.6.5)$$

where $d = d/dx$ and the $v_i(x)$ are simple functions.

Our strategy is to make this commutation relation, Eq. (13.6.4), the basis for the generalized KdV hierarchies [26]. We solve this commutation relation by constructing a new pth-order operator out of the operator Q_i, using the theory of pseudodifferential operators. We demand that this new operator satisfies the same commutation relations as before. We then identify this operator with P_i.

The theory of pseudodifferential operators tells us that it is possible to create a pth-order operator from a qth-order operator. Given the q-th order operator Q, we can define:

$$Q^{1/q} \equiv d + \sum_{i=1}^{\infty} \{e_i, d^{-i}\} \tag{13.6.6}$$

where we define the operator d^{-1} to satisfy:

$$d^{-1}f = \sum_{j=0}^{\infty} (-1)^j f^{(j)} d^{-j-1} \tag{13.6.7}$$

(With this convention, it is easy to check that $d(d^{-1}f) = f$).

From the operator $Q^{1/q}$, we can construct the pth-order operator $Q^{p/q}$, such that:

$$[Q_+^{p/q}, Q] = 1 \tag{13.6.8}$$

where the plus (+) subscript means we take only the nonnegative powers of d in the expansion. A minus ($-$) subscript means taking negative powers of d.

Generally, we define:

$$Q = d^q + \sum_{i=0}^{q-2} \{v_i, d^i\}$$
$$Q_-^{p/q} = \sum_{i=1}^{\infty} \{e_i, d^{-i}\} \tag{13.6.9}$$

Evaluating the commutator yields:

$$[P, Q] = [Q_+^{p/q}, Q] = [Q, Q_-^{p/q}] = \sum_{i=0}^{q-2} \{r_i, d^i\} = 1 \tag{13.6.10}$$

where:

$$r_i = q e'_{q-1-i} + \cdots \tag{13.6.11}$$

Last, imposing that the commutator yield one, sets all r_i equal to zero, except for $r_0 = \frac{1}{2}$. Let us use this rather abstract formalism to first rederive the results for the one-matrix model and then to derive the results for the multimatrix model.

Example: One-Matrix Model

For the one-matrix model, the theory reproduces two-dimensional gravity coupled to (p, q) conformal matter, where $p = 2l - 1$ and $q = 2$. For $q = 2$, we take the second-order hermitian operator for Q:

$$Q = d^2 - u(x) \tag{13.6.12}$$

Our strategy is to construct the operator $P = Q_+^{p/q} = Q_+^{l-1/2}$, where:

$$Q^{l-1/2} = d^{2l-1} - \frac{2l-1}{4}\{u, d^{2l-3}\} + \cdots \tag{13.6.13}$$

We break this up into two pieces, containing positive and negative orders,

$$Q^{l-1/2} = Q_+^{l-1/2} + Q_-^{l-1/2} \tag{13.6.14}$$

with:

$$Q_-^{l-1/2} = \{R_l, d^{-1}\} + O(d^{-3}) \tag{13.6.15}$$

where R_l is yet undetermined, and:

$$
\begin{aligned}
Q_+^{1/2} &= d; \qquad Q_+^{3/2} = d^3 - \frac{3}{4}\{u, d\} \\
Q_+^{5/2} &= d^5 - 5/4\{u, d^3\} + \frac{5}{16}\{(3u^2 + u''), d\} \\
Q_+^{7/2} &= d^7 - \frac{7}{4}\{u, d^5\} + \frac{35}{16}\{(u^2 u''), d^3\} \\
&\quad - \frac{7}{64}\{[13u^{(4)} + 10uu'' + 10u^3 + 25(u')^2], d\}
\end{aligned} \tag{13.6.16}
$$

The R_l, which appears in $Q_-^{l-1/2}$, can be calculated by taking repeated commutators of the various Q's. By evaluating:

$$Q^{l+1/2} = QQ^{l-1/2} = Q^{l-1/2}Q \tag{13.6.17}$$

we find:

$$Q_+^{l+1/2} = \frac{1}{2}\left(Q_+^{l-1/2}Q + QQ_+^{l-1/2}\right) + \{R_l, d\} \tag{13.6.18}$$

Commuting both sides with Q, we find the recursion relation:

$$R'_{l+1} = \frac{1}{4}R''_l - uR'_l - \frac{1}{2}u'R_l \tag{13.6.19}$$

which is the same recursion relation satisfied by the R_l found previously in Eq. (13.5.13). Thus, we have now made contact with the KdV equations.

Last, we can show:

$$[Q_+^{l-1/2}, Q] = 4R'_l \tag{13.6.20}$$

The right-hand side must equal one. Integrating that relation to remove the prime, we now have (after rescaling):

$$\left(l + \frac{1}{2}\right) R_l(u) = x \tag{13.6.21}$$

which is the string equation found earlier in Eq. (13.5.28).

It is then straightforward to generalize the method to higher matrix models.

Example: Two-Matrix Model

For the two-matrix model [27, 28], we are interested in the coupling between two-dimensional gravity and $(4,3)$ conformal matter, or the critical Ising model. We find:

$$
\begin{aligned}
Q &= d^3 - \frac{3}{4}\{u, d\} + \frac{3}{2}w = K_+^{3/2} + \frac{3}{2}w \\
P &= Q_+^{p/q} = Q_+^{4/3} = K^2 + \{w, d\} + v
\end{aligned}
\tag{13.6.22}
$$

where $v = -8R_2/3$, the operator K is the Q found in the one-matrix model ion Eq. (13.6.12), and w is a breaking field resulting from coupling to a magnetic field. The commutation relations yield:

$$
\begin{aligned}
1 = [P, Q] = &- \left\{(4R_2' + \frac{3}{2}v'), d^2\right\} + \frac{1}{2}\left\{(w''' - 3(uw)', d\right\} \\
&+ 8uR_2' + \frac{1}{2}v''' + \frac{3}{2}uv' + \frac{3}{2}(w^2)'
\end{aligned}
\tag{13.6.23}
$$

Solving the equation and then integrating (so the number one becomes x), we find:

$$x = -8R_3 + \frac{3}{2}uv - \frac{1}{4}v'' + \frac{3}{2}w^2 \tag{13.6.24}$$

which is the solution for the two-matrix model. The higher matrix models can be done in a similar matter.

Example: Three-Matrix Model

For the three-matrix model [27, 28], where we couple to $(5,4)$ conformal matter (tricritical Ising model), we have:

$$
\begin{aligned}
Q &= K^2 + \{w, d\} + v \\
P &= Q_+^{p/q} = Q_+^{5/4} = K_+^{5/4} + \frac{5}{4}\{w, d^2\} + \frac{5}{8}\{v, d\} - \frac{5}{4}uw
\end{aligned}
\tag{13.6.25}
$$

We repeat the same steps. The commutation relations yield:

$$1 = [P, Q] = \sum_{i=0}^{2}\{r_i, d^i\} \tag{13.6.26}$$

where:

$$0 = r_2 = 4R_3' + \frac{5}{8}v''' - \frac{5}{4}(uv)' + \frac{5}{4}(w^2)'$$

$$0 = r_1 = -\frac{1}{2}\left[\frac{1}{2}w'''' - \frac{5}{4}(uw)'' - \frac{5}{4}uw'' - \frac{5}{2}vw + \frac{5}{4}u^2w\right]'$$

$$1 = 2r_0 = -8R_3'u - \frac{1}{4}v^{(5)} + \frac{5}{8}\left(2u'v'' + u''v' + 2vv' + 3u^2v' + 4uu'v\right)$$

$$- \frac{5}{2}ww''' - 5w'w'' + \frac{5}{2}w^2u' + \frac{5}{2}uww'$$

$$(13.6.27)$$

Solving the equation and integrating, we have the final result:

$$x = 8R_4 + \frac{1}{16}v^{(4)} + \frac{5}{8}v^2 + \frac{15}{8}u^2v - \frac{5}{8}(uv'' + u'v' + vu'')$$

$$- \frac{5}{4}ww'' + \frac{5}{4}w^2u + Tu$$

$$(13.6.28)$$

where T is an integration constant.

In summary, using the method of quasi-differential operators, the fundamental relationships, Eqs. (13.6.4) and (13.6.11), have given us the differential equations that define the nonperturbative behavior of the scaling functions.

13.7. D = 1 Matrix Models

Up to now, we have only treated the case of two-dimensional gravity coupled to conformal matter. Although the exact results involving the KdV hierarchy are encouraging, we do not expect many of the features to survive when we enter the region $D > 1$, where the KPZ formalism breaks down. In fact, as we approach $D = 1$ with $k \to \infty$, the potential becomes infinitely steep, the exponents become complex, and the formalism that we have developed breaks down.

What is remarkable, however, is that the case of $D = 1$ is still solvable and serves as a testing ground for many of our ideas concerning nonperturbative strings [29–32]. For example, we expect the perturbation theory to break down as we pass $D = 1$ because the ground state particle becomes a tachyon. Evaluating the zero-point contribution to the string, we find that the mass squared of the ground state is:

$$m^2 = \frac{1}{12}(1 - D) \qquad (13.7.1)$$

showing that the ground state becomes massless at $D = 1$ and tachyonic beyond that limit ($D - 1$, not $D - 2$, appears in this equation because the longitudinal mode does not decouple in noncritical string theory). Thus, we expect the exact solution for the $D = 1$ case should reflect this fact, which will destabilize the perturbation theory but perhaps allow a nonperturbative interpretation.

For the $D = 1$ case, the matrix field M is now one dimensional, that is, depends on a time variable, and we can write the equivalent Schrödinger equation in this time variable:

$$H\psi = N^2 E^{(1)}(g)\psi \qquad (13.7.2)$$

where

$$H = -\frac{1}{2}\Delta + V$$

$$\Delta = \sum_i \frac{\partial^2}{\partial M_{ii}^2} + \frac{1}{2}\sum_{i<j}\frac{\partial^2}{\partial \operatorname{Re} M_{ij}^2} + \frac{\partial^2}{\partial \operatorname{Im} M_{ij}^2} \qquad (13.7.3)$$

$$V = \frac{1}{2}\operatorname{Tr} M^2 + \frac{g}{N}\operatorname{Tr} M^4$$

Now, let us repeat the same trick that we introduced before, decomposing the matrix M in terms of its eigenvalues λ_i and its diagonalizing matrix U, which can be trivially integrated over. We find that the expression for the energy is given by:

$$E^{(1)}(g) = \lim_{N\to\infty} \frac{1}{N^2} \operatorname{Min} \frac{\int \prod_i d\lambda_i \prod_{i<j}^2 \left[(1/2)\sum_i(\partial\psi/\partial\lambda_i)^2 + V(\lambda_i)\psi^2\right]}{\int \prod_i d\lambda_i \prod_{i<j}(\lambda_i - \lambda_j)^2\psi^2} \qquad (13.7.4)$$

Now, let us make the key observation, which will render this problem almost trivial. We will redefine a new wave function ϕ as follows:

$$\phi(\lambda_1,\ldots,\lambda_N) \equiv \left[\prod_{i<j}(\lambda_i - \lambda_j)\right]\psi(\lambda_1,\ldots,\lambda_N) \qquad (13.7.5)$$

The first remarkable feature now emerges. Notice that the presence of this new factor has rendered the wave function ϕ to be antisymmetric, as if the theory were based on *fermions* rather than bosons. The second remarkable feature is that the highly coupled Schrödinger equation now reduces to a noninteracting theory when we eliminate the U matrix for the ground state:

$$\sum_i \left(-\frac{1}{2}\frac{\partial^2}{\partial\lambda_i^2} + \frac{1}{2}\lambda_i^2 + \frac{g}{N}\lambda_i^4\right)\phi = N^2 E^{(1)}\phi \qquad (13.7.6)$$

This result shows the power of the substitution. Notice that the theory, which at first seemed intractable, has now been reduced to an almost trivial problem, the theory of a Fermi gas interacting via a central potential given by:

$$\frac{\lambda^2}{2} + \frac{g}{N}\lambda^4 \qquad (13.7.7)$$

This theory can now be solved using standard techniques. For example, we know that in a Fermi gas, we can let $e_1 \leq e_2 \leq \cdots$ represent the energies of each fermion subject to the Hamiltonian:

$$h = -\frac{1}{2}\frac{\partial^2}{\partial\lambda^2} + \frac{\lambda^2}{2} + \frac{g}{N}\lambda^4 \qquad (13.7.8)$$

Let μ_F be the Fermi level. Then the total energy is the sum of the individual energies below the Fermi level or:

$$N^2 E^{(1)} = \sum k e_k \theta(\mu_F - e_k)$$
$$N = \sum_k \theta(\mu_F - e_k) \qquad (13.7.9)$$

So far, everything has been exact. Now, let us introduce the large N approximation, so that the eigenvalues λ_i become a smooth function $\lambda(x)$. Then, e_k can be written in terms of p, the momentum of the particle. We can then integrate over p:

$$N^2 E^{(1)} = N\mu_F - \int \frac{d\lambda dp}{2\pi} \theta\left(\mu_F - \frac{p^2}{2} - \frac{\lambda^2}{2} - \frac{g\lambda^4}{N}\right)$$
$$\times \left(\mu_F - \frac{p^2}{2} - \frac{\lambda^2}{2} - \frac{g\lambda^4}{N}\right) \qquad (13.7.10)$$
$$N = \int \frac{d\lambda dp}{2\pi} \theta\left(\mu_F - \frac{p^2}{2} - \frac{\lambda^2}{2} - \frac{g\lambda^4}{N}\right)$$

Let us now integrate over p and then rescale by $\lambda \to \sqrt{N}\lambda$ and $\mu_F \to N\epsilon$. Then, we have:

$$E^{(1)}(g) = \epsilon - \int \frac{d\lambda}{3\pi}\left(2\epsilon - \lambda^2 - 2g\lambda^4\right)^{3/2}\theta(2\epsilon - \lambda^2 - 2g\lambda^4)$$
$$1 = \int \frac{d\lambda}{2\pi}\left(2\epsilon - \lambda^2 - 2g\lambda^4\right)^{1/2}\theta(2\epsilon - \lambda^2 - 2g\lambda^4) \qquad (13.7.11)$$

This simple large N approximation has yielded a solution for the energy that, compared with numerical results of the anharmonic oscillator, yields results that are only off by at most 12%. For example, asymptotically, the planar approximation yields [17]:

$$E^{(1)}(g) \sim .58993g^{1/3} \qquad (13.7.12)$$

while the exact result is:

$$E^{(1)}(g) \sim .66799g^{1/3} \qquad (13.7.13)$$

We will now approach the $D = 1$ from an entirely new direction. We will exploit its resemblance to a system of harmonic oscillators. This, in turn, will give us the ability to perform both weak and strong coupling approximations which will reveal the non-Borel summability of the weak coupling approximation. In this approximation, we will solve for the exact solution for the ground state energy. We will find it convenient to introduce ρ, the density of states, as follows:

$$\rho(e) = \frac{1}{\beta} \sum_n \delta(e_n - e) \qquad (13.7.14)$$

Then, the ground state energy E_{gs} in Eq. (13.7.9) can be written as:

$$E_{gs} = \beta^2 \int_0^{\mu_F} \rho(e) e \, de; \qquad g = \frac{N}{\beta} = \int_0^{\mu_F} \rho(e) \, de \qquad (13.7.15)$$

where μ_F is the highest energy level, the Fermi level.

In the critical limit, we will take the limit $N/\beta \equiv g \to g_c$. We will adjust the potential so that $U = \mu_c$ as it approaches the Fermi level from above. Then, we define the cosmological constants Δ and μ as $\Delta \equiv g_c - g$ and $\mu \equiv \mu_c - \mu_F$

We will find it convenient to work with derivatives of Eq. (13.7.15):

$$\frac{\partial g}{\partial \mu} = -\rho(\mu_F)$$
$$\frac{\partial E_{gs}}{\partial \mu} = -\beta^2 \mu_F \rho(\mu_F) = \beta^2 \mu_F \frac{\partial g}{\partial \mu} \qquad (13.7.16)$$

Our strategy will be to solve for ρ and then invert to find the ground state energy as we perturb in $1/\beta\mu$, which is the weak coupling limit, or $\beta\mu$, which is the strong coupling limit.

The key observation is that, in the scaling limit, the system resembles an inverted harmonic oscillator.

To see this, we define:

$$\rho(\mu_F) = \frac{1}{\pi\beta} \operatorname{Im} \operatorname{Tr} \frac{1}{h - \mu_F - i\epsilon} \qquad (13.7.17)$$

where h is the Hamiltonian. If we expand near criticality: $y \sim x_c - x \sim 0$, so $h - \mu_F \sim -(1/2\beta^2)\partial_y^2 - \mu_F - 2y^2 + 0(y^3)$. For small y, we can ignore the cubic and higher terms, so the theory is dominated by the y^2 term, that is, it corresponds to an inverted harmonic oscillator. This is a great simplification, because we have now reduced a rather complicated system to solving a known system where we can use well-established methods to find its solutions.

We begin by remarking that the density of states of the normal harmonic oscillator, of frequency ω, is:

$$\rho(E) = \frac{1}{\pi} \operatorname{Im} \sum_n \frac{1}{[n + (1/2)]\omega\hbar - E - i\epsilon} \qquad (13.7.18)$$

When we continue ω to imaginary frequency, we have:

$$\rho(\mu_F) = \frac{1}{\pi} \operatorname{Re} \sum_n \frac{1}{2n + 1 + i\beta\mu} \qquad (13.7.19)$$

This is singular, but the divergent part is μ dependent and hence will disappear in the critical limit. We thus have:

$$
\begin{aligned}
\rho(\mu_F) &= (1/2\pi) \operatorname{Re}\{\zeta[1, (1/2)(1 + i\beta\mu)]\} \\
&= (1/2\pi)\{ \ln(\beta/2) - \operatorname{Re} \psi[(1 + i\beta\mu)/2] \}
\end{aligned}
\qquad (13.7.20)
$$

where:

$$
\begin{aligned}
\zeta(z, q) &= \sum_{n=0}^{\infty} \frac{1}{(n + q)^z} \\
\psi(z) &= \frac{\Gamma'(z)}{\Gamma(z)}
\end{aligned}
\qquad (13.7.21)
$$

Let us now take the weak coupling limit (which is an expansion in $1/\beta\mu$). Power expanding the ζ function in Eq. (13.7.20), we find:

$$
\begin{aligned}
\rho(\mu_F) &= -\frac{1}{2\pi} \ln\mu + \frac{1}{\pi\beta\mu} \operatorname{Im} \sum_{n=0}^{\infty} \frac{1}{1 - i(2n + 1)/\beta\mu} \\
&= \frac{1}{2\pi} \left[-\ln\mu + 2\sum_{k=1}^{\infty} (-1)^k (\beta\mu)^{-2k} (2^{2k-1} - 1)\zeta(1 - 2k) \right] \\
&= \frac{1}{2\pi} \left[-\ln\mu + \sum_{m=1}^{\infty} (2^{2m-1} - 1)\frac{|B_{2m}|}{m(\beta\mu)^{2m}} \right]
\end{aligned}
\qquad (13.7.22)
$$

where B_{2m} are the Bernoulli numbers.

Now, we invert. We integrate the expression $\partial g/\partial \mu = -\rho(\mu_F)$ in Eq. (13.7.17) to find:

$$
\Delta = \frac{\mu}{2\pi} \left[-\ln\mu - \sum_{m=1}^{\infty} (2^{2m-1} - 1)\frac{|B_{2m}|}{m(2m - 1)(\beta\mu)^{2m}} \right] \qquad (13.7.23)
$$

Inverting again to find the energy, we find [29]:

$$
E_{\text{gs}} = \frac{1}{g_{\text{st}}^2} \left[1 + \sum_{n=1}^{\infty} \sum_{m=1}^{n} \epsilon_{n,m} g_{\text{st}}^{2n} (-\ln\Delta)^m \right] \qquad (13.7.24)
$$

where $\epsilon_{n,m}$ are constants, and the string coupling constant is:

$$
g_{\text{st}}^2 = -\frac{\ln\Delta}{2\pi\beta^2\Delta^2} \qquad (13.7.25)
$$

The important point is that we have logarithmic singularities appearing in the series, which spoil the perturbation theory. In fact, no matter how small we make g_{st}^2, we still find these singularities.

These logarithmic singularities probably reflect the presence of the massless mode in the theory (which becomes tachyonic for $D > 1$). This is due to the fact that we can always attach a tadpole with the massless particle to any Riemann surface. Thus, this is an infrared problem. It also means that the nice picture of summing over Riemann surfaces breaks down perturbatively. In addition, the perturbation theory also suffers from the fact that diagrams diverge as $(2n)!$, meaning that it is not Borel summable in the weak coupling limit.

Now, let us analyze the strong coupling limit as a power expansion in $\beta\mu$, where everything is well behaved. We power expand Eq. (13.7.19):

$$\rho(\mu_F) = -\frac{1}{2\pi} \ln \mu + \frac{1}{\pi} \operatorname{Re} \sum_{n=0}^{\infty} \frac{1}{(2n+1)(1+i\beta\mu/(2n+1))} \qquad (13.7.26)$$

We repeat the same steps, inverting a series of equations. We find:

$$\frac{\partial \Delta}{\partial \mu} = \rho(\mu_F) = -\frac{1}{2\pi} \ln \mu + \frac{1}{\pi} \sum_{k=1}^{\infty} (-1)^k [1 - 2^{-(2k+1)}] \zeta(2k+1)(\beta\mu)^{2k}$$

$$(13.7.27)$$

Inverting, we find for μ:

$$\mu = -\frac{2\pi\Delta}{\ln \Delta} \left[1 + \sum_{n=1}^{\infty} \sum_{m=n+1}^{\infty} d_{n,m} \left(\frac{\beta^2 \Delta^2}{\ln \Delta} \right)^n (\ln \Delta)^{-m} \right] \qquad (13.7.28)$$

which gives us the final answer [29]:

$$E_{\mathrm{gs}} = \frac{1}{g_{\mathrm{gs}}^2} \left(1 + \sum_{n=1}^{\infty} \sum_{m=n+1}^{\infty} \theta_{n,m} \frac{1}{g_{\mathrm{gs}}^{2n} \ln^m \Delta} \right) \qquad (13.7.29)$$

where d_{nm} and θ_{nm} are constants. In contrast to the weak coupling limit, we find that the strong coupling limit is well behaved.

So far, we have been able to solve for the energy of the $D = 1$ theory because of a key observation, that the problem reduces to that of an inverted harmonic oscillator. We need to find a more comprehensive formalism, however, if we are to understand the physical origin of the curious logarithmic divergences and the non-Borel summability. The formalism that we will introduce is string field theory [33–35].

Certain features which seem obscure in the previous discussion have a natural explanation from the point of view of string field theory. For example, the ground state energies found earlier emerge directly as eigenstates of the string field Hamiltonian. Also, the curious logarithmic infinities found earlier, which apparently ruin Borel summability, can be interpreted as the length of one of the compactified dimensions. The mysterious Liouville mode also has a nice re-interpretation in string field theory; it appears as the space of eigenvalues of the matrix M, and combines with the single dimension of $D = 1$ string theory to give an effective $D = 2$ theory. Furthermore, the fact that $D = 1$ is solvable, which appears obscure in the previous section, appears almost obvious because the string field theory action is integrable.

We saw earlier that string field theory enables us to describe all the states of the string in a single field Ψ, whose decomposition yields all the states in the Fock space, and whose products yields the interactions of the theory.

We will introduce string field theory via our earlier observation that the $D = 1$ theory reduces to a fermionic theory of uncoupled particles. We noticed that, if we diagonalize our matrix field M into eigenvalues via $M = \Omega \Lambda \Omega^\dagger$, then the Hamiltonian reduces to:

$$H = -\frac{1}{2\beta^2 \Delta(\lambda)} \sum_i \frac{d^2}{d\lambda_i^2} + \sum_i U(\lambda_i) + \sum_{i \le j} \frac{\Pi_{ij}^2}{\beta^2 (\lambda_i - \lambda_j)^2} \qquad (13.7.30)$$

where Π_{ij} depends on the angular part of the M matrix. Because this Π_{ij} factor is difficult to work with, we will only consider $SU(N)$ singlet states in which the angular part of the Hamiltonian is exactly zero, so that this term vanishes. We can further reduce the system by introducing $\psi(\lambda_i) = \Delta(\lambda_i)\phi(\lambda_i)$, which yields an anti-symmetric, fermionic wave function.

In the singlet sector, the matrix model reduces to ordinary quantum mechanics with N non-interacting fermions moving in a potential $U(\lambda)$, with Planck's constant $\hbar$ given by $1/\beta \sim 1/N$. In this new basis, the Hamiltonian reduces to:

$$h = -\frac{1}{2\beta^2} \frac{d^2}{d\lambda^2} + U(\lambda) \qquad (13.7.31)$$

Let us now introduce a fermionic string field Ψ, which in turn is a linear superposition of an infinite number of states ψ_i with energy e_i:

$$\Psi(\lambda, t) = \sum_i \alpha_i \psi_i(\lambda) e^{-ie_i t} \qquad (13.7.32)$$

In this representation, we can easily re-write our original Hamiltonian in terms of this string field:

$$h = \int d\lambda \left[\frac{1}{2\beta^2} \frac{\partial \Psi^\dagger}{\partial \lambda} \frac{\partial \Psi}{\partial \lambda} + U(\lambda)\Psi^\dagger \Psi - \mu_F(\Psi^\dagger \Psi - N) \right] \qquad (13.7.33)$$

(we adjust the Lagrange multiplier μ_F to equal the Fermi level).

This Hamiltonian, in turn, can be derived from the following action:

$$S = \int dt d\lambda \left[i\Psi^\dagger \dot{\Psi} - \frac{1}{2\beta^2} \frac{\partial \Psi^\dagger}{\partial \lambda} \frac{\partial \Psi}{\partial \lambda} - U(\lambda)\Psi^\dagger \Psi + \mu_F(\Psi^\dagger \Psi - N) \right] \qquad (13.7.34)$$

This string field theory action contains all the information of the $D = 1$ matrix model for the $SU(N)$ singlet sector. Notice that the action appears to be defined in *two* dimensions if we treat λ and t as two space-time coordinates. This is how the Liouville mode appears in our formulation.

Notice that the previous action appeared to be non-relativistic. However, with some modification, we can re-write this in a more relativistic fashion. Let us define Ψ_L and Ψ_R:

$$
\Psi(\lambda, t) = \frac{e^{i\mu_F t}}{\sqrt{2v(\lambda)}} \left[\exp\left(-i\beta \int^\lambda d\lambda' v(\lambda') + i\pi/4 \right) \Psi_L(\lambda, t) \right.
$$
$$
\left. + \exp\left(i\beta \int^\lambda d\lambda' v(\lambda') - i\pi/4 \right) \Psi_R(\lambda, t) \right]
$$
(13.7.35)

where $v(\lambda)$ is the velocity of a classical particle at the Fermi level in the potential $U(\lambda)$, that is, $v(\lambda) = d\lambda/d\tau = \sqrt{2(\mu_F - U(\lambda))}$.

In terms of these new variables, the action becomes:

$$
H = \int_0^{T/2} d\tau \left[i\Psi_R^\dagger \partial_\tau \Psi_R - i\Psi_L^\dagger \partial_\tau \Psi_L + \frac{1}{2\beta v^2}(\partial_\tau \Psi_L^\dagger \partial_\tau \Psi_L + \partial_\tau \Psi_R^\dagger \partial_\tau \Psi_R) \right.
$$
$$
\left. + \frac{1}{4\beta}(\Psi_L^\dagger \Psi_L + \Psi_R^\dagger \Psi_R)\left(\frac{v''}{v^3} - \frac{5(v')^2}{2v^4} \right) \right]
$$
(13.7.36)

The equations of motion can be read off the action:

$$
i\gamma^\mu \partial_\mu \Psi = \mathbf{K}\Psi
$$
(13.7.37)

where $\partial_\mu = (\partial_t, \partial_\tau)$ and

$$
\mathbf{K} = \gamma_0 \left[\partial_\tau \frac{1}{2\beta v^2} \partial_\tau - \frac{1}{4\beta}\left(\frac{v''}{v^3} - \frac{5(v')^2}{2v^4} \right) \right]
$$
(13.7.38)

One advantage of string field theory is that we can see that the system is integrable. Since the field theory is based on free fermionic fields, we can construct an infinite number of conserved currents:

$$
J_\mu = \bar{\Psi}(\tau + a, t)\gamma_\mu T \left[\exp\left(-i \int_\tau^{\tau+a} \gamma_1 \mathbf{K}(\tau, \partial_\tau)d\tau' \right) \right] \Psi(\tau, t)
$$
(13.7.39)

The existence of this infinite set of conserved currents, in turn, indicates that the system is integrable.

In fact, it also indicates that the system is topological. In the next chapter, we will see that the Green's functions found in matrix models is identical to the Green's functions found for topological field theory.

13.8. Summary

Matrix models provide the first nonperturbative information concerning string theory. They serve as a theoretical laboratory in which to test many of our ideas about string theory. However, there seems to be a qualitative problem in extending the beautiful results of matrix models beyond $D = 1$.

Matrix models begin with string theory defined at below the critical dimension, where we must be careful to calculate the contribution of the function measure. After a scale transformation, both the string and the ghost parts contribute to the Liouville action:

$$S_{\mathrm{L}}(\sigma, g) = \int d^2 z \sqrt{g} \left(\frac{1}{2} g^{ab} \, \partial_a \sigma \, \partial_b \sigma + R\sigma + \mu e^{\sigma} \right) \qquad (13.8.1)$$

which can be eliminated only in 26 dimensions.

By scaling arguments, KPZ showed how to calculate the string susceptibility. One can show that the partition function diverges as:

$$Z(A) \sim A^{-3+\gamma} \exp(kA) \qquad (13.8.2)$$

where γ is the string susceptibility and is given by:

$$\gamma(h) = (1/12)(1 - h) \left(D - 25 - \sqrt{25 - D} \sqrt{1 - D} \right) + 2 \qquad (13.8.3)$$

where h is the genus number. One of the early successes of matrix models was their ability to reproduce this result for the susceptibility.

Matrix models are based on the old assumption that Feynman graphs for the matrix model, when viewed as a power expansion in $1/N^2$, become planar to lowest order. Hence, we can view them as approximations to two-dimensional gravity. For higher powers of $1/N^2$, the power expansion becomes identical to a power expansion in the genus of the two-dimensional surface.

We will take the action:

$$L = \sum_{\mu=1}^{D} \mathrm{Tr}(\partial_\mu M \, \partial^\mu M^\dagger) + \mathrm{Tr}(MM^\dagger) + \frac{g}{N} \mathrm{Tr}(MM^\dagger MM^\dagger) \qquad (13.8.4)$$

More generally, we can have the interaction:

$$U = \sum_{i=3} V_i = g_3 \, \mathrm{Tr}\, M^3 + g_4 \, \mathrm{Tr}\, M^4 + \cdots \qquad (13.8.5)$$

The generating function is given by:

$$Z(\beta) = \int D\Phi \, \exp\left[-\beta \, \mathrm{Tr}\, U(\Phi) \right] \qquad (13.8.6)$$

with potential U. We are interested in the double scaling limit

$$N \to \infty, \qquad \beta/N \to \text{const.} \qquad (13.8.7)$$

where we will find universality, that is, most of the particular properties of the potential are washed out.

There is an old trick that allows us to solve most of these models exactly and that is to decompose M in terms of its eigenvalues and its angular part. The angular part, in fact, decouples, leaving us with the measure:

$$\prod_{ij} dM_{ij} = \prod_i d\lambda_i \prod_{i<j} (\lambda_i - \lambda_j)^2 \, d\Pi_{ij} \qquad (13.8.8)$$

where $d\Pi$ can be trivially integrated over for our case.

Next, we evaluate the integration over the eigenvalues λ using the method of orthogonal polynomials, which are defined via:

$$\int d\lambda \exp\left[-\beta U(\lambda)\right] P_n(\lambda) P_m(\lambda) = \delta_{n,m} h_n \qquad (13.8.9)$$

It is possible to construct these polynomials so that they satisfy a recursion relation for any potential $U(\lambda)$:

$$\lambda P_n(\lambda) = P_{n+1}(\lambda) + S_n P_n(\lambda) + R_n P_{n-1}(\lambda) \qquad (13.8.10)$$

Using the recursion relation, one can also show:

$$
\begin{aligned}
h_{n+1} &= \int d\lambda \exp\left[-\beta U(\lambda)\right] P_{n+1} \lambda P_n \\
&= \int d\lambda \exp\left[-\beta U(\lambda)\right] \left(P_{n+2} + S_{n+1} P_{n+1} + R_{n+1} P_n\right) P_n \qquad (13.8.11) \\
&= R_{n+1} h_n
\end{aligned}
$$

The point of using these orthogonal polynomials is that we can write the partition function simply in terms of them:

$$Z_N = N! \prod_{i=0}^{N-1} h_i \qquad (13.8.12)$$

Using these recursion relations, we arrive at the equation:

$$\frac{n}{\beta} = 2R_n \sum_{p=0}^{n} (p+1) g_{g+1} \sum_{paths} R_{\alpha_1} \cdots R_{\alpha_p} \qquad (13.8.13)$$

where the sum over *paths* is rather complicated. To lowest order, we have:

$$t = f^2 + f'' \qquad (13.8.14)$$

where f is the scaling function found in the free energy. This solution is a Painlevé transcendental of the first kind.

It is tedious, however, to sum over *paths* as we go to higher and higher levels. It is much more convenient to convert to an operator language and treat the problem from a different perspective. As an operator expression, our basic recursion relations, in terms of the potential U, are:

$$\langle n|U'(\hat{\lambda})|n\rangle = 0$$
$$\langle n-1|U'(\hat{\lambda})|n\rangle = n/\beta\sqrt{R_n} \qquad (13.8.15)$$

We will introduce the KdV hierarchy through the operator:

$$K[f(t), \nabla_t] \equiv -\frac{1}{2}\frac{d^2}{dt^2} + f(t) + \frac{1}{\nabla_t}f(t)\nabla_t \qquad (13.8.16)$$

and:

$$R(t, t'; \xi) \equiv \left\langle t \left| \frac{1}{\xi + \hat{H}} \right| t' \right\rangle \qquad (13.8.17)$$

Then, we can show that the matrix element of the inverse of the Hamiltonian can be given as:

$$R(t, t; \xi) = \sum_{l=0}^{\infty} \frac{R_l(f)}{\xi^{l+1/2}}$$
$$R_l(f) = \left\{ -\frac{1}{2}K[f(t), \nabla_t] \right\}^l \cdot \frac{1}{2} \qquad (13.8.18)$$

We now have all the tools necessary to solve for t. By inserting the expression for the resolvant $R(t, t; \xi)$ into the expression for t, we find that Eq. (13.8.15) reduces to:

$$t = \frac{k!}{(2k-1)!!}\{K[f(t), \nabla_t]\}^k \cdot 1 \qquad (13.8.19)$$

This is the final answer, which includes the earlier solution. The lowest order solutions include:

$$
\begin{aligned}
k = 1: &\quad t = f \\
k = 2: &\quad t = f^2 - \frac{1}{3}f'' \\
k = 3: &\quad t = f^3 - ff'' - \frac{1}{2}(f')^2 + \frac{1}{10}f^{(4)} \\
k = 4: &\quad t = f^4 - 2f(f')^2 - 2f^2f'' + \frac{3}{5}(f'')^2 \\
&\quad + \frac{4}{5}f'f''' + \frac{2}{5}ff^{(4)} - \frac{1}{35}f^{(6)}
\end{aligned}
\qquad (13.8.20)
$$

By comparing our results with the KPZ equation, we can find which two-dimensional theory the matrix model is approximating. For $k = 2$, we have the case of pure two-dimensional gravity (that is, zero dimensional strings). For $k = 3$, although initially suspected to be the coupling of 2D gravity with the unitary minimal series, it is now known to approximate 2D gravity coupled to nonunitary conformal matter.

To see how to get 2D gravity coupled to the unitary minimal series and other forms of (p, q) conformal matter, we must now generalize our approach to the multimatrix approach, where we introduce several types of matrices M_i obeying:

$$Z = \ln \int \prod_{i=1}^{q-1} dM_i \, \exp\left\{ -\mathrm{Tr}\left[\sum_{i=1}^{q-1} V_i(M_i) - \sum_{i=1}^{q-2} c_i M_i M_{i+1} \right] \right\} \quad (13.8.21)$$

The solution to these equations can be solved via the theory of pseudodifferential operators.

Define Q_i and P_i to be operators that represent insertions of λ_i and $d/d\lambda_i$, respectively, into the functional integral. Although these operators are actually quite complicated, they must obey the simple relation:

$$[P_i, Q_i] = 1 \quad (13.8.22)$$

because λ and $d/d\lambda$ have the same commutation relations.

A solution to these equations, in terms of Q_i, can be found once one introduces the definition of Q^{-n} via pseudodifferential operators. We find the solution:

$$[Q_+^{p/q}, Q] = 1 \quad (13.8.23)$$

where the plus $(+)$ subscript means we take only the nonnegative powers of d in the expansion. A minus $(-)$ subscript means taking negative powers of d.

By solving for these equations, we can reproduce the one-matrix result. For higher matrix models, we find the coupling of 2D gravity to the minimal unitary series, as well as (p, q) conformal matter. Our goal, however, is to model string interactions in higher dimensions. Let us analyze the $D = 1$ case, which may be the limit for matrix models.

We begin with the density function for states in the matrix model:

$$\rho(e) = \frac{1}{\beta} \sum_n \delta(e_n - e) \quad (13.8.24)$$

Then, the ground state energy E_{gs} can be written as:

$$g = \frac{N}{\beta} = \int_0^{\mu_F} \rho(e) \, de; \qquad E_{\mathrm{gs}} = \beta^2 \int_0^{\mu_F} \rho(e) e \, de \quad (13.8.25)$$

where μ_F is the highest energy level, the Fermi level, and:

$$\begin{aligned} \rho(\mu_F) &= (1/2\pi) \, \mathrm{Re}\{\zeta[1, (1/2)(1 + i\beta\mu)]\} \\ &= (1/2\pi)\{ \ln(\beta/2) - \mathrm{Re}\,\psi[(1 + i\beta\mu)/2] \} \end{aligned} \quad (13.8.26)$$

Perturbing in the weak coupling limit (expanding in $1/\beta\mu$), we find that the energy is not well behaved:

$$E_{\text{gs}} = \frac{1}{g_{\text{st}}^2} \left[1 + \sum_{n=1}^{\infty} \sum_{m=1}^{n} \epsilon_{n,m} g_{\text{st}}^{2n} (-\ln \Delta)^m \right] \tag{13.8.27}$$

that is, the theory behaves badly, and is not Borel summable. In the strong coupling limit (expanding in $1/\beta\mu$), we find perfectly acceptable results for the energy:

$$E_{\text{gs}} = \frac{1}{g_{\text{gs}}^2} \left(1 + \sum_{n=1}^{\infty} \sum_{m=n+1}^{\infty} \theta_{n,m} \frac{1}{g_{\text{gs}}^{2n} \ln^m \Delta} \right) \tag{13.8.28}$$

This seems to confirm our earlier conjecture about the non-Borel summability of the perturbation theory.

References

1. V. Kazakov, *Phys. Lett.* **60B** 2105 (1988).
2. V. Kazakov, I. Kostov, and A. Migdal, *Phys. Lett.* **157B** 295 (1985).
3. F. David, *Nucl. Phys.* **B257** [**FS14**] 45, 543 (1985).
4. J. Ambjorn, B. Durhuus, and J. Frohlich, *Nucl. Phys.* **B257** [**FS14**], 433 (1985).
5. J. Ambjorn, B. Durhuus, J. Frohlich, and P. Orland, *Phys. Lett. Nucl. Phys.* **B270** [**FS16**], 457 (1986).
6. J. Jurkievic, A. Krzywicki, and B. Peterson, *Phys. Lett.* **168B**, 273 (1986).
7. A. Billoire and F. David, *Phys. Lett.* **186B**, 279 (1986).
8. D. V. Boulatov, V. A. Kazakov, I. K. Kostov, and A. A. Migdal, *Nucl. Phys.* **B275**, 641 (1986).
9. I. Kostov and M. Mehta, *Phys. Lett.* **189B**, 247 (1987).
10. V. Kazakov and A. Migdal, *Nucl. Phys.* **B311**, 171 (1989).
11. B. Sakita and M.A. Virasoro, *Phys. Rev. Lett.* **24**, 1146 (1970); H.B. Nielsen and P. Olesen, *Phys. Lett.* **32B**, 203 (1970).
12. G. 't Hooft, *Nucl. Phys.* **B72**, 461 (1974).
13. A. M. Polyakov, *Mod. Phys. Lett.* **A2**, 899 (1987).
14. V. G. Knizhnik, A. M. Polyakov, and A. A. Zamolodchikov, *Mod. Phys. Lett.* **A3**, 819 (1988).
15. F. David, *Mod. Phys. Lett.* **A3**, 207 (1988).
16. J. Distler and H. Kawaii, *Nucl. Phys.* **B231**, 509 (1989); see also: J. L. Gervais and A. Neveu, *Nucl. Phys.* **B238**, 125 (1984).
17. E. Brezin, C. Itzykson, G. Parisi, and J.-B. Zuber, *Comm. Math. Phys.* **59**, 35 (1978).
18. D. Bessis, C. Itzykson, and J.-B. Zuber, *Adv. Appl. Math.* **1**, 109 (1980).
19. D. Bessis, *Comm. Math. Phys.* **69**, 147 (1979).
20. C. Itzykson and J. B. Zuber, *J. Math. Phys.* **21**, 411 (1980).
21. E. Brezin and V. A. Kazakov, *Phys. Lett.* **236B**, 144 (1989).
22. D. Gross and A. Midgal, *Phys. Rev. Lett.* **64**, 127 (1990); Princeton preprint (1989).
23. M. Douglas and S. H. Shenker, *Nucl. Phys.* **B335**, 635 (1990).
24. T. Banks, M. R. Douglas, N. Seiberg, and S. H. Shenker, *Phys. Lett.* **238B**, 279 (1989).
25. I. Gelfand and L. Dikii, *Usp. Matem. Nauk.* **30**, 5 (1975).
26. M. R. Douglas, *Phys. Lett.* **238B**, 176 (1989).

27. P. Ginsparg, M. Goulian, M. R. Plesser, and J. Zinn-Justin, HUTP-90/AO15 (1990).
28. M. Kreuzer and R. Schimmrigk, Santa Barbara preprint NSF-ITP-90-30 (1990); H. Kunitomo and S. Odake, U. Tokyo preprint UT-558 (1990).
29. D. Gross and N. Miljkovic, Princeton preprint PUPT-1160 (1990).
30. P. Ginsparg and J. Zinn-Justin, Harvard preprint PUPT-1160 (1990).
31. G. Parisi, Rome preprint ROM2F-90/2 (1990).
32. E. Brezin, V. Kazakov, and Al. Zamolodchikov, École Normale preprint LPS-ENS-89-182 (1989).
33. S.R. Das and A. Jevicki, Brown-Het-750 (1990).
34. J. Polchinski, UTTG-15-90 (1990).
35. D. J. Gross and I. R. Klebanov, PUPT-1198 (1990).

Chapter 14

Topological Field Theory

14.1. Unbroken Phase of String Theory

The fundamental problem facing string theory at present is our inability to select its true vacuum nonperturbatively. Until the true string vacuum can be discovered, it is impossible to determine whether the theory predicts nonsense, and must be discarded as yet another failed attempt at a unified field theory, or gives a valid description of our universe and a unification of all known quantum forces. The frustration is that string theory has been evolving backward, ever since its accidental discovery in 1968 by Veneziano and Suzuki, so its underlying geometry is totally unknown.

By contrast, the "natural home" for Yang–Mills theory and the general theory of relativity are well known. Their "natural home" lies in the realm of unbroken local $SU(N)$ symmetry or general covariance. Even if we study these theories in a domain where all their symmetries have been severely broken, we know that the near-miraculous properties that persist in the broken theory arise because of its underlying geometry. Likewise, perhaps the key to understanding the underlying geometry of string theory is to understand its natural home.

There is a useful analogy that illustrates this problem. Hypothetically, one can ask the question of what might have happened to the evolution of physics if Einstein did not discover the work of Riemann and write the general theory of relativity in 1915. The most pessimistic scenario would be that relativity would not have been discovered until decades later, in the 1950s, as field theorists began a systematic search for higher spin field equations. The successes of spin-0 meson theories, spin-$\frac{1}{2}$ Dirac theories, and spin-1 Maxwell theories might have led to the study of purely hypothetical spin-2 systems in flat space.

As Feynman independently discovered, gauge invariance is necessary to kill the ghosts of a spin-two particle and maintain unitarity, but this necessarily complicates the search for the action. A simple cubic action for gravitons can be used to construct four-point scattering amplitudes, but these fail to maintain gauge invariance. A fundamental four-point contact term is necessary. But then, the five-point scattering amplitude fails to be gauge invariant, requiring the addition of a fundamental five-point interaction, and so on. The final result is a nontrivial nonpolynomial action.

However, it might be discovered that this ugly and contrived nonpolynomial action possessed mysterious, near-miraculous properties. After a long and difficult calculation, it might be recognized that the theory was independent of the classical background metric. Thus, a hunt might begin to find the natural home for the nonpolynomial theory. However, even though the action was completely known as a power expansion, it might be a leap of logic to postulate that general covariance was the actual origin of these miracles.

Likewise, we are still searching for the natural home for string theory. However, there are strong indications that the natural home for string theory does not lie in the low-energy realm of perturbation theory around conformal field theories.

As we have seen in Chapter 9, there are several indications that surprises await us at high energies and high temperatures. First, the high-energy behavior of the multiloop amplitudes shows that the perturbation theory is not Borel summable, with the genus g amplitudes growing as $g!$. Also, there is a strong indication that a new "symmetry" of some type is appearing at high energies beyond the Planck length. Second, the high-temperature behavior of multiloop string amplitudes shows that there may be a first-order phase transition occurring near the Hagedorn temperature.

These are indirect indications that, in analogy with ordinary gauge theories, an "unbroken phase" of string theory may be opening up at high energies and high temperatures.

In some sense, the natural home of string theory is not the perturbation theory based on Riemann surfaces at all, but an entirely new domain. This is crucially important for the ultimate isolation of the true vacuum of the theory.

At first, this may sound confusing, because for the past 75 years physicists have studied Einstein's general theory of relativity, where we make the important approximation:

$$g_{\mu\nu} = g_{\mu\nu}^{(o)} + \kappa h_{\mu\nu} \qquad (14.1.1)$$

where $g_{\mu\nu}^{(o)}$ is a solution to the classical equations of motion, usually taken to be the Minkowski metric of flat space. However, this approximation breaks local general covariance explicitly, leaving only global Poincaré covariance in the action. Thus, most quantum mechanical approaches to general relativity inherently break general covariance.

What would a theory look like in which general covariance was not broken, where we did not power expand around some background metric? Such a theory would look strange indeed. In a scheme where $g_{\mu\nu}$ is power expanded around zero (without ever refering to the Minkowsky metric $\delta_{\mu\nu}$), there is no light cone, no propagation of waves, no meter sticks, and no motion. In other words, physics as we know it apparently ceases to exist in a quantum theory where local general covariance is preserved exactly, in this "unbroken phase" of general relativity!

An important step in probing the "unbroken phase" of string theory was the development of topological field theory. Witten originally created topological field theory [1] as an attempt to use the sigma model as a tool to construct topological invariants for manifolds, using the input of physics to solve problems in pure topology. Specifically, quantum field theory was used to correct certain weaknesses in Morse theory.

Given the success of the sigma model as a new mathematical tool, Floer [2–3] was then able to generalize Witten's formulation to include the Chern–Simons Yang–Mills theory in three dimensions. This, in turn, gave rise to a powerful formalism by which to construct new topological invariants in three dimensions.

Independently, working in four dimensions, Donaldson [4] startled the world of mathematics by creating new topological invariants in four dimensions using the input of physics: exploiting the instanton solutions of four-dimensional Yang–Mills theory. It had been known for decades that manifolds in $D = 3, 4$ behaved in qualitatively different ways than in higher dimensions $D \geq 5$. For example, in Smale's celebrated proof, the Poincaré conjecture could be demonstrated for $D \geq 5$, but attempts to understand the nature of the Poincaré conjecture for lower dimensions met with frustration. Thus, Donaldson's use of Yang–Mills theory to settle a long-standing problem in mathematics created quite a sensation in the world of mathematics. His instanton formulation not only showed that the Poincaré conjecture failed in four dimensions and that "exotic four-spheres" existed, his new polynomial invariants allowed one to distinguish between new classes of four-manifolds.

At about the same time, Jones [5] was able to write new polynomial invariants for knots, making the first significant advance in knot theory in decades. Topology in lower dimensions was thus experiencing a renaissance, but these advances were occurring in a variety of scattered, unrelated directions, without any unifying theme or picture.

Given this renewed interest in topological invariants, Atiyah [6] then asked several questions: Could quantum field theory be used to give a unifying approach to all these disparate results? In particular, could a quantum field theory be found in four dimensions to explain the new topological invariants of Donaldson that generalizes the three-dimensional theory of Floer? Also, could a quantum field theory be found in three dimensions that generates the polynomial invariants of Jones?

If so, then quantum field theory would also solve a nagging defect in these purely topological formulations, that is, the inability to solve for explicit, analytic forms for the various topological invariants. Quantum field theory, however, might give a specific algorithm by which to calculate the numerical value of these invariants and perhaps generate new classes of them.

Witten showed that the answer to all these questions was yes [7, 8]. In fact, the theories of Jones, Floer, and Donaldson gave rise to two classes of topological field theories:

1. metric-free topological models, where the theory is manifestly free of any metric dependence; and
2. cohomological topological field theories, where a background field may be present but the energy-momentum tensor is BRST trivial.

We now turn to a discussion of these two approaches.

14.2. Topology and Morse Theory

Over the decades, it has become common knowledge that generally covariant theories can be created by introducing a metric tensor and then functionally integrating over all possible metric tensors in the functional measure. This is the way general covariance is implemented in relativity. However, this neglects important classes of theories in which general covariance is implemented in an entirely different fashion, such as theories that lack a metric tensor entirely or theories with a metric in which the Green's functions are independent of the metric.

These theories are called topological field theories, and have novel properties that separate them from all other quantum field theories. Because the Green's functions are independent of the choice of metric, it means that they must be purely topological, generating numerical values of topological invariants for certain manifolds. But, this also means that they might possess a finite number of degrees of freedom, unlike the infinite number of degrees of freedom found even for the simplest point particle quantum field theory.

The first class of topological field theories, which are manifestly metric free, include the three-dimensional Chern–Simons theory found in Chapter 8 in our discussion of knot theory, where the Lagrangian is given by:

$$L = \frac{k}{8\pi} \int \epsilon^{ijk} \operatorname{Tr}\left[A_i(\partial_j A_k - \partial_k A_j) + \frac{2}{3} A_i[A_j, A_k] \right] \qquad (14.2.1)$$

The action is manifestly locally generally covariant (because ϵ^{ijk} transforms as a density under coordinate transformations), yet it contains no metric whatsoever to tell us how to define the light cone of Minkowski space. Therefore, as we have seen in Chapter 8, the correlation functions are given by Wilson loops. By numerically evaluating these correlation functions, we generate new classes of knot polynomials, generalizing the Jones' polynomials.

Another example of a metric-free topological model is 2+1 gravity [9], whose action is given by:

$$L = \epsilon^{abc}\epsilon^{ijk}e_i^a R_{jk}^{bc}(\omega)$$
$$R_{jk}^{bc} = \partial_j \omega_k^{bc} + \omega_j^{be}\omega_k^{ec} - (j \leftrightarrow k) \tag{14.2.2}$$

This model is usually thought to be both trivial and non-renormalizable. However, upon closer inspection, we see that these two attributes are actually mutually exclusive.

In fact, upon closer examination, one can also show that $2+1$ dimensional gravity, reinterpreted in this fashion, is equivalent to Chern–Simons $2+1$ Yang–Mills theory [9]. The theory is hence exactly solvable. The correlation functions of the theory, not surprisingly, are knot invariants, which are topological and do not depend upon the metric of space–time.

The second type of topological field theory, in which a metric explicitly appears but where the Green's functions are independent of the metric, is much more complicated but also much richer in mathematical content. To understand cohomological topological field theories, it is first important to understand how they evolved out of an attempt to use quantum field theory to solve problems in pure topology and Morse theory.

To understand the significance of applying quantum field theory to Morse theory, let us quickly (and not very rigorously) review some of the highlights of de Rahm cohomology, which is how topological invariants can be constructed for real manifolds. Usually, when analyzing the topological invariants of a real manifold M, we traditionally begin with de Rahm cohomology, rather than Morse theory. On the manifold, we define p forms as follows:

$$\omega = \omega_{\mu_1,\mu_2,\ldots,\mu_p}\, dx^{\mu_1} \wedge dx^{\mu_2} \wedge \cdots \wedge dx^{\mu_p} \tag{14.2.3}$$

The operator d acting on this form is defined as follows:

$$d\omega = \frac{\partial \omega_{\mu_1,\mu_2,\ldots,\mu_p}}{\partial x^j}\, dx^j \wedge dx^{\mu_1} \wedge dx^{\mu_2} \wedge \cdots \wedge dx^{\mu_p} \tag{14.2.4}$$

where d is nilpotent, that is, $d^2 = 0$.

We then define topological invariants on the manifold. The space of independent forms that are annihilated by d is denoted by $\ker d$. However, we wish to subtract those forms that are themselves expressed as $d\tilde{\omega}$ for some $\tilde{\omega}$, that is, we wish to remove those states that are the image of d. Thus, we define the nth cohomology as:

$$H^n = \ker d / \operatorname{im} d \tag{14.2.5}$$

for n forms.

The dimension of the de Rahm cohomology is given by the Betti numbers:

$$b_n = \dim H^n \tag{14.2.6}$$

Notice that the de Rahm cohomology depends on the local, differential properties of the manifold. On this smooth manifold, we can also define a homology, which depends on the global properties of the manifold. Let us define ∂ to be the boundary operator, that is, it maps a manifold M into its boundary manifold ∂M. Then, we can also show that the boundary operator is also nilpotent: $\partial^2 = 0$.

We can similarly define a homology on the manifold as follows:

$$H_n = \ker \partial / \operatorname{im} \partial \qquad (14.2.7)$$

The correspondence between cohomology and homology is made by showing that the boundary operator is dual to the operator d. This is done via the Stoke's theorem:

$$\int_M d\omega = \int_{\partial M} \omega \qquad (14.2.8)$$

To make this correspondence transparent, let us define the "scalar product" between a manifold M and a form ω:

$$\langle M|\omega \rangle \equiv \int_M \omega \qquad (14.2.9)$$

Then, Stoke's theorem can be rewritten as:

$$\langle M|d\omega \rangle = \langle \partial M|\omega \rangle \qquad (14.2.10)$$

that is, by moving the d operator from the left-hand to the right-hand side of the scalar product, it has become ∂. Thus, under very general conditions, these two operators can be shown to be dual to each other. We call d the coboundary operator. This, in turn, means that the two spaces, homology and cohomology, are dual to each other and have the same dimension.

Thus, the Betti number can be written as:

$$b_n = \dim H_n = \dim H^n \qquad (14.2.11)$$

From this, we can write the Euler characteristic, a topological invariant:

$$\chi(M) = \sum_{q=0} (-1)^q b_q(M) \qquad (14.2.12)$$

Last, one can also define the Laplacian $dd^* + d^*d$. A form ω is called *harmonic* if it satisfies:

$$(dd^* + d^*d)\omega = 0 \qquad (14.2.13)$$

Then, it can be shown that the Betti number b_q is also equal to the number of independent harmonic q forms that one can write on the manifold. In summary, the de Rahm theory allows us to write topological invariants of manifolds based upon examining either the local or global properties of the manifold via the kernel of nilpotent operators d or ∂.

However, there is also another, less powerful method by which to analyze the topological invariants of a manifold, and this is Morse theory, which differs markedly from de Rahm theory. Morse theory is not based on nilpotent operators, but on analyzing the critical points of a certain function defined on a manifold.

To be a little more precise, let h be a function defined on the manifold, that is, a mapping of M onto the real numbers. Let P_i be the critical points of this function, that is,

$$\partial h(P_i)/\partial x_k = 0 \qquad (14.2.14)$$

for $k = 1, 2, \ldots, \dim M$. Let us now define the Hessian as the matrix:

$$\partial^2 h(x)/\partial x_i\, \partial x_j \qquad (14.2.15)$$

We define the *Morse index* $\mu(P_i)$ at the critical point P_i as the number of negative eigenvalues of the Hessian of h. We then define M_p as the number of critical points with the Morse index p. Then, one of the results of Morse theory is the inequality relating M_q of Morse theory and b_q of de Rahm theory:

$$M_p \geq b_p \qquad (14.2.16)$$

In terms of Morse theory, the Euler characteristic can be shown to equal:

$$\chi(M) = \sum_{q=0}(-1)^q M_q \qquad (14.2.17)$$

We see, therefore, that Morse theory differs qualitatively from de Rahm theory. Morse theory depends on the properties of a manifold at its critical points, rather than the cohomology or homology of nilpotent boundary operators.

Unfortunately, Morse theory is not powerful enough to calculate the Betti numbers in terms of operators defined on Morse theory. Morse theory is weaker than de Rahm theory, yielding mainly inequalities. Let us illustrate this new approach with an example [10].

Example: Gravitational Potential

The simplest example of Morse theory is a two-manifold M placed in the earth's gravitational field. We then choose $h(x)$ to be the height of the point x (that is, the gravitational potential). At every point x on the surface of the manifold, we can assign a real number $h(x)$, its height off the ground.

For example, imagine a torus with two handles so that it appears upright, as in a figure eight. This torus has six critical points at which the derivative of the height function at these points is zero. Each loop has two critical points, and there is a critical point at the very top of the figure eight and one at the very bottom. By taking the second derivative of the height function at these critical points, we obtain the Hessian and can then calculate the number of negative eigenvalues at each critical point. Physically, this corresponds to finding the points on the surface where a marble could be placed in stable or unstable equilibrium.

The lowest critical point at the bottom is stable and has Morse index 0. The number of critical points with index 0 is 1, so $M_0 = 1$. The highest point on the surface is unstable and has Morse index 2. The number of critical points with index 2 is also 1, so $M_2 = 1$. The other 4 critical points, located in the holes, are saddle points, with one stable and one unstable direction, so they have Morse index 1. Since there are 4 such saddle points, $M_1 = 4$. Putting these all together, we then find that the Euler characteristic is:

$$\chi = M_0 - M_1 + M_2 = 1 - 4 + 1 = -2 \qquad (14.2.18)$$

which is indeed the Euler characteristic for a torus of genus g found from de Rahm theory.

This can also be easily generalized to two-dimensional surfaces of arbitrary genus g. We know that the Betti numbers for this surface are given by:

$$b_0 = b_2 = 1, \qquad b_1 = 2g \qquad (14.2.19)$$

so that the Euler characteristic is given by

$$\chi(M) = 2(1 - g) \qquad (14.2.20)$$

If we now compare this to the negative eigenvalues of the height function, we find an exact correspondence. We find that the Betti numbers b_q equal the M_q for this surface.

For more complicated manifolds, however, this exact correspondence between Betti numbers and M_q breaks down and only the weak Morse inequalities apply. In particular, we are unable to calculate exact expressions for the Betti numbers via Morse theory.

However, we will now greatly expand the power of Morse theory by using techniques from an unexpected source: supersymmetric quantum field theory. Using the input of physics, we will now show how to improve upon the old Morse theory by calculating the exact expression for the Betti numbers.

We begin by defining a modified set of coboundary operators as a function of some fictitious "time" parameter t:

$$\begin{aligned} d_t &\equiv e^{th} \, d \, e^{-th} \\ d_t^* &\equiv e^{th} \, d^* e^{-th} \end{aligned} \qquad (14.2.21)$$

for some Morse function $h(x)$. We then define $b_q(t)$ to be a t-dependent Betti number:

$$b_q(t) \equiv \dim \ker(d_t d_t^* + d_t^* d_t) \qquad (14.2.22)$$

Although $b_q(t)$ is a t-dependent number, it is also a discrete function and is therefore independent of t. Thus, we also have $b_q(t) = b_q(0)$.

Last, we also define the t-dependent Hamiltonian as half of the t-dependent Laplacian:

$$H_t = \frac{1}{2}(d_t d_t^* + d_t^* d_t) \qquad (14.2.23)$$

Notice that the number of zero energy states of this Hamiltonian at $t = 0$ is also the number of harmonic forms and hence equals the Betti number. Thus, the Betti number counts the number of zero energy states of the Hamiltonian at $t = 0$.

To see how the number of zero energy states changes for finite t, let us power expand the operator d_t as a function of t:

$$d_t = d + ta^{*i}(\partial h/\partial x^i) + \cdots$$
$$d_t^* = d^* + ta^i(\partial h/\partial x^i) + \cdots \qquad (14.2.24)$$

where we have introduced a^i via:

$$dw = a^{*i} \frac{\partial w}{\partial x^i}$$
$$d^* w = a^i \frac{\partial w}{\partial x^i} \qquad (14.2.25)$$

and:

$$\{a^i, a^{*j}\} = g^{ij} \qquad (14.2.26)$$

Putting this back into the Hamiltonian acting on a form w, we find the expansion in t:

$$2H_t w = (dd^* + d^* d)w + t^2 g^{ij} \frac{\partial h}{\partial x^i} \frac{\partial h}{\partial x^j} w + t[a^{*i}, a^j]D_i D_j hw \qquad (14.2.27)$$

where D_i is the covariant derivative with respect to g^{ij}.

To lowest order in t, we have shown that the t-dependent Hamiltonian contains a term proportional to the square of the gradient of h. The minima of the Hamiltonian therefore correspond to the critical points of the Morse function. If we power expand around one of these critical points, we find:

$$h(x) = h(0) + \lambda_i x_i^2/2 + O(x^2)$$
$$2H_t = \sum_i \left(-\frac{\partial^2}{\partial x_i^2} + t^2 \lambda_i^2 x_i^2 + t\lambda_i [a_i^*, a_i] \right) \qquad (14.2.28)$$

where λ_i is the Morse index at the critical point.

The first two terms define the usual harmonic oscillator theory. The second term is also easily analyzed by noticing that:

$$[a_i^*, a_i]dx^{\mu_1} \wedge \cdots \wedge dx^{\mu_p} = \pm dx^{\mu_1} \wedge \cdots \wedge dx^{\mu_p} \qquad (14.2.29)$$

(The eigenvalue is $+1$ if i is one of the indices μ_j appearing in the volume element and -1 if it is not.)

The energy is therefore given by the energy of a series of uncoupled harmonic oscillators plus a correction factor:

$$E_t = \frac{1}{2} t \sum_i [|\lambda_i|(1 + 2N_i) + \lambda_i n_i] + O(t^0) \qquad (14.2.30)$$

where $n_i = \pm 1$.

We are interested in the number of zero-energy solutions. The energy is zero if $N_i = 0$ and $n_i = -\text{sign}\,\lambda_i$. However, we recall that the number of negative eigenvalues λ_i is the Morse index p. Each critical point of the Hamiltonian thus defines a wave function whose energy is zero to order t, and there are M_p such wave functions.

However, in the limit as t approaches zero, some of these zero-energy states receive positive energy contributions, and hence the number of zero-energy states decreases. But, in the zero t limit, the number of zero-energy states equals b_q, as we saw earlier. Thus, the number of zero-energy states at zero t (the Betti number) is less than the number of zero-energy states at small but finite t (M_q):

$$b_q \leq M_q \tag{14.2.31}$$

which is the Morse inequality derived from quantum mechanics.

In summary, this simple quantum mechanical model reproduces a proof of the known Morse inequality conditions. However, using the power of supersymmetry, we can do even better than this. We can go beyond the standard inequalities of Morse theory and generate new information, that is, we can calculate the Betti numbers in terms of Morse theory.

14.3. Sigma Models and Floer Theory

To be specific, we will start with a supersymmetric nonlinear sigma model, with a metric $g_{ij}(\phi)$, which is a function of a scalar field ϕ^i. We also introduce the Morse function $h(\phi)$. In supersymmetric language, we introduce the superfield Φ^i:

$$\Phi^i = \phi^i + \bar\theta\psi^i + \bar\theta\theta F^i/2 \tag{14.3.1}$$

The action is then [1]:

$$
\begin{aligned}
S &= \int d^2x\, d^2\theta \left[g_{ij}(\Phi)\bar D\Phi^i\, D\Phi^j + h(\Phi) \right] \\
&= \frac{1}{2} \int d^2x \left[g_{ij}(\phi)\partial_\mu\phi^i\, \partial^\mu\phi^j + i g_{ij}(\phi)\bar\psi^i\gamma^\mu\, D_\mu\psi^j \right. \\
&\quad \left. + \frac{1}{12}R_{iklj}(\phi)\bar\psi^i\psi^l\bar\psi^k\psi^j - g^{ij}(\phi)\frac{\partial h}{\partial\phi^i}\frac{\partial h}{\partial\phi^j} - D_iD_j\, h(\phi)\bar\psi^i\psi^j \right] \\
D_\mu\psi^i &= \partial_\mu\psi^i + \Gamma^i_{jk}(\phi)\partial_\mu\phi^j\psi^k
\end{aligned}
$$
$$\tag{14.3.2}$$

We can make contact with Morse theory by making the following identification between cohomology operators d acting on p forms and the supersymmetric operator Q of the sigma model acting on states with fermion number F:

$$(-1)^p \leftrightarrow (-1)^F$$
$$d \leftrightarrow Q$$
$$d^* \leftrightarrow Q^*$$
$$dd^* + d^*d \leftrightarrow 2H = \{Q, Q^*\}$$

$$(14.3.3)$$

The supersymmetric sigma model thus gives us a specific realization of Morse theory. With this identification of Q as a cohomology operator d, we can use the previous discussion to prove the Morse inequalities.

Our discussion so far has been perturbative. Now, however, we will take the formalism one step further, by analyzing tunneling between different critical points. Supersymmetry will guarantee that the energy of the Hamiltonian will vanish to all orders in perturbation theory; however, tunneling will in general lift the some of the degeneracies among the zero-energy states. Tunneling via instantons can remove some of the critical points. The key point is that if we can calculate the number of critical points removed by tunneling, then we can calculate corrections to Eq. (14.2.31) and hence b_p itself.

Let $|P_i\rangle$ represent the perturbative vacuum defined at one of the critical points P_i. We wish to calculate the matrix element between different critical points $\langle P_i|d_t|P_j\rangle$. Normally, this matrix element is zero (because of the presence of fermionic zero modes). However, nonperturbative instanton effects can render the matrix element nonzero if the instanton effect produces a fermion zero mode, which is absorbed by d_t.

At this point, we use standard instanton arguments to calculate this amplitude. We wish to find solutions $x(\tau)$ that take us from one critical point to another, parameterized by some τ. Thus, we wish to find the equation for $x(\tau)$ connecting two critical points, that is, $x(-\infty) = P_i$ and $x(\infty) = P_j$.

The equation for the instanton is:

$$\frac{dx^i(\tau)}{d\tau} = -g^{ij}[x(\tau)]\frac{\partial h[x(\tau)]}{\partial x^j}$$

$$(14.3.4)$$

Given the instanton solution connecting two critical points, we can use tunneling arguments to show that the matrix element connecting the two critical points is given by:

$$\langle P_i|d_t|P_j\rangle = n(P_i, P_j)\exp\left\{-t[h(P_i) - h(P_j)]\right\}$$

$$(14.3.5)$$

where $n(P_i, P_j)$ is an integer that is computable once we are given the Morse function h.

So far, our discussion has been rather general. Now, we come to the key point of our discussion: we will define a new cohomology operator δ that will establish the link between the Betti numbers and Morse theory.

Let us first define W_p to be the set of eigenstates $|P\rangle$ such that $\mu(P) = p$ for some integer p. Let us define a new cohomology operator δ, defined in terms of the integer $n(P_i, P_j)$, which is given by:

$$\delta|Q\rangle = \sum_{P\in W_{p+1}} n(Q,P)\,|P\rangle \qquad (14.3.6)$$

for $Q \in W_p$. Notice that the operator δ takes us from W_p to W_{p+1}. We can also show that $\delta^2 = 0$ and that it satisfies all the properties of a standard cohomology.

Given this new cohomology operator δ, we can then define the Betti number as:

$$b_p = \dim\left([\ker\delta/\operatorname{im}\delta]\cap W_p\right) \qquad (14.3.7)$$

We have now succeeded in our goal [1] of defining the Betti number of the manifold totally in terms of Morse theory. Using the tool of the super-symmetric sigma model, we have converted the old Morse inequalities into precise identities.

Let us now generalize our discussion to the case of more complicated manifolds in three and four dimensions. By now, we see a strategy emerging for using quantum field theory to calculate the Morse invariants W_q.

1.   First, start with a supersymmetric quantum field theory where we can define a Morse function $h(\Phi)$ and identify the supersymmetric operator Q with the nilpotent coboundary operator d.
2.   Construct the t-dependent Hamiltonian as a function of the Laplacian. At $t = 0$, the number of zero energy states equals the number of harmonic forms, or the Betti number.
3.   Calculate the Morse number M_p as the number of zero-energy states and compare it to the zero-energy states of the $t = 0$ Hamiltonian, which are the Betti numbers.
4.   Using instanton methods, calculate the transition matrix element between different critical points, and from this, define a new nilpotent δ whose cohomology gives us the Betti numbers directly in terms of Morse theory.

Let us now follow these simple steps and sketch how this approach can be applied to the case of the Yang–Mills theory. This will, in turn, give us entirely new topological invariants defined in three and four dimensions.

We begin by taking the h function to be the Chern–Simons action in three dimensions [2]:

$$h(A) = \int \operatorname{Tr}\left(A\wedge dA + \frac{2}{3}A\wedge A\wedge A\right) \qquad (14.3.8)$$

The critical points of h are found by taking functional derivatives of the action:

$$\partial h(A)/\partial A_i^a(x) = -\epsilon^{ijk}F_{jk}^a(x)/2 = -B_i^a(x) \qquad (14.3.9)$$

Because the derivatives are proportional to the curvature tensor, the critical points correspond to the space of connections where the curvature vanishes, that is, the space of flat connections. We will call the space of flat connections, modulo gauge transformations, *Floer complexes*.

Repeating the same steps as before, we find that the cohomology operators are given by:

$$d = \int d^3x \, \psi_i^a \, \frac{\delta}{\delta A_i^a(x)}$$
$$d^* = -\int d^3x \, \bar{\psi}_i^a(x) \, \frac{\delta}{\delta A_i^a(x)} \qquad (14.3.10)$$

Generalizing Eq. (14.2.21), we now define e-dependent cohomology operators as:

$$d_e \equiv e^{-h/e^2} \, d \, e^{h/e^2}$$
$$d_e^* \equiv e^{h/e^2} \, d^* \, e^{-h/e^2} \qquad (14.3.11)$$

We introduce the Hamiltonian via:

$$2e^{-2} H_e \equiv d_e d_e^* + d_e^* d_e \qquad (14.3.12)$$

Written out explicitly, we have:

$$H = \int d^3x \, \mathrm{Tr} \left(e^2 \Pi_i^{a\,2} + e^{-2} B_i^{a\,2} + \epsilon^{ijk} \psi_i D_j \bar{\psi}_k \right) \qquad (14.3.13)$$

and:

$$\Pi_i^a(x) = -i \, \frac{\delta}{\delta A_i^a(x)} \qquad (14.3.14)$$

As before, we can construct W_q as the number of zero-energy eigenfunctions of the Hamiltonian for finite e and b_q as the number of zero-energy eigenfunctions at $e = \infty$.

We must now analyze whether nonperturbative effects play an important role. Let us analyze tunneling effects that connect different critical points. As before, the tunneling effects are determined by a differential equation in a fictitious parameter τ connecting the different critical points. The analog of Eq. (14.3.4) is:

$$\partial A_i^a(x, \tau)/\partial \tau = B_i^a(x, \tau) \qquad (14.3.15)$$

(This equation has a simple meaning. If we identify τ as a time coordinate, then it is easy to see that the previous equation sets the electric field proportional to the magnetic field, that is, $F = -F^*$ and the curvature is self-dual. Thus, the tunneling effects are described by the standard instanton effects found in ordinary Yang–Mills theory.)

We conclude this discussion by noting that Floer [2] was able to use the Yang–Mills theory to construct a new boundary operator ∂ that was nilpotent and to define a new homology group, called the Floer group. From these, he was able to construct new topological invariants for three-manifolds, using physics as the crucial input in a purely topological formulation.

Floer's discussion, however, was incomplete because it left open the possibility of a generalization to a fully four-dimensional-type formulation. Atiyah then conjectured [6], and Witten later proved [7], that a four-dimensional generalization of Floer theory should give the topological polynomial invariants of Donaldson. We now turn to a discussion of how to generalize this formulation to four dimensions and to a wide variety of other theories.

14.4. Cohomological Topological Field Theories

Other topological models, in which general covariance is exact, are the cohomological models [7], where a background metric may be present, but where the Lagrangian is given by:

$$L = 0 \qquad (14.4.1)$$

or a topological invariant, such as:

$$L \sim F \wedge \tilde{F} \qquad (14.4.2)$$

This may appear strange, because then the system appears to be empty. Although the action is zero, the "physics" of the theory is to be found entirely in the field content and its gauge variation. We will find that, after gauge fixing the gauge fields, a nilpotent BRST operator Q arises, and the gauge fixed action, with its Faddeev–Popov term, is given by:

$$L_{\text{GF+FP}} = \{Q, V\} \qquad (14.4.3)$$

where V is some field composed out of the original fields and their ghosts.

Most important, we will then take the variation of the Lagrangian with respect to the background metric to derive the energy-momentum tensor. Because the gauge fixed action is itself BRST invariant, we find that the variation of the action with respect to the background metric yields the energy-momentum tensor $T^{\alpha\beta}$ via:

$$\delta L = \frac{1}{2} \int \sqrt{g}\, \delta g^{\alpha\beta} T_{\alpha\beta} \qquad (14.4.4)$$

and:

$$T_{\alpha\beta} = \{Q, V_{\alpha\beta}\} \qquad (14.4.5)$$

for some field $V_{\alpha\beta}$.

This last statement, that the energy-momentum tensor is BRST trivial, is one of the most important features of the cohomological topological field theories. Since BRST trivial operators vanish when inserted into a correlation function, it means that we are free to vary the background metric $g^{\mu\nu}$ without changing the theory, that is, the theory is locally generally covariant.

For example, let us insert an operator O into a path integral:

$$Z_0 = \int D\phi\, e^{i \int L d^4 x} \{O\} \qquad (14.4.6)$$

Let us now make a small BRST variation of the path integral labeled by ϵ, where the action and the measure are both BRST invariant. The path integral becomes Z_ϵ, which equals Z. Then, we find:

$$
\begin{aligned}
0 = Z_\epsilon - Z_0 &= \int D\phi\, e^{\epsilon Q} e^{i \int L d^4 x} \{O\} - Z_0 \\
&= \int D\phi\, e^{i \int L d^4 x} \left(O + \epsilon\{Q, O\}\right) - Z_0 \qquad (14.4.7) \\
&= \int D\phi\, e^{i \int L d^4 x} \epsilon\{Q, O\}
\end{aligned}
$$

Thus,

$$\langle\{Q, O\}\rangle = 0 \qquad (14.4.8)$$

for any field O. (Another, more intuitive way in which to see that cohomological topological field theories are independent of the choice of metric is to notice that the metric tensor enters the theory through BRST gauge fixing. Since the metric is introduced as a gauge artifact, and since the physical properties of a field theory are always independent of gauge fixing, the cohomological topological theories must also be independent of the metric.)

One of the most important examples of such a cohomological theory is a four-dimensional topological Yang–Mills theory, which resembles a twisted version of supersymmetric $N = 2$ gauge theory and is the four-dimensional extension of the three-dimensional Floer theory. There are many ways to approach the quantization of this topological Yang–Mills theory. We will explore just a few of them.

We begin with a theory of zero action, but then postulate that this action is invariant under the following gauge transformation:

$$\delta A_\mu^a = \psi_\mu^a \qquad (14.4.9)$$

Upon first glance, this is a highly unusual gauge transformation, much larger than the usual $SU(N)$ gauge transformation. Because ψ_μ^a has the same number of indices as A_μ^a, it implies that we can use ψ_μ^a to eliminate all the fields contained within A_μ^a, that is, the theory is vacuous. However, several nontrivial features begin to emerge when we gauge fix this seemingly trivial theory with zero action [11–13].

First, let us choose a gauge so that the $F_{\mu\nu}^a$ is self-dual:

$$\text{gauge choice}: \quad F_{\mu\nu}^a - \frac{1}{2}\epsilon_{\mu\nu\alpha\beta} F^{\alpha\beta a} = 0 \qquad (14.4.10)$$

Naively, we might believe that this gauge completely determines all fields of the theory, and hence, the theory is again vacuous. However, there is a subtle point here that will prove crucial in our later discussion. Although demanding that the curvature be self-dual fixes the infinite degrees of freedom within our fields, it is not sufficient to fix all *finite* degrees of freedom. As is well known in gauge theory, there are nontrivial solutions to the self-dual equation, given by *instantons*. In other words, after gauge fixing there are still finite degrees of freedom left in the theory given by the space of instantons.

The space of parameters necessary to label one-to-one the space of instantons is called the "moduli space" of instantons (which in turn is intimately linked to the Donaldson polynomials). In fact, for each cohomological topological field theory, we will find that gauge fixing leaves finite degrees of freedom labeled by some moduli space.

There is also a second complication, however. Since our theory is locally gauge invariant, we demand that our field transform under local $SU(N)$, so that the total variation is

$$\delta A_\alpha^a = (D_\alpha \phi)^a + \psi_\alpha^a \qquad (14.4.11)$$

where ϕ^a is a $SU(N)$ gauge parameter. Notice, however, that it is possible to absorb the ϕ^a term completely into the ψ_α^a term. This means that the gauge parameter ψ_α^a has its own hidden gauge symmetry given by:

$$\delta \psi_\alpha^a = (D_\alpha \phi)^a \qquad (14.4.12)$$

Because of the tight relationship between the two gauge parameters ϕ^a and ψ_α^a, the Faddeev–Popov ghosts arising from gauge fixing will themselves have ghosts. In general, for complicated gauge choices, we will have to use the BV "ghosts-for-ghosts" quantization method described in Appendix 1. However, we will choose a simple enough gauge so that only second generation ghosts are required, so that the Faddeev–Popov prescription is adequate.

With these preliminaries, let us begin the quantization of the topological theory with zero action. The usual prescription gives us the gauge fixing term and the Faddeev–Popov ghost term:

$$L_{\text{GF+FP}} = (i/8)\alpha_0 B^{\alpha\beta} B_{\alpha\beta} + (i/4) B^{\alpha\beta}(F_{\alpha\beta} + \tilde{F}_{\alpha\beta}) - i\chi^{\alpha\beta} D_\alpha \psi_\beta \quad (14.4.13)$$

where α_0 is a constant, χ and ψ are the standard Faddeev–Popov ghosts, such that χ is self-dual, and $B_{\alpha\beta}$ is a self-dual auxiliary field.

Notice that we can, in turn, write this action as the off-shell, nilpotent BRST variation of the following term:

$$L_{\text{GF+FP}} = (i/4)\delta_1 \left[\chi^{\alpha\beta}(F_{\alpha\beta} + \tilde{F}_{\alpha\beta} + (1/2)\alpha_0 B_{\alpha\beta}) \right] \qquad (14.4.14)$$

where:

$$\delta_1 A_\alpha^a = \psi_\alpha^a$$
$$\delta_1 \psi_\alpha^a = 0$$
$$\delta_1 \chi^{\alpha\beta a} = B^{\alpha\beta a} \qquad (14.4.15)$$
$$\delta_1 B^{\alpha\beta a} = 0$$

(At this point, we have the option of solving for the $B_{\alpha\beta}$ field via its equations of motion, so the action reduces to the the square of the self-dual condition on the Yang–Mills field. However, the BRST invariance only holds on-shell, and we must use the quantization procedure described in the Appendix 1.)

As we noted earlier, there is still a hidden symmetry in the theory because the ghost field ψ_α^a has its own remaining ghost symmetry, parameterized by the ghost field ϕ^a. The action possesses a hidden symmetry:

$$\delta_G \psi_\alpha^a = i(D_\alpha \phi)^a$$
$$\delta_G B^{\alpha\beta a} = -ie_0[\phi, \chi^{\alpha\beta}]^a \qquad (14.4.16)$$

The field ψ has four degrees of freedom, but χ has only three. We find that the "ghosts" themselves require more Faddeev–Popov fixing.

To fix the remaining symmetry within the anticommuting Faddeev–Popov fields, we must introduce a set of commuting Faddeev–Popov fields, ϕ and λ, and an anticommuting field η. Let us therefore introduce a new gauge fixing action with more Faddeev–Popov terms:

$$L'_{\text{GF+FP}} = \delta_{\text{BRST}}\Big[ic_0\lambda(D_\alpha\psi^\alpha + sb) + c_1\chi^{\alpha\beta}B_{\alpha\beta}\Big] \qquad (14.4.17)$$

where $\delta_{\text{BRST}} = \delta_1 + \delta_G$.

Let us now write the full action:

$$
\begin{aligned}
(L + L')_{\text{GF+FP}} = {}& -i\chi^{\alpha\beta}D_\alpha\psi_\beta - i\eta D^\alpha\psi_\alpha + (1/2)\lambda D^\alpha D_\alpha\phi \\
& - (i/2)e_0\lambda[\psi^\alpha, \psi_\alpha] - (i/8)e_0\phi[\chi^{\alpha\beta}, \chi_{\alpha\beta}] \\
& + se_0\big[i\phi[\eta, \eta] + (e_0/4)[\phi, \lambda]^2\big] + (1/8)B^{\alpha\beta}B_{\alpha\beta} \\
& + (i/4)B^{\alpha\beta}(F_{\alpha\beta} + \tilde{F}_{\alpha\beta}) \\
= {}& (1/8)(F + \tilde{F})^2 - i\chi^{\alpha\beta}D_\alpha\psi_\beta - i\eta D^\alpha\psi_\alpha \\
& + (\lambda/2)D^\alpha D_\alpha\phi - (i/2)e_0\lambda[\psi^\alpha, \psi_\alpha] - (i/8)e_0\phi[\chi^{\alpha\beta}, \chi_{\alpha\beta}] \\
& + se_0\big[i\phi[\eta, \eta] + (e_0/4)[\phi, \lambda]^2\big]
\end{aligned}
$$
$$(14.4.18)$$

In this action, we have replaced the b field in terms of a new field η, given by $b \equiv e_0[\phi, \eta]$. The gauge symmetry is maintained by having $\delta_G\lambda = 2\eta$ and $\delta_G\eta = -(i/2)e_0[\psi, \lambda]$. We have made this replacement in order to preserve a symmetry arising from the Floer three dimensional action, which is called U symmetry. With this replacement, the scaling dimensions and U weights of the fields $(A, \phi, \lambda, \psi, \chi)$ are given as (1, 0, 2, 1, 2) and (0, 2, -2, 1, -1), respectively. Also, we have made the choice $c_0 = -(1/2)$ and $c_1 = 1/8$. (We note that the action of Eq. (14.4.18) is not unique. By adding in a BRST variation of some arbitrary collection of fields, the theory remains the same. For example, the ϕ field is inert under the BRST variation.)

As we mentioned earlier, the hallmark of a cohomological topological theory is that its energy-momentum tensor is BRST trivial. By direct calculation, we can show that the energy momentum tensor is equal to:

$$
\begin{aligned}
T_{\alpha\beta} &= \{Q, \lambda_{\alpha\beta}\} \\
\lambda_{\alpha\beta} &= \frac{1}{2}\operatorname{Tr}\left(F_{\alpha\sigma}\chi_{\beta}^{\sigma} + F_{\beta\sigma}\chi_{\alpha}^{\sigma} - \frac{1}{2}g_{\alpha\beta}F_{\sigma\tau}\chi^{\sigma\tau}\right) \\
&\quad + \frac{1}{2}\operatorname{Tr}\left(\psi_{\alpha}D_{\beta}\lambda + \psi_{\beta}D_{\alpha}\lambda - g_{\alpha\beta}\psi_{\sigma}D^{\sigma}\lambda\right) \\
&\quad + \frac{1}{4}g_{\alpha\beta}\operatorname{Tr}\left(\eta[\phi, \lambda]\right)
\end{aligned}
\tag{14.4.19}
$$

By direct BRST gauge-fixing, we have therefore converted the original action, which was zero, into a BRST variation, Eq. (14.1.19). There is, however, yet another way in which to construct a topological theory, and this is through its relationship to $N = 2$ supersymmetry.

If we analyze the field content of the previous theory, we see that it is equivalent to an $N = 2$ supersymmetric Yang–Mills theory, but with an important difference. An $N = 2$ theory possesses two spinorial supersymmetry generators, Q_{α}^{i}, where α is a spinor index. Our goal is to rewrite this theory such that we extract a single fermionic, nilpotent Lorentz scalar Q out of the supersymmetric generators.

Normally, this is impossible, because an irreducible Lorentz spinor does not contain any scalars. But, this can be changed if we add a twist to the theory, such that the energy-momentum tensor is altered so that one component of Q_{α}^{i} becomes a Lorentz scalar.

This twisting process is most easily represented on a two-dimensional theory. We begin with an $N = 2$ supersymmetry with an R symmetry associated with it, whose generator is R_{μ}. Then, the supersymmetric generator is $Q_{\alpha\pm}$, with commutation relations:

$$
\begin{aligned}
\{Q_{\alpha+}, Q_{\beta-}\} &= \gamma_{\alpha\beta}^{\mu}P_{\mu} \\
\{Q_{\alpha+}, Q_{\beta+}\} &= \{Q_{\alpha-}, Q_{\beta-}\} = 0
\end{aligned}
\tag{14.4.20}
$$

In two dimensions, a spinor has only two components, also labeled $\pm$, so that we have four nilpotent supercharges $Q_{\pm\pm}$.

The key step is that we will now modify the theory so that the energy-momentum tensor becomes:

$$
T'_{\mu\nu} = T_{\mu\nu} + \epsilon_{\mu\sigma}\,\partial^{\sigma}R_{\nu} + \epsilon_{\nu\sigma}\,\partial^{\sigma}R_{\mu}
\tag{14.4.21}
$$

Altering the energy-momentum tensor means that we are also altering the rotation group generator J, which only has one component, by $J' = J + R$.

In this new basis, with an altered rotation group generator, we find that Q_{-+} and Q_{+-} now transform as scalars. Since both are nilpotent, we now define the new BRST generator as:

$$Q_{\text{BRST}} \equiv Q_{-+} + Q_{+-} \qquad (14.4.22)$$

In this way, an $N = 2$ theory has now been modified so that a new scalar, nilpotent Q_{BRST} operator can be constructed, such that the action becomes a BRST commutator. It is thus not surprising that topological field theories have the same field content as $N = 2$ superfield theories, but with a different realization of supersymmetry and Lorentz invariance.

14.5. Correlation Functions

Because the underlying action is zero, one is tempted to conclude that the correlation functions must also be zero. However, this is not true. Because topological field theories are totally independent of the choice of background metric by construction, we find that the correlation functions reproduce the known topological invariants found by topologists. In fact, this was one of the original motivations for studying these topological field theories: to provide a quantum field theoretical framework in which to generate topological invariants.

Perhaps the most intriguing of these topological invariants are the Donaldson polynomials [4]. Surprisingly enough, Donaldson first analyzed instanton solutions to the Yang–Mills equation and their moduli space in order to construct his invariants.

Because topological gauge theory is based on self-dual fields, one can show that the correlation functions of the theory are precisely the Donaldson polynomials. To see this, we want to construct correlation functions among fields that are BRST invariant, but not BRST trivial. In other words, we want a field whose BRST variation is zero, but cannot be written as a BRST commutator (in which case, as we have seen, its correlation functions are exactly zero).

Examining the list of fields found in the topological gauge theory, we find that the only invariant field is ϕ. For a gauge invariant combination, we are led to choose the following gauge invariant, BRST invariant field at point P:

$$\begin{aligned} W_0(P) &= \frac{1}{2}\,\text{Tr}\,\phi^2 \\ \delta W_0(P) &= 0 \\ W_0 &\neq \{Q, V\} \end{aligned} \qquad (14.5.1)$$

for the group $SU(2)$. (For higher groups, there are obviously more gauge invariants one can construct via ϕ^a, corresponding to the number of Casimir invariants of the group.) The correlation functions we are interested in are:

$$Z(k) = \int D\phi\, e^{-I} \prod_{i=1}^{k} W_0(P_i) = \langle W_0(P_1) \cdots W_0(P_k) \rangle \qquad (14.5.2)$$

The next step is to show that this is really a topological invariant, that is, it is independent of the location of the points P_k. This is easily shown. Let us move the point P a small distance. Then, the variation of the W_0 is given by:

$$\frac{\partial}{\partial x^\alpha} W_0 = \frac{1}{2} \frac{\partial}{\partial x^\alpha} \left(\text{Tr}\, \phi^2 \right) = \text{Tr}\, \phi\, D_\alpha \phi = i\{Q, \text{Tr}\, \phi\psi_\alpha\} \qquad (14.5.3)$$

that is, the variation of W_0 is equal to a BRST commutator. Then,

$$W_0(P) - W_0(P') = \int_{P'}^{P} \frac{\partial W_0}{\partial x^\alpha}\, dx^\alpha = \{Q, \int_{P'}^{P} W_1\} \qquad (14.5.4)$$

where $W_1 = \text{Tr}(\phi\psi_\alpha)dx^\alpha$.

Then, the variation of the correlation function by moving the point P_1 is equal to the matrix element of a BRST commutator, so it vanishes:

$$\delta Z(k) = \langle \delta W_0(P_1) \cdots W_0(P_k) \rangle = \left\langle \left\{ Q, i \int_{P'}^{P} W_1 \prod_{j=1}^{k} W_0(P_j) \right\} \right\rangle$$

$$= 0$$

$$(14.5.5)$$

as desired.

Now that we have shown that the correlation functions composed of W_0 are topological invariants, let us construct a sequence of these BRST invariant (but BRST nontrivial) operators. We notice that:

$$0 = i\{Q, W_0\}, \qquad dW_0 = i\{Q, W_1\} \qquad (14.5.6)$$

Let us now extract, from W_1, a new BRST invariant operator W_2, which is BRST non-trivial and so on [7]:

$$\begin{aligned} dW_1 &= i\{Q, W_2\} \\ dW_2 &= i\{Q, W_3\} \\ dW_3 &= i\{Q, W_4\} \\ dW_4 &= 0 \end{aligned} \qquad (14.5.7)$$

where:

$$W_2 = \text{Tr}\left(\frac{1}{2} \psi \wedge \psi + i\phi \wedge F \right)$$

$$W_3 = i\,\text{Tr}\,(\psi \wedge F) \qquad (14.5.8)$$

$$W_4 = -\frac{1}{2}\,\text{Tr}\,(F \wedge F)$$

(Here, $dW_4 = 0$ because it is a five-form, which in four dimensions equals zero by the antisymmetry of dx^α.)

Let us now generalize these results for an arbitrary topological theory, where the invariant fields W_k obey $dW_k = i\{Q, W_{k+1}\}$. Then, the integral:

$$I(\gamma) = \int_\gamma W_k \qquad (14.5.9)$$

around a k-dimensional cycle γ is BRST invariant, that is,

$$\{Q, I\} = \int_\gamma \{Q, W_k\} = -i \int_\gamma dW_{k-1} = 0 \qquad (14.5.10)$$

This integral is only sensitive to the homology class of γ, that is, if we add to γ a boundary term $\partial\beta$, then it remains unaltered up to a BRST commutator:

$$\begin{aligned}
I(\gamma + \partial\beta) &= \int_{\gamma+\partial\beta} W_k = I(\gamma) + \int_\beta dW_k \\
&= I(\gamma) + i \int_\beta \{Q, W_{k+1}\} = I(\gamma)
\end{aligned} \qquad (14.5.11)$$

Now, we construct the following topological invariant:

$$Z(\gamma_1, \ldots, \gamma_r) = \int DX\, e^{-I} \prod_{i=1}^r \int_{\gamma_i} W_{k_1} = \left\langle \prod_{i=1}^r \int_{\gamma_i} W_{k_i} \right\rangle \qquad (14.5.12)$$

The above correlation function is gauge invariant, BRST invariant, and independent of the location of the points where W_{k_i} are located. It is only dependent on the homology cycles γ_i.

The four-dimensional invariants defined in Eq. (14.5.12) correspond to the Donaldson polynomials. Hence, quantum field theory yields a straightforward way in which to generate analytical expressions for these complicated polynomials.

To motivate this identification, let us note the following. Donaldson originally found his polynomials by examing the moduli space of instanon solutions to the Yang–Mills equation. The dimension of the moduli space of instantons is given by:

$$\dim M = 8p_1(M) - \frac{3}{2}[\chi(M) + \sigma(M)] \qquad (14.5.13)$$

where p_1 is the first Pontryagin index, χ is the Euler index, and σ is the signature index of the manifold M.

The moduli space of instantons arises when one looks for the solutions to the self-dual equation $F = -\tilde{F}$ in Yang–Mills theory. To find the moduli space, let us make small variations in the field A appearing in the self-dual equation:

$$\delta(F + \tilde{F}) \sim D_\alpha \delta A_\beta - D_\beta \delta A_\alpha + \epsilon_{\alpha\beta\gamma\delta} D^\gamma \delta A^\delta = 0 \qquad (14.5.14)$$

Of course, we want a solution to these equations modulo gauge transformations. Thus, to eliminate this redundancy, we break gauge invariance by imposing:

$$D_\alpha \delta A^\alpha = 0 \qquad (14.5.15)$$

But, notice that these equations, which define the moduli space of solutions to the self-dual equation, are equivalent to the equations for the ψ equation arising from our action in Eq. (14.4.18). The χ equation for ψ is:

$$D_\alpha \psi_\beta - D_\beta \psi_\alpha + \epsilon_{\alpha\beta\gamma\delta} D^\gamma \psi^\delta \qquad (14.5.16)$$

while the η equation for ψ is given by:

$$D_\alpha \psi^\alpha = 0 \qquad (14.5.17)$$

These equations are identical to the equations defining the moduli space of instantons. More precisely, the numer of zero modes of ψ minus the number of zero modes of η and χ equals the dimension of the space spanned by these solutions, which is identical to the moduli space of instantons. So, the sums of the fermion zero modes must equal $\dim M$.

However, the U number of ψ is equal to $+1$, while the U number of η and χ equals -1. Thus, we can alternatively count the U number of various operators to calculate the number of zero modes of these fermion fields and hence the dimension of moduli space.

The U number of W_{k_i} is equal to $4 - k_i$. Thus, the correlation function appearing in Eq. (14.5.12) must satisfy:

$$\sum_i (4 - k_i) = \dim M \qquad (14.5.18)$$

(If this equation is not satisfied, then the number fermion zero modes in the integration measure does not match the number of fermion zero modes in the integrand, so the correlation function vanishes.)

Last, we can functionally integrate out the nonzero modes appearing in the integration measure, consisting of the integration over the various fields appearing in the model. Once all the nonzero modes have been integrated out, we have:

$$Z(\gamma_1, \gamma_2, \ldots, \gamma_r) = \int \Phi^{(\gamma_1)} \wedge \Phi^{(\gamma_2)} \wedge \cdots \wedge \Phi^{(\gamma_r)} \qquad (14.5.19)$$

where each of the $\Phi^{(\gamma_i)}$ which remain after integrating over nonzero modes, is a $4 - k_i$ form. We have now written the correlation function in terms of an integral over the moduli space of instantons, which is the desired form for comparison to the Donaldson polynomials.

14.6. Topological Sigma Models

One of the objectives of topological field theory is to analyze the possible "unbroken phase" of string theory, where general covariance is unbroken. As a consequence, we will now apply our knowledge of topological gauge theory to write topological sigma models and topological gravity in two dimensions. Perhaps these will give us prototypes for the true theory that we are seeking.

As before, we start with a vanishing action, or an action that is purely topological, and proceed to quantize it. We start with the topological action [11, 14] defined for the two-dimensional field X_μ, where μ is a space–time index:

$$\int_\Sigma d^2 z\, \epsilon^{ab}\, J_{\mu\nu}\, \partial_a X^\mu\, \partial_b X^\nu = \int_T J \qquad (14.6.1)$$

where Σ is the two-dimensional world sheet, and T is space–time. We set

$$J = \frac{1}{2} J_{\mu\nu}\, dX^\mu \wedge dX^\nu \qquad (14.6.2)$$

$J_{\mu\nu}$ describes the almost complex structure of space–time, such that $dJ = 0$ and:

$$J_\mu^\nu J_\nu^\lambda = -\delta_\mu^\lambda \qquad (14.6.3)$$

As before, the entire field X_μ is a gauge field and can be eliminated. We now fix the gauge for the theory. Let us choose the self-dual condition:

$$\partial_a X^\mu + \epsilon_a^b J_\nu^\mu\, \partial_b X^\nu = 0 \qquad (14.6.4)$$

which eliminates all infinite degrees of freedom contained within the gauge fields (leaving only finite degrees of freedom).

We now add the gauge fixing term and the ghost contribution to the action in the usual way. There are no "ghosts-for-ghosts" to complicate our discussion. We add to the action:

$$L_{\mathrm{GF+FP}} = \frac{i}{2} \delta \left[\rho_\mu^a \left(\partial_a X^\mu + \epsilon_a^b J_\nu^\mu\, \partial_b X^\nu - \frac{1}{2} H_a^\mu \right) \right] \qquad (14.6.5)$$

where:

$$\rho_\mu^a = \epsilon_b^a J_\mu^\nu \rho_\nu^b \qquad (14.6.6)$$

and H_a^μ is a self-dual field. The transformations of the fields (in flat world sheet space) are given by:

$$\begin{aligned}
\delta_0 X^\mu &= i\epsilon \chi^\mu \\
\delta_0 \chi^\mu &= 0 \\
\delta_0 \rho_a^\mu &= \epsilon \left(H_a^\mu - i \Gamma_{\nu\kappa}^\mu \chi^\nu \rho_a^\kappa \right) \\
\delta_0 H_a^\mu &= -\epsilon \left[\frac{1}{4} \chi^\nu \chi^\lambda \left(R_{\nu\lambda\lambda'}^\mu + R_{\nu\lambda\mu'\mu''} J^{\mu'\mu} J_{\lambda'}^{\mu''} \right) \rho_a^{\lambda'} - i\Gamma_{\nu\kappa}^\mu \chi^\nu H_a^\kappa \right]
\end{aligned} \qquad (14.6.7)$$

Notice that H occurs quadratically in the action, and that it does not propagate. If we eliminate H via its equation of motion, we find:

$$I = \int d^2 z \left[\frac{1}{2} (g_{\mu\nu} \, \partial_a X^\mu \, \partial^a X^\nu + \epsilon^{ab} J_{\mu\nu} \, \partial_a X^\mu \, \partial_b X^\nu) \right.$$
$$\left. - i \rho^a_\mu \, \partial_a X^\mu - \frac{1}{8} \chi^\kappa \chi^\lambda \rho^a_\mu \rho_{a\nu} R^{\mu\nu}_{\kappa\lambda} \right] \qquad (14.6.8)$$

The BRST current is:

$$J_a = g_{\mu\kappa} (\partial_a X^\mu + \epsilon^b_a J^\mu_\nu \, \partial_b X^\nu) \chi^\kappa \qquad (14.6.9)$$

The elimination of the H field has resulted in the standard propagator of the X^μ field. However, the metric of space–time is almost complex and simplifies further if it is Kähler. However, the physical interpretation of this is obscure.

Once again, we see the same features emerging for the topological theory, that is, the action is zero and the gauge fixed action is a BRST commutator. The energy-momentum tensor is also BRST invariant, meaning that the theory is topological. In contrast to the previous case, however, there is now a restriction on the background metric.

14.7. Topological 2D Gravity

Normally, in critical string theory, the two-dimensional gravitational metric can be entirely gauged away. In noncritical dimensions, such as those found in matrix models, we find that two-dimensional gravity plays an essential role. Thus, the next theory we will examine is topological gravity in two dimensions [11, 16, 17].

As usual, we begin the action being zero or a topological term. (In two dimensions, the Einstein action itself is topological.) However, since there are many ways in which the theory can be formulated, we will take the one that most closely resembles the usual conformal field theory found in ordinary string theory.

We begin by specifying the gauge fields. $\omega^{ab} = \omega^{ab}_\mu dx^\mu$ is the gauge field associated with the Lorentz group. However, in two dimensions, there is only one component to an antisymmetric second rank tensor, so we will simply describe this gauge field as ω. Second, there is the gauge field of the translations for the Poincaré group in two dimensions, the vierbein $e^a = e^a_\mu dx^\mu$. The fields are therefore $\{\omega, e^+, e^-\}$.

Then, we will need superpartners for these fields in order to construct the BRST operator out of the supersymmetric operator. These superfields will be represented by $\{\psi^0, \psi^+, \psi^-\}$.

The gauge choice we will choose is to set the curvatures to zero:

$$S = \int (\pi_0 d\omega + \pi_+ De^+ + \pi_- De^-) + \int (\chi_0 d\psi^0 + \chi_+ D\psi^+ + \chi_- D\psi^-) \quad (14.7.1)$$

where π and χ are Lagrange multipliers, which enforce the gauge constraints:

$$
\begin{aligned}
De^+ &= de^+ - \omega \wedge e^+ \\
De^- &= de^- + \omega \wedge e^- \\
D\psi^+ &= d\psi^+ - \omega \wedge \psi^+ + e^+ \wedge \psi^0 \\
D\psi^- &= d\psi^- + \omega \wedge \psi^- - e^- \wedge \psi^0
\end{aligned}
\quad (14.7.2)
$$

The action is still invariant under local Lorentz transformations, parameterized by α, and by diffeomorphisms in two dimensions, parameterized by ξ:

$$
\begin{aligned}
\delta \omega &= d\alpha + \xi \cdot d\omega \\
\delta e^\pm &= \pm \alpha e^\pm + D(\xi \cdot e^\pm) + \xi \cdot De^\pm
\end{aligned}
\quad (14.7.3)
$$

The supersymmetry interchanges the fields as follows:

$$\delta_s \omega = \psi^0, \qquad \delta_s e^\pm = \psi^\pm \quad (14.7.4)$$

If we now make the supersymmetry transformation on the transformation of ω and $e^\pm$, we find the transformation of ψ under local Lorentz transformations and diffeomorphisms:

$$
\begin{aligned}
\delta \psi^0 &= \xi \cdot d\psi^0 \\
\delta \psi^\pm &= \pm \alpha e^\pm + D(\xi \cdot \psi^\pm) + \xi \cdot \psi^\pm
\end{aligned}
\quad (14.7.5)
$$

We can now write the BRST structure of the theory. Let c_0 and c represent the anticommuting ghosts associated with local Lorentz transformations and diffeomorphisms, and let their supercounterparts be represented by γ_0 and γ.

In terms of these ghost parameters, we can now write the BRST transformations for the fields ω, ψ_0 as:

$$
\begin{aligned}
\delta \omega &= \psi_0 + dc_0 \\
\delta \psi_0 &= d\gamma_0 \\
\delta c_0 &= \gamma_0 \\
\delta \gamma_0 &= 0
\end{aligned}
\quad (14.7.6)
$$

The other variations are given by:

$$
\begin{aligned}
\delta e^\pm &= \psi^\pm \pm c_0 e^\pm + D(c \cdot e^\pm) \\
\delta \psi^\pm &= \pm c_0 \psi^\pm + D(c \cdot \psi^\pm) \pm \gamma_0 e^\pm + D(\gamma \cdot e^\pm) \\
\delta c &= \gamma + c \cdot \partial c \\
\delta \gamma &= c \cdot \partial \gamma - \gamma \cdot \partial c
\end{aligned}
\quad (14.7.7)
$$

There are some interesting features of this transformation. First, notice that correlation functions are made of BRST invariant fields; however, the likely candidate for a BRST invariant field is γ_0, which in turn can be expressed as the BRST transformation of another field. This means that, at first glance, the theory seems totally empty. The set of BRST invariant operators that are not in turn BRST variations of another field seems to be the null set. This is an example of what is called "equivariant cohomology," the BRST is nilpotent up to a gauge-dependent parameter c_0. Thus, as we shall see, the theory is not totally empty.

Let us now break the symmetries of the theory. We can choose the conformal gauge, so we can define a complex structure on the theory, so that:

$$e^+ = e^{\phi+} \, dz, \qquad e^- = e^{\phi-} \, d\bar{z} \tag{14.7.8}$$

We now break local Lorentz transformations by setting:

$$\phi^+ = \phi^- \tag{14.7.9}$$

Let us make a BRST variation of the previous gauge fixing condition. Then, we find:

$$c_0 = \frac{1}{2}(\partial c + c \, \partial\phi - \bar{\partial}\bar{c} - \bar{c} \, \bar{\partial}\phi) \tag{14.7.10}$$

If we let $\phi = \phi_+ + \phi_-$, then we can simplify the action and express it entirely in terms of ϕ and its Lagrange multiplier. Also, we have, as in ordinary string theory, the gauge fixing of the vierbein, which creates a Faddeev–Popov ghost action. Adding both parts, we find that the bosonic part of the action becomes:

$$S_B = \int \pi \bar{\partial}\partial\phi + \int (b \, \bar{\partial}c + \bar{b} \, \partial\bar{c}) \tag{14.7.11}$$

We now repeat all these steps for the fermionic part of the action. The superconformal gauge yields:

$$\begin{aligned} \psi_+ &= e^{\phi+} \psi_+ \, dz \\ \psi^- &= e^{\phi-} \psi_- \, d\bar{z} \end{aligned} \tag{14.7.12}$$

while breaking the super local Lorentz invariance yields:

$$\psi_+ = \psi_- \tag{14.7.13}$$

Making a supersymmetric variation on c_0 in Eq. (14.7.10), we also find an equation for γ_0:

$$\gamma_0 = \frac{1}{2}(\partial\gamma + \gamma \, \partial\phi + c \, \partial\psi - \bar{\partial}\bar{\gamma} - \bar{\gamma} \, \bar{\partial}\phi - \bar{c} \, \bar{\partial}\psi) \tag{14.7.14}$$

Thus, the fermionic part of the action now becomes:

$$S_F = \int \chi \bar{\partial} \, \partial\psi + \int (\beta \, \bar{\partial}\gamma + \bar{\beta} \, \partial\bar{\gamma}) \tag{14.7.15}$$

The final action is the sum of S_B and S_F.

We are now in a position to write the complete BRST operator for the theory. The energy-momentum tensor is equal to the sum of two pieces:

$$T_L = \partial\pi\,\partial\phi + \partial^2\pi + \partial\chi\,\partial\psi$$
$$T_{gh} = c\,\partial b + 2\partial c\,b + \gamma\,\partial\beta + 2\,\partial\gamma\,\beta$$

$$(14.7.16)$$

while the superconformal generator is also given by the sum of two pieces:

$$G_L = \partial\chi\,\partial\phi + \partial^2\chi$$
$$G_{gh} = c\,\partial\beta + 2\,\partial c\,\beta$$

$$(14.7.17)$$

We also have the supersymmetry charge:

$$Q_s = \oint (\partial\pi\,\psi + b\gamma)$$

$$(14.7.18)$$

Putting everything together, the total BRST operator is now given by:

$$Q_{BRST} = Q_s + \oint \left[c\left(T_L + T_{\beta\gamma} + \frac{1}{2}T_{bc} \right) + \gamma G_L \right]$$

$$(14.7.19)$$

14.8. Correlation Functions for 2D Topological Gravity

Now that we have an explicit representation of the action and BRST operator in terms of familiar conformal fields, we can write correlation functions, as in Eq. (14.5.12). As before, the correlation functions must be over BRST invariant fields that are not in turn BRST variation of other fields.

Here, γ_0 (like ϕ appearing in the topological Yang–Mills theory) is a BRST invariant operator. However, unlike our previous case, it can be written as the BRST variation of another field c_0. Normally, this means that all correlation functions made out of γ_0 vanish. This means that the entire formalism collapses, and the theory is empty. However, topological gravity is an example of an "equivariant cohomology," that is, the BRST operator is nilpotent, modulo a gauge-dependent field such as c_0. In practice, when constructing correlation functions, we will find that they do not vanish, because of contact interactions.

As in topological Yang–Mills theory, we have a sequence of operators that we can construct from γ_0. Repeating the sequence of manipulations developed in Eqs. (14.5.6)–(14.5.8), we can define the following sequence of operators:

$$\sigma_n^{(0)} \equiv \gamma_0^n$$
$$d\sigma_n^{(0)} = \delta\sigma_n^{(1)}$$
$$d\sigma_n^{(1)} = \delta\sigma_n^{(2)}$$

$$(14.8.1)$$

where the superscript (n) represents an n form on the Riemann surface. Since the differential of $\sigma_n^{(i)}$ is the BRST variation of $\sigma_n^{(i+1)}$, correlation functions composed out of $\sigma_n^{(i)}$ are independent of the points at which they are defined. Explicitly, we have the following:

$$\sigma_n^{(1)} = n\psi_0\gamma_0^{n-1}$$
$$\sigma_n^{(2)} = n\,d\omega\,\gamma_0^{n-1} + \frac{1}{2}n(n-1)\psi_0 \wedge \psi_0\gamma_0^{n-2} \qquad (14.8.2)$$

Given the form of the correlation functions, we can write recursion relations [18–21] between them that will be shown to be precisely the same as those found in the matrix model.

The goal for the genus zero recursion relation is to find a relation between correlation functions involving σ_{d_i} and correlation functions where d_i is replaced by $d_i - 1$. Then, by repeatedly applying this recursion relation, eventually σ_{d_i} can be reduced to correlation functions involving only $\sigma_0 \equiv P$, which are known. Thus, this recursion relation will be able to reduce all possible correlation functions to the known correlation functions of the puncture operator P.

Let $M_{g,s}$ be the moduli space of a Riemann surface with genus g and s punctures. The dimension of moduli space is $6g - 6 + 2s$. For every σ_{d_i}, we can associate a $2d_i$ form $\lambda_{(i)}$ defined in the moduli space $M_{g,s}$ as in Eq. (14.8.2). Then, we have, as in the topological Yang–Mills theory:

$$\left\langle \sigma_{d_1}\sigma_{d_2}\cdots\sigma_{d_s} \right\rangle = \int_{M_{g,s}} \lambda_{(1)} \wedge \lambda_{(2)} \wedge \cdots \wedge \lambda_{(s)} \qquad (14.8.3)$$

where twice the sum of the d_i must equal the dimension of moduli space:

$$2\sum_i d_i = 6g - 6 + 2s \qquad (14.8.4)$$

Fortunately, because these theories are purely topological and hence independent of the location of the operators σ_n, the correlation functions must be topologically defined. Thus, the evaluation of these correlation functions can be accomplished by using counting arguments.

Using purely topological arguments, we see that the correlation function is a function of the integration region of each $\lambda_{(i)}$, which we will call $H_{(i)}$. These $H_{(i)}$ can be taken to be cycles defined on moduli space. Since the correlation function is purely topological, we can take these $H_{(i)}$ to be homology cycles.

Fortunately, it is possible to define a simple topological invariant out of these homology cycles. The intersection of these cycles is a number, which is topologically defined. Thus, we find that:

$$\left\langle \sigma_{d_1}\sigma_{d_2}\cdots\sigma_{d_n} \right\rangle = \#\left(H_{(1)} \cap H_{(2)} \cap \cdots \cap H_{(n)} \right) \prod_i d_i! \qquad (14.8.5)$$

where we have normalized each $\lambda_{(i)}$ to be a $2d_i$ form times $d_i!$.

Our goal is to calculate the correlation functions of topological gravity and compare them to the correlation functions found in matrix models. We must, therefore, find a way of reducing the d_i appearing in the correlation functions. The key to constructing recursion relations for matrix models is to find the topologically defined operator for $\lambda_{(i)}$ appearing in the correlation function. It can be shown that the correct expression for this is given by:

$$\lambda_{(i)} = c_1 \left[L_{(i)} \right]^{d_i} \tag{14.8.6}$$

where c_1 is the first Chern class defined for the line bundle $L_{(i)}$ defined on the moduli space $M_{g,s}$ (or actually the moduli space of stable curves, which is obtained by compactifying $M_{g,s}$ by adjoining curves with double points).

Our strategy is to reduce all the d_i appearing in the correlation function until we are left with the correlation function of products of σ_0, which we identify as the puncture operator. The way to do this is to split off one of the c_1:

$$\lambda_{(i)} = c_1 \left[L_{(i)} \right] \left\{ c_1 \left[L_{(i)} \right] \right\}^{d_i - 1} \tag{14.8.7}$$

and then perform the complex integration over the two variables appearing in $L_{(i)}$. In this way, we eliminate two moduli and reduce d_i by one.

Let us consider the case of a genus one surface. Then, the integral over the two moduli appearing in c_1 is defined with certain poles and zeros. We must be careful to examine the case when the points z_i approach each other.

Let us pick three points on the Riemann surface Σ, labeled z_1, z_{s-1}, and z_s, and perform the integral over the moduli associated with $L_{(1)}$. Let S be the set of all other points. In general, we find that nodes appear in the integrand, resulting in the Riemann surface Σ splitting into two disjoint pieces, Σ_1 and Σ_2.

The only case of interest is when z_1 appears on Σ_1 and z_{s-1} and z_s appear on the other (the other possibilities do not contribute to the integral). In this case, the points appearing in the set S can be distributed over Σ_1 or Σ_2. In fact, we have to sum over all possible distributions of the set of points S into the set X on Σ_1 and the set Y on Σ_1. By performing the integration over two moduli, we find the recursion relation (for genus zero only):

$$\left\langle \sigma_{d_1} \sigma_{d_2} \cdots \sigma_{d_n} \right\rangle = d_1 \sum_{S = X \cup Y} \left\langle \sigma_{d_1 - 1} \prod_{j \in X} \sigma_{d_j} P \right\rangle \times \left\langle P \prod_{k \in Y} \sigma_{d_k} \sigma_{d_s - 1} \sigma_{d_s} \right\rangle \tag{14.8.8}$$

where we have summed over all possible ways in which the points in S can be distributed over Σ_1 and Σ_2, and where we have inserted the puncture operator P at the node separating the Riemann surfaces Σ_1 and Σ_2. (The puncture operator does nothing but give us a marked point on the Riemann surface.)

Notice that the number of moduli matches. The correlation function on the left-hand side satisfies:

$$2\sum_i d_i = 2n - 6 \tag{14.8.9}$$

since the genus equals zero. On the right-hand side, we have integrated over two moduli (which produced the node that split the Riemann surface Σ into two pieces). These two correlation functions satisfy:

$$2(d_1 - 1) + 2\sum_{i \in X} d_i = 2(n_1) - 6 + 2$$
$$2d_{s-1} + 2d_s + \sum_{i \in Y} d_i = 2(n_2) - 6 + 2 \tag{14.8.10}$$

where n_1 and n_2 are the number of z_i on each Riemann surface. By adding these two equations, we arrive at Eq. (14.8.9), as expected. The presence of the puncture operator P, which does nothing but give us a marked point at the node separating the two Riemann surfaces, was essential to get the counting of moduli correct.

Let us use this recursion relation in order to calculate the matrix elements for the various product of operators. We begin by constructing the generating function for correlation functions. We assume that the action, which is zero, is supplemented by ϵP, where P is the puncture operator. (An analogous assumption was made for the matrix model case.) Let us make scaling arguments to assume that the matrix elements are powers of ϵ. Then,

$$\langle \sigma_n \rangle = a_n \epsilon^{b_n} \tag{14.8.11}$$

for some a_n and b_n, which are as yet undetermined.

We know that taking derivatives of this expression by ϵ pulls down a P operator, that is,

$$\langle \sigma_n P \rangle = (\partial/\partial\epsilon)\langle \sigma_n \rangle = a_n b_n \epsilon^{b_n - 1}$$
$$\langle \sigma_n PP \rangle = (\partial^2/\partial\epsilon^2)\langle \sigma_n \rangle = a_n b_n (b_n - 1)\epsilon^{b_n - 2} \tag{14.8.12}$$

Now, let us insert these values into the recursion relations, which, for this simple case, read:

$$\langle \sigma_n PP \rangle = n\langle \sigma_{n-1}P \rangle\langle PPP \rangle$$
$$\langle \sigma_n \sigma_m P \rangle = n\langle \sigma_{n-1}P \rangle\langle P\sigma_m P \rangle \tag{14.8.13}$$
$$\langle \sigma_n \sigma_m \sigma_p \rangle = n\langle \sigma_{n-1}P \rangle\langle \sigma_m \sigma_p \rangle$$

where we have normalized $\langle PPP \rangle = (1/k)\epsilon^{(1/k)-1}$.

Plugging in the values for the various correlation functions, we find a simple recursion relation in the a_n and b_n, which allows us to calculate all of them. We find:

$$a_n = \frac{1}{(n+1)[1+(1+n)/k]}$$

$$b_n = 1 + \frac{n+1}{k}$$

(14.8.14)

This then allows us to calculate the following expectation values:

$$\langle \sigma_n \rangle = \frac{\epsilon^{1+(n+1)/k}}{(n+1)[1+(n+1)/k]}$$

$$\langle \sigma_n \sigma_m \rangle = \frac{\epsilon^{(n+m+1)/k}}{n+m+1}$$

(14.8.15)

$$\langle \sigma_n \sigma_m \sigma_p \rangle = \frac{1}{k}\epsilon^{(n+m+p+1)/k-1}$$

These values (for the sphere) are precisely what we found for the matrix model in Eq. (13.5.27), showing that topological gravity, in some sense, is identical to ordinary two-dimensional gravity, at least for low genus. Repeating the same arguments for the arbitrary case, we find:

$$\langle \sigma_{d_1} \sigma_{d_2} \cdots \sigma_{d_s} \rangle = \frac{1}{k}\left(\frac{\partial}{\partial \epsilon}\right)^{s-3} \epsilon^{(n+1)/k-1}$$

(14.8.16)

which reproduces the result of the matrix models.

14.9. Virasoro Constraint, W-algebras, and KP Hierarchies

We still, however, have not used the full power of this formalism, which is capable of deriving the complete set of constraints satisfied by the generating functional to all orders in perturbation theory. These constraints, in turn, are equivalent to the string equation and the recursion relations found in the KdV formalism. We thus obtain a way to conveniently reformulate all the constraints found in matrix models in a more familiar language.

At first, this may seem strange because the Green's functions of topological gravity appear to be zero. This is because the physical operators σ_n are all BRST trivial, and hence the Green's functions should all be zero, and the S-matrix vanishes. However, a careful analysis of the Green's functions shows that there is indeed a source of non-trivial contributions, and this comes from whenever σ_n approaches σ_m. In other words, the entire contribution to the Green's functions comes from *contact terms*. This vastly simplifies the calculation of N-point functions, since all we have to calculate are the contributions of contact terms, which can be isolated using operator identities.

For example, consider what happens when σ_n approaches σ_m. By explicit calculation, using the conformal field theory identities we have written down, we can show the following identity which isolates the contact terms:

$$\int_{D_\epsilon} \sigma_n|\sigma_m\rangle = \frac{1}{3}(2n+1)|\sigma_{n+m-1}\rangle \qquad (14.9.1)$$

where D_ϵ is a propagator representing an infinitesimal neighborhood separating the two operators. (This identity can be proven by inserting the propagator between these two operators and then moving it to the right, where it annihilates on the vacuum.)

In the same way, by carefully isolating contact terms, we can construct the entire set of identities satisfied by N-point Green's functions. The calculation is rather intricate, so we will just present final the result of this calculation. Let S be a collection of fields σ_m. This set can be broken up into two smaller sets, X and Y. Then the general recursion relation on genus g Green's functions is given by [21]:

$$\left\langle \sigma_{n+1} \prod_{m \in S} \sigma_m \right\rangle = x \left\langle \sigma_n \prod_{m \in S} \sigma_m \right\rangle_g$$

$$+ \sum_{j \in S}(2j+1)\left\langle \sigma_{j+n} \prod_{m \neq j} \sigma_m \right\rangle_g + \sum_{j=1}^{n}\left\{ \left\langle \sigma_{j-1}\sigma_{n-j} \prod_{m \in S} \sigma_m \right\rangle_{g-1} \right.$$

$$\left. + \frac{1}{2} \sum_{\substack{S=X\cup Y \\ g=g_1+g_2}} \left\langle \sigma_{j-1} \prod_{m \in X} \sigma_m \right\rangle_{g_1} \left\langle \sigma_{n-j} \prod_{m \in Y} \sigma_m \right\rangle_{g_2} \right\}$$

$$(14.9.2)$$

where x is the cosmological constant, and we have divided up the Riemann surface of genus g into two smaller Riemann surfaces of genus g_1 and g_2.

This equation, although formidable looking, can be broken down into simpler components. The essence of this identity is that the major contribution to the Green's function comes when σ_{n+1} approaches the other σ_j. We saw earlier in Eq. (14.9.1) that the effect of this contact term is to generate $(2j+1)\sigma_{j+n}$. That is the contribution found in the first term on the second line of the equation. The last two terms in the equation must be added because the N-point function may develop nodes (e.g. the Riemann surface may fission into two pieces with genus g_1 and g_2) and we must insert operators at the nodes.

This identity, although it summarizes all the information contained within the Green's functions, is still unwieldy. It simplifies enormously, however, if we re-express this in terms of operators. Let the generating functional be represented as $Z(t_0, t_1, \cdots)$, where the t_i are the sources for the σ_i. Taking a derivative with respect to t_i simply pulls down the operator σ_i into the functional integration. Let us now re-write Eq. (14.9.1) and (14.9.2) in terms of operators acting on $Z(t_0, t_1, \cdots)$. Let us define L_n as the operator which pulls down σ_{n+1}.

Now apply this operator L_n on the generating function twice. Then the essence of Eq. (14.9.1) is that the commutator of two L's acting on the generating functional yields the algebra $[L_n, L_m] = (n - m)L_{n+m} + \cdots$, which resembles the Virasoro algebra! Encouraged by this result, we suspect that the complete algebra contained within Eq. (14.9.2) is precisely the Virasoro algebra. It is gratifying to note that we can now summarize the entire content of Eq. (14.9.2), which in turn contains all the constraints of the matrix models, in one equation:

$$L_n \tau = 0; \quad n \geq -1 \tag{14.9.3}$$

where τ is the square root of the generating function:

$$Z(t_0, t_1, \cdots) = \tau^2(t_0, t_1, \cdots) \tag{14.9.4}$$

and the explicit expression for the L_n's is given by:

$$L_{-1} = \sum_{m=1}^{\infty} (m + \frac{1}{2}) t_m \frac{\partial}{\partial t_{m-1}} + \frac{1}{8}\lambda^{-2}t_0^2$$

$$L_0 = \sum_{m=0}^{\infty} (m + \frac{1}{2}) t_m \frac{\partial}{\partial t_m} + \frac{1}{16} \tag{14.9.5}$$

$$L_n = \sum_{m=0}^{\infty} (m + \frac{1}{2}) t_m \frac{\partial}{\partial t_{m+n}} + \frac{1}{2}\lambda^2 \sum_{m=1}^{n} \frac{\partial^2}{\partial t_{m-1}\partial t_{n-m}}$$

where λ is the string coupling constant.

If we let the operator expression for L_n operate on τ, we obtain the constraints (14.9.2). Remarkably, the entire content of the recursion relations can now be expressed compactly in one equation!

Furthermore, we see that the L_n operators, which pull down σ_{n+1} into the generating function, satisfy precisely the algebra of the usual Virasoro operators. However, these are not the ordinary Virasoro operators; instead of acting on the space of parametrizations of the world sheet of the string, these operators act on the space of physical operators appearing within the generating functional. Although the algebra is the same, the physical content appears to be entirely different.

We stress that these recursion relations for the topological field theory are identical to the recursion relations found in matrix models using the Schwinger–Dyson equations [21–22]. Thus, the equivalence of the matrix model to the topological field theory approach is established to all orders in perturbation theory. Both approaches (the contact algebra of topological field theory and the Schwinger–Dyson equations of matrix models) yield the same Virasoro conditions.

So far, we have only treated pure topological gravity, which has only one primary field, given by the puncture operator P. Encouraged by our surprising success in reformulating the recursion relations of matrix models in terms of a Virasoro constraint, we are led to examine whether the recursion relations for the higher matrix models can also be reproduced in this way.

We can generalize the Virasoro constraint by coupling topological gravity to topological minimal models [23]. The advantage is that we can expand the number of primary fields and hence derive a larger set of identities satisfied by the generating function.

Let us call the primary fields of the topological minimal models V_k, and the expanded set of physical operators is given by the product of V_k with the old physical operators σ_n:

$$\sigma_{n,k} = V_k \gamma_0^n P \qquad (14.9.6)$$

where P is the puncture operator.

As in the purely gravitational case, we find that the Green's functions are all zero, because the physical operators are all BRST trivial, except for the possibility of contact terms. By explicit calculation, we find that the contact algebra is given by:

$$\int_{D_\epsilon} \sigma_{m,k}|\sigma_n\rangle - \int_{D_\epsilon} \sigma_n|\sigma_{m,k}\rangle = \frac{h(kn-m)}{h+1}|\sigma_{m+n-1,k}\rangle \qquad (14.9.7)$$

The important point is to notice that the right hand side contains the factor $(kn - m)$. This means that the algebra generated by this extended theory is not the usual Virasoro algebra. This extended algebra must therefore be an extension of the Virasoro algebra. If the operator algebra L_n pulls down the term σ_{n+1} and a new operator $W_m^{(k+1)}$ pulls down the term $\sigma_{m+1,k}$, then the commutator between these two operators must include:

$$[L_n, W_m^{(k+1)}] = (kn - m)W_{n+m}^{(k+1)} + \cdots \qquad (14.9.8)$$

The only algebra of this type which includes the Virasoro algebra as a subset is the W-algebra of Zamolodchikov [24].

Arguing from purely theoretical grounds, Zamolodchikov investigated generalizations of the Virasoro algebra with operators of various conformal spins. He found that the addition of a conformal spin 3 operator generated a larger algebra than the usual Virasoro algebra. If we define $W_n^{(2)} = L_n$ and $W_n^{(3)} = W_n$, then the algebra becomes:

$$[L_m, L_n] = (m - n)L_{m+n} + \frac{c}{12}m(m^2 - 1)$$

$$[L_m, W_n] = (2m - n)W_{m+n}$$

$$[W_m, W_n] = (m - n)\left[\frac{1}{15}(m + n + 3)(m + n + 2)\right.$$

$$\left. - \frac{1}{6}(m+2)(n + 2)\right]L_{m+n} + \frac{c}{360}m(m^2 - 4)\delta_{m+n,0} + \frac{16}{22 + 5c}\Lambda_m$$

$$(14.9.9)$$

where:

$$\Lambda_m = \sum_{n}^{\infty} L_{m-n}L_n - \frac{3}{10}(m+2)(m+2)L_m \qquad (14.9.10)$$

The last commutator contains terms which are bilinear in the Virasoro generators. This means that the algebra is *not* a standard Lie algebra. However, one can show that the Jacobi identities still close.

Although the details for the full theory have yet to be worked out in detail, we conjecture that the full set of constraints for the $p-1$ matrix model are given by:

$$W_n^{(k)}\tau = 0; \quad 2 \le k \le p, \, n \ge -k+1 \qquad (14.9.11)$$

For the case $p = 1$, the one matrix model, this reduces to the ordinary Virasoro condition found earlier. However, for higher matrix models, we find a series of non-trivial constraints. Since the Jacobi identities are satisfied by the $W^{(k)}$, we are guaranteed to have a set of self-consistent equations. Similarly, for the (p,q) matrix model, it is also possible to use the generalized W algebras which can reproduce the constraints for these models as well.

We can also take this formalism one step further. In the same way that we found the KdV hierarchy emerging from the one matrix model, these higher identities should emerge as a generalization of the KdV hierarchy. This generalization is called the Kadomtsev-Petviasvili (KP) hierarchy .

To see how the higher W-algebra constraints emerge from the KP hierarchy, let us first define the pseudo-differential operator:

$$L = d + u_2(t_i)d^{-1} + u_3(t_i)d^{-2} + \cdots \qquad (14.9.12)$$

where the u's are functions of t_i (which will be linearly related to the t_i discussed earlier), d is d/dt_0, and $(L^n)_+$ represents taking the positive differential part of L^n.

To constrain the u functions, we impose:

$$\frac{\partial}{\partial t_n}L = [(L^n)_+, L] \quad n = 1, 2, 3, \cdots \qquad (14.9.13)$$

Once this constraint is placed, the remarkable feature of this formalism is that the entire system can be re-written in terms of a new function, called the Hirota τ function, which satisfies:

$$\frac{\partial^2}{\partial t_0 \partial t_n} \ln \tau = (L^{n+1})_{-1}$$
$$\frac{\partial^2}{\partial t_1 \partial t_n} \ln \tau = 2(L^{n+1})_{-2} + \frac{\partial}{\partial t_0}(L^n)_{-1} \qquad (14.9.14)$$

and so on.

Lastly, to reduce this KP hierarchy to simpler hierarchies, we will impose one more additional constraint. The p reduction of the KP hierarchy is defined by stating that the u's have no dependence on $t_{p-1}, t_{2p-1}, t_{3p-1}, \cdots$, and:

$$(L^p)_- = 0 \qquad (14.9.15)$$

where $(L^p)_-$ stands for taking the part negative in d. The 2-reduction of the KP hierarchy is called the KdV hierarchy, encountered earlier for the one matrix model, and the 3-reduction is called the Boussinesq hierarchy, which describes the two matrix model. In this way, we can incorporate the p matrix model as part of a KP hierarchy.

The last step is to compare the constraint $W_n^{(k)}\tau = 0$ with the constraints coming from the KP hierarchy. By carefully expanding both constraints in terms of t_i, we find that the τ function appearing in the W algebra constraint is precisely the Hirota τ function, and that the two sets of t_i are linearly related to each other.

Example: One Matrix Model

To see how this happens, it is useful to take a specific example, the one matrix model. There is a conjecture that the Virasoro constraint $L_n\tau = 0$ for $n \geq -1$ is equivalent to the constraint $L_{-1}\tau = 0$ with the added condition that τ satisfy the KP hierarchy.

Assuming that this is true, let us take the constraint $L_{-1}\tau = 0$ and apply d to it. Then this constraint reduces to the following:

$$\left(\frac{1}{4}\lambda^{-2}t_0 + \sum_{m=1}^{\infty}(m + \frac{1}{2})t_m \frac{\partial^2}{\partial t_0 \partial t_{m-1}} \right) \tau = 0 \qquad (14.9.16)$$

Now assume that τ satisfies once reduced KP hierarchy. This means that $(L^2)_- = 0$, which in turn can be shown to lead to:

$$(L^{2k-1})_{-1} = 2R_k(-2u_2); \quad k \geq 1 \qquad (14.9.17)$$

where R_k are the familiar Gelfand-Dikii polynomials.

Because τ satisfies the KP hierarchy, it also satisfies: $\partial^2/\partial t_0 \partial t_m \ln \tau = (L^{m+1})_{-1}$. Now substitute this expression into (14.9.16), and we are finally left with:

$$\frac{1}{4}\lambda^{-2}t_0 + \sum_{m=1}^{\infty}(2m + 1)t_m R_m = 0 \qquad (14.9.18)$$

which is the string equation found earlier using matrix models.

In summary, it is quite remarkable that so much non-trivial, non-perturbative information can be so compactly represented in terms of a

W algebra constraint on the generating functional. A vast amount of information, summarizing the complex interactions over Riemann surfaces of all genus, is succinctly compressed into these W algebra constraints.

Lastly, we remark that these constraints, in turn, look suspiciously like Ward identities on a string field theory. In fact, it is possible to re-express topological field theory in second quantized language and use the gauge invariance of the theory to write down the Ward identities for the theory. When this is done, we find that we reproduce precisely the W algebra constraints [25]. In other words, the fact that the W algebra constraints are all self-consistent is due to the fact that the Ward identities are manifestations of the gauge invariance of the theory. Since the theory is invariant under multiple gauge transformations, we find that the Ward identities must also be self-consistent among each other, which is guaranteed by the Jacobi identities satisfied by the W algebra.

Thus, the remarkable self-consistency of the tower of W algebra constraints can now be seen to be the self-consistency of the gauge transformations of string field theory.

14.10. Summary

Previously, we analyzed the high-energy and high-temperature behavior of string amplitudes, which demonstrated the possibility of a phase transition and the restoration of vast symmetries. These new symmetries may indicate the reemergence of general covariance as an exact symmetry of the system. Usually, we break general covariance in the quantization scheme by power expanding the metric around some classical solution. However, if we were to quantize the theory without making such an unnatural split, then we might see the "topological phase" of the underlying theory. Thus, Witten's topological field theory is an attempt to construct the "unbroken phase" of string theory, where general covariance is unbroken.

Two types of topological field theories exist. The first type involves covariant theories without a metric tensor, such as the Chern–Simons gauge theory or the $2 + 1$ gravity theory, which can be written as:

$$L = \epsilon^{abc}\epsilon^{ijk}e_i^a R_{jk}^{bc}(\omega) \sim e \wedge R$$
$$R_{jk}^{bc} = \partial_j \omega_k^{bc} + \omega_j^{be}\omega_k^{ec} - (j \leftrightarrow k)$$

$$(14.10.1)$$

The second type of topological field theory involves the cohomological theories. In general, they have an explicit dependence on the metric tensor, but the final correlation functions are independent of the geometry of space–time. Thus, their correlation functions must be topological invariants. This means that we can use cohomological topological field theories to generate analytic expressions for the various topological invariants that have been recently written in three and four dimensions. Quantum field theory is then being used to answer difficult questions in pure mathematics, such as Morse theory, Floer complexes, and Donaldson polynomials.

In the cohomological theories, the action is zero,

$$L = 0 \tag{14.10.2}$$

or a topological invariant, such as:

$$L \sim F \wedge \tilde{F} \tag{14.10.3}$$

In the cohomological theory, the content of the theory lies in the field variations, which give us a Faddeev–Popov ghost contribution and gauge fixing part. The important fact is that the total gauge fixed action is a BRST commutator:

$$L_{\text{GF+FP}} = \{Q, V\} \tag{14.10.4}$$

More important, the energy-momentum tensor is BRST invariant:

$$T_{\alpha\beta} = \{Q, V_{\alpha\beta}\} \tag{14.10.5}$$

which means that, even if the background metric occurs after gauge fixing, the correlation functions are independent of the choice of metric. Since the background metric arises as a by-product of BRST quantization and since the physics should not be altered by the details of BRST quantization, the final theory should be background independent.

For example, in the four-dimensional Yang–Mills case, the symmetry of the action is large enough to completely eliminate the connection field. We will fix the gauge by demanding that the curvature be self-dual:

$$\text{gauge choice} \quad F^a_{\mu\nu} - \frac{1}{2}\epsilon_{\mu\nu\alpha\beta}F^{\alpha\beta a} = 0 \tag{14.10.6}$$

Normally, this is enough to make the theory empty, since the solutions of this gauge condition include zero. However, although this gauge is strong enough to eliminate the infinite degrees of freedom of the connection field, it is not strong enough to eliminate the finite degrees of freedom. As is well known, instantons are finite-dimensional solutions of this constraint. Thus, it is not surprising that the moduli space of instantons should play a key role in topological Yang–Mills theory. This is indeed the case, because the correlation functions will be topological invariants defined on four-dimensional manifolds; for example we will find an analytic derivation of the celebrated Donaldson polynomials.

The gauge fixed action, in the BRST formalism, becomes a BRST commutator:

$$L_{\text{GF+FP}} = \delta_1 \left\{ \chi^{\alpha\beta} \left[F_{\alpha\beta} + \tilde{F}_{\alpha\beta} - (1/2)\alpha_0 B_{\alpha\beta} \right] \right\} \tag{14.10.7}$$

where:

$$\delta_1 A_\alpha^a = \psi_\mu^a$$
$$\delta_1 \psi_\alpha^a = 0$$
$$\delta_1 \chi^{\alpha\beta a} = B^{\alpha\beta a} \qquad (14.10.8)$$
$$\delta_1 B^{\alpha\beta a} = 0$$

The quantization is not yet over, because the theory possesses a hidden symmetry:

$$\delta_G \psi_\alpha^a = i(D_\alpha \phi)^a$$
$$\delta_G B^{\alpha\beta a} = -ie_0[\phi, \chi^{\alpha\beta}]^a \qquad (14.10.9)$$

Since the ghosts themselves have a gauge degree of freedom, we must introduce "ghosts-for-ghosts" to completely eliminate all local invariances. Once this remaining symmetry is gauge fixed, we can obtain the final action for topological Yang–Mills theory.

Alternatively, we could have used the quantization method of Batalin–Vilkovisky (see Appendix 1), or we could have observed that the $N = 2$ supersymmetric Yang–Mills theory has a field content almost identical to the topological version. By "twisting" the $N = 2$ supersymmetric Yang–Mills theory, we can convert one of the supersymmetry generators Q_α^i into a genuine nilpotent Lorentz scalar, which we can then define to be Q_{BRST}. In this way, we arrive at the identical action.

To see this, observe that in $N = 2$ superconformal theories we have the supersymmetry generators:

$$\{Q_{\alpha+}, Q_{\beta-}\} = \gamma^\mu_{\alpha\beta} P_\mu$$
$$\{Q_{\alpha+}, Q_{\beta+}\} = \{Q_{\alpha-}, Q_{\beta-}\} = 0 \qquad (14.10.10)$$

In two dimensions, a spinor has only two components, also labeled $\pm$, so that we have four nilpotent supercharges $Q_{\pm\pm}$. The key step is that we will now modify the theory so that the energy-momentum tensor becomes:

$$T'_{\mu\nu} = T_{\mu\nu} + \epsilon_{\mu\sigma}\, \partial^\sigma R_\nu + \epsilon_{\nu\sigma}\, \partial^\sigma R_\mu \qquad (14.10.11)$$

where R_μ is the current associated with R symmetry. Then, because the structure of the Lorentz group has been altered, we can extract a genuine Lorentz scalar out of the supersymmetry charges and call it the BRST charge:

$$Q_{\text{BRST}} \equiv Q_{-+} + Q_{+-} \qquad (14.10.12)$$

To find the correlation functions of the theory (which will generate the Donaldson polynomials) we observe that there is a BRST invariant scalar ϕ in the Yang–Mills theory. Defining $W_0 = \frac{1}{2}\text{Tr}\,\phi^2$, we can develop a chain of BRST invariant operators:

$$0 = i\{Q, W_0\}$$
$$dW_0 = i\{Q, W_1\} \qquad (14.10.13)$$

Let us now extract from W_1 a new BRST invariant operator W_2, which is BRST nontrivial, and so:

$$\begin{aligned} dW_1 &= i\{Q, W_2\} \\ dW_2 &= i\{Q, W_3\} \\ dW_3 &= i\{Q, W_4\} \\ dW_4 &= 0 \end{aligned}$$
(14.10.14)

where:

$$\begin{aligned} W_2 &= \mathrm{Tr}\left(\frac{1}{2}\psi \wedge \psi + i\phi \wedge F\right) \\ W_3 &= i\,\mathrm{Tr}\,(\psi \wedge F) \\ W_4 &= -\frac{1}{2}\,\mathrm{Tr}\,(F \wedge F) \end{aligned}$$
(14.10.15)

In this way, we can construct operators W_n, which can be inserted into the correlation function to obtain topological invariants.

A wide array of cohomological topological field theories exist, depending on whether topological invariants can be constructed for them. For the σ model, for example, the action can be taken to be:

$$\int_\Sigma d^2 z \, \epsilon^{ab} J_{\mu\nu} \partial_a X^\mu \, \partial_b X^\nu = \int_T J$$
(14.10.16)

where Σ is the two-dimensional world sheet, and T is space–time. We set

$$J = \frac{1}{2} J_{\mu\nu}\, dX^\mu \wedge dX^\nu$$
(14.10.17)

The gauge fixing condition is that the derivative of X is self-dual:

$$\partial_a X^\mu + \epsilon_a^b J_\nu^\mu \, \partial_b X^\nu = 0$$
(14.10.18)

The action is then obtained by straightforward gauge fixing:

$$L_{\mathrm{GF+FP}} = \frac{i}{2}\delta\left[\rho_\mu^a\left(\partial_a X^\mu + \epsilon_a^b J_\mu^\nu \, \partial_b X^\nu - \frac{1}{2}H_a^\mu\right)\right]$$
(14.10.19)

where:

$$\rho_\mu^a = \epsilon_b^a J_\mu^\nu \rho_\nu^b$$
(14.10.20)

and H_a^μ is a self-dual, commuting ghost. The transformations of the fields (in flat world sheet space) are given by:

$$\begin{aligned} \delta_0 X^\mu &= i\epsilon\chi^\mu \\ \delta_0 \chi^\mu &= 0 \\ \delta_0 \rho_a^\mu &= \epsilon(H_a^\mu - i\Gamma_{\nu\kappa}^\mu \chi^\nu \rho_a^\kappa) \\ \delta_0 H_a^\mu &= -\epsilon\left[\frac{1}{4}\chi^\nu\chi^\lambda(R_{\nu\lambda\lambda'}^\mu + R_{\nu\lambda\mu'\mu''}J^{\mu'\mu}J_{\lambda'}^{\mu''})\rho_a^{\lambda'} - i\Gamma_{\nu\kappa}^\mu\chi^\nu H_a^\kappa\right] \end{aligned}$$
(14.10.21)

Here, H occurs quadratically in the action and it does not propagate. If we eliminate H via its equation of motion, we find:

$$I = \int d^2 z \left[\frac{1}{2} (g_{\mu\nu} \partial_a X^\mu \partial^a X^\nu + \epsilon^{ab} J_{\mu\nu} \partial_a X^\mu \partial_b X^\nu) \right.$$
$$\left. - i\rho_\mu^a \partial_a X^\mu - \frac{1}{8} \chi^\kappa \chi^\lambda \rho_\mu^a \rho_{a\nu} R_{\kappa\lambda}^{\mu\nu} \right] \tag{14.10.22}$$

Similarly, two-dimensional gravity can be made into a topological theory. The gauge choice we will choose is to set the curvatures to zero:

$$S = \int (\pi_0 d\omega + \pi_+ De^+ + \pi_- De^-) + \int (\chi_0 d\psi^0 + \chi_+ D\psi^+ + \chi_- D\psi^-) \tag{14.10.23}$$

where π and χ are Lagrange multipliers, which enforce the gauge constraints:

$$\begin{aligned} De^+ &= de^+ - \omega \wedge e^+ \\ De^- &= de^- + \omega \wedge e^- \\ D\psi^+ &= d\psi^+ - \omega \wedge \psi^+ + e^+ \wedge \psi^0 \\ D\psi^- &= d\psi^- + \omega \wedge \psi^- - e^- \wedge \psi^0 \end{aligned} \tag{14.10.24}$$

To fix local Lorentz invariance, we decompose the zweibein:

$$e^+ = e^{\phi^+} dz, \qquad e^- = e^{\phi^-} d\bar{z} \tag{14.10.25}$$

and set the gauge:

$$\phi^+ = \phi^- \tag{14.10.26}$$

The final action, for both the bosonic and fermion parts, reduces to:

$$S_B = \int \pi \bar{\partial} \partial \phi + \int (b \bar{\partial} c + \bar{b} \partial \bar{c}) \tag{14.10.27}$$

and:

$$S_F = \int \chi \bar{\partial} \partial \psi + \int (\beta \bar{\partial} \gamma + \bar{\beta} \partial \bar{\gamma}) \tag{14.10.28}$$

The final action is the sum of S_B and S_F.

We are now in a position to write the complete BRST operator for the theory. The energy-momentum tensor is equal to the sum of two pieces:

$$\begin{aligned} T_L &= \partial\pi \, \partial\phi + \partial^2 \pi + \partial\chi \, \partial\psi \\ T_{gh} &= c \, \partial b + 2 \, \partial c \, b + \gamma \, \partial\beta + 2 \, \partial\gamma \, \beta \end{aligned} \tag{14.10.29}$$

while the superconformal generator is also given by the sum of two pieces:

$$\begin{aligned} G_L &= \partial\chi \, \partial\phi + \partial^2 \chi \\ G_{gh} &= c \, \partial\beta + 2 \, \partial c \, \beta \end{aligned} \tag{14.10.30}$$

We also have the supersymmetry charge:

$$Q_s = \oint (\partial \pi \, \psi + b\gamma) \tag{14.10.31}$$

Putting everything together, the total BRST operator is now given by:

$$Q_{\text{BRST}} = Q_s + \oint \left[c \left(T_{\text{L}} + T_{\beta\gamma} + \frac{1}{2} T_{bc} \right) + \gamma G_{\text{L}} \right] \tag{14.10.32}$$

We can form BRST invariant objects σ_n by iterating the field γ_0, the ghost associated with super-Lorentz transformations. The matrix elements of these operators are:

$$\left\langle \sigma_{d_1} \sigma_{d_2} \cdots \sigma_{d_n} \right\rangle = \# \left(H_{(1)} \cap H_{(2)} \cap \cdots \cap H_{(n)} \right) \prod_i d_i! \tag{14.10.33}$$

where we have normalized each $\lambda_{(i)}$ to be a $2d_i$ form times $d_i!$ and where H_i are homology cycles on the moduli space of flat connections.

Because the correlation function is defined topologically, we can, by reducing the index n of the BRST invariant field, develop recursion relations for the matrix elements. The recursion relation is (for genus zero only):

$$\left\langle \sigma_{d_1} \sigma_{d_2} \cdots \sigma_{d_n} \right\rangle = d_1 \sum_{S=X \cup Y} \left\langle \sigma_{d_1-l} \prod_{j \in X} \sigma_{d_j} P \right\rangle \left\langle P \prod_{k \in Y} \sigma_{d_k} \sigma_{d_s-1} \sigma_{d_s} \right\rangle \tag{14.10.34}$$

Last, using these recursion relations, we can completely calculate the matrix elements for the BRST operators:

$$\left\langle \sigma_n \right\rangle = \frac{\epsilon^{1+(n+1)/k}}{(n+1)[1+(n+1)/k]}$$

$$\left\langle \sigma_n \sigma_m \right\rangle = \frac{\epsilon^{(n+m+1)/k}}{n+m+1} \tag{14.10.35}$$

$$\left\langle \sigma_n \sigma_m \sigma_p \right\rangle = \frac{1}{k} \epsilon^{(n+m+p+1)/k-1}$$

Comparing these with the matrix elements found in the matrix models approach, we find they are the same. Thus, it can be shown, by calculating the recursion relation for correlation functions, that matrix models for two-dimensional gravity and topological gravity are the same. In retrospect, this may not be too surprising because both theories have finite degrees of freedom and both theories are tolological.

The equivalence between matrix models and topological field theories can also be generalized to include all orders in perturbation theory. It is possible to write down generalized recursion relations in topological field theory for the generating functional $Z(t_0, t_1 \cdots)$. When re-written in terms of Z, these recursion relations for two-dimensional gravity assume the remarkable form:

$$L_n \tau(t_0, t_1, \cdots) = 0; \quad n \geq 1 \qquad (14.10.36)$$

where $\tau \equiv \sqrt{Z}$ and where L_n, although they obey the same relations as the usual Virasoro algebra, is defined in the source space $\{t_i\}$ rather than on the world sheet. Thus, these new Virasoro operators have an entirely different meaning than the usual ones.

When written for two-dimensional gravity coupled to topological conformal matter, we find the constraints given by an extension of the Virasoro algebra, which is the W-algebra:

$$W_n^{(k)} \tau = 0; \quad 2 \leq k \leq p; \quad n \geq -k + 1 \qquad (14.10.37)$$

We stress that these two remarkable equations were derived independently using matrix model techniques [22].

We find that we can summarize a vast amount of non-perturbative information very succinctly in these constraint equations using the W algebra.

Lastly, we point out that it is possible, in turn, to re-express these W algebra constraints as Ward identities on a second quantized topological string field theory. Thus, the W algebra constraints are nothing but an expression of the gauge invariance inherent within string field theory [25].

14.11. Conclusion

As we have stressed throughout this book, string theory has an enormously powerful gauge group which allows us to eliminate the divergences and anomalies which riddle point particle theories of quantum gravity. Thus, at present string theory has no rival; no other theory can claim to self-consistently unify all known fundamental forces.

We have seen how string theory has unified not only the four fundamental interactions into a coherent picture, it has also unified previously unrelated branches of mathematics. Powerful mathematical methods, especially coming from topology, have been essential in eliminating the anomalies and divergences associated with quantum gravity and particle interactions. In addition, topology has played an essential role in extracting phenomenologically acceptable results from the theory.

However, in spite of all of the remarkable theoretical successes of string theory in eliminating these anomalies and divergences, we must always keep in mind that the theory is difficult, if not impossible, to test experimentally. Since string theory is really a theory of creation, we must eventually re-create the energies found at the Big Bang in order to fully test its consequences.

Although this is often singled out as the main criticism of string theory, in our opinion, the main problem facing string theory is not experimental at all, but theoretical. If we were clever enough, we would be able to solve the theory and calculate its true ground state. In this way, we should be able to determine whether the theory accurately predicts the properties of our universe, or whether it predicts nonsense. Since a theory whose "natural home" lies beyond the Planck length is also a theory of everyday phenomena, we should be able to theoretically extract the masses and couplings of the familiar particles found in nature at low energies.

The central theme of this book, therefore, is the search for the true vacuum of string theory. This search, in turn, has opened up startling new areas of mathematical research and has revealed highly non-linear relationships between seemingly unrelated physical systems. In fact, the fact that entirely new areas of mathematical physics are being broken open by string theory is nothing less than breathtaking.

It is gratifying to realize that whole areas of mathematics, which previously had no relationship to physics, are now seen as essential ingredients in solving the theory non-perturbatively. In particular, conformal invariance and topology have proven to be useful tools by which to probe the true vacuum of string theory. By making certain simplifying assumptions, we have been able to use these two powerful tools to completely solve string theory non-perturbatively.

For example, by assuming that we have a finite number of primary fields, conformal methods have proven to be so powerful that we now have a relatively good understanding of all possible rational conformal field theories.

Also, by assuming that the string vibrates in $D \leq 1$ dimensions, we now have a complete non-perturbative understanding of the bosonic theory. Remarkably, beginning with a purely topological theory, we can solve the constraints on the Green's functions exactly and write them succinctly as W algebra constraints. Or, in the language of string field theory, we can write them as Ward identities.

Although the simplifying assumptions we have made are unphysical, we can, even from our limited vantage point in lower dimensions, begin to see certain fascinating features emerge, especially the highly non-trivial nature of string theory in 26 or 10 dimensions. Clearly, much more work has to be done before we can get a handle on string theory at the critical dimension. However, it is encouraging that we have been able to make such progress in calculating the true vacuum of string theory with the limited mathematical tools at our disposal. The hope is that the rich lessons we have learned in exploring low dimensional topological systems and systems with finite primary fields will give us the insight necessary to probe string theory in the critical dimension.

In conclusion, this book will have served its purpose if it has captured some of the excitement of string theory and conveyed the richness of the theory. At present, we have only scratched the surface of the theory with our primitive mathematical tools. More surprises await us as we probe into the non-perturbative regime. The goal of this book, therefore, has been to communicate some of the remarkable properties of string theory which has propelled it to the forefront of research.

It is our hope that some of the readers of this book, inspired by the intensive research in string theory, will go on to make original contributions in superstring research. Perhaps one of the readers will eventually shed light on the fundamental problem facing string theory: determining its true vacuum state.

References

1. E. Witten, *Nucl. Phys.* **B202**, 253 (1982); *J. Diff. Geom.* **17**, 661 (1982).
2. A. Floer, *Bull. A.M.S.* **126**, 335 (1987); *Comm. Math. Phys.* **118**, 215 (1988).
3. M. Gromov, *Inven. Math.* **82**, 307 (1985).
4. S. Donaldson, *J. Diff. Geom.* **18**, 269(1983); **26**, 397 (1987).
5. V. F. R. Jones, *Bull. A.M.S.* **12**, 103 (1986); *Ann. Math.* **126**, 335 (1987).
6. M. F. Atiyah, in *Symposium on the Mathematical Heritage of Hermann Weyl*, University of North Carolina Press, Chapel Hill (1987).
7. E. Witten, *Comm. Math. Phys.* **117**, 353 (1988).
8. E. Witten, *Comm. Math. Phys.* **121**, 351 (1989).
9. E. Witten, *Nucl. Phys.* **B311**, 46 (1988/1989).
10. P. Van Baal, CERN-TH.5453/89 (1989).
11. D. Montano and J. Sonnenschein, *Nucl. Phys.* **B313**, 258 (1989); *Nucl. Phys.* **B324** , 348 (1989).
12. S. Ouvry, R. Stora, and P. Van Baal, *Phys. Lett.* **220B**, 159 (1988).
13. J. M. F. Labastida and M. Pernici, *Phys. Lett.* **B212**, 56 (1988).
14. L. Baulieu and I. M. Singer, *Nucl. Phys. Suppl.* **5B**, 12 (1988); see also L. Baulileu and B. Grossman, *Phys. Lett.* **212B**, 319 (1988).
15. E. Witten, *Comm. Math. Phys.* **118**, 411 (1988)
16. J. Labastida, M. Pernici, and E. Witten, *Nucl. Phys.* **B310**, 611 (1988).
17. J. Distler, PUPT-1161 (1989).
18. E. Witten, IASSNS-HEP 89/66 (1989).
19. R. Dijkgraaf and E. Witten, IASSNS-HEP 90/18 (1990).
20. E. Verlinde and. H. Verlinde, IASSNS-HEP 90/40 (1990).
21. R. Dijkgraaf, H. Verlinde, and E. Verlinde, Princeton preprint PUPT-1194 (1990).
22. M. Fukuma, H. Kawai, and R. Nakayama, Tokyo preprint UT-562 (1990).
23. K. Li, CALT-68-1662.
24. A. B. Zamolodchikov, *Theor. Math. Phys.* **65**, 1205 (1986); see also A. Bilal and J.L. Gervais, *Phys. Lett.* **206B**, 412 (1988).
25. M. Kaku, CCNY preprint (1991).

Appendix I

Batalin–Vilkovisky Quantization

Historically, quantizing a gauge invariant action has posed conceptual problems because of the presence of negative metric states, which arise from the Lorentz metric. The usual canonical quantization method, although rigorous and unitary, is noncovariant and very difficult to use.

A great breakthrough was the Faddeev–Popov method, which introduced ghost fields in order to cancel the negative metric states and also maintain covariance. This, in turn, led to the development of modern BRST methods, which have significantly reduced the effort necessary to quantize covariant gauge theories.

However, the recent actions found in supersymmetric theories and in superstring theory are beyond the realm of the standard Faddeev–Popov quantization.

Specifically, two types of covariant actions pose problems for the standard Faddeev–Popov approach.

1. There are theories in which there are "ghosts-for-ghosts," that is, where the Faddeev–Popov ghosts possess a local gauge invariance, requiring a second generation of ghosts to cancel them. In superstring theories, there can be an infinite tower of such ghosts-for-ghosts.
2. There are theories in which the algebra defined by the gauge variation of the fields does not close properly except when we use the equations of motion. For example, in ordinary $N = 1$ supergravity, the gauge variations of the fields do not close off-shell, and the naive Faddeev–Popov quantization fails to produce a four-ghost interaction, which is necessary to maintain the unitarity of the theory.

The most recent example of a gauge theory in which the standard Faddeev–Popov quantization fails is the Green–Schwarz (GS) superstring, where we have both an infinite tower of ghosts-for-ghosts and a gauge algebra that does not close off-shell. In fact, a naive counting of the GS theory shows that we need 8-dimensional spinorial representations of the Lorentz group in 10 dimensions in order to separate out the first and second class constraints. However, 8-dimensional spinorial representations in 10 dimensions do not exist, which makes the quantum theory of the GS string problematic.

Because of the growing realization that supersymmetric and super-string actions require more sophisticated tools for covariant quantization, we will now present the method of Batalin–Vilkovisky (BV) [1], the most powerful method so far devised in which to quantize gauge invariant actions. Although the BV quantization method, for most ordinary point particle systems, is much too cumbersome for practical use, we will find that for many supersymmetric and superstring applications, there is no other covariant quantization method.

The point of the BV quantization method is to make sense out of the functional integral:

$$Z = \int d\phi_i \, \exp\left[\frac{i}{\hbar} S_0(\phi_i)\right] \tag{A.1.1}$$

which diverges if ϕ_i is a gauge field. To render this functional integral meaningful, the BV quantization method can be broken down into several steps.

1. *Antifields*: Let Φ collectively represent the original set of gauge fields ϕ_i plus the complete set of ghosts C_j necessary for Faddeev–Popov quantization. Now, define Φ^* to be the *antifield*, which is similar to the field Φ except that it is a collection of gauge fields ϕ^{i*} and ghost fields C_j^*, which have the opposite statistics $\epsilon$to the ordinary fields.

2. *Antibrackets*: Next, we define a commutator called the *antibracket*:

$$(X, Y) = \frac{\partial_r X}{\partial \Phi^A} \frac{\partial_l Y}{\partial \Phi_A^*} - \frac{\partial_r X}{\partial \Phi_A^*} \frac{\partial_l Y}{\partial \Phi^A} \tag{A.1.2}$$

where the subscripts r and l refer to right and (left) derivatives of the various fields. Some of the antibracket's elementary properties are:

$$(B, B) = 2 \frac{\partial_r B}{\partial \Phi^A} \frac{\partial_l B}{\partial \Phi_A^*}, \qquad B \text{ bosonic}$$

$$(F, F) = 0, \qquad F \text{ fermionic} \tag{A.1.3}$$

$$\left[(G, G), G\right] = 0, \qquad \text{any } G$$

It also satisfies the cyclic identity:

$$(-1)^{(\epsilon_F+1)(\epsilon_H+1)}\left((F, G), H\right) + \text{cyclic permutation} = 0 \tag{A.1.4}$$

where ϵ represents the Grassmann number of the field.

3. *Action and gauge fermion*: Here, we introduce two new functions. The first is the action $W(\Phi, \Phi^*)$, which reduces to the usual action in the limit that the antifields vanish:

$$W(\Phi, 0) = S_0(\Phi) \tag{A.1.5}$$

The second is the *gauge fermion* Ψ, which will contain all the information concerning gauge fixing. We define the antifields in terms of this gauge fermion:

$$\Phi^* = \partial \Psi / \partial \Phi \qquad (A.1.6)$$

The gauge fermion is a function of Φ, plus a second set of ghosts denoted symbolically by $\bar{C}$ and C'. Once Ψ is fixed, then the gauge conditions that we impose are:

$$\frac{\partial \Psi}{\partial \bar{C}} = 0; \qquad \frac{\partial \Psi}{\partial C'} = 0 \qquad (A.1.7)$$

To implement this choice in the action, we must introduce a Lagrange multiplier for each $\bar{C}$ and C', so that we add to the action:

$$L_{\text{gauge}} = \frac{\partial \Psi}{\partial \bar{C}} \bar{\pi} + \frac{\partial \Psi}{\partial C'} \pi' \qquad (A.1.8)$$

4. *Master equation*: So far, we have done nothing but make empty definitions. But, now we come to the heart of the BV quantization method. We demand that the functional integral:

$$Z = \int d\Phi \, \exp\left[\frac{i}{\hbar} W\left(\Phi, \frac{\partial \Psi}{\partial \Phi}\right)\right] \qquad (A.1.9)$$

be independent of the choice of the gauge fermion Ψ, but that it maintain the same physical content as the original functional. For example, if we set the gauge fermion $\Psi = 0$, then we retrieve the original functional integral, which was infinite. With nontrivial choices of Ψ, we can break the gauge invariance and obtain finite results without changing the physical content of the theory.

The trick is to extract the conditions on W such that the functional integral is independent of Ψ, that is, it is independent of the gauge choice needed to render the functional integral finite. Let us make a small variation of Ψ and use the chain rule:

$$
\begin{aligned}
\delta Z &= \int D\Phi \left[\frac{\delta}{\delta \Phi_B^*} e^{iW(\Phi,\Phi^*)/\hbar}\right]\Bigg|_{\Phi^*=\partial\Psi/\partial\Phi} \frac{\partial \delta \Psi}{\partial \Phi} \\
&= \int D\Phi \delta\Psi \sum_A (-1)^{\Phi_A} \frac{\partial}{\partial \Phi_A} \frac{\partial}{\partial \Phi_A^*} e^{\frac{iW}{\hbar}}
\end{aligned}
\qquad (A.1.10)
$$

where we have integrated by parts (and eliminated a second term proportional to the double derivative of Ψ, which vanishes because Φ and Φ^* have opposite statistics).

Independence from the choice of Ψ means, therefore,

$$\Delta \exp\left(\frac{i}{\hbar} W\right) = 0; \qquad \Delta \equiv \frac{\partial_r}{\partial \Phi^A} \frac{\partial_l}{\partial \Phi_A^*} \qquad (A.1.11)$$

Our master equation is then:

$$\frac{1}{2}(W, W) = i\hbar \Delta W \qquad (A.1.12)$$

The master equation is usually too difficult to solve, so we will power expand in $\hbar$:

$$W = S + \sum_{p=1}^{\infty} \hbar^p W_p \qquad (A.1.13)$$

So, the master equation actually becomes a series of equations:

$$
\begin{aligned}
(S, S) &= 0 \\
(W_1, S) &= i\Delta S \\
(W_p, S) &= i\Delta M_{p-1} - \frac{1}{2} \sum_{q=1}^{p-1} (W_q, W_{p-q})
\end{aligned}
\qquad (A.1.14)
$$

[For most systems, the master equation can be solved without this power expansion, so the master equation simply reads $(S, S) = 0$.]

5. *BRST Invariance*: Last, we define the BRST variation as:

$$\delta \Phi^A = (-1)^{\Phi_A} \frac{\partial W}{\partial \Phi_A^*} \bigg|_{\Phi^* = \partial \Psi / \partial \Phi} \qquad (A.1.15)$$

Using the previous identities, we can show that the action is invariant under this BRST transformation, which is nilpotent.

To show this, let us take the variation of the exponential of the action:

$$
\begin{aligned}
\delta e^{iW/\hbar} &= \frac{i}{\hbar} e^{iW/\hbar} \delta \Phi_A \left(\frac{\partial W}{\partial \Phi_A} + \sum_B \frac{\partial^2 \Psi}{\partial \Phi_A \partial \Phi_B} \frac{\partial W}{\partial \Phi_B^*} \right) \\
&= \frac{i}{\hbar} e^{iW/\hbar} (-1)^{\Phi_A} \frac{\partial W}{\partial \Phi_A^*} \frac{\partial W}{\partial \Phi_A}
\end{aligned}
\qquad (A.1.16)
$$

where we have dropped a term proportional to the second derivative of Ψ because Φ and Φ^* come in opposite statistics and cancel against each other.

Now, let us calculate the variation of the functional measure, which contributes a Jacobian (which is actually a superdeterminant). Let θ be a constant Grassmann number. Then:

$$s \det \left[\partial \Phi_A + \frac{\theta \delta \Phi_A}{\partial \Phi_B} \right] = 1 + \theta (-1)^{\Phi_A} \frac{\partial^2 W}{\partial \Phi_A \partial \Phi_A^*} \qquad (A.1.17)$$

If we add the contribution from the variation of the action functional with the variation of the measure, we find an exact cancellation if we use the master equation, so the action is therefore invariant. [For systems where $W_n = 0$, the proof of nilpotency is trivial. We write:

$$
\begin{aligned}
\delta \Phi^A &= (-1)^{\epsilon_A} (\Phi^A, S) \\
\delta \Phi_A^* &= (-1)^{\epsilon_A + 1} (\Phi_A^*, S)
\end{aligned}
\qquad (A.1.18)
$$

Then the proof of nilpotency follows immediately, since $(S, S) = 0$.]

Example: Antisymmetric Tensor Field

The BV quantization method, as detailed above, is deceptively simple. Its strength is that it is the most versatile covariant quantization method so far devised for quantum field theory. However, it has a weak point. Most of the work is now concentrated in constructing $W(\Phi, \Phi^*)$, which satisfies the master equation Eq. (A.1.12). Although the BV method gives a rough prescription for finding $W(\Phi, \Phi^*)$ in most simple cases, the method unfortunately does not give a universal prescription for finding such a function in all cases. A certain amount of guesswork is still needed to find $W(\Phi, \Phi^*)$ for complicated actions.

To appreciate the power of the method, let us examine a few simple cases, such as the case of a spin-0 antisymmetric tensor field, and also a spin-$\frac{5}{2}$ symmetric spin field, both of which require "ghosts-for-ghosts." An antisymmetric tensor field $A_{\mu\nu}$ has the following action:

$$L = -\frac{1}{12} \int F_{\mu\nu\rho} F^{\mu\nu\rho} \sqrt{g} \, d^4x$$

$$F_{\mu\nu\rho} \equiv \partial_\mu A_{\nu\rho} + \partial_\rho A_{\mu\nu} + \partial_\nu A_{\rho\mu} \tag{A.1.19}$$

which is invariant under the following gauge transformation:

$$\delta A_{\mu\nu} = \partial_\mu \theta_\nu - \partial_\nu \theta_\mu \tag{A.1.20}$$

Notice that the gauge parameter θ_μ is itself a gauge field, that is, we can always make the following transformation on the parameter:

$$\delta\theta_\mu = \partial_\mu \Lambda \tag{A.1.21}$$

which leaves the variation of the original θ_μ invariant.

Notice that the "ghosts-for-ghosts" stops here. One cannot find a gauge variation of Λ that preserves the previous variations. We say, therefore, that this theory is *first-stage reducible*. Obviously, if the ghost field Λ was itself a gauge field, then we would have a *second-stage reducible* theory, and so on.

Example: Spin-$\frac{5}{2}$ Field

Another example of a reducible theory is a spin-$\frac{5}{2}$ symmetric Majorana field, given by the action

$$L = \int \left(-\frac{1}{2} \bar{\psi}_{\mu\nu} \gamma^\alpha \, \partial_\alpha \psi^{\mu\nu} - \bar{\psi}_{\mu\nu} \gamma^\nu \gamma^\alpha \, \partial_\alpha \gamma_\lambda \psi^{\lambda\mu} + 2\bar{\psi}_{\mu\nu} \gamma^\nu \, \partial_\lambda \psi^{\lambda\mu} \right.$$

$$\left. + \frac{1}{4} \bar{\psi}^\lambda_\lambda \gamma^\alpha \, \partial_\alpha \psi^\mu_\mu - \bar{\psi}^\lambda_\lambda \partial_\mu \gamma_\nu \psi^{\mu\nu} \right) d^4x$$

$$\tag{A.1.22}$$

which is invariant under the following gauge transformation:

$$\delta\psi_{\mu\nu} = \partial_\mu\epsilon_\nu + \partial_\nu\epsilon_\mu; \qquad \gamma_\mu\epsilon^\mu = 0 \qquad (A.1.23)$$

where ϵ is a spin $[\frac{3}{2}$ Majorana spinor. As before, this gauge transformation, in turn, possesses a gauge transformation for the gauge parameter. Because of the last constraint, we can always make the variation:

$$\delta\epsilon_\mu = \left(\delta_{\mu\nu} - \frac{1}{4}\gamma_\mu\gamma_\nu\right)\theta^\nu \qquad (A.1.24)$$

Thus, the ghost field ϵ_μ itself has a ghost field.

Example: Reducible Theories

The previous discussion was specific to the antisymmetric tensor field or a spin-$\frac{5}{2}$ field. Now, let us generalize this discussion and represent a general gauge field by ϕ^i and rewrite the original gauge transformation more abstractly as:

$$\delta\phi^i = R^i_{\alpha_0}\theta^{\alpha_0} \qquad (A.1.25)$$

where θ^{α_0} is the gauge parameter and α_0 is a collection of Lorentz indices. Because this is a symmetry of the action, the variation of the action with respect to this gauge transformation is zero:

$$\frac{\delta L}{\delta\phi^i}R^i_{\alpha_0} = 0 \qquad (A.1.26)$$

For the case of the antisymmetric tensor field in Eq. (A.1.20), the explicit representation of $R^i_{\alpha_0}$ is given by:

$$R^i_{\alpha_0} = \partial_\nu\delta^\alpha_\mu\delta(x - x_0) \qquad (A.1.27)$$

where $i = (\mu, \nu, x)$ and $\alpha_0 = (\alpha, x_0)$. For the spin-$\frac{5}{2}$ theory in Eq. (A.1.23), the $R^i_{\alpha_0}$ is given by:

$$R^i_{\alpha_0} = (\partial_\mu\delta^\beta_\nu + \partial_\nu\delta^\beta_\mu)\left(\delta_{\beta\alpha} - \frac{1}{4}\gamma_\beta\gamma_\alpha\right)\delta(x - x_0) \qquad (A.1.28)$$

Notice that we have done nothing. We have merely rewritten the original gauge variation for $A_{\mu\nu}$ and $\psi_{\mu\nu}$ in a more abstract notation. The advantage of this notation, however, is that we can now state precisely the criterion for a "ghosts-for-ghosts" theory.

We say that a theory is *irreducible* when the $R^i_{\alpha_0}$ have no zero eigenvalues. Examples of irreducible theories the familiar Yang–Mills theory or the general theory of relativity. However, a theory is defined to be *reducible* when $R^i_{\alpha_0}$ has zero eigenvalues, that is,

$$R^i_{\alpha_0}Z^{\alpha_0}_{1\alpha_1} = 0 \qquad (A.1.29)$$

If the eigenvectors Z_1 are linearly independent, then the "ghosts-for-ghosts" stops here, and we say that the theory is *first-stage* reducible.

Our previous examples of the antisymmetric tensor field and the spin-$\frac{5}{2}$ field represent first-stage reducible theories. In fact, their Z_1 are represented by, respectively,

$$
\begin{aligned}
Z^{\alpha_0}_{1\alpha_1} &= \partial_\alpha \delta(x_0 - x_1) \\
Z^{\alpha_0}_{1\alpha_1} &= \gamma_\sigma \delta(x_0 - x_1)
\end{aligned}
\tag{A.1.30}
$$

If the Z_1 themselves are linearly dependent, then the theory is still reducible. A second-stage reducible theory, for example, has zero eigenvalues of Z_1 given by:

$$
Z^{\alpha_0}_{1\alpha_1} Z^{\alpha_1}_{2\alpha_2} = 0
\tag{A.1.31}
$$

but Z_2 does not have any zero eigenvalues.

Example: Ghost Structure of Yang–Mills Theory

Now, let us analyze how to add in the ghosts into the various theories. For the familiar case of the Yang–Mills theory, which is irreducible, let us rederive the known Faddeev–Popov action. We can solve the master equation by adding terms that are linear in the antifields:

$$
S(\Phi, \Phi^*) = L(\phi) + \phi^* R^i_{\alpha_0} C^{\alpha_0}_0
\tag{A.1.32}
$$

Inserting the fields A^{*a}_μ and C^{*a} into the equation and using $R^i_{\alpha_0}$ for both A^a_μ and C^a, we have:

$$
S = -\frac{1}{4} F^a_{\mu\nu} F^{\mu\nu a} + A^{*a}_\mu D^{\mu ab} C^b + \frac{1}{2} C^{*a} f^{abc} C^c C^b
\tag{A.1.33}
$$

Invariance under the BRST transformation can also be read by taking the antibracket of the various fields with respect to S:

$$
\begin{aligned}
\delta A^a_\mu &= D^{ab}_\mu C^b \\
\delta C^a &= -\frac{1}{2} F^{abc} C^c C^b \\
\delta A^{*a}_\mu &= \frac{\partial S_0}{\partial A^{\mu a}} - A^{*b}_\mu f^{bac} C^c \\
\delta C^{*a} &= D^{\mu ab} A^{*b}_\mu + c^{*b} F^{bac} C^c
\end{aligned}
\tag{A.1.34}
$$

To break the gauge invariance of the theory, we choose the usual gauge $\partial_\mu A^\mu = 0$, which yields the following gauge fermion:

$$
\Psi = \int \bar{C}_{0\alpha_0} \partial_\mu A^\mu
\tag{A.1.35}
$$

Now, using the fact that $\Phi^* = \delta \Psi / \delta \Phi$, we can enforce the gauge constraint by adding one more equation:

$$
L_{\text{gauge}} = \frac{\partial \Psi}{\partial \bar{C}_{0\alpha_0}} \pi_{0\alpha_0} = \partial_\mu A^\mu \pi_{0\alpha_0}
\tag{A.1.36}
$$

Adding the action coming from Faddeev–Popov fixing of the gauge and the term coming from enforcing the constraint, we find the complete Faddeev–Popov action for the Yang–Mills theory, with $\bar{C}_0^{\alpha_0}$ and $C_{\alpha_0}^0$ being the usual Faddeev–Popov ghosts.

Example: First-Stage Reducible Theory

For a first-stage reducible theory, such as the antisymmetric tensor field or spin-$\frac{5}{2}$ field, the quantization is straightforward, except that there are more ghost fields. The new solution of the master equation now contains both $R_{\alpha_0}^i$ and Z_1:

$$S(\Phi, \Phi^*) = L(\phi) + \phi_i^* R_{\alpha_0}^i C_0^{\alpha_0} + C_{0\alpha_0}^* Z_{1\alpha_1}^{\alpha_0} C_1^{\alpha_1} \tag{A.1.37}$$

Now, we now introduce the gauge fermion Ψ, in which case the above action (including the gauge part) now becomes:

$$\begin{aligned}
L_{\text{ghost}} &= \frac{\partial \Psi}{\partial \phi^i} R_{\alpha_0}^i C_0^{\alpha_0} + \frac{\partial \Psi}{\partial C_0^{\alpha_0}} Z_{1\alpha_1}^{\alpha_0} C_1^{\alpha_1} \\
L_{\text{gauge}} &= \frac{\partial \Psi}{\partial \bar{C}_{0\alpha_0}} \pi_{0\alpha_0} + \frac{\partial \Psi}{\partial \bar{C}_{1\alpha_1}} \pi_{1\alpha_1} + \frac{\partial \Psi}{\partial C_1'^{\alpha_1}} \pi_1'^{\alpha_1}
\end{aligned} \tag{A.1.38}$$

where the Lagrange multipliers are $\pi_{0\alpha_0}$, $\pi_{1\alpha_1}$, $\pi_1'^{\alpha_1}$.

We can now solve for the action for the antisymmetric tensor field and the spin-$\frac{5}{2}$ field by an appropriate choice of Ψ. For the antisymmetric field, we wish to impose the gauge choice $\nabla^\mu A_{\mu\nu}$, as well as $\nabla^\mu C_{0\mu}$ to fix the gauge. So the choice of the gauge fermion is:

$$\Psi = \int \left(\bar{C}_{0\alpha} g^{\alpha\beta} \nabla^\mu A_{\mu\beta} + \bar{C}_1 \nabla^\mu C_{0\mu} + C_{0\alpha} \nabla^\alpha C_1' \right) \tag{A.1.39}$$

Example: General Case of Lth-Stage Reducible Theories

Now, let us present the solution for the general case. An Lth-stage reducible theory is defined via the following conditions:

$$\begin{aligned}
\frac{\delta L}{\delta \phi^i} R_{\alpha_0}^i &= 0; & R_{\alpha_o}^i Z_{1\alpha_1}^{\alpha_0}|_{\phi_0} &= 0 \\
Z_{s-1,\alpha_{s-1}}^{\alpha_{s-2}} Z_{s\alpha_s}^{\alpha_{s-1}}|_{\phi_0} &= 0; & s &= 2, \dots, L
\end{aligned} \tag{A.1.40}$$

where the last equation terminates at the Lth level.

For each level of reducibility, we introduce a set of ghost fields $C_s^{\alpha_s}$:

$$\Phi_{\text{min}} = \{\phi^i, C_s^{\alpha_s}\}; \qquad s = 0, \dots, L \tag{A.1.41}$$

The ghosts have statistics ϵ and ghost numbers gh given by:

$$\begin{aligned}
\epsilon(C_s^{\alpha_s}) &= \epsilon_{\alpha_s} + s + 1 \\
\text{gh}(C_s^{\alpha_2}) &= s + 1
\end{aligned} \tag{A.1.42}$$

This, in turn, allows us to introduce the antifields Φ^*, with opposite statistics to each field in Φ.

The action $S(\Phi, \Phi^*)$ satisfies the master equation $(S, S) = 0$ (assuming that the higher W_n vanish) leaving us with:

$$S(\Phi, 0) = L(\Phi)$$

$$\frac{\partial_l}{\partial \phi_i^*} \frac{\partial_r}{\partial C_0^{\alpha_0}} S(\Phi, \Phi^*) \Big|_{\Phi^*=0} = R_{\alpha_0}^i(\phi)$$

$$\frac{\partial_l}{\partial C_{s-1}^{*\alpha_s}} \frac{\partial_r}{\partial C_s^{\alpha_s}} S(\Phi, \Phi^*) \Big|_{\Phi^*=0} = Z_{s\alpha_s}^{\alpha_{s-1}}(\phi) \tag{A.1.43}$$

The solution of these equations gives us $S(\Phi_{\min}, \Phi_{\min}^*)$.

Now, let us introduce the gauge fermion Ψ, which in turn introduces the second set of ghost fields $\bar{C}, C'$. In fact, the precise number of new ghosts introduced with each new stage is given by:

$$\begin{aligned}
C_0^{\alpha_0} &\to \bar{C}_{0\alpha_0} \\
C_1^{\alpha_1} &\to \bar{C}_{1\alpha_1}, \, C_1'^{\alpha_1} \\
C_2^{\alpha_2} &\to \bar{C}_{2\alpha_2}, \, C_2'^{\alpha_2}, \, C_2''^{\alpha_2} \\
C_3^{\alpha_3} &\to \bar{C}_{3\alpha_3}, \, C_3'^{\alpha_3}, \, \bar{C}_{3\alpha_3}'', \, C_3'''^{\alpha_3} \\
C_n^{\alpha_n} &\to \cdots
\end{aligned} \tag{A.1.44}$$

The gauge choice is given by the variation of the Ψ fields with respect to the $\bar{C}$ and C'. The gauge fixing in accomplished by introducing Lagrange multipliers. The solution for the action is therefore given by:

$$S(\Phi, \Phi^*) = S(\Phi_{\min}, \Phi_{\min}^*) + \frac{\partial \Psi}{\partial \bar{C}} \bar{\pi} + \frac{\partial \Psi}{\partial C'} \pi' \tag{A.1.45}$$

summed over all the various $\bar{C}$ and C' fields.

Let us now assemble all pieces, both the L_{ghost} and L_{gauge} parts, into one expression. Let us suppress all subscripts and superscripts, for convenience. Then, the final gauge fixed action can be simply represented as:

$$L_{\text{gauge fixed}} = L(\phi^i) + \delta\Psi \tag{A.1.46}$$

where:

$$\Psi = \bar{C}\chi(\Phi) \tag{A.1.47}$$

and the BRST variations are:

$$\begin{aligned}
\delta C' &= \pi \\
\delta \pi &= 0 \\
\delta \bar{C} &= \bar{\pi} \\
\delta \bar{\pi} &= 0
\end{aligned} \tag{A.1.48}$$

Inserting the value of Ψ into the action and taking the BRST variation, we now have the final result:

$$L_{\text{gauge fixed}} = L(\phi^i) + \bar{\pi}\chi + \bar{C}\frac{\partial\chi}{\partial C'}\pi + \bar{C}\frac{\partial\chi}{\partial\Phi}\delta\Phi \qquad (A.1.49)$$

We can now easily identify all the pieces on the right-hand side of the equation. The first two terms, involving the Lagrange multipliers $\bar{\pi}$ and π', are the gauge fixing terms. The last term is the Faddeev–Popov ghost term. In summary, we have now condensed all the terms arising in the action from gauge fixing a reducible quantum system into one term: $\delta\Psi$.

Open Algebras

So far, we have only dealt with the case where the algebra generated by the variation of the fields closes off-shell, that is, without making any mention of the equations of motion. However, in many quantum systems (such as supergravity and superstrings), we find that the algebra often does not close off-shell. In the proof of BRST invariance, there was one step where we actually used the fact that the algebra formed by the variation of the fields was closed off-shell. In this case, the naive BRST quantization method actually gives us the incorrect answer, leaving out important terms such as the four-ghost interaction.

Let us now generalize our discussion to the case of open algebras, which only close on-shell. In particular, we will find that the square of a BRST variation of a field is not zero:

$$\delta^2\Phi^A = S^{Ai}(\Phi)\frac{\partial L_{\text{classical}}}{\partial\phi^i} \qquad (A.1.50)$$

Notice that the BRST operator is nilpotent only if we can set the variation of the classical Lagrangian to zero, that is, if we are on-shell. In this case, one convenient method is to add a new term to the variation of the fields and then add a new term to the gauge fixed action, such that the sum of the two contributions exactly cancels.

Let us modify the gauge variation of the field:

$$(\delta + \delta')\Phi^A = S^A + \frac{\partial\Psi}{\partial\Phi^B}S^{AB} \qquad (A.1.51)$$

where S^A is the usual BRST variation of the field and S^{AB} is the new term arising from the nonclosure of the algebra. Then, the modified action becomes:

$$\begin{aligned}
L_{\text{gauge fixed}} &= L(\phi^i) + (\delta + \frac{1}{2}\delta')\Psi \\
&= L(\phi^i) + \bar{C}\chi + \bar{\pi}\frac{\partial\chi}{\partial C'} + \bar{\pi}\frac{\partial\chi}{\partial\Phi^A}S^A \qquad (A.1.52) \\
&\quad + \frac{1}{2}\bar{C}\frac{\partial\chi}{\partial\Phi^A}\bar{C}\frac{\partial\chi}{\partial\Phi^B}S^{AB}
\end{aligned}$$

Notice that the last term in the gauge fixed action is the four-ghost inter-action that was missing in the naive BRST quantization of supergravity. By taking the variation of this modified action with respect to the modified variation of the fields, we can show that the action is fully invariant under the modified BRST variation.

For more complicated actions, however, it may be necessary to add a succession of terms to the variation of the fields and to the action. This occurs in the BV quantization program if the master equation requires the addition of higher order antifields:

$$
\begin{aligned}
S_{\text{gauge fixed}} = S_{\text{cl}} &+ \frac{\partial \Psi}{\partial \Phi^A} S^A + \frac{1}{2} \frac{\partial \Psi}{\partial \Phi^A} \frac{\partial \Psi}{\partial \Phi^B} S^{AB} \\
&+ \frac{1}{3} \frac{\partial \Psi}{\partial \Phi^A} \frac{\partial \Psi}{\partial \Phi^B} \frac{\partial \Psi}{\partial \Phi^C} S^{ABC} + \cdots
\end{aligned}
\tag{A.1.53}
$$

In this case, the presence of S^{ABC} and higher terms spoils the BRST invari-ance mentioned above. We must continue to add new correction terms to the variation of fields, proportional to the various $S^{ABCD\cdots}$ and new terms to the action until the modified action is fully BRST invariant. Fortunately, we will find that supergravity and the superstring require only one more modification to the naive BRST rules in order to construct BRST invariant actions. (The only known quantum systems in which these higher order terms are required are the super-p-dimensional membranes, or p-branes.)

Appendix 2

Covariant Quantization of the Green–Schwarz String

We now apply the BV quantization method to the Green–Schwarz (GS) string [2–6], which requires the full power of the method. Although the light cone quantization of the GS string is trivial, the covariant string has posed considerable problems for three reasons.

1. The first and second class constraints of the theory cannot be separated covariantly.
2. The supersymmetry algebra does not close off-shell.
3. The system is infinitely reducible.

Because the precise details of the quantization are quite involved (and some delicate technical points still must be resolved rigorously), we will only sketch the quantization procedure.

For simplicity, we will only discuss the heterotic string in the GS formalism. The heterotic string action we are interested in is:

$$ L = e \left(\Pi_z^\mu \Pi_{\bar{z}}^\mu + i\partial_z X^\mu \bar{\theta}\gamma_\mu \partial_{\bar{z}}\theta - i\partial_{\bar{z}} X^\mu \bar{\theta}\gamma_\mu \partial_z \theta - \frac{i}{2}\Psi^I \partial_{\bar{z}}\Psi^I \right) \quad (A.2.1) $$

where the superstring coordinates are given by X^μ, θ; the zweibeins by $e_z^a, e_{\bar{z}}^a$; and the left-handed chiral fermions by Ψ^I, where I labels the isospin index for $E_8 \otimes E_8$, and where $\Pi_z^\mu = \partial_z X^\mu - i\bar{\theta}\gamma^\mu \partial_{\bar{z}}\theta$.

Let us now carefully analyze the field/antifield structure of the theory [4]. Our fields/antifields are as follows:

$$ \phi_0 = \{X^\mu, e_z^a, e_{\bar{z}}^a, \Psi^I, c^a, \rho, \Lambda, \theta_{p,0}\} \quad (A.2.2) $$

where c^a, ρ, and Λ are the ghosts for reparameterizations, conformal transformations, and Lorentz rotations. Notice that we have our first sequence of ghosts in $\theta_{p,0}$. We will define $\theta_{0,0} \equiv \theta$, because the cascading sequence of "ghosts-for-ghosts" will produce the series of ghosts $\theta_{p,q}$.

The antighosts are given by:

$$ \bar{\phi}_0 = \left\{ \bar{c}_a^z, \bar{c}_a^{\bar{z}}, \bar{\theta}_{\bar{z}}^{p,0} \right\} \quad (A.2.3) $$

where $\bar{c}_a^z$ and $\bar{c}_a^{\bar{z}}$ are the antighosts associated with c^a, ρ, Λ and $\bar{\theta}^{p,0}$ are the antighosts associated with $\theta_{p,0}$. The next level fields are given by:

$$\phi_q = \{\theta_{p,q}\}, \qquad p \geq q \tag{A.2.4}$$

while their antighosts are given by:

$$\bar{\phi}^p = \{\bar{\theta}_{\bar{z}}^{p,q}\} \tag{A.2.5}$$

Finally, the Lagrange multipliers for the infinite κ symmetry are given by:

$$\begin{aligned} \bar{\mu}^q &= \{\bar{\mu}_a^z, \bar{\mu}_a^{\bar{z}}, \lambda_{\bar{z}}^{p,q}\} \quad q \geq 0 \\ \mu_q &= \{\lambda_{p,q}\}; \qquad (q \geq 1) \end{aligned} \tag{A.2.6}$$

To write the gauge fixed action, let us now group the fields together and make the following symbolic definitions [4]:

$$\begin{aligned} \phi &\equiv \{\phi_0, \phi_q\} \\ \phi_{extra} &\equiv \{\phi_q\} \\ \bar{\phi} &\equiv \{\bar{\phi}^p\} \\ \mu &\equiv \{\mu_q\} \\ \bar{\mu} &\equiv \{\bar{\mu}^q\} \end{aligned} \tag{A.2.7}$$

Then, the gauge fixed action is given by [see Eq. (A.1.46)]:

$$L_{GF} = L_0 + \delta\Psi(\phi, \bar{\phi}) \tag{A.2.8}$$

If we choose the gauge fermion:

$$\Psi = \bar{\phi}\chi(\phi) \tag{A.2.9}$$

then the final action is given by:

$$\begin{aligned} L_{GF} &= L_0 + L_1 + L_2 + L_3 \\ L_1 &= \bar{\mu}\chi \\ L_2 &= \bar{\phi}\,\frac{\partial\chi}{\partial\phi_{extra}}\,\mu \\ L_3 &= \bar{\phi}\,\frac{\partial\chi}{\partial\phi_0}\,S(\phi_0) \end{aligned} \tag{A.2.10}$$

where the BRST variations are given by [see Eq. (A.1.48)]:

$$\begin{aligned} \delta\phi_0 &= S(\phi_0) \\ \delta\phi_{extra} &= \mu \\ \delta\bar{\phi} &= \bar{\mu} \\ \delta\mu &= 0 \\ \delta\bar{\mu} &= 0 \end{aligned} \tag{A.2.11}$$

Now, we wish to apply the BV quantization method to the Green–Schwarz string by choosing the gauge fermion condition. We will choose:

$$\Psi = \bar{c}_a^z \left[e_z^a - (e_z^a)^0 \right] + \bar{c}_a^{\bar{z}} \left[e_{\bar{z}}^a - (e_{\bar{z}}^a)^0 \right] + \sum_{q=0}^{\infty} \sum_{p=0}^{\infty} \bar{\theta}_{\bar{z}}^{p,q} \, \nabla_z \tilde{\theta}_{p,q} \qquad (A.2.12)$$

where:

$$\tilde{\theta}_{p,q} \equiv \theta_{p,q} + \theta_{p+1,q+1}$$

$$\theta_{p,q} = \sum_{r=0}^{\infty} (-1)^r \tilde{\theta}_{p+r,q+r} \qquad (A.2.13)$$

We have chosen this particular gauge fermion because it will simply set the zweibein equal to its background value (which can be taken to be conformally flat). It also simply sets the divergence of the $\tilde{\theta}_{p,q}$ to zero. (It can be shown that, because there are no X_μ appearing in the gauge fermion, a four-ghost term due to nonclosure of the algebra, as appears in supergravity, is absent.)

Although the action seems to be very complicated, we find that the action collapses down to an extremely simple structure because most of the terms can be eliminated by redefining the fields. For example, all of L_2 and part of L_3 (the part coming from the $\theta_{p,0}$ term) can be simply recombined as follows:

$$\sum_{q=1}^{\infty} \sum_{p=q}^{\infty} \frac{\partial \Psi}{\partial \theta_{p,q}} \lambda_{p,q} + \sum_{p=0}^{\infty} \frac{\partial \Psi}{\partial \theta_{p,0}} \delta \theta_{p,0} = \sum_{q=0}^{\infty} \sum_{p=q}^{\infty} \bar{\theta}_{\bar{z}}^{p,q} \nabla_z \pi_{p,q} \qquad (A.2.14)$$

where we have defined:

$$\pi_{p,0} \equiv \delta \theta_{p,0} + \lambda_{p+1,1}$$

$$\pi_{p,q} \equiv \lambda_{p,q} + \lambda_{p+1,q+1} \qquad (A.2.15)$$

Likewise, it is possible to drop other terms and recombine the rest of L_3. To do this, we will define:

$$b^{zz} \equiv e_{\bar{z}}^a \frac{\partial \Psi}{\partial e_z^a}; \qquad b^{\bar{z}z} \equiv e_z^a \frac{\partial \Psi}{\partial e_{\bar{z}}^a}$$

$$b^{z\bar{z}} \equiv e_z^a \frac{\partial \Psi}{\partial e_z^a}; \qquad b^{\bar{z}\bar{z}} \equiv e_{\bar{z}}^a \frac{\partial \Psi}{\partial e_{\bar{z}}^a} \qquad (A.2.16)$$

By using the field equations for $\rho, \Lambda, b^{z\bar{z}}, b^{\bar{z}z}$, we can set them to zero:

$$\rho = \Lambda = b^{z\bar{z}} = b^{\bar{z}z} = 0 \qquad (A.2.17)$$

Next, we can collapse the terms in L_3 by noting that Ψ does not depend on the string variable X_μ and that the remaining terms in L_3 reduce to:

$$L_3 \rightarrow K_{\bar{z}} \partial_z \theta \qquad (A.2.18)$$

where:

$$K_{\bar{z}} \equiv -2ie\,\partial_{\bar{z}}X^{\mu}\bar{\theta}\gamma_{\mu} - e(\bar{\theta}\gamma_{\mu}\,\partial_{\bar{z}}\theta)\bar{\theta}\gamma^{\mu} + 4ieb^{zz}\bar{\theta}_{1,0} \tag{A.2.19}$$

We can further redefine terms by combining L_3 and part of L_1 that depends on $\bar{\lambda}_{\bar{z}}^{p,q}$:

$$K_{\bar{z}}\partial_z\theta + \sum_{q=0}^{\infty}\sum_{p=q}\bar{\lambda}_{\bar{z}}^{p,q}\nabla_z\tilde{\theta}_{p,q} = \sum_{q=0}^{\infty}\sum_{p=q}^{\infty}\tilde{\bar{\pi}}_{\bar{z}}^{p,q}\nabla_z\tilde{\theta}_{p,q} \tag{A.2.20}$$

where:

$$\tilde{\bar{\pi}}_{\bar{z}}^{p,q} \equiv \bar{\lambda}_{\bar{z}}^{p,q} + (-1)^q K_{\bar{z}}\delta^{p,q} \tag{A.2.21}$$

Finally, because so many terms have been dropped or absorbed into redefinitions of fields, we can now write the final action, which miraculously enough is a free theory [4]:

$$\begin{aligned}
L = {}& e\partial_z X\,\partial_{\bar{z}}X - \left(\frac{ie}{2}\right)\Psi^I\partial_{\bar{z}}\Psi^I \\
& + \sum_{q=0}^{\infty}\sum_{p=q}^{\infty}\left(\bar{\theta}_{\bar{z}}^{p,q}\nabla_z\pi_{p,q} + \tilde{\bar{\pi}}^{p,q}\nabla_z\tilde{\theta}_{p,q}\right) \\
& - eb^{zz}\nabla_z c_z - eb^{\bar{z}\bar{z}}\nabla_{\bar{z}}c_{\bar{z}} \\
& + \bar{\mu}_a^z\left[e_z^a - (e_z^a)^0\right] + \bar{\mu}_a^{\bar{z}}\left[e_{\bar{z}}^a - (e_{\bar{z}}^a)^0\right]
\end{aligned} \tag{A.2.22}$$

The first line is the usual free part of the string, while the second line consists of an infinite sequence of conformal spinors. The third line consists of the usual b, c ghost fields. The last line simply sets the zweibein equal to the background metric, which can be chosen to be flat. In summary, the second and third lines have the familiar form:

$$L = \frac{1}{\pi}\sum_i b_i\,\bar{\partial}c_i + \text{c.c.} \tag{A.2.23}$$

except that now we have an infinite set of these b, c fields with varying conformal weights, with varying spins, satisfying either commutation or anticommutation relations.

In some sense, an infinite sequence of ghosts was to be expected. A naive quantization of the GS string would have required separating first- and second-class constraints, in which case we need 8 dimensional spinorial representations of the 10-dimensional Lorentz group. However, such representations do not exist. (They exist in the light cone case only because Lorentz invariance is broken.)

Thus, the final, gauge fixed action of the GS string must utilize a new irreducible representation of the supergroup. The problem is evaded by using an entirely new infinite-dimensional representation of the group. The fact that we have an infinite number of fields in the system means that we are somehow utilizing a new, unexpected representation of the supergroup.

As a check, let us carefully count the contribution to the degrees of freedom and to the anomaly by summing over all p, q. Since the b, c contribution to the degrees of freedom and the central charge c are known for all conformal spins, we can show that the total degrees of freedom are given by:

$$\sum_{m=0}^{\infty}(-1)^m \sum_{n=0}^{\infty}(-1)^n f_n \qquad (A.2.24)$$

Since each ghost contributes one degree of freedom, $f_n = 1$ for the counting of ghost degrees of freedom.

At this point, we notice that the sum diverges. Therefore, there exists a certain ambiguity in whether or not the theory is really anomaly free. However, truncation methods are possible (for example, summing the BV ghosts according to their position along the "ghost-for-ghost" tree).

If, however, we use the regularization:

$$\sum_{m=0}^{\infty}\sum_{n=0}^{\infty}(-y)^n f_n \qquad (A.2.25)$$

then the sum converges. For the case of counting the degrees of freedom, the sum can be performed for $f_n = 1$ to arrive at:

$$N = \frac{16}{(1+x)(1+y)} \to 4 \qquad (A.2.26)$$

The contribution to the anomaly can also be calculated. A b, c system, with arbitrary conformal weight, contributes a central charge:

$$f_n = 6n^2 + 6n + 1 \qquad (A.2.27)$$

for arbitrary n. For the anomaly, the summation can also be performed, and we have:

$$c = -2(16)\left[\frac{12}{(1+x)^4} - \frac{12}{(1+x)^3} + \frac{1}{(1+x)^2}\right] \to -16 \qquad (A.2.28)$$

which are exactly the correct values one needs. Similarly, one can perform the sum with different regularizations, with the same result, which tends to justify the method (but one must point out that this does not prove that the methods are regularization independent).

References

1. I. Batalin and G. Vilkovisky, *Phys. Rev.* **D28**, 2567 (1983); **D30**, 508 (1984); *Phys. Lett.* **120B**, 166 (1983); *J. Math. Phys.* **26**, 172 (1985).
2. R. Kallosh, *Phys. Lett.* **224B**, 273; **225B**, 49 (1989).
3. R. Kallosh and M. Rahmanov, *Phys. Lett.* **209B**, 133 (1988); *Phys. Lett.* **214B**, 549 (1988).
4. E. A. Bergshoef and R. E. Kallosh, *Nucl. Phys.* **B333**, 605 (1990).
5. M. B. Green and C. M. Hull, *Phys. Lett.* **225B**, 57 (1989).
6. U. Lindstrom, M. Rocek, W. Siegel, P. van Nieuwenhuizen, and A. E. van de Ven, *Nucl. Phys.* **B330**, 19 (1989).

Index

1-MONTH